TRACKatlas
of Mainland Britain

2nd edition

Edited by Mike Bridge

TRACKatlas
of Mainland Britain

2nd edition

First published 2009

Second edition published 2012

Published by Platform 5 Publishing Ltd, Wyvern House, Sark Road, Sheffield S2 4HG, England.

ISBN: 978 1902336 97 8

Printed in England by Berforts Information Press, Eynsham, Oxford.

Cover Photograph: A view from above the entrance to Broad Street Tunnel overlooking the north end of Sheffield station, taken on 15 September 2012. **Robert Pritchard**

TRACKatlas
of Mainland Britain

Contents

Foreword to the Second Edition

Welcome to the second edition of TRACKatlas of Mainland Britain.

In the three years that have elapsed since publication of the first edition, the number of passengers carried on Britain's railways has continued to grow. Consequently, many locations on the railway network have been modified to meet new traffic levels and projected future growth. Several major junctions have been extensively remodelled and new stations have opened around the country. The maps give as much information as possible, subject to the limitations of scale and include some major changes that come into effect between the date of publication and the end of 2012.

As well as incorporating all of these alterations and additions, a number of enhancements have been made to the content of this edition. Most notably, the track layouts of 44 major heritage railways have been included. For the purposes of this atlas we have defined `major' as being over 8 miles in length, operating a timetabled service every day in the peak summer season, or having a permanent connection to the national rail network. Included are several narrow gauge lines.

Other improvements for this edition have been the re-drawing of the London maps which now provide a complete layout with consistent scale throughout and an effort to ensure that each signal box throughout the atlas has its code prefix appended.

Acknowledgements

This edition is edited by Mike Bridge and the cartography has been provided by Dave Padgett with assistance from John Szwenk and Bob Lewis.

For readers who do not have the earlier edition, it is also appropriate to acknowledge the work of the original cartographers Chris Fry, Dave Jones, Bob Lewis, John Gill, Jo Grant, Ann Rayski and Kathy and John Szwenk, led by Dave Padgett, who gave form to an idea by Mike Bridge.

We are also grateful for the contributions, great and small, from David Allen, Pat Birkett, Allan Brackenbury, Chris Bushell, Jim Davies, Barry Doe, Colin Duff, John English, Nigel Farebrother, Jim Ferguson, Neil Ferguson, John Gashion, John Gilbert, Steven Gustafson, Simon Hodges, B.R. Jones, T. S. Keep, Ray Liffen, David Longley, Matthew Lupton, John Macnab, David Markham, Ian McKenzie, John McNab, Andrew Millward, Myles Munsey, William Murchsion, David Muxworthy, Roger Newman, A. D. Nicholson, Gareth Parry, John Patston, Malcolm Pickering, Michael Reade, P.J. Rodgers, Bob Sambrook, Stephen Sangwine, John Savage, Graham Spencer, Mike Stone, David Thomson, Dr. F. G. Tomlins and John Yellowlees.

Contact Details

This atlas has been compiled with care to be as accurate as possible, but official information is not always available and the publisher cannot be held responsible for any errors or omissions. In particular, it should be noted that layout information for heritage railways has been compiled using the best available knowledge, but the nature of these lines means minor track alterations are often less well documented than their equivalents on the national rail network. We would welcome notification of any corrections or updates to this atlas of which readers have first-hand knowledge and also any additional information which may be used to enhance future editions.

Please send any comments or amendments to the publisher's address opposite the contents page or by e-mail to updates@platform5.com (telephone 0114 255 2625).

About this Atlas

The object of this atlas is to show today's railway network in mainland Britain in its geographical context and at a level of detail which allows readers to understand where they might be in railway terms and the implications of the trackwork in front of them.

The detail shown here started life as geo-referenced centrelines and from there, using industry sources for the track layout detail, the line work was drawn over. However, if drawn at true scale, it is likely that an atlas would require several hundred pages to show the detail at readable size. Indeed, many of those pages would show very little information as countryside lines can be both long and devoid of infrastructure.

To overcome this problem, a schematic approach has been taken to detail and one could call the result a Geo-schematic Map. It does have an underlying scale from the original geo-referenced centrelines but the infrastructure has been "pushed and pulled" to enable sufficient detail to appear. Therefore, the maps can be considered to have a notional scale only which, for the main part, is 1:210,000. To help the reader understand if such pushing and pulling has been used, a 5 mile marker symbol has been added to the maps. If, in any run, the markers are wider apart than elsewhere, it means detail at the beginning and end of the section has been moved to make space for detail in-between.

There are one or two other aspects that the reader needs to be aware of. For example, to help with the geographic context, rivers and urban sprawl have been shown. These are essentially true-scale. However, the points at which rivers cross track routes should not be taken as evidence of the actual crossing point.

Sidings have also been an issue due to scale. To give some substance to the information, the commonest treatment has been to describe them by name rather than by detail but it is admitted there is no consistency about this.

In principle, the maps show the passenger and freight routes which represent the national rail network together with certain heritage lines. The track formations are shown, including connections, crossovers and infrastructure. The maps show stations, signal boxes, junctions and tunnels including their names and railway mileages. Level crossings that affect signalling are included together with information describing their type. Where the maps become very complex in conurbations, specific areas are shown as larger scale insets. These appear towards the back of the book.

The reader will note that, unlike most other atlases but following industry practice, the Trackatlas indicates not only the name of an asset but also gives it a reference in miles and chains; the railway mileage. Railway mileages are subject in their own right and the article published in the 1st edition appears again in the following pages.

RAILWAY MILEAGES

A Brief History

One of the key aspects of managing Railway Infrastructure is detailed knowledge of the location of the assets. Throughout the UK rail system, principal assets are given a location along each route and this is expressed as the Railway Mileage. Generally, this mileage is given in miles and chains and the infrastructure assets identified in this Atlas have this information appended in each case (eg, 146m 39ch). The use of miles and chains may seem archaic to some but is worthy of explanation as these mileages are as old as the Railway itself.

In the beginning
In the early to mid-1800s, when the first railway lines were being built, Parliamentary approval was required. The various legal instruments used to gain this approval included plans which were intended to show not only the routes but also the major constructions on the way. These major constructions and other fixed items all had to be referenced in some way and initially, their positions would be related to items of lineside furniture such as telegraph poles, plaques, statutory obelisks and posts. While many of these features have disappeared now, they were used to provide unique identities for signal posts, bridges, tunnels and viaducts and the gradient posts marking the changes in the gradient of the track.

As the methods or markers used by each applicant tended to vary and to institute some commonality, Parliament enacted the Railways Clauses Consolidation Act of 1845 (section 94) which stated that railway companies "*...shall cause the length of the railway to be measured, and milestones, posts, or other conspicuous objects to be set up and maintained along the whole line thereof, at the distance of one quarter of a mile from each other, with numbers or marks inscribed thereon denoting such distances*". As a consequence, the idea of measuring the railway linearly along its route became the standard (as opposed to the eastings and northings used elsewhere for geographic spatial position) and the railway milepost, including quarter, half and three-quarter posts, became a statutory measure point for railway locations.

For increased accuracy, a smaller unit of measure was required and it became the practice to use chains to sub-divide miles and quarter miles. The Chain (22 yards, 20 chains per quarter mile, 80 chains to the mile), as the most commonly used fraction of a mile, had been in use in land surveying for more than 200 years before the early railways arrived and it was, therefore, natural that it came into use for railway measurement. Once established as the method of measurement, the whole railway system for the next 150 years followed this format. While major projects in more recent times may have been constructed to metric measurement, with the exception of the Channel Tunnel Rail Link (CTRL), once the construction comes into use, miles and chains are applied to maintain a level of consistency across the network. As a consequence, the present day use of miles and chains measure, even though apparently archaic and with such distant origins, has become rooted in railway history and geography.

Applying the Mileage
Railway measurement in this form would have been familiar to everyone before the 1923 Grouping of Railways. In the simplest terms, it required the 'planting' of a milepost (other than a Zero) along a line of route to indicate a 'measured' distance from a critical point of origin. There was no standard practice governing whether the mileposts were placed on one side of the line or the other and each company normally adopted one or the other throughout their system. Some companies, the L&YR for example, placed full mileposts on each side of the line.

Distances were usually measured from the buffer stops at the terminal station or from a junction of origin. Distances attributed to intermediate stations were normally calculated at a mean point from the extremities of the down and up platforms. A point of origin may generate miles which proceed along more than one route.

Mileages in Practice
Notwithstanding the apparent simplicity in the method of measurement, following a route by mileage can occasionally mislead the reader. Take start or zero points, for example. Following the 1923 Grouping, each of the 'Big Four' acquired multiple routes with different start points. This is probably best illustrated by the table below with reference to the principal pre-grouping companies.

Great Western	London, Paddington
Southern Railway	
L&SWR	London, Waterloo
LB&SCR	London Bridge (Central)
SER	London, Charing Cross
LCDR	London, Victoria
LNER	
GNR	London, Kings Cross
GER	London, Liverpool Street
GCR	Manchester, London Road
NBR	Edinburgh, Waverley
GNofS	Aberdeen

LMSR

L&NWR	London, Euston
Midland	London, St. Pancras
L&YR	Manchester, Victoria
Caledonian	Carlisle
G&SWR	Glasgow, Bridge Street
Highland	Perth

With expansions and amalgamations of railways, re-measurements were inevitable. There are various examples of the effect on mileages. For example, the South Eastern line would have been measured in its earliest form from London Bridge to Dover via Redhill; currently this route is measured from London, Charing Cross via Sevenoaks to Dover.

On the former Midland Railway, all distances were measured from Derby; this still applies to the routes which head to Birmingham and the south west but with the opening of their London, St. Pancras terminal all northbound routes were re-measured from that point. When different routes from London met, usually the route with the lesser mileage went forward. As the Midland Railway was a fascinating tapestry of routes, interpreting mileposts can be a little bit bewildering. However, there is a trail which links St. Pancras right through to Carlisle, even if there are gaps where tracks have been removed.

On the former Great Western Railway, the primary route continues to be known as the Main Line and has a mileage sequence which runs from Paddington to Penzance via Bristol. The mileage of the later alternate route via Newbury, however, expires at Cogload Junction, near Taunton.

Longer distances create more opportunities for confusion and neither the East Coast nor the West Coast Main lines have continuous mileage sequences. On the WCML, the mileage originating from Euston starts with a negative value because of platform extensions and finishes at 187m 76ch at Golborne Junction, north of Warrington. A new mileage which has originated from Newton-le-Willows, on the former Grand Junction line, goes forward as far Preston from where another sequence runs to Lancaster. Another sequence then runs to Carlisle to link with the Caledonian mileage which goes forward to Glasgow and Aberdeen.

On the ECML, the GNR mileage runs from London, Kings Cross to York, albeit the final 28 miles are over the metals of the former North Eastern Railway. York is a mileage origin point for a number of routes, including one that takes the ECML to Newcastle and then another from Newcastle to just north of Berwick where a mileage which has originated at Edinburgh and running south to Berwick completes the route.

Other peculiarities become obvious when studying the sequences of miles. When a new line was built and made a junction with a former one, it is the latter's mileage which usually goes forward. Gaps in sequences may occur when a 'joint' line arises, that is one managed by more than one company and only one sequence of miles will be evident. Sometimes mileages are reversed at a junction and may then either increase or decrease over the branch line. In a rare case, two mileages may actually apply over a common section of line.

All of these possibilities are evident in the Atlas. The reader will see where a mileage run begins, how it may end at a junction or carry on over the adjoining line. It will be seen where a long route can be a continuous mileage over its length or where it may be broken into various sections each with their own origin (not necessarily a new zero!). Where possible, such changes have been indicated by the abbreviation COM or 'change of mileage'. Mileages may also change on open lengths of route. This often points to the existence of a junction now removed but leaving evidence of its earlier location.

It is also necessary to make a distinction when apparently conflicting mileages may appear in print. In the context of this article and in this Atlas, the mileage for any location is that most recognised in the Railway Industry, usually having its origin in track engineering mileages. Other compilations of mileage tables and information have been constructed by others to suit passenger and freight requirements or personal survey but do not appear except in the absence of any other.

Future Probabilities

Not everything stays the same. Eventually miles and chains will change to kilometres as new lines develop or outside forces come into play but at the moment this is limited. CTRL was mentioned earlier and uses kilometres as the primary measurement mixed with miles and chains in the immediate area of St. Pancras station. London Underground lines (which do not appear in this Atlas) were converted to kilometres in 1972 and measured from zero at Ongar. Kilometre 'overlays' relating to the overhead electrification exist on both the WCML and ECML, also between Newcastle and Sunderland on the Metro lines. However, there is little evidence at the moment that full scale conversion to kilometre measurement is a planned project

In summary, railway mileages form an interesting sub-division of railway knowledge. They can inform, they can confuse and they can infuriate. Steeped in history and with logic of their own, they are likely to be in regular use for the foreseeable future.

Editor's note: a table equating chains with yards and an equivalent kilometre measurement is provided at the end of the Index. This will also help the reader calculate tunnel lengths where the maps give only the portal mileages.

Key to Map Pages

In this Atlas, the UK Rail Network has been detailed over 104 maps set out as shown on this Key. Certain parts of the country do not benefit from network lines and these have not been included. Complex areas have been extracted from the main maps and are shown on a further 42 area and inset maps which have their own Contents page at 106. The Key opposite describes the conventions and abbreviations used in the Atlas.

The user can either identify his or her area of interest from this Key Map or from the Index of locations which appears at the end of the numbered map section. Some maps have a grey tone boundary inside the frame. This represents an overlap with the next adjacent page. Adjacent map pages are indicated by a direction arrow.

The main map pages have a notional scale of 1:210,000. However, the scales of the area and inset maps vary widely from this figure and the values have not been calculated.

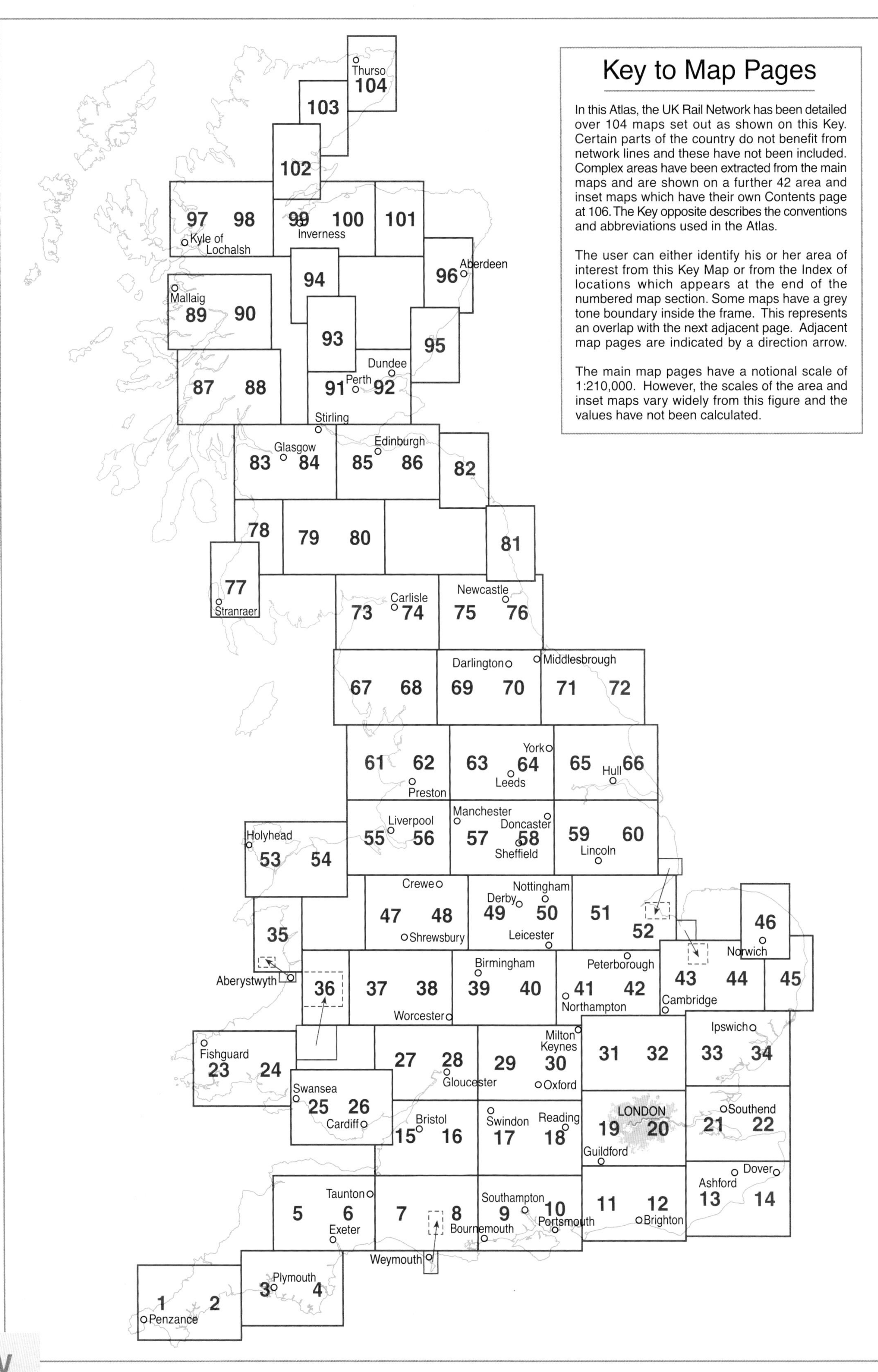

Key to Map Symbols

Non Electrified line
Electrified Overhead line
Electrified 3rd Rail line
Electrified 4th Rail line
Electrified Overhead & Conductor Rail line
Light Rail line
Line proposed or under construction
Tunnel
LC Level Crossing
see table for abbreviations
oou Line out of use
Connection to Depot, Siding, Private lines or other off-network locations
Bridge over or viaduct
5 Mile Marker

NB: The 5 mile marker is used on main map lines as an indicator of the true distance between two points which may have been varied in scale in schematic.

54m 36ch .... Distance in miles and chains from specified zero
57.60km .... Distance in Kilometres
24m 51ch/39.65km .... Station Mileage/ Signal Box Mileage
27m 00ch/43.45km .... Junction Mileage
12m 06ch/19.43km .... Crossing &Tunnel Mileage
NB: Mileages in brackets indicate an approximation

Station Platform
2 Station Platform (with platform number)
Platform out of use
Platform proposed or under construction
Signal Box (or ASC, IECC, SC or SCC)
Signal Room (CTRL)
Heritage Line SB
Gatebox

Level Crossing and Other Abbreviations

ABCL .... Automatic Barrier Crossing, locally monitored
AHBC .... Automatic Half-Barrier Crossing
AHBC-X .... Automatic Half-Barrier Crossing which works automatically for movements in the wrong direction
AOCL .... Automatic Open Crossing, locally monitored
CCTV .... Closed Circuit Television (normally MCB or MCG)
FP .... Footpath
OPEN (or OC) .... Open
LC .... Un-controlled Level Crossing
MCB .... Manually Controlled Barriers by Signaller or Keeper
MCB-OD .... Manually Controlled Barriers by Signaller or Keeper with obstruction detector
MCG .... Manually Controlled Gates by Signaller or Keeper
MGH .... Manually Controlled Gates, worked by hand
MSL .... Miniature Stop Lights
RC .... Remote Controlled
R/G .... Miniature Red/Green warning lights
R/G-X .... Miniature Red/Green warning lights which work automatically for movements in the wrong direction
TMO .... Traincrew Operated
UWC .... User Worked (with telephone for user to contact signaller)

NB: Any crossings without abbreviations are locally operated by signaller, crossing keeper or others.

ASC .... Area Signalling Centre
CE .... Civil Engineer's
COM .... Change of Mileage
CTRL .... Channel Tunnel Rail Link (HS1)
EGF .... Emergency Ground Frame
EMD .... Electric Maintenance Depot
GF .... Ground Frame
GSP .... Ground Switch Panel
IECC .... Integrated Electronic Control Centre
Jn .... Junction
LC .... Level Crossing (manned or open)
LUL .... London Underground
MOD .... Ministry of Defence
NR .... Network Rail
oou .... Out of Use
PGF .... Power Ground Frame
SB .... Signal Box
SC .... Signalling Centre
SCC .... Signalling Control Centre
Sdg(s) .... Siding or sidings
SR .... Signal Room (CTRL)
TMD .... Traction Maintenance Depot
UG/DG .... Up/Down Goods
UGL/DGL .... Up/Down Goods Loop

*NB: (**) appearing after the name of an ASC, IECC, SB, SC or SCC (where * is a letter) indicates the prefix applicable to the identification numbers of assets controlled from the relevant Box or Centre. (#) indicates there are 5 or more prefixes associated with the relevant SC and not itemised.*

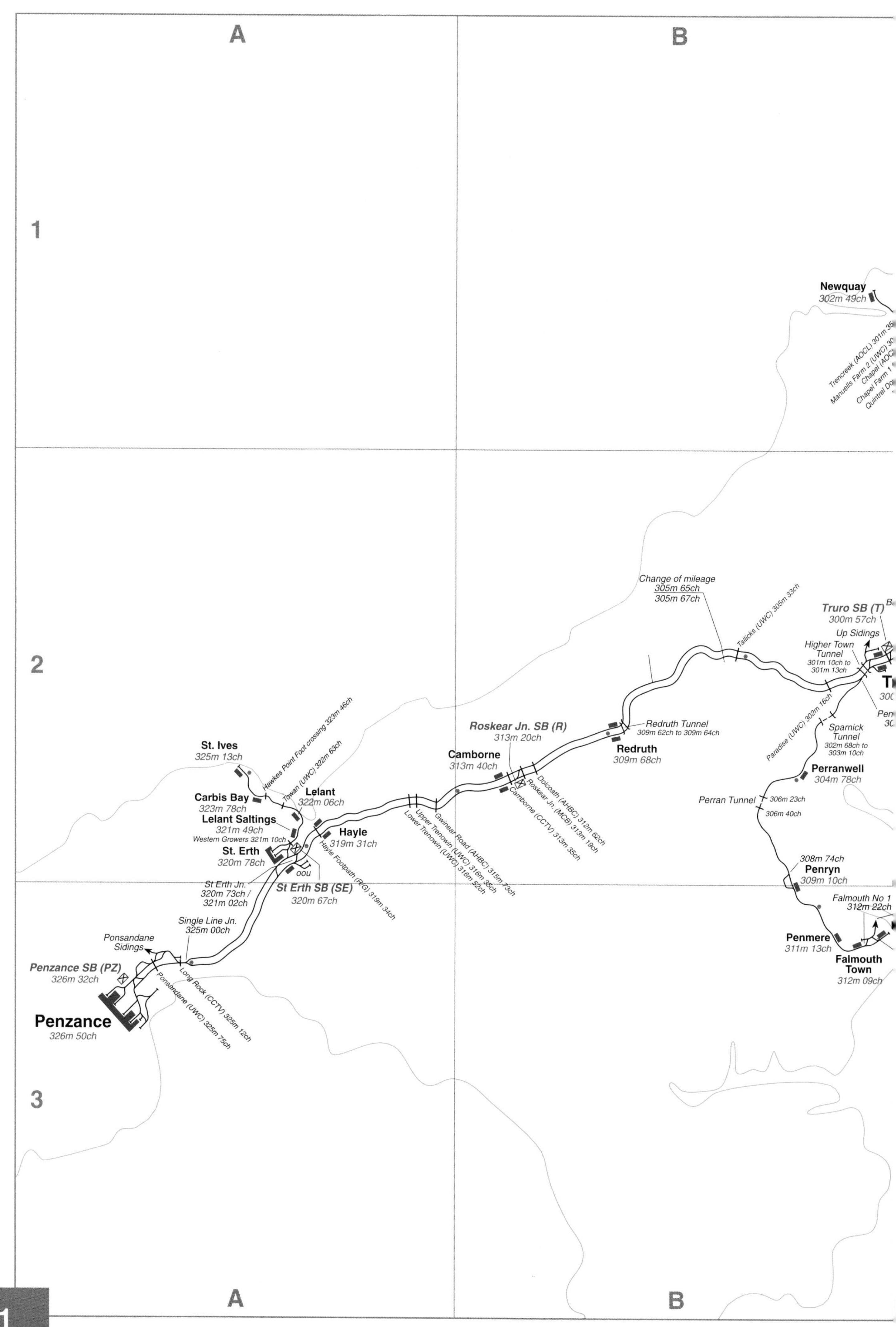

A
B
1
2
3
Newquay
302m 49ch
Trencreek (AOCL) 301m 35
Manuells Farm 2 (UWC) 30
Chapel (AOC
Chapel Farm 1
Quintrel Do
Change of mileage
305m 65ch
305m 67ch
Tallicks (UWC) 305m 33ch
Truro SB (T)
300m 57ch
Up Sidings
Higher Town Tunnel
301m 10ch to
301m 13ch
Paradise (UWC) 302m 16ch
Sparnick Tunnel
302m 68ch to
303m 10ch
Perranwell
304m 78ch
Perran Tunnel
306m 23ch
306m 40ch
308m 74ch
Penryn
309m 10ch
Falmouth No 1
312m 22ch
Penmere
311m 13ch
Falmouth Town
312m 09ch
Redruth Tunnel
309m 62ch to 309m 64ch
Redruth
309m 68ch
Roskear Jn. SB (R)
313m 20ch
Camborne
313m 40ch
Dolcoath (AHBC) 312m 62ch
Roskear Jn. (MCB) 313m 19ch
Camborne (CCTV) 313m 35ch
Gwinear Road (AHBC) 315m 73ch
Upper Trenowin (UWC) 316m 35ch
Lower Trenowin (UWC) 316m 52ch
St. Ives
325m 13ch
Hawkes Point Foot crossing 323m 46ch
Towan (UWC) 322m 63ch
Carbis Bay
323m 78ch
Lelant
322m 06ch
Lelant Saltings
321m 49ch
Western Growers 321m 10ch
St. Erth
320m 78ch
Hayle
319m 31ch
Hayle Footpath (FG) 319m 34ch
OOU
St Erth Jn.
320m 73ch /
321m 02ch
St Erth SB (SE)
320m 67ch
Single Line Jn.
325m 00ch
Ponsandane Sidings
Penzance SB (PZ)
326m 32ch
Long Rock (CCTV) 325m 12ch
Ponsandane (UWC) 325m 75ch
Penzance
326m 50ch
A
B

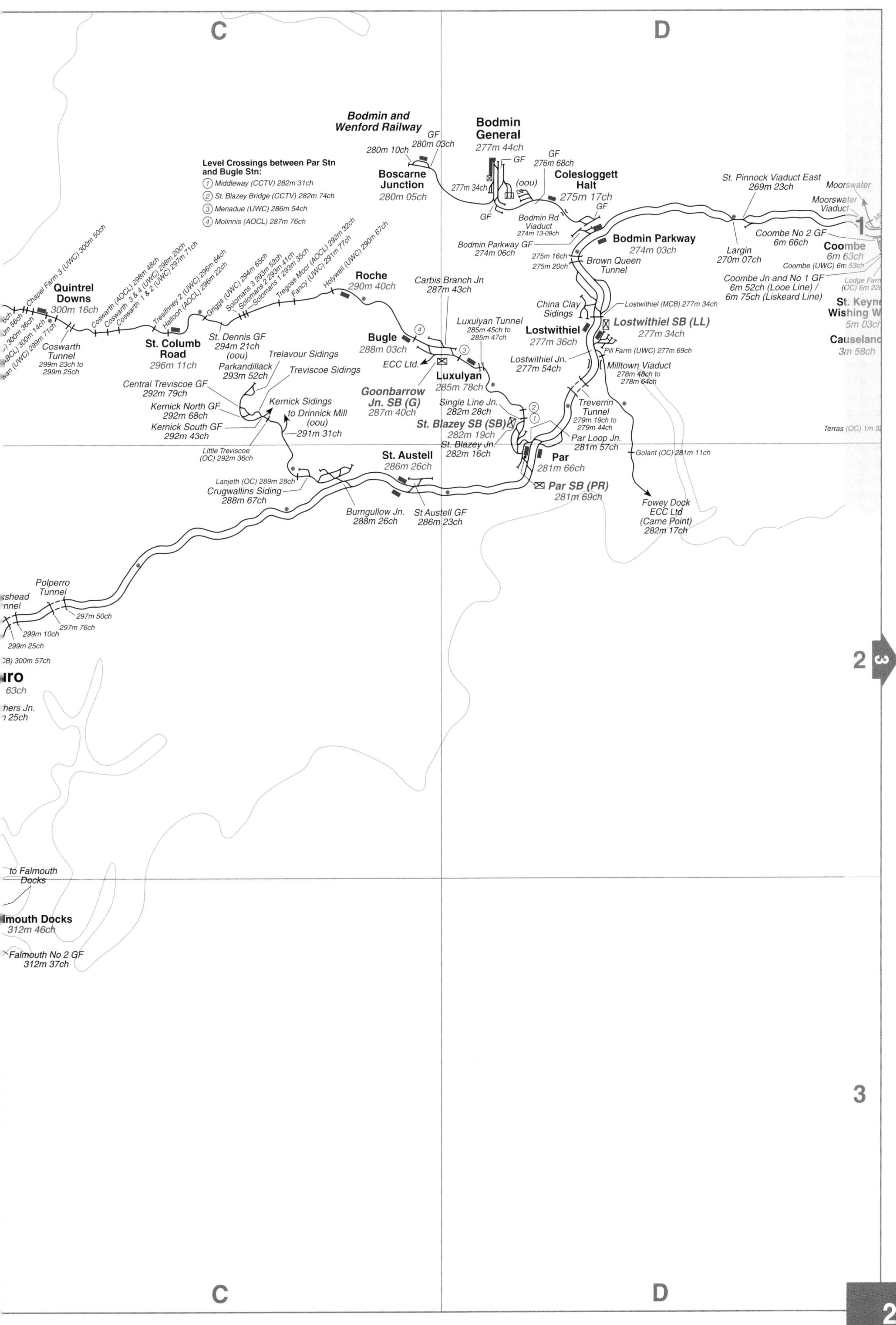

C
D
Bodmin and Wenford Railway
Bodmin General 277m 44ch
Boscarne Junction 280m 05ch
Coleslogett Halt 275m 17ch
Level Crossings between Par Stn and Bugle Stn:
1 Middleway (CCTV) 282m 31ch
2 St. Blazey Bridge (CCTV) 282m 74ch
3 Menadue (UWC) 286m 54ch
4 Molinnis (AOCL) 287m 76ch
Bodmin Parkway 274m 03ch
St. Pinnock Viaduct East 269m 23ch
Largin 270m 07ch
Coombe No 2 GF 6m 66ch
Coombe Jn and No 1 GF 6m 52ch (Looe Line) / 6m 75ch (Liskeard Line)
Brown Queen Tunnel
Quintrel Downs 300m 16ch
Roche 290m 40ch
Carbis Branch Jn 287m 43ch
Luxulyan Tunnel 285m 45ch to 285m 47ch
China Clay Sidings
Lostwithiel 277m 36ch
Lostwithiel SB (LL) 277m 34ch
St. Columb Road 296m 11ch
Bugle 288m 03ch
Luxulyan 285m 78ch
Goonbarrow Jn. SB (G) 287m 40ch
St. Blazey SB (SB) 282m 19ch
Lostwithiel Jn. 277m 54ch
Milltown Viaduct 278m 48ch to 278m 64ch
Treverrin Tunnel 279m 19ch to 279m 44ch
Par Loop Jn. 281m 57ch
Par 281m 66ch
Par SB (PR) 281m 69ch
St. Austell 286m 26ch
Burngullow Jn. 288m 26ch
St Austell GF 286m 23ch
Fowey Dock ECC Ltd (Carne Point) 282m 17ch
Golant (OC) 281m 11ch
Coswarth Tunnel 299m 23ch to 299m 25ch
Trelavour Sidings
Treviscoe Sidings
Kernick Sidings
to Drinnick Mill (oou)
Polperro Tunnel
to Falmouth Docks
Falmouth No 2 GF 312m 37ch
2
3
C
D

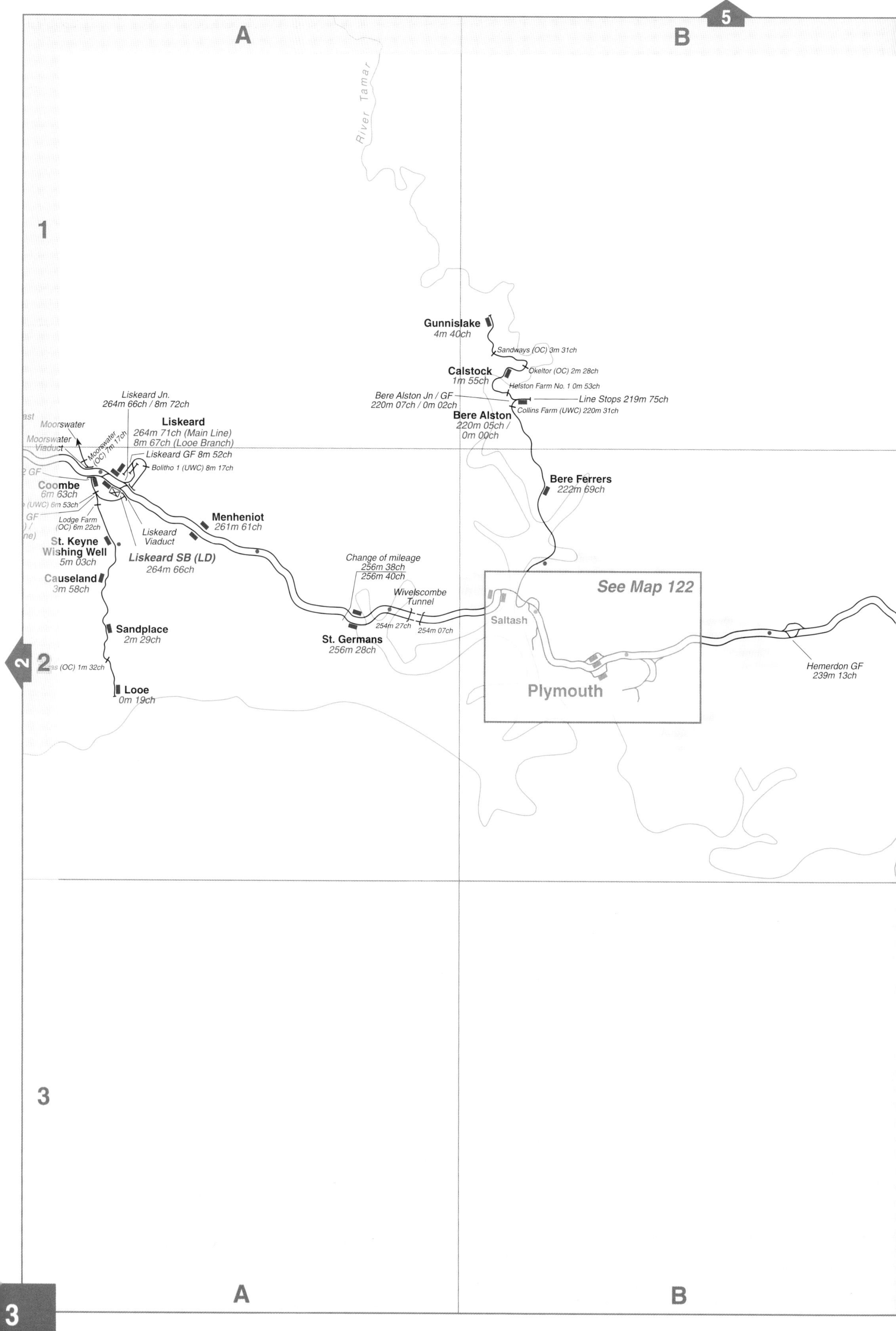

A
B
5
River Tamar
1
Gunnislake
4m 40ch
Sandways (OC) 3m 31ch
Okeltor (OC) 2m 28ch
Calstock
1m 55ch
Helston Farm No. 1 0m 53ch
Bere Alston Jn / GF
220m 07ch / 0m 02ch
Line Stops 219m 75ch
Collins Farm (UWC) 220m 31ch
Bere Alston
220m 05ch /
0m 00ch
Liskeard Jn.
264m 66ch / 8m 72ch
Moorswater
Moorswater Viaduct
Moorswater (OC) 7m 17ch
Liskeard
264m 71ch (Main Line)
8m 67ch (Looe Branch)
Liskeard GF 8m 52ch
Bolitho 1 (UWC) 8m 17ch
Coombe
6m 63ch
Lodge Farm (OC) 6m 22ch
Liskeard Viaduct
Menheniot
261m 61ch
St. Keyne Wishing Well
5m 03ch
Liskeard SB (LD)
264m 66ch
Causeland
3m 58ch
Sandplace
2m 29ch
(OC) 1m 32ch
Looe
0m 19ch
Bere Ferrers
222m 69ch
Change of mileage
256m 38ch
256m 40ch
Wivelscombe Tunnel
254m 27ch
254m 07ch
St. Germans
256m 28ch
See Map 122
Saltash
Plymouth
Hemerdon GF
239m 13ch
2
3

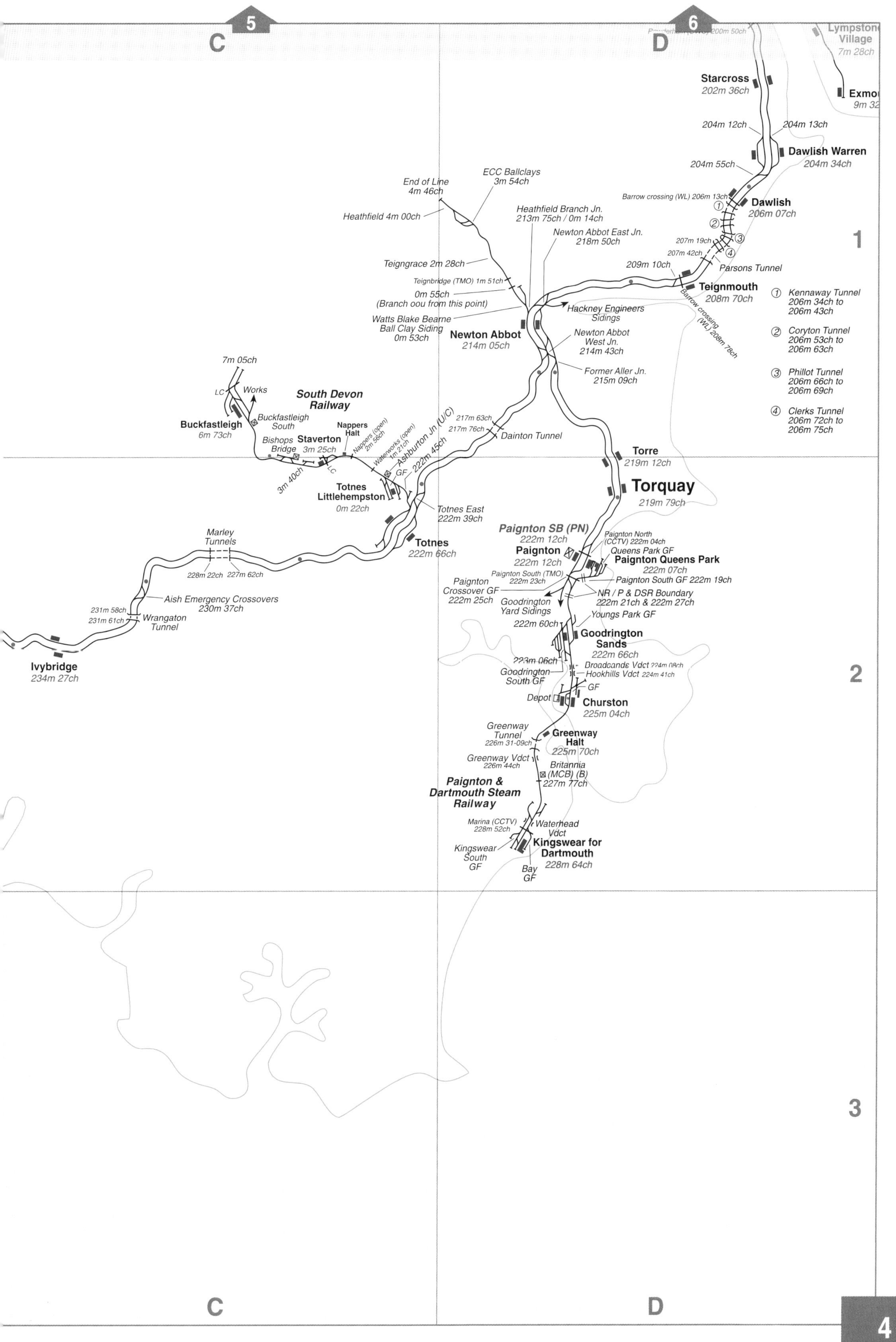

5
6
C
D
Lympstone Village
7m 28ch
Exmo
9m 32
Starcross
202m 36ch
204m 12ch
204m 13ch
Dawlish Warren
204m 34ch
204m 55ch
Barrow crossing (WL) 206m 13ch
Dawlish
206m 07ch
207m 19ch
207m 42ch
Parsons Tunnel
209m 10ch
Teignmouth
208m 70ch
Barrow crossing (WL) 208m 78ch
① Kennaway Tunnel 206m 34ch to 206m 43ch
② Coryton Tunnel 206m 53ch to 206m 63ch
③ Phillot Tunnel 206m 66ch to 206m 69ch
④ Clerks Tunnel 206m 72ch to 206m 75ch
1
End of Line
4m 46ch
ECC Ballclays
3m 54ch
Heathfield 4m 00ch
Heathfield Branch Jn.
213m 75ch / 0m 14ch
Newton Abbot East Jn.
218m 50ch
Teigngrace 2m 28ch
Teignbridge (TMO) 1m 51ch
0m 55ch
(Branch oou from this point)
Watts Blake Bearne
Ball Clay Siding
0m 53ch
Hackney Engineers Sidings
Newton Abbot
214m 05ch
Newton Abbot West Jn.
214m 43ch
Former Aller Jn.
215m 09ch
7m 05ch
LC
Works
South Devon Railway
Buckfastleigh
6m 73ch
Buckfastleigh South
Bishops Bridge
Staverton
3m 25ch
Nappers Halt
Nappers (open) 2m 56ch
Waterworks (open) 1m 21ch
Ashburton Jn (U/C)
222m 45ch
217m 63ch
217m 76ch
Dainton Tunnel
3m 40ch
LC
GF
Totnes Littlehempston
0m 22ch
Totnes East
222m 39ch
Totnes
222m 66ch
Torre
219m 12ch
Torquay
219m 79ch
Paignton SB (PN)
222m 12ch
Paignton
222m 12ch
Paignton North (CCTV) 222m 04ch
Queens Park GF
Paignton Queens Park
222m 07ch
Paignton South (TMO)
222m 23ch
Paignton South GF 222m 19ch
Paignton Crossover GF
222m 25ch
Goodrington Yard Sidings
NR / P & DSR Boundary
222m 21ch & 222m 27ch
Youngs Park GF
222m 60ch
Goodrington Sands
222m 66ch
223m 06ch
Goodrington South GF
Broadsands Vdct 224m 08ch
Hookhills Vdct 224m 41ch
GF
Depot
Churston
225m 04ch
Marley Tunnels
228m 22ch
227m 62ch
Aish Emergency Crossovers
230m 37ch
231m 58ch
231m 61ch
Wrangaton Tunnel
Ivybridge
234m 27ch
2
Greenway Tunnel
226m 31-09ch
Greenway Halt
225m 70ch
Greenway Vdct
226m 44ch
Britannia (MCB) (B)
227m 77ch
Paignton & Dartmouth Steam Railway
Marina (CCTV)
228m 52ch
Waterhead Vdct
Kingswear for Dartmouth
228m 64ch
Kingswear South GF
Bay GF
3
C
D

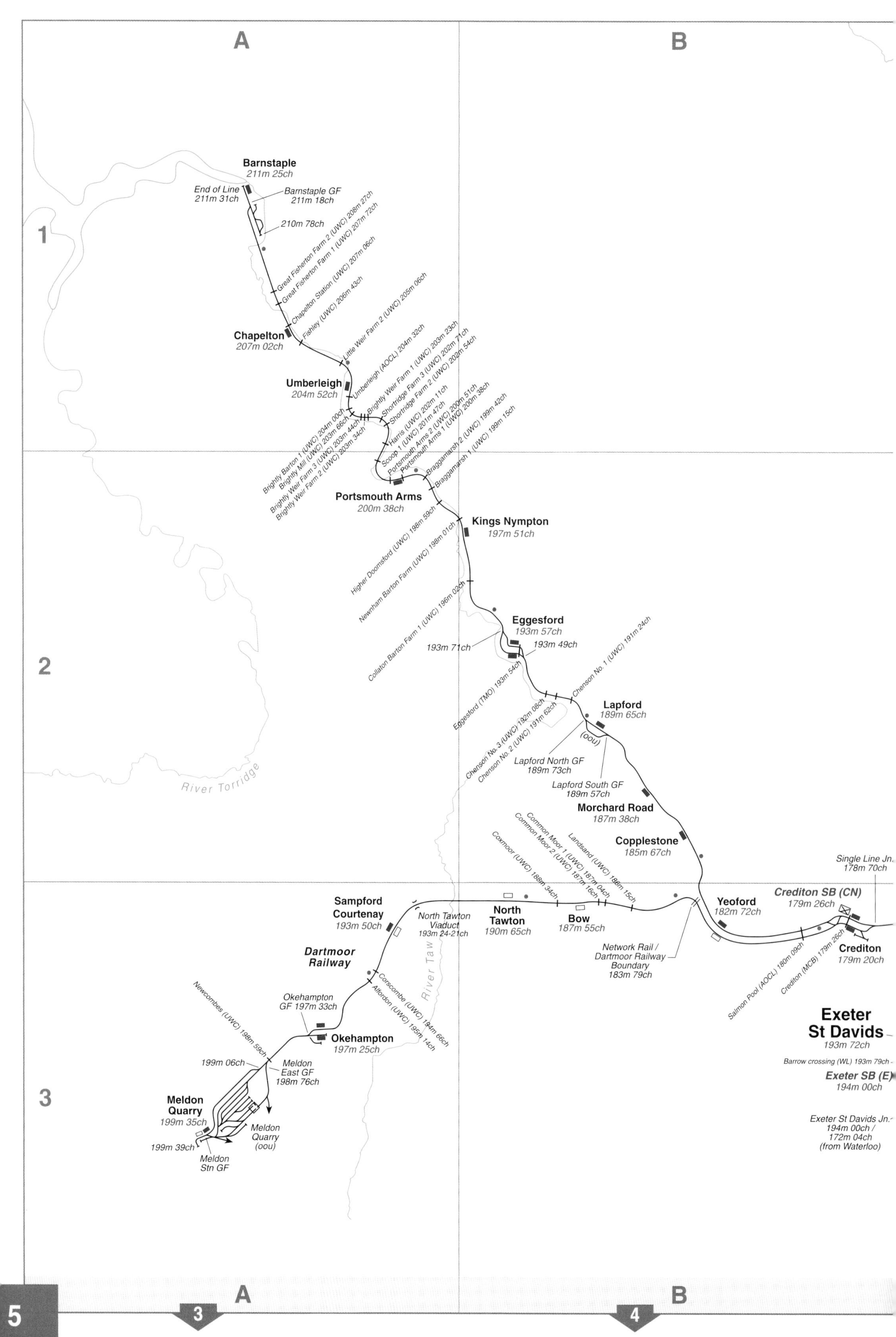
A
B
1
2
3
Barnstaple
211m 25ch
End of Line
211m 31ch
Barnstaple GF
211m 18ch
210m 78ch
Great Fisherton Farm 2 (UWC) 208m 27ch
Great Fisherton Farm 1 (UWC) 207m 72ch
Chapelton Station (UWC) 207m 06ch
Fishley (UWC) 206m 43ch
Little Weir Farm 2 (UWC) 205m 06ch
Chapelton
207m 02ch
Umberleigh
204m 52ch
Umberleigh (AOCL) 204m 32ch
Brightly Weir Farm 1 (UWC) 203m 23ch
Shortridge Farm 3 (UWC) 202m 71ch
Shortridge Farm 2 (UWC) 202m 54ch
Harris (UWC) 202m 11ch
Scoop 1 (UWC) 201m 47ch
Portsmouth Arms 2 (UWC) 200m 51ch
Portsmouth Arms 1 (UWC) 200m 38ch
Braggamarsh 2 (UWC) 199m 42ch
Braggamarsh 1 (UWC) 199m 15ch
Brightly Barton 1 (UWC) 204m 00ch
Brightly Mill (UWC) 203m 66ch
Brightly Weir Farm 3 (UWC) 203m 44ch
Brightly Weir Farm 2 (UWC) 203m 34ch
Portsmouth Arms
200m 38ch
Kings Nympton
197m 51ch
Higher Doomsford (UWC) 198m 59ch
Newnham Barton Farm (UWC) 198m 01ch
Collaton Barton Farm 1 (UWC) 196m 02ch
Eggesford
193m 57ch
193m 71ch
193m 49ch
Eggesford (TMO) 193m 54ch
Chenson No. 1 (UWC) 191m 24ch
Chenson No. 3 (UWC) 192m 08ch
Chenson No. 2 (UWC) 191m 62ch
Lapford
189m 65ch
(oou)
Lapford North GF
189m 73ch
Lapford South GF
189m 57ch
Morchard Road
187m 38ch
Copplestone
185m 67ch
River Torridge
River Taw
Common Moor 1 (UWC) 187m 04ch
Common Moor 2 (UWC) 187m 16ch
Landsand (UWC) 186m 15ch
Coxmoor (UWC) 188m 34ch
Single Line Jn.
178m 70ch
Crediton SB (CN)
179m 26ch
Sampford Courtenay
193m 50ch
North Tawton Viaduct
193m 24-21ch
North Tawton
190m 65ch
Bow
187m 55ch
Yeoford
182m 72ch
Crediton
179m 20ch
Network Rail / Dartmoor Railway Boundary
183m 79ch
Salmon Pool (AOCL) 180m 09ch
Crediton (MCB) 179m 26ch
Dartmoor Railway
Corscombe (UWC) 194m 66ch
Alfordon (UWC) 195m 14ch
Okehampton GF 197m 33ch
Okehampton
197m 25ch
Newcombes (UWC) 198m 59ch
199m 06ch
Meldon East GF
198m 76ch
Meldon Quarry
199m 35ch
Meldon Quarry
(oou)
199m 39ch
Meldon Stn GF
Exeter St Davids
193m 72ch
Barrow crossing (WL) 193m 79ch
Exeter SB (E)
194m 00ch
Exeter St Davids Jn.
194m 00ch /
172m 04ch
(from Waterloo)
A
B
3
4

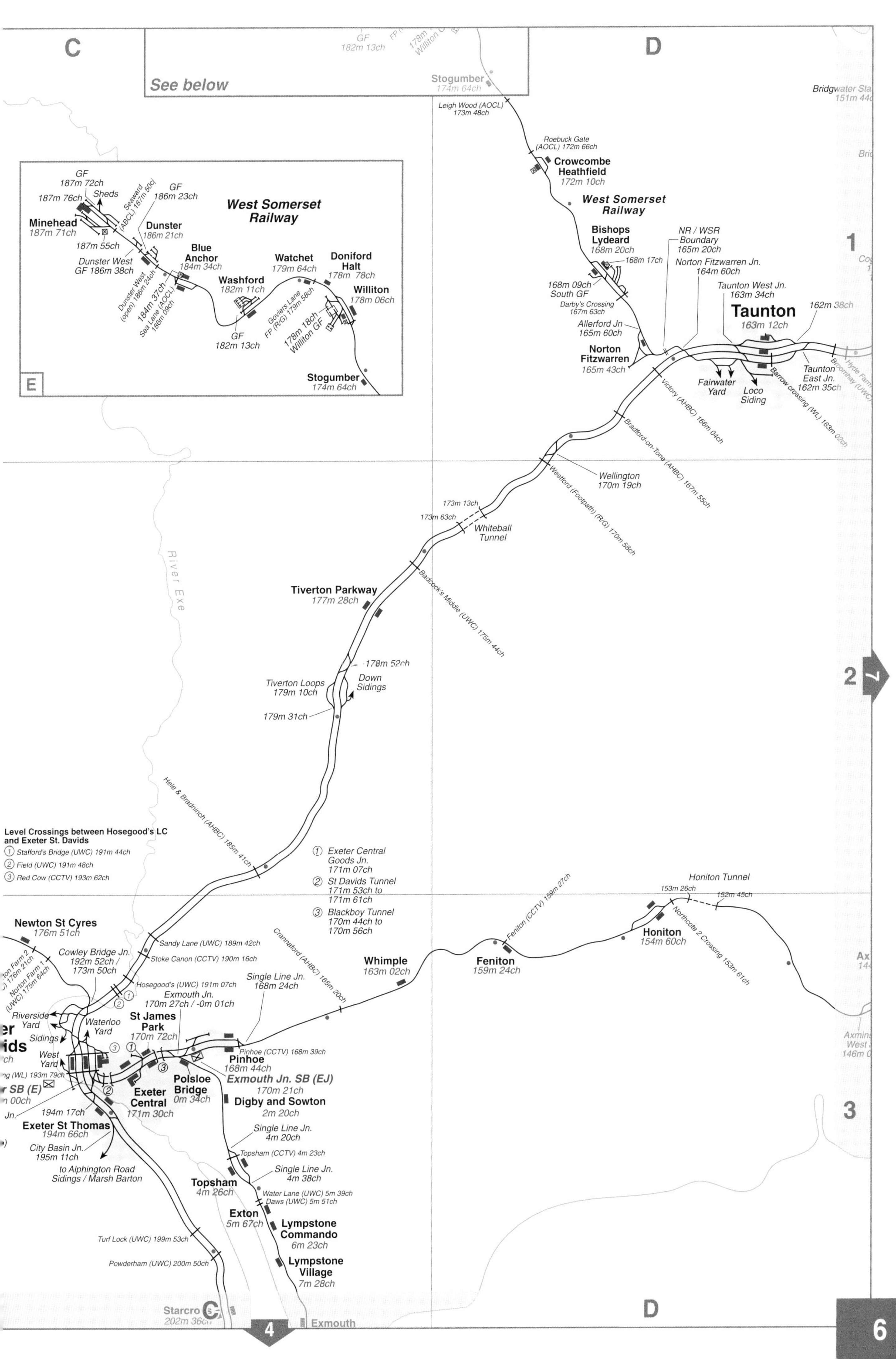
C
D
See below
Stogumber
174m 64ch
Leigh Wood (AOCL)
173m 48ch
Roebuck Gate
(AOCL) 172m 66ch
Crowcombe
Heathfield
172m 10ch
West Somerset
Railway
Bishops
Lydeard
168m 20ch
168m 17ch
168m 09ch
South GF
Darby's Crossing
167m 63ch
Allerford Jn
165m 60ch
Norton
Fitzwarren
165m 43ch
NR / WSR
Boundary
165m 20ch
Norton Fitzwarren Jn.
164m 60ch
Taunton West Jn.
163m 34ch
Taunton
163m 12ch
162m 38ch
Taunton
East Jn.
162m 35ch
Fairwater
Yard
Loco
Siding
Victory (AHBC) 166m 04ch
Barrow crossing (WL) 163m 02ch
Bradford-on-Tone (AHBC) 167m 55ch
Wellington
170m 19ch
Westford (Footpath) (R/G) 170m 58ch
173m 13ch
173m 63ch
Whiteball
Tunnel
Badcock's Middle (UWC) 175m 44ch
Tiverton Parkway
177m 28ch
178m 52ch
Tiverton Loops
179m 10ch
Down
Sidings
179m 31ch
River Exe
Hele & Bradninch (AHBC) 185m 41ch
Bridgwater
151m 44
GF
187m 72ch
187m 76ch
Sheds
Seaward (ABCL) 187m 50ch
GF
186m 23ch
Minehead
187m 71ch
Dunster
186m 21ch
187m 55ch
Dunster West
GF 186m 38ch
Blue
Anchor
184m 34ch
Dunster West (open) 186m 24ch
184m 37ch
Sea Lane (AOCL) 186m 09ch
Washford
182m 11ch
GF
182m 13ch
Watchet
179m 64ch
Doniford
Halt
178m 78ch
Goviers Lane FP (R/G) 179m 58ch
178m 18ch
Williton GF
Williton
178m 06ch
Stogumber
174m 64ch
E
1
2
3
Level Crossings between Hosegood's LC
and Exeter St. Davids
(1) Stafford's Bridge (UWC) 191m 44ch
(2) Field (UWC) 191m 48ch
(3) Red Cow (CCTV) 193m 62ch
(1) Exeter Central
Goods Jn.
171m 07ch
(2) St Davids Tunnel
171m 53ch to
171m 61ch
(3) Blackboy Tunnel
170m 44ch to
170m 56ch
Newton St Cyres
176m 51ch
Cowley Bridge Jn.
192m 52ch /
173m 50ch
Sandy Lane (UWC) 189m 42ch
Stoke Canon (CCTV) 190m 16ch
Hosegood's (UWC) 191m 07ch
Exmouth Jn.
170m 27ch / -0m 01ch
Single Line Jn.
168m 24ch
Crannaford (AHBC) 165m 20ch
Whimple
163m 02ch
Feniton (CCTV) 159m 27ch
Feniton
159m 24ch
Honiton Tunnel
153m 26ch
152m 45ch
Honiton
154m 60ch
Northcote 2 Crossing 153m 61ch
Riverside
Yard
Waterloo
Yard
Sidings
West
Yard
St James
Park
170m 72ch
Pinhoe (CCTV) 168m 39ch
Pinhoe
168m 44ch
Exmouth Jn. SB (EJ)
170m 21ch
Digby and Sowton
2m 20ch
Polsloe
Bridge
0m 34ch
Exeter
Central
171m 30ch
194m 17ch
Exeter St Thomas
194m 66ch
City Basin Jn.
195m 11ch
to Alphington Road
Sidings / Marsh Barton
Single Line Jn.
4m 20ch
Topsham (CCTV) 4m 23ch
Single Line Jn.
4m 38ch
Topsham
4m 26ch
Water Lane (UWC) 5m 39ch
Daws (UWC) 5m 51ch
Exton
5m 67ch
Lympstone
Commando
6m 23ch
Lympstone
Village
7m 28ch
Turf Lock (UWC) 199m 53ch
Powderham (UWC) 200m 50ch
Starcross
202m 36ch
Exmouth
4
7
6

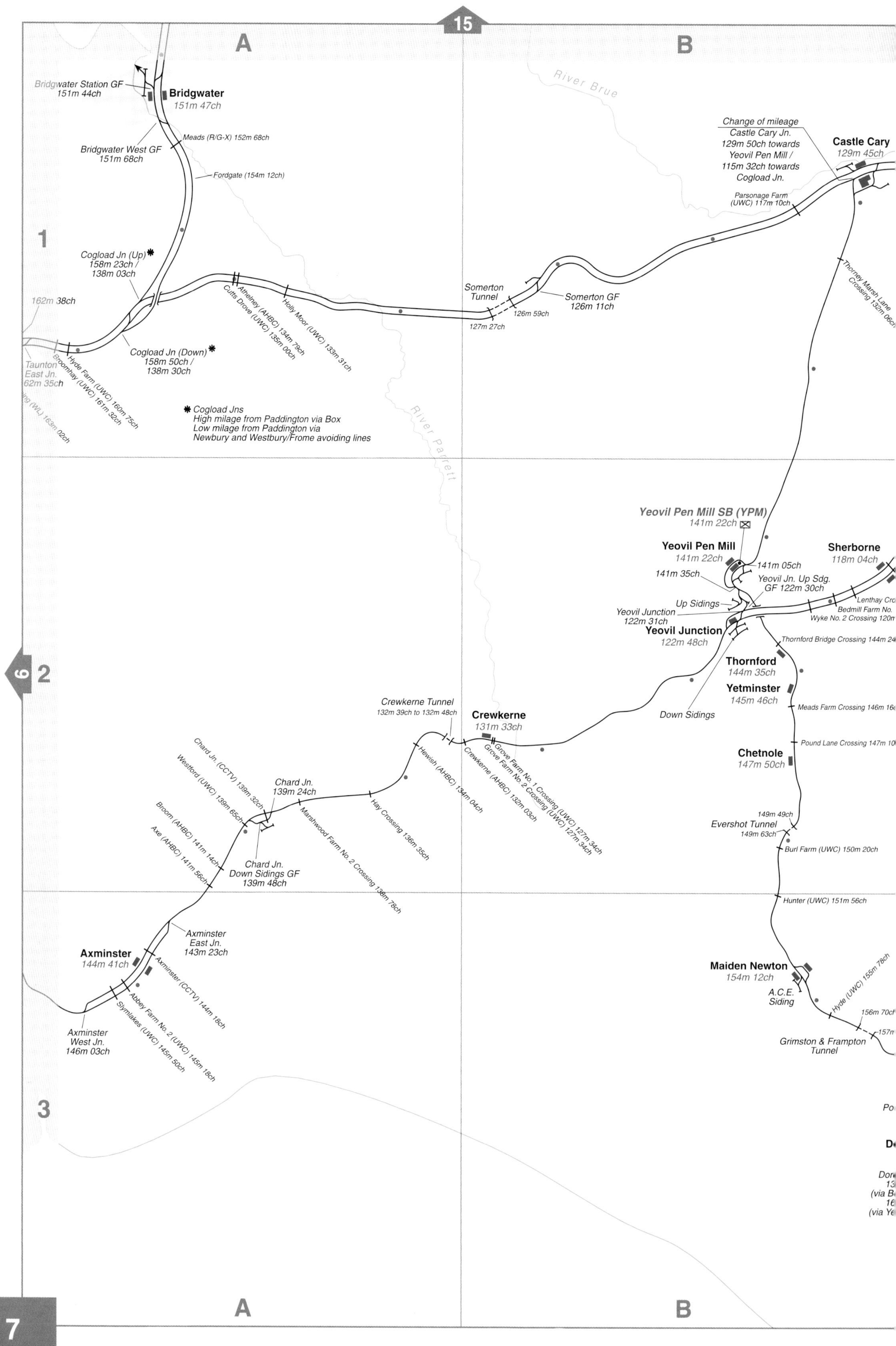
A
B
1
2
3
River Brue
River Parrett
Bridgwater Station GF 151m 44ch
Bridgwater
151m 47ch
Meads (R/G-X) 152m 68ch
Bridgwater West GF 151m 68ch
Fordgate (154m 12ch)
Cogload Jn (Up) ✱
158m 23ch / 138m 03ch
162m 38ch
Athelney (AHBC) 134m 79ch
Cutts Drove (UWC) 135m 00ch
Holly Moor (UWC) 133m 31ch
Cogload Jn (Down) ✱
158m 50ch / 138m 30ch
Taunton East Jn. 62m 35ch
Hyde Farm (UWC) 160m 75ch
Broomhay (UWC) 161m 32ch
✱ Cogload Jns
High milage from Paddington via Box
Low milage from Paddington via Newbury and Westbury/Frome avoiding lines
Somerton Tunnel
126m 59ch
127m 27ch
Somerton GF 126m 11ch
Change of mileage
Castle Cary Jn.
129m 50ch towards Yeovil Pen Mill /
115m 32ch towards Cogload Jn.
Castle Cary
129m 45ch
Parsonage Farm (UWC) 117m 10ch
Thorney Marsh Lane Crossing 132m 06ch
Yeovil Pen Mill SB (YPM)
141m 22ch
Yeovil Pen Mill
141m 22ch
141m 05ch
141m 35ch
Yeovil Jn. Up Sdg. GF 122m 30ch
Up Sidings
Yeovil Junction 122m 31ch
Yeovil Junction
122m 48ch
Sherborne
118m 04ch
Lenthay Cro
Bedmill Farm No.
Wyke No. 2 Crossing 120m
Thornford Bridge Crossing 144m 24
Thornford
144m 35ch
Yetminster
145m 46ch
Down Sidings
Meads Farm Crossing 146m 16
Pound Lane Crossing 147m 10
Chetnole
147m 50ch
149m 49ch
Evershot Tunnel
149m 63ch
Burl Farm (UWC) 150m 20ch
Hunter (UWC) 151m 56ch
Maiden Newton
154m 12ch
A.C.E. Siding
Hyde (UWC) 155m 78ch
156m 70ch
157m
Grimston & Frampton Tunnel
Crewkerne Tunnel
132m 39ch to 132m 48ch
Crewkerne
131m 33ch
Hewish (AHBC) 134m 04ch
Crewkerne (AHBC) 132m 03ch
Grove Farm No. 1 Crossing (UWC) 127m 34ch
Grove Farm No. 2 Crossing (UWC) 127m 34ch
Hay Crossing 136m 35ch
Marshwood Farm No. 2 Crossing 138m 78ch
Chard Jn.
139m 24ch
Chard Jn. (CCTV) 139m 32ch
Westford (UWC) 139m 65ch
Chard Jn. Down Sidings GF 139m 48ch
Broom (AHBC) 141m 14ch
Axe (AHBC) 141m 56ch
Axminster East Jn. 143m 23ch
Axminster
144m 41ch
Axminster (CCTV) 144m 18ch
Abbey Farm No. 2 (UWC) 145m 18ch
Slymlakes (UWC) 145m 50ch
Axminster West Jn. 146m 03ch
6

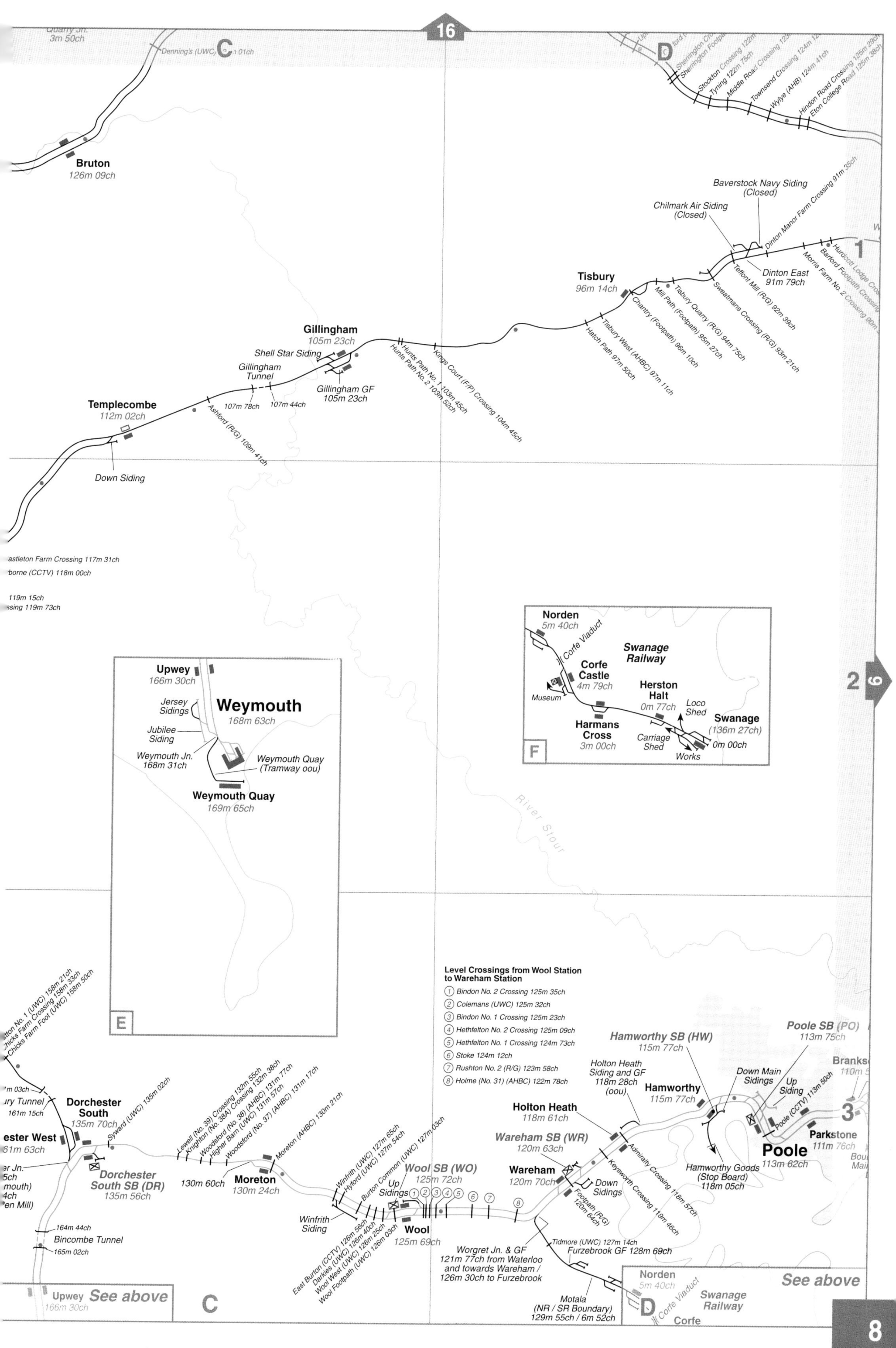
Bruton
126m 09ch
Tisbury
96m 14ch
Gillingham
105m 23ch
Shell Star Siding
Gillingham Tunnel
Gillingham GF
105m 23ch
Templecombe
112m 02ch
Down Siding
Baverstock Navy Siding (Closed)
Chilmark Air Siding (Closed)
Dinton East
91m 79ch
Norden
5m 40ch
Corfe Viaduct
Swanage Railway
Corfe Castle
4m 79ch
Museum
Herston Halt
0m 77ch
Loco Shed
Swanage
(136m 27ch)
0m 00ch
Harmans Cross
3m 00ch
Carriage Shed
Works
F
Upwey
166m 30ch
Jersey Sidings
Weymouth
168m 63ch
Jubilee Siding
Weymouth Jn.
168m 31ch
Weymouth Quay (Tramway oou)
Weymouth Quay
169m 65ch
E
River Stour
Level Crossings from Wool Station to Wareham Station
1 Bindon No. 2 Crossing 125m 35ch
2 Colemans (UWC) 125m 32ch
3 Bindon No. 1 Crossing 125m 23ch
4 Hethfelton No. 2 Crossing 125m 09ch
5 Hethfelton No. 1 Crossing 124m 73ch
6 Stoke 124m 12ch
7 Rushton No. 2 (R/G) 123m 58ch
8 Holme (No. 31) (AHBC) 122m 78ch
Dorchester South
135m 70ch
Dorchester South SB (DR)
135m 56ch
Moreton
130m 24ch
130m 60ch
Wool SB (WO)
125m 72ch
Up Sidings
Winfrith Siding
Wool
125m 69ch
Bincombe Tunnel
Worgret Jn. & GF
121m 77ch from Waterloo and towards Wareham / 126m 30ch to Furzebrook
Furzebrook GF 128m 69ch
Motala
(NR / SR Boundary)
129m 55ch / 6m 52ch
Holton Heath
118m 61ch
Wareham SB (WR)
120m 63ch
Wareham
120m 70ch
Down Sidings
Hamworthy SB (HW)
115m 77ch
Holton Heath Siding and GF
118m 28ch (oou)
Hamworthy
115m 77ch
Hamworthy Goods (Stop Board)
118m 05ch
Poole SB (PO)
113m 75ch
Down Main Sidings
Up Siding
Poole
113m 62ch
Parkstone
111m 76ch
Upwey See above
166m 30ch
See above
C
D
8

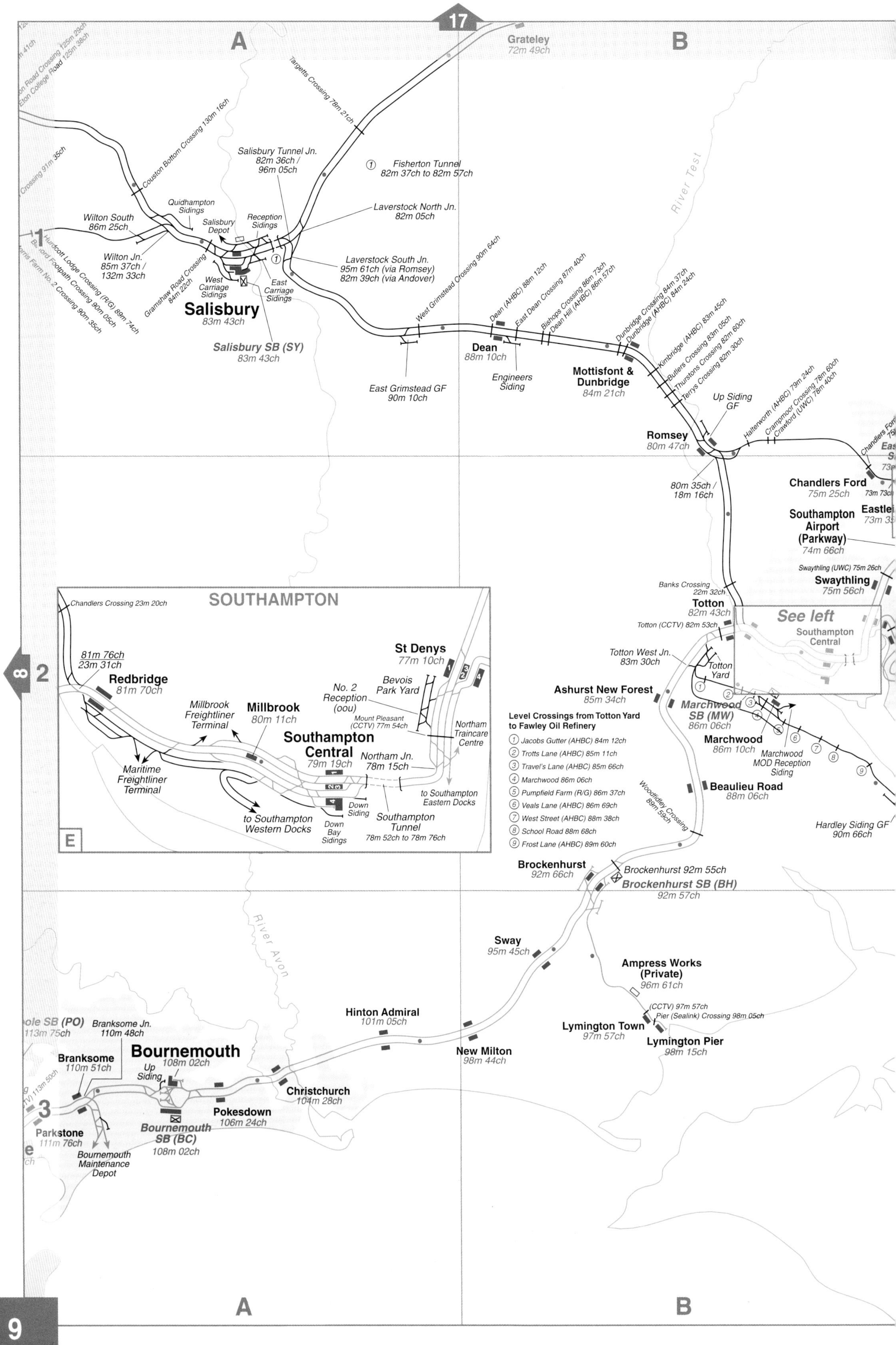
17
A
B
Grateley
72m 49ch
Targetts Crossing 78m 21ch
Couston Bottom Crossing 130m 16ch
Salisbury Tunnel Jn.
82m 36ch /
96m 05ch
1 Fisherton Tunnel
82m 37ch to 82m 57ch
River Test
Laverstock North Jn.
82m 05ch
Quidhampton Sidings
Wilton South
86m 25ch
Salisbury Depot
Reception Sidings
Wilton Jn.
85m 37ch /
132m 33ch
Laverstock South Jn.
95m 61ch (via Romsey)
82m 39ch (via Andover)
Gramshaw Road Crossing 84m 22ch
West Carriage Sidings
East Carriage Sidings
Salisbury
83m 43ch
Salisbury SB (SY)
83m 43ch
West Grimstead Crossing 90m 64ch
Dean (AHBC) 88m 12ch
East Dean Crossing 87m 40ch
Bishops Crossing 86m 73ch
Dean Hill (AHBC) 86m 57ch
Dunbridge Crossing 84m 37ch
Dunbridge (AHBC) 84m 24ch
Dean
88m 10ch
East Grimstead GF
90m 10ch
Engineers Siding
Mottisfont & Dunbridge
84m 21ch
Kimbridge (AHBC) 83m 45ch
Butlers Crossing 83m 05ch
Thurstons Crossing 82m 60ch
Terrys Crossing 82m 30ch
Up Siding GF
Halterworth (AHBC) 79m 24ch
Crampmoor Crossing 78m 60ch
Crawford (UWC) 78m 40ch
Romsey
80m 47ch
80m 35ch /
18m 16ch
Chandlers Ford
75m 25ch
Southampton Airport (Parkway)
74m 66ch
Swaythling (UWC) 75m 26ch
Swaythling
75m 56ch
Banks Crossing
22m 32ch
Totton
82m 43ch
Totton (CCTV) 82m 53ch
See left
Southampton Central
Totton West Jn.
83m 30ch
Totton Yard
Ashurst New Forest
85m 34ch
Marchwood SB (MW)
86m 06ch
Marchwood
86m 10ch
Marchwood MOD Reception Siding
Level Crossings from Totton Yard to Fawley Oil Refinery
1 Jacobs Gutter (AHBC) 84m 12ch
2 Trotts Lane (AHBC) 85m 11ch
3 Travel's Lane (AHBC) 85m 66ch
4 Marchwood 86m 06ch
5 Pumpfield Farm (R/G) 86m 37ch
6 Veals Lane (AHBC) 86m 69ch
7 West Street (AHBC) 88m 38ch
8 School Road 88m 68ch
9 Frost Lane (AHBC) 89m 60ch
Beaulieu Road
88m 06ch
Woodfidley Crossing 89m 59ch
Hardley Siding GF
90m 66ch
SOUTHAMPTON
Chandlers Crossing 23m 20ch
81m 76ch
23m 31ch
Redbridge
81m 70ch
Millbrook Freightliner Terminal
Millbrook
80m 11ch
Maritime Freightliner Terminal
St Denys
77m 10ch
Bevois Park Yard
No. 2 Reception (oou)
Mount Pleasant (CCTV) 77m 54ch
Northam Traincare Centre
Southampton Central
79m 19ch
Northam Jn.
78m 15ch
to Southampton Eastern Docks
to Southampton Western Docks
Down Siding
Down Bay Sidings
Southampton Tunnel
78m 52ch to 78m 76ch
E
Brockenhurst
92m 66ch
Brockenhurst 92m 55ch
Brockenhurst SB (BH)
92m 57ch
River Avon
Sway
95m 45ch
Ampress Works (Private)
96m 61ch
(CCTV) 97m 57ch
Pier (Sealink) Crossing 98m 05ch
Lymington Town
97m 57ch
Lymington Pier
98m 15ch
Hinton Admiral
101m 05ch
New Milton
98m 44ch
Branksome Jn.
110m 48ch
Branksome
110m 51ch
Bournemouth
108m 02ch
Up Siding
Christchurch
104m 28ch
Pokesdown
106m 24ch
Parkstone
111m 76ch
Bournemouth SB (BC)
108m 02ch
Bournemouth Maintenance Depot
8
1
2
3
A
B

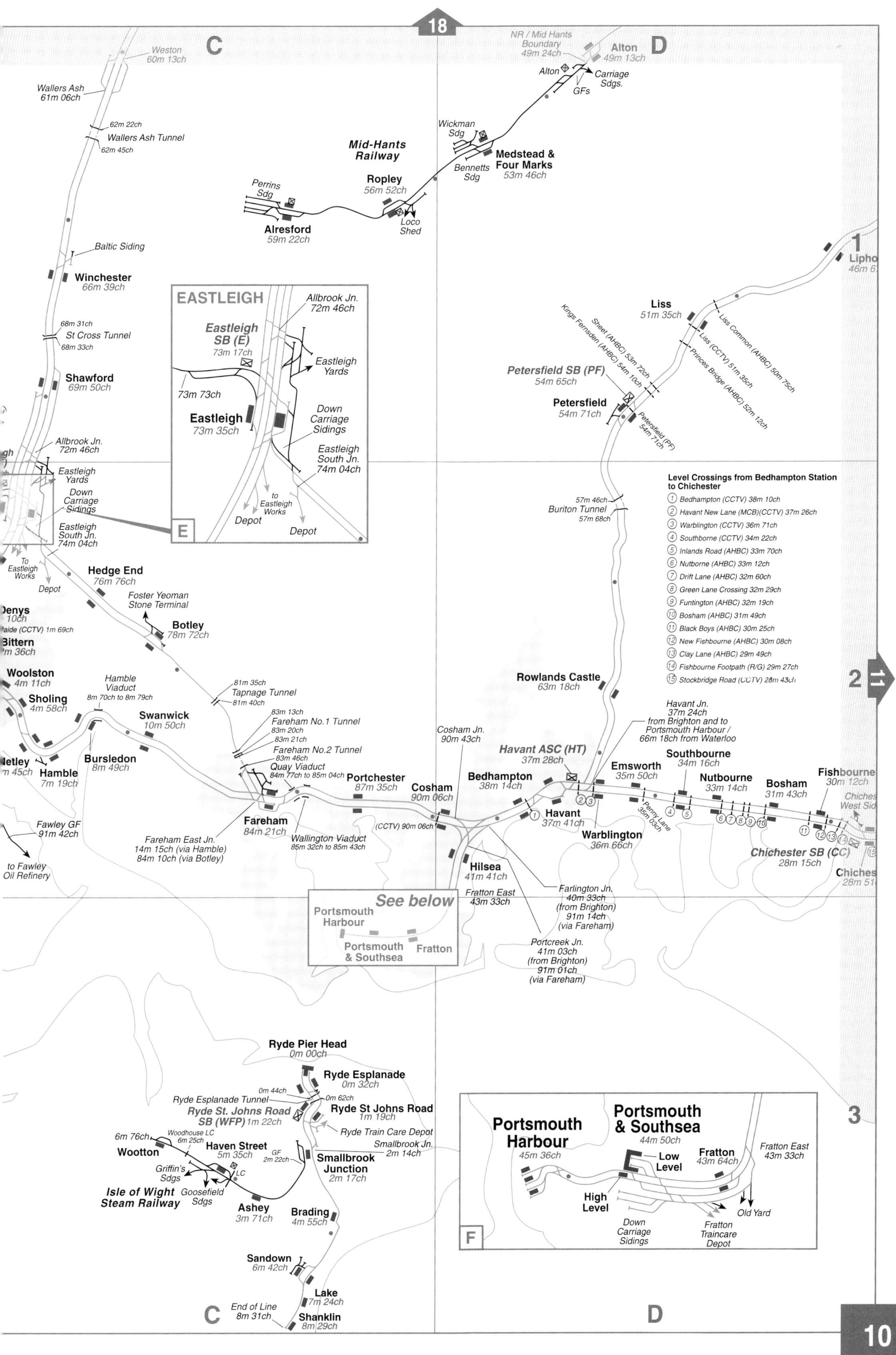
C
D
Weston
60m 13ch
Wallers Ash
61m 06ch
62m 22ch
Wallers Ash Tunnel
62m 45ch
Baltic Siding
Winchester
66m 39ch
68m 31ch
St Cross Tunnel
68m 33ch
Shawford
69m 50ch
Allbrook Jn.
72m 46ch
Eastleigh
Yards
Down
Carriage
Sidings
Eastleigh
South Jn.
74m 04ch
To
Eastleigh
Works
Depot
Hedge End
76m 76ch
Foster Yeoman
Stone Terminal
Botley
78m 72ch
Woolston
4m 11ch
Sholing
4m 58ch
Hamble
Viaduct
8m 70ch to 8m 79ch
Swanwick
10m 50ch
Bursledon
8m 49ch
Hamble
7m 19ch
81m 35ch
Tapnage Tunnel
81m 40ch
83m 13ch
Fareham No.1 Tunnel
83m 20ch
83m 21ch
Fareham No.2 Tunnel
83m 46ch
Quay Viaduct
84m 77ch to 85m 04ch
Portchester
87m 35ch
Fareham
84m 21ch
Fareham East Jn.
14m 15ch (via Hamble)
84m 10ch (via Botley)
Wallington Viaduct
85m 32ch to 85m 43ch
Fawley GF
91m 42ch
to Fawley
Oil Refinery
Mid-Hants
Railway
Perrins
Sdg
Alresford
59m 22ch
Ropley
56m 52ch
Loco
Shed
Wickman
Sdg
Bennetts
Sdg
Medstead &
Four Marks
53m 46ch
NR / Mid Hants
Boundary
49m 24ch
Alton
49m 13ch
Alton
Carriage
Sdgs.
GFs
EASTLEIGH
Eastleigh
SB (E)
73m 17ch
Allbrook Jn.
72m 46ch
Eastleigh
Yards
73m 73ch
Eastleigh
73m 35ch
Down
Carriage
Sidings
Eastleigh
South Jn.
74m 04ch
to
Eastleigh
Works
Depot
Depot
E
1
Liss
51m 35ch
Liss Common (AHBC) 50m 75ch
Liss (CCTV) 51m 35ch
Princes Bridge (AHBC) 52m 12ch
Sheet (AHBC) 53m 72ch
Kings Fernsden (AHBC) 54m 10ch
Petersfield SB (PF)
54m 65ch
Petersfield
54m 71ch
Petersfield (PF)
54m 71ch
57m 46ch
Buriton Tunnel
57m 68ch
Level Crossings from Bedhampton Station to Chichester
① Bedhampton (CCTV) 38m 10ch
② Havant New Lane (MCB)(CCTV) 37m 26ch
③ Warblington (CCTV) 36m 71ch
④ Southborne (CCTV) 34m 22ch
⑤ Inlands Road (AHBC) 33m 70ch
⑥ Nutborne (AHBC) 33m 12ch
⑦ Drift Lane (AHBC) 32m 60ch
⑧ Green Lane Crossing 32m 29ch
⑨ Funtington (AHBC) 32m 19ch
⑩ Bosham (AHBC) 31m 49ch
⑪ Black Boys (AHBC) 30m 25ch
⑫ New Fishbourne (AHBC) 30m 08ch
⑬ Clay Lane (AHBC) 29m 49ch
⑭ Fishbourne Footpath (R/G) 29m 27ch
⑮ Stockbridge Road (CCTV) 28m 43ch
2
Rowlands Castle
63m 18ch
Havant Jn.
37m 24ch
from Brighton and to
Portsmouth Harbour /
66m 18ch from Waterloo
Cosham Jn.
90m 43ch
Havant ASC (HT)
37m 28ch
Emsworth
35m 50ch
Southbourne
34m 16ch
Nutbourne
33m 14ch
Bosham
31m 43ch
Fishbourne
30m 12ch
Cosham
90m 06ch
(CCTV) 90m 06ch
Bedhampton
38m 14ch
Havant
37m 41ch
Warblington
36m 66ch
Penny Lane
35m 03ch
Chichester SB (CC)
28m 15ch
Hilsea
41m 41ch
Fratton East
43m 33ch
Farlington Jn.
40m 33ch
(from Brighton)
91m 14ch
(via Fareham)
Portcreek Jn.
41m 03ch
(from Brighton)
91m 01ch
(via Fareham)
See below
Portsmouth
Harbour
Portsmouth
& Southsea
Fratton
Ryde Pier Head
0m 00ch
Ryde Esplanade
0m 32ch
0m 44ch
Ryde Esplanade Tunnel
0m 62ch
Ryde St. Johns Road
SB (WFP) 1m 22ch
Ryde St Johns Road
1m 19ch
Ryde Train Care Depot
Smallbrook Jn.
2m 14ch
6m 76ch
Woodhouse LC
6m 25ch
Wootton
Haven Street
5m 35ch
GF
2m 22ch
Smallbrook
Junction
2m 17ch
Griffin's
Sdgs
LC
Isle of Wight
Steam Railway
Goosefield
Sdgs
Ashey
3m 71ch
Brading
4m 55ch
Sandown
6m 42ch
Lake
7m 24ch
End of Line
8m 31ch
Shanklin
8m 29ch
Portsmouth
Harbour
45m 36ch
Portsmouth
& Southsea
44m 50ch
Low
Level
High
Level
Fratton
43m 64ch
Fratton East
43m 33ch
Old Yard
Down
Carriage
Sidings
Fratton
Traincare
Depot
F
3

11

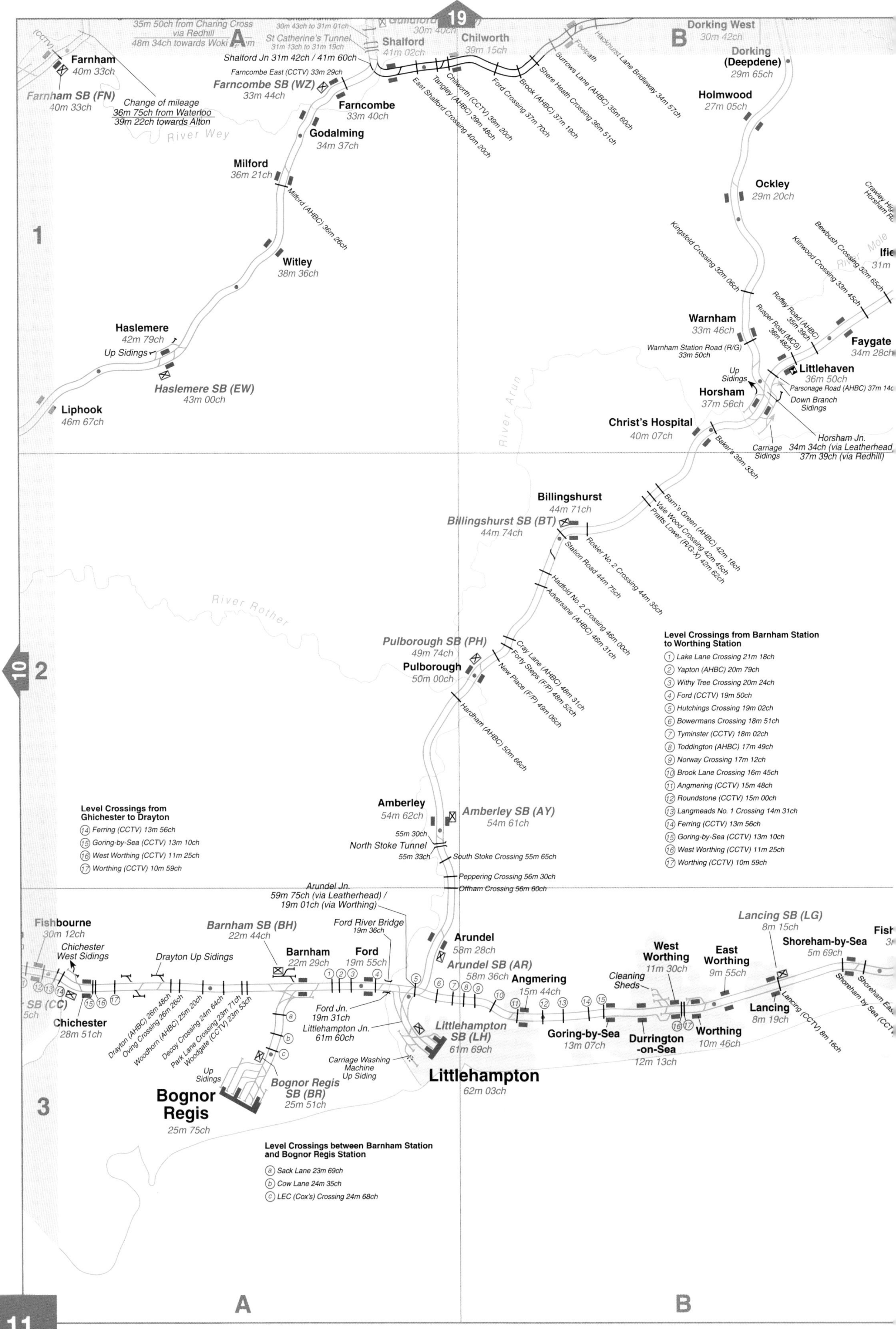

19
10
A
B
1
2
3
35m 50ch from Charing Cross via Redhill
48m 34ch towards Woking
30m 43ch to 31m 01ch
St Catherine's Tunnel
31m 13ch to 31m 19ch
Shalford
41m 02ch
Chilworth
39m 15ch
Dorking West
30m 42ch
Dorking (Deepdene)
29m 65ch
Farnham
40m 33ch
Farnham SB (FN)
40m 33ch
Shalford Jn 31m 42ch / 41m 60ch
Farncombe East (CCTV) 33m 29ch
Farncombe SB (WZ)
33m 44ch
Farncombe
33m 40ch
Change of mileage
36m 75ch from Waterloo
39m 22ch towards Alton
River Wey
Godalming
34m 37ch
East Shalford Crossing 40m 20ch
Tangley (AHBC) 39m 48ch
Chilworth (CCTV) 39m 20ch
Ford Crossing 37m 70ch
Brook (AHBC) 37m 19ch
Shere Heath Crossing 36m 51ch
Surrows Lane (AHBC) 35m 60ch
Footpath
Hackhurst Lane Bridleway 34m 57ch
Holmwood
27m 05ch
Milford
36m 21ch
Milford (AHBC) 36m 26ch
Ockley
29m 20ch
Witley
38m 36ch
Kingsfold Crossing 32m 06ch
Kilnwood Crossing 33m 45ch
Bewbush Crossing 32m 65ch
River Mole
Haslemere
42m 79ch
Up Sidings
Haslemere SB (EW)
43m 00ch
Liphook
46m 67ch
Warnham
33m 46ch
Warnham Station Road (R/G) 33m 50ch
Roffey Road (AHBC) 35m 39ch
Rusper Road (MCG) 36m 48ch
Faygate
34m 28ch
Littlehaven
36m 50ch
Parsonage Road (AHBC) 37m 14ch
Up Sidings
Horsham
37m 56ch
Down Branch Sidings
River Arun
Christ's Hospital
40m 07ch
Baker's 39m 33ch
Carriage Sidings
Horsham Jn.
34m 34ch (via Leatherhead)
37m 39ch (via Redhill)
Billingshurst
44m 71ch
Billingshurst SB (BT)
44m 74ch
Barn's Green (AHBC) 42m 18ch
Vale Wood Crossing 42m 45ch
Pratts Lower (R/G-X) 42m 62ch
Station Road 44m 75ch
Rosier No. 2 Crossing 44m 35ch
Hadfold No. 2 Crossing 46m 00ch
Adversane (AHBC) 46m 31ch
River Rother
Pulborough SB (PH)
49m 74ch
Pulborough
50m 00ch
Cray Lane (AHBC) 48m 31ch
Forty Steps (F/P) 48m 52ch
New Place (F/P) 49m 06ch
Hardham (AHBC) 50m 66ch
Level Crossings from Barnham Station to Worthing Station
1 Lake Lane Crossing 21m 18ch
2 Yapton (AHBC) 20m 79ch
3 Withy Tree Crossing 20m 24ch
4 Ford (CCTV) 19m 50ch
5 Hutchings Crossing 19m 02ch
6 Bowermans Crossing 18m 51ch
7 Tyminster (CCTV) 18m 02ch
8 Toddington (AHBC) 17m 49ch
9 Norway Crossing 17m 12ch
10 Brook Lane Crossing 16m 45ch
11 Angmering (CCTV) 15m 48ch
12 Roundstone (CCTV) 15m 00ch
13 Langmeads No. 1 Crossing 14m 31ch
14 Ferring (CCTV) 13m 56ch
15 Goring-by-Sea (CCTV) 13m 10ch
16 West Worthing (CCTV) 11m 25ch
17 Worthing (CCTV) 10m 59ch
Amberley
54m 62ch
Amberley SB (AY)
54m 61ch
Level Crossings from Ghichester to Drayton
14 Ferring (CCTV) 13m 56ch
15 Goring-by-Sea (CCTV) 13m 10ch
16 West Worthing (CCTV) 11m 25ch
17 Worthing (CCTV) 10m 59ch
55m 30ch
North Stoke Tunnel
55m 33ch
South Stoke Crossing 55m 65ch
Peppering Crossing 56m 30ch
Offham Crossing 56m 60ch
Arundel Jn.
59m 75ch (via Leatherhead) /
19m 01ch (via Worthing)
Fishbourne
30m 12ch
Chichester West Sidings
Barnham SB (BH)
22m 44ch
Drayton Up Sidings
Barnham
22m 29ch
Ford
19m 55ch
Ford River Bridge
19m 36ch
Arundel
58m 28ch
Arundel SB (AR)
58m 36ch
Lancing SB (LG)
8m 15ch
Shoreham-by-Sea
5m 69ch
West Worthing
11m 30ch
East Worthing
9m 55ch
Cleaning Sheds
Angmering
15m 44ch
Chichester
28m 51ch
Drayton (AHBC) 26m 48ch
Oving Crossing 26m 26ch
Woodhorn (AHBC) 25m 20ch
Decoy Crossing 24m 64ch
Park Lane Crossing 23m 71ch
Woodgate (CCTV) 23m 53ch
Ford Jn.
19m 31ch
Littlehampton Jn.
61m 60ch
Littlehampton SB (LH)
61m 69ch
Goring-by-Sea
13m 07ch
Durrington-on-Sea
12m 13ch
Worthing
10m 46ch
Lancing
8m 19ch
Lancing (CCTV) 8m 16ch
Shoreham by Sea (CCTV)
Carriage Washing Machine Up Siding
Up Sidings
Bognor Regis SB (BR)
25m 51ch
Littlehampton
62m 03ch
Bognor Regis
25m 75ch
Level Crossings between Barnham Station and Bognor Regis Station
a Sack Lane 23m 69ch
b Cow Lane 24m 35ch
c LEC (Cox's) Crossing 24m 68ch

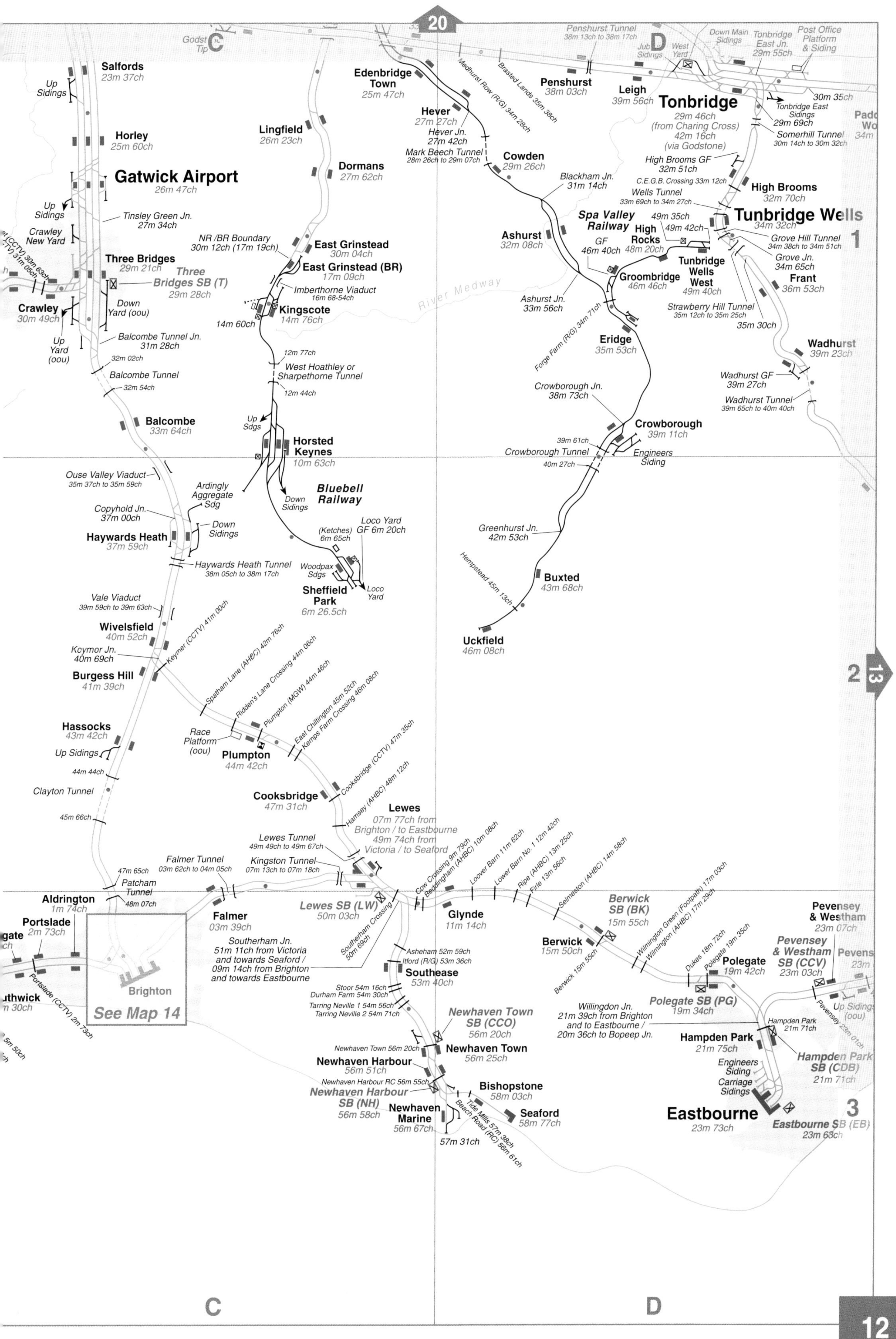

20
Salfords
23m 37ch
Up Sidings
Horley
25m 60ch
Gatwick Airport
26m 47ch
Up Sidings
Tinsley Green Jn.
27m 34ch
Crawley New Yard
Three Bridges
29m 21ch
Three Bridges SB (T)
29m 28ch
Crawley
30m 49ch
Down Yard (oou)
Up Yard (oou)
Balcombe Tunnel Jn.
31m 28ch
32m 02ch
Balcombe Tunnel
32m 54ch
Balcombe
33m 64ch
Ouse Valley Viaduct
35m 37ch to 35m 59ch
Ardingly Aggregate Sdg
Copyhold Jn.
37m 00ch
Down Sidings
Haywards Heath
37m 59ch
Haywards Heath Tunnel
38m 05ch to 38m 17ch
Vale Viaduct
39m 59ch to 39m 63ch
Wivelsfield
40m 52ch
Keymer Jn.
40m 69ch
Keymer (CCTV) 41m 00ch
Burgess Hill
41m 39ch
Hassocks
43m 42ch
Up Sidings
44m 44ch
Clayton Tunnel
45m 66ch
47m 65ch
Patcham Tunnel
48m 07ch
Aldrington
1m 74ch
Portslade
2m 73ch
Portslade (CCTV) 2m 73ch
Brighton
See Map 14
Godstone Tip
Lingfield
26m 23ch
Dormans
27m 62ch
NR /BR Boundary
30m 12ch (17m 19ch)
East Grinstead
30m 04ch
East Grinstead (BR)
17m 09ch
Imberthorne Viaduct
16m 68-54ch
Kingscote
14m 76ch
14m 60ch
12m 77ch
West Hoathley or Sharpethorne Tunnel
12m 44ch
Up Sdgs
Horsted Keynes
10m 63ch
Down Sidings
Bluebell Railway
(Ketches) 6m 65ch
Loco Yard GF 6m 20ch
Woodpax Sdgs
Sheffield Park
6m 26.5ch
Loco Yard
Spatham Lane (AHBC) 42m 76ch
Ridden's Lane Crossing 44m 06ch
Plumpton (MGW) 44m 46ch
East Chiltington 45m 52ch
Kemps Farm Crossing 46m 08ch
Race Platform (oou)
Plumpton
44m 42ch
Cooksbridge (CCTV) 47m 35ch
Cooksbridge
47m 31ch
Hamsey (AHBC) 48m 12ch
Lewes
07m 77ch from Brighton / to Eastbourne
49m 74ch from Victoria / to Seaford
Lewes Tunnel
49m 49ch to 49m 67ch
Kingston Tunnel
07m 13ch to 07m 18ch
Falmer Tunnel
03m 62ch to 04m 05ch
Falmer
03m 39ch
Lewes SB (LW)
50m 03ch
Southerham Jn.
51m 11ch from Victoria and towards Seaford /
09m 14ch from Brighton and towards Eastbourne
Southerham Crossing 50m 69ch
Cow Crossing 9m 79ch
Beddingham (AHBC) 10m 08ch
Glynde
11m 14ch
Loover Barn 11m 62ch
Lower Barn No. 1 12m 42ch
Ripe (AHBC) 13m 25ch
Firle 13m 56ch
Selmeston (AHBC) 14m 58ch
Asheham 52m 59ch
Itford (R/G) 53m 36ch
Southease
53m 40ch
Stoor 54m 16ch
Durham Farm 54m 30ch
Tarring Neville 1 54m 56ch
Tarring Neville 2 54m 71ch
Newhaven Town SB (CCO)
56m 20ch
Newhaven Town 56m 20ch
Newhaven Town
56m 25ch
Newhaven Harbour
56m 51ch
Newhaven Harbour RC 56m 55ch
Newhaven Harbour SB (NH)
56m 58ch
Newhaven Marine
56m 67ch
57m 31ch
Bishopstone
58m 03ch
Tide Mills 57m 38ch
Beach Road (RC) 56m 61ch
Seaford
58m 77ch
Edenbridge Town
25m 47ch
Hever
27m 27ch
Hever Jn.
27m 42ch
Mark Beech Tunnel
28m 26ch to 29m 07ch
Medhurst Row (R/G) 34m 28ch
Brasted Lands 35m 38ch
Cowden
29m 26ch
Penshurst Tunnel
38m 13ch to 38m 17ch
Penshurst
38m 03ch
Jub Sidings
West Yard
Leigh
39m 56ch
Down Main Sidings
Tonbridge East Jn.
29m 55ch
Post Office Platform & Siding
Tonbridge
29m 46ch
(from Charing Cross)
42m 16ch
(via Godstone)
30m 35ch
Tonbridge East Sidings
29m 69ch
Somerhill Tunnel
30m 14ch to 30m 32ch
High Brooms GF
32m 51ch
C.E.G.B. Crossing 33m 12ch
High Brooms
32m 70ch
Wells Tunnel
33m 69ch to 34m 27ch
Blackham Jn.
31m 14ch
Spa Valley Railway
High Rocks
48m 20ch
49m 35ch
49m 42ch
GF
46m 40ch
Tunbridge Wells
34m 32ch
Grove Hill Tunnel
34m 38ch to 34m 51ch
Grove Jn.
34m 65ch
Ashurst
32m 08ch
Groombridge
46m 46ch
Tunbridge Wells West
49m 40ch
Frant
36m 53ch
River Medway
Ashurst Jn.
33m 56ch
Strawberry Hill Tunnel
35m 12ch to 35m 25ch
35m 30ch
Forge Farm (R/G) 34m 71ch
Eridge
35m 53ch
Wadhurst
39m 23ch
Wadhurst GF
39m 27ch
Wadhurst Tunnel
39m 65ch to 40m 40ch
Crowborough Jn.
38m 73ch
Crowborough
39m 11ch
39m 61ch
Crowborough Tunnel
40m 27ch
Engineers Siding
Greenhurst Jn.
42m 53ch
Buxted
43m 68ch
Hempstead 45m 13ch
Uckfield
46m 08ch
Berwick SB (BK)
15m 55ch
Berwick
15m 50ch
Berwick 15m 55ch
Wilmington Green (Footpath) 17m 03ch
Wilmington (AHBC) 17m 29ch
Dukes 18m 72ch
Polegate 19m 35ch
Polegate
19m 42ch
Polegate SB (PG)
19m 34ch
Willingdon Jn.
21m 39ch from Brighton and to Eastbourne /
20m 36ch to Bopeep Jn.
Pevensey & Westham
23m 07ch
Pevensey & Westham SB (CCV)
23m 03ch
Up Sidings (oou)
Pevensey 23m 01ch
Hampden Park
21m 71ch
Hampden Park
21m 75ch
Hampden Park SB (CDB)
21m 71ch
Engineers Siding
Carriage Sidings
Eastbourne
23m 73ch
Eastbourne SB (EB)
23m 63ch
1
2
13
3
C
D

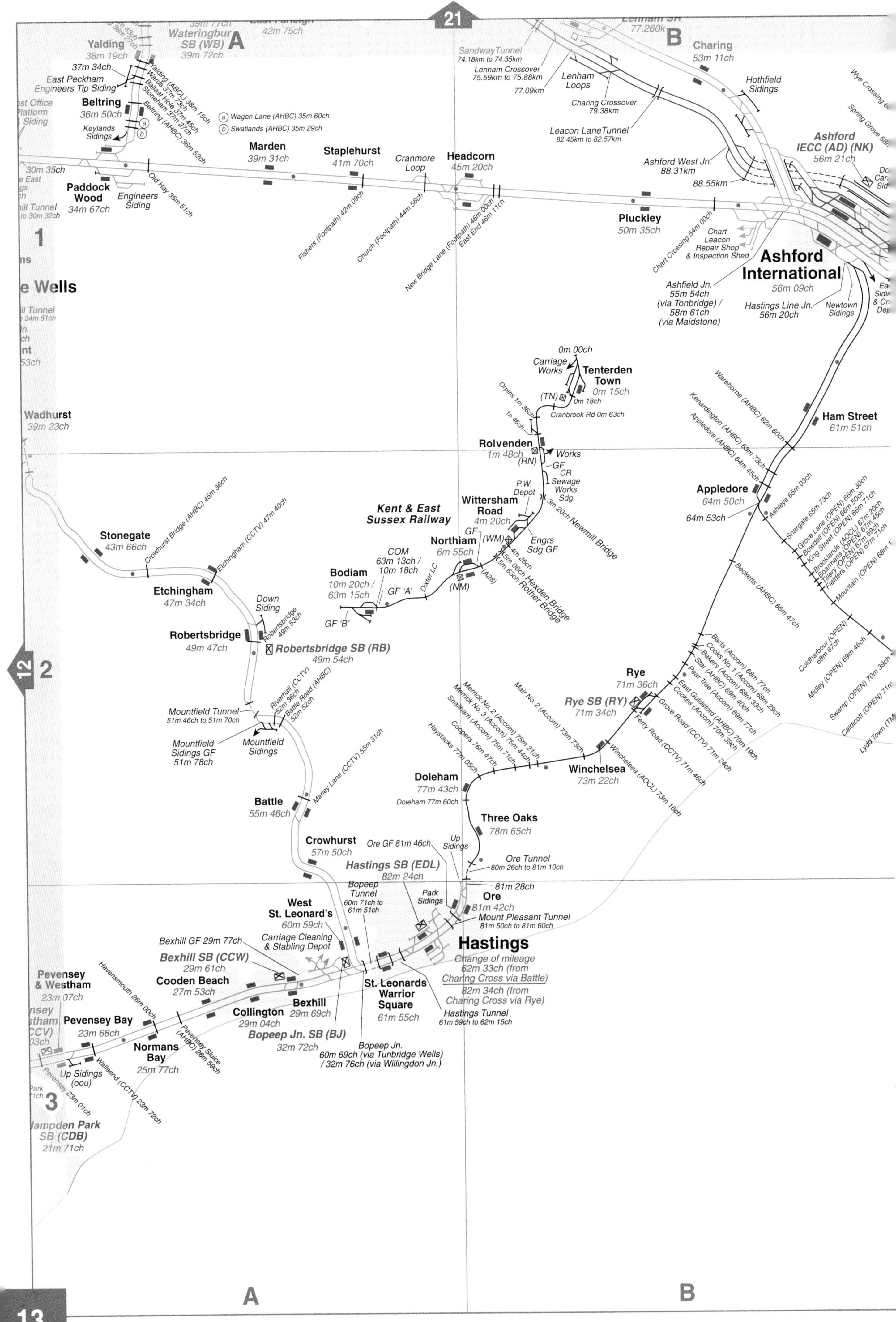

21
A
B
1
2
3
12
Wateringbury SB (WB)
39m 72ch
Yalding
38m 19ch
37m 34ch
East Peckham Engineers Tip Siding
Beltring
36m 50ch
Keylands Sidings
Yalding (ABCL) 38m 15ch
Wards 37m 73ch
Ballast Hole 37m 45ch
Stoneham 37m 27ch
Beltring (AHBC) 36m 52ch
a Wagon Lane (AHBC) 35m 60ch
b Swatlands (AHBC) 35m 29ch
Paddock Wood
34m 67ch
Engineers Siding
Old Hay 35m 51ch
30m 35ch
Marden
39m 31ch
Staplehurst
41m 70ch
Cranmore Loop
Headcorn
45m 20ch
Fishers (Footpath) 42m 09ch
Church (Footpath) 44m 56ch
New Bridge Lane (Footpath) 46m 00ch
East End 46m 11ch
SandwayTunnel
74.18km to 74.35km
Lenham Crossover
75.59km to 75.88km
77.09km
Lenham Loops
Charing Crossover
79.38km
Leacon LaneTunnel
82.45km to 82.57km
Charing
53m 11ch
Hothfield Sidings
Ashford West Jn.
88.31km
88.55km
Ashford IECC (AD) (NK)
56m 21ch
Pluckley
50m 35ch
Chart Crossing 54m 00ch
Chart Leacon Repair Shop & Inspection Shed
Ashford International
56m 09ch
Ashfield Jn.
55m 54ch (via Tonbridge) / 58m 61ch (via Maidstone)
Hastings Line Jn.
56m 20ch
Newtown Sidings
Wadhurst
39m 23ch
0m 00ch
Carriage Works
Tenterden Town
0m 15ch
(TN)
0m 18ch
Orpins 1m 36ch
1n 46ch
Cranbrook Rd 0m 63ch
Rolvenden
1m 48ch
(RN)
Works
GF
CR Sewage Works Sdg
P.W. Depot
3m 20ch Newmill Bridge
Kent & East Sussex Railway
Wittersham Road
4m 20ch
GF
(WM)
Engrs Sdg GF
4m 26ch
5m 06ch Hexden Bridge
5m 63ch Rother Bridge
Northiam
6m 55ch
(NM)
(A28)
COM
63m 13ch / 10m 18ch
Dixter LC
Bodiam
10m 20ch / 63m 15ch
GF 'A'
GF 'B'
Warehorne (AHBC) 62m 60ch
Kenardington (AHBC) 63m 73ch
Appledore (AHBC) 64m 45ch
Ham Street
61m 51ch
Appledore
64m 50ch
64m 53ch
Ashleys 65m 03ch
Snargate 65m 73ch
Grove Lane (OPEN) 66m 30ch
Bowdell (OPEN) 66m 50ch
King Street (OPEN) 66m 71ch
Brooklands (AOCL) 67m 20ch
Boarmans (OPEN) 67m 45ch
Tillery (OPEN) 67m 59ch
Fielders (OPEN) 67m 71ch
Mountain (OPEN) 68m
Beckets (AHBC) 66m 47ch
Coldharbour (OPEN) 68m 67ch
Midley (OPEN) 69m 46ch
Swamp (OPEN) 70m 39ch
Stonegate
43m 66ch
Crowhurst Bridge (AHBC) 45m 36ch
Etchingham (CCTV) 47m 40ch
Etchingham
47m 34ch
Down Siding
Robertsbridge 49m 53ch
Robertsbridge
49m 47ch
Robertsbridge SB (RB)
49m 54ch
Mountfield Tunnel
51m 46ch to 51m 70ch
Riverhall (CCTV) 52m 36ch
Battle Road (AHBC) 52m 52ch
Mountfield Sidings GF
51m 78ch
Mountfield Sidings
Marley Lane (CCTV) 55m 31ch
Battle
55m 46ch
Crowhurst
57m 50ch
Barts (Accom) 68m 77ch
Cooks No. 1 (Accom) 69m 29ch
Bakers (Accom) 69m 33ch
Star (AHBC) 69m 40ch
Pear Tree (Accom) 69m 77ch
East Guldeford (AHBC) 70m 19ch
Cookes (Accom) 70m 39ch
Rye
71m 36ch
Rye SB (RY)
71m 34ch
Grove Road (CCTV) 71m 24ch
Ferry Road (CCTV) 71m 46ch
Merrick No. 2 (Accom) 75m 21ch
Merrick No. 3 (Accom) 75m 44ch
Snailham (Accom) 75m 71ch
Coopers 76m 47ch
Haystacks 77m 05ch
Mair No. 2 (Accom) 73m 73ch
Winchelsea (AOCL) 73m 16ch
Winchelsea
73m 22ch
Doleham
77m 43ch
Doleham 77m 60ch
Three Oaks
78m 65ch
Ore GF 81m 46ch
Up Sidings
Ore Tunnel
80m 26ch to 81m 10ch
81m 28ch
Ore
81m 42ch
Hastings SB (EDL)
82m 24ch
Bopeep Tunnel
60m 71ch to 61m 51ch
Park Sidings
Mount Pleasant Tunnel
81m 50ch to 81m 60ch
West St. Leonard's
60m 59ch
Carriage Cleaning & Stabling Depot
Hastings
Change of mileage
62m 33ch (from Charing Cross via Battle)
82m 34ch (from Charing Cross via Rye)
Hastings Tunnel
61m 59ch to 62m 15ch
St. Leonards Warrior Square
61m 55ch
Bexhill GF 29m 77ch
Bexhill SB (CCW)
29m 61ch
Cooden Beach
27m 53ch
Bexhill
29m 69ch
Collington
29m 04ch
Bopeep Jn. SB (BJ)
32m 72ch
Bopeep Jn.
60m 69ch (via Tunbridge Wells) / 32m 76ch (via Willingdon Jn.)
Havensmouth 26m 00ch
Pevensey Sluice (AHBC) 26m 59ch
Pevensey & Westham
23m 07ch
Pevensey Bay
23m 68ch
Normans Bay
25m 77ch
Up Sidings (oou)
Pevensey 23m 01ch
Wallsend (CCTV) 23m 72ch
Hampden Park SB (CDB)
21m 71ch

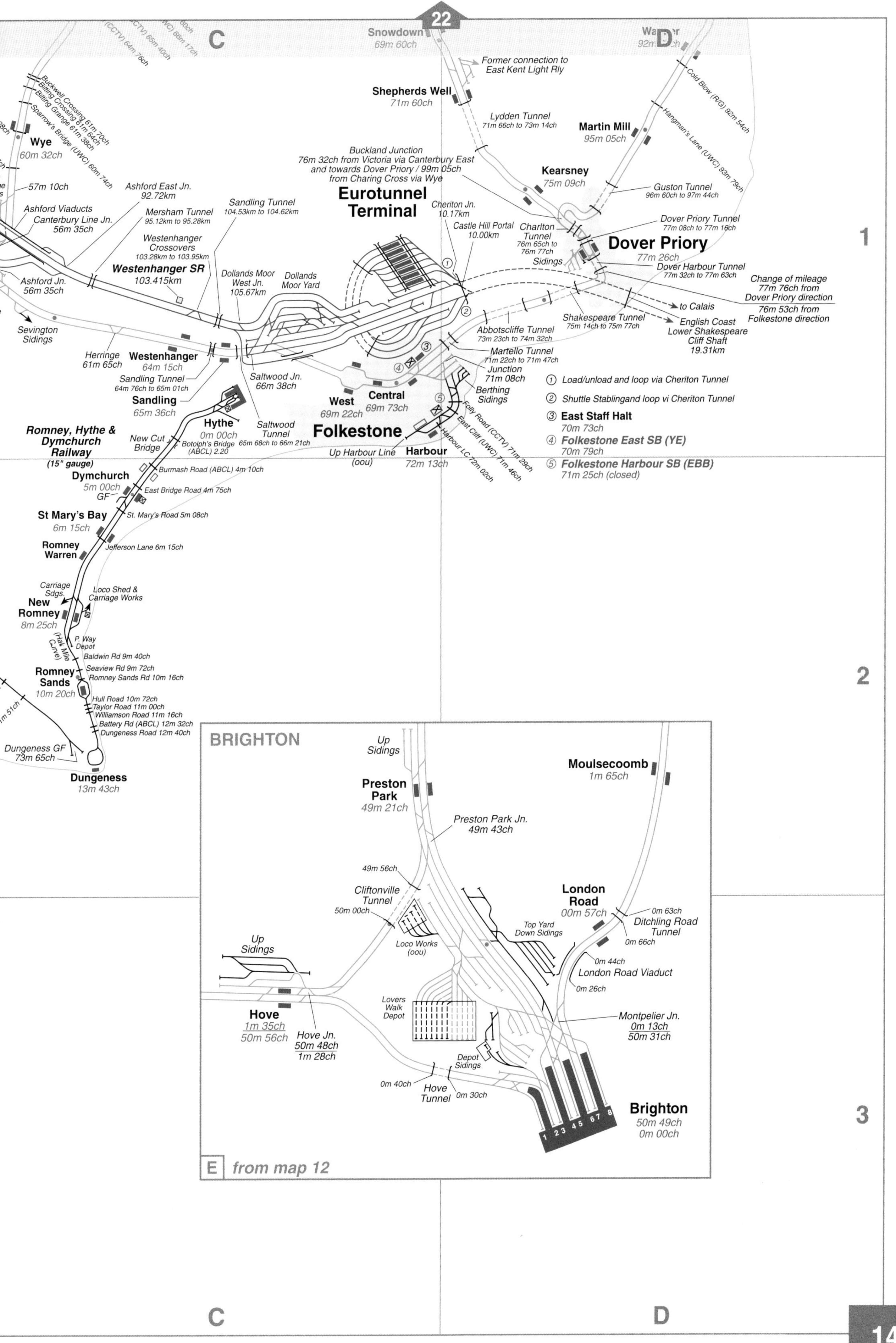

22
C
D
1
2
3
Snowdown
69m 60ch
Former connection to
East Kent Light Rly
Shepherds Well
71m 60ch
Lydden Tunnel
71m 66ch to 73m 14ch
Cold Blow (R/G) 92m 54ch
Hangman's Lane (UWC) 93m 79ch
Martin Mill
95m 05ch
Kearsney
75m 09ch
Guston Tunnel
96m 60ch to 97m 44ch
Buckwell Crossing 61m 70ch
Bilting Crossing 61m 64ch
Bilting Grange 61m 38ch
Sparrow's Bridge (UWC) 60m 74ch
Wye
60m 32ch
57m 10ch
Ashford Viaducts
Canterbury Line Jn.
56m 35ch
Ashford East Jn.
92.72km
Mersham Tunnel
95.12km to 95.28km
Sandling Tunnel
104.53km to 104.62km
Westenhanger
Crossovers
103.28km to 103.95km
Westenhanger SR
103.415km
Buckland Junction
76m 32ch from Victoria via Canterbury East
and towards Dover Priory / 99m 05ch
from Charing Cross via Wye
Eurotunnel
Terminal
Cheriton Jn.
10.17km
Castle Hill Portal
10.00km
Charlton
Tunnel
76m 65ch to
76m 77ch
Sidings
Dover Priory Tunnel
77m 08ch to 77m 16ch
Dover Priory
77m 26ch
Dover Harbour Tunnel
77m 32ch to 77m 63ch
Change of mileage
77m 76ch from
Dover Priory direction
76m 53ch from
Folkestone direction
to Calais
English Coast
Lower Shakespeare
Cliff Shaft
19.31km
Shakespeare Tunnel
75m 14ch to 75m 77ch
Abbotscliffe Tunnel
73m 23ch to 74m 32ch
Martello Tunnel
71m 22ch to 71m 47ch
Junction
71m 08ch
Berthing
Sidings
Ashford Jn.
56m 35ch
Sevington
Sidings
Dollands Moor
West Jn.
105.67km
Dollands
Moor Yard
Herringe
61m 65ch
Westenhanger
64m 15ch
Sandling Tunnel
64m 76ch to 65m 01ch
Sandling
65m 36ch
Saltwood Jn.
66m 38ch
West
69m 22ch
Central
69m 73ch
Folkestone
Hythe
0m 00ch
Saltwood
Tunnel
65m 68ch to 66m 21ch
Romney, Hythe &
Dymchurch
Railway
(15" gauge)
New Cut
Bridge
Botolph's Bridge
(ABCL) 2.20
Up Harbour Line
(oou)
Harbour
72m 13ch
Folly Road (CCTV) 71m 29ch
East Cliff (UWC) 71m 46ch
Harbour LC 72m 02ch
① Load/unload and loop via Cheriton Tunnel
② Shuttle Stablingand loop vi Cheriton Tunnel
③ East Staff Halt
70m 73ch
④ Folkestone East SB (YE)
70m 79ch
⑤ Folkestone Harbour SB (EBB)
71m 25ch (closed)
Dymchurch
5m 00ch
GF
Burmash Road (ABCL) 4m 10ch
East Bridge Road 4m 75ch
St Mary's Bay
6m 15ch
St. Mary's Road 5m 08ch
Romney
Warren
Jefferson Lane 6m 15ch
Carriage
Sdgs.
Loco Shed &
Carriage Works
New
Romney
8m 25ch
(Half Mile Curve)
P. Way
Depot
Baldwin Rd 9m 40ch
Seaview Rd 9m 72ch
Romney Sands Rd 10m 16ch
Romney
Sands
10m 20ch
Hull Road 10m 72ch
Taylor Road 11m 00ch
Williamson Road 11m 16ch
Battery Rd (ABCL) 12m 32ch
Dungeness Road 12m 40ch
Dungeness GF
73m 65ch
Dungeness
13m 43ch
BRIGHTON
Up
Sidings
Preston
Park
49m 21ch
Preston Park Jn.
49m 43ch
Moulsecoomb
1m 65ch
49m 56ch
Cliftonville
Tunnel
50m 00ch
Loco Works
(oou)
Top Yard
Down Sidings
London
Road
00m 57ch
0m 63ch
Ditchling Road
Tunnel
0m 66ch
0m 44ch
London Road Viaduct
0m 26ch
Up
Sidings
Hove
1m 35ch
50m 56ch
Hove Jn.
50m 48ch
1m 28ch
Lovers
Walk
Depot
Montpelier Jn.
0m 13ch
50m 31ch
Depot
Sidings
0m 40ch
Hove
Tunnel
0m 30ch
1 2 3 4 5 6 7 8
Brighton
50m 49ch
0m 00ch
E from map 12
C
D

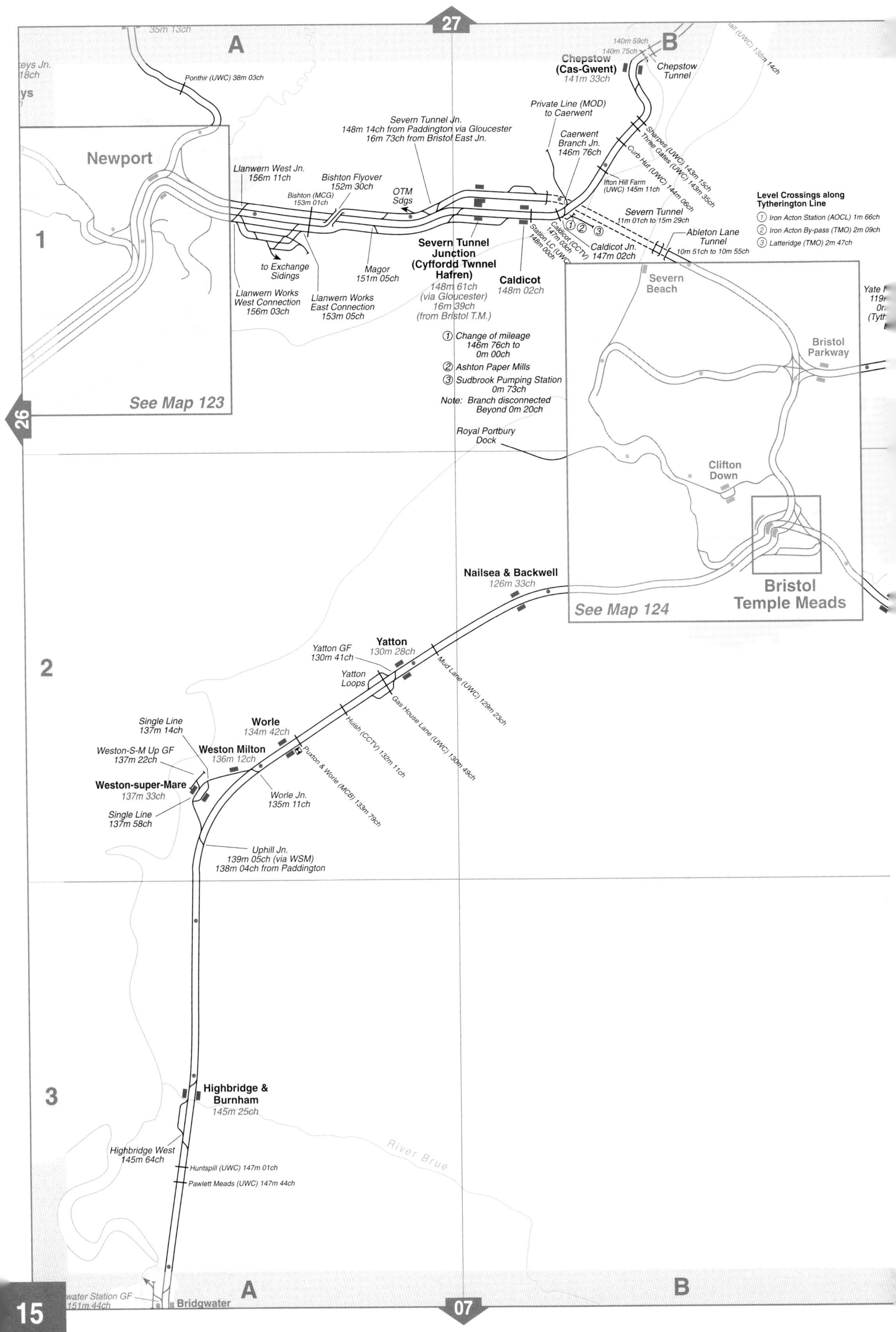
A
B
Ponthir (UWC) 38m 03ch
Chepstow
(Cas-Gwent)
141m 33ch
Chepstow
Tunnel
140m 59ch
140m 75ch
Private Line (MOD)
to Caerwent
Severn Tunnel Jn.
148m 14ch from Paddington via Gloucester
16m 73ch from Bristol East Jn.
Caerwent
Branch Jn.
146m 76ch
Newport
Llanwern West Jn.
156m 11ch
Bishton Flyover
152m 30ch
Bishton (MCG)
153m 01ch
OTM
Sdgs
Sharpes (UWC) 143m 15ch
Three Gates (UWC) 143m 35ch
Curb Hut (UWC) 144m 06ch
Ifton Hill Farm
(UWC) 145m 11ch
Level Crossings along
Tytherington Line
① Iron Acton Station (AOCL) 1m 66ch
② Iron Acton By-pass (TMO) 2m 09ch
③ Latteridge (TMO) 2m 47ch
Severn Tunnel
11m 01ch to 15m 29ch
Ableton Lane
Tunnel
10m 51ch to 10m 55ch
Caldicot Jn.
147m 02ch
Caldicot (CCTV)
147m 03ch
Station LC (UWC)
148m 00ch
1
Severn Tunnel
Junction
(Cyffordd Twnnel
Hafren)
148m 61ch
(via Gloucester)
16m 39ch
(from Bristol T.M.)
Caldicot
148m 02ch
to Exchange
Sidings
Magor
151m 05ch
Llanwern Works
West Connection
156m 03ch
Llanwern Works
East Connection
153m 05ch
Severn
Beach
Bristol
Parkway
① Change of mileage
146m 76ch to
0m 00ch
② Ashton Paper Mills
③ Sudbrook Pumping Station
0m 73ch
Note: Branch disconnected
Beyond 0m 20ch
See Map 123
Royal Portbury
Dock
Clifton
Down
Nailsea & Backwell
126m 33ch
Bristol
Temple Meads
See Map 124
2
Yatton
130m 28ch
Yatton GF
130m 41ch
Yatton
Loops
Mud Lane (UWC) 129m 23ch
Gas House Lane (UWC) 130m 49ch
Huish (CCTV) 132m 11ch
Puxton & Worle (MCB) 133m 79ch
Single Line
137m 14ch
Worle
134m 42ch
Weston-S-M Up GF
137m 22ch
Weston Milton
136m 12ch
Weston-super-Mare
137m 33ch
Worle Jn.
135m 11ch
Single Line
137m 58ch
Uphill Jn.
139m 05ch (via WSM)
138m 04ch from Paddington
3
Highbridge &
Burnham
145m 25ch
Highbridge West
145m 64ch
River Brue
Huntspill (UWC) 147m 01ch
Pawlett Meads (UWC) 147m 44ch
Bridgwater
A
B

26
07

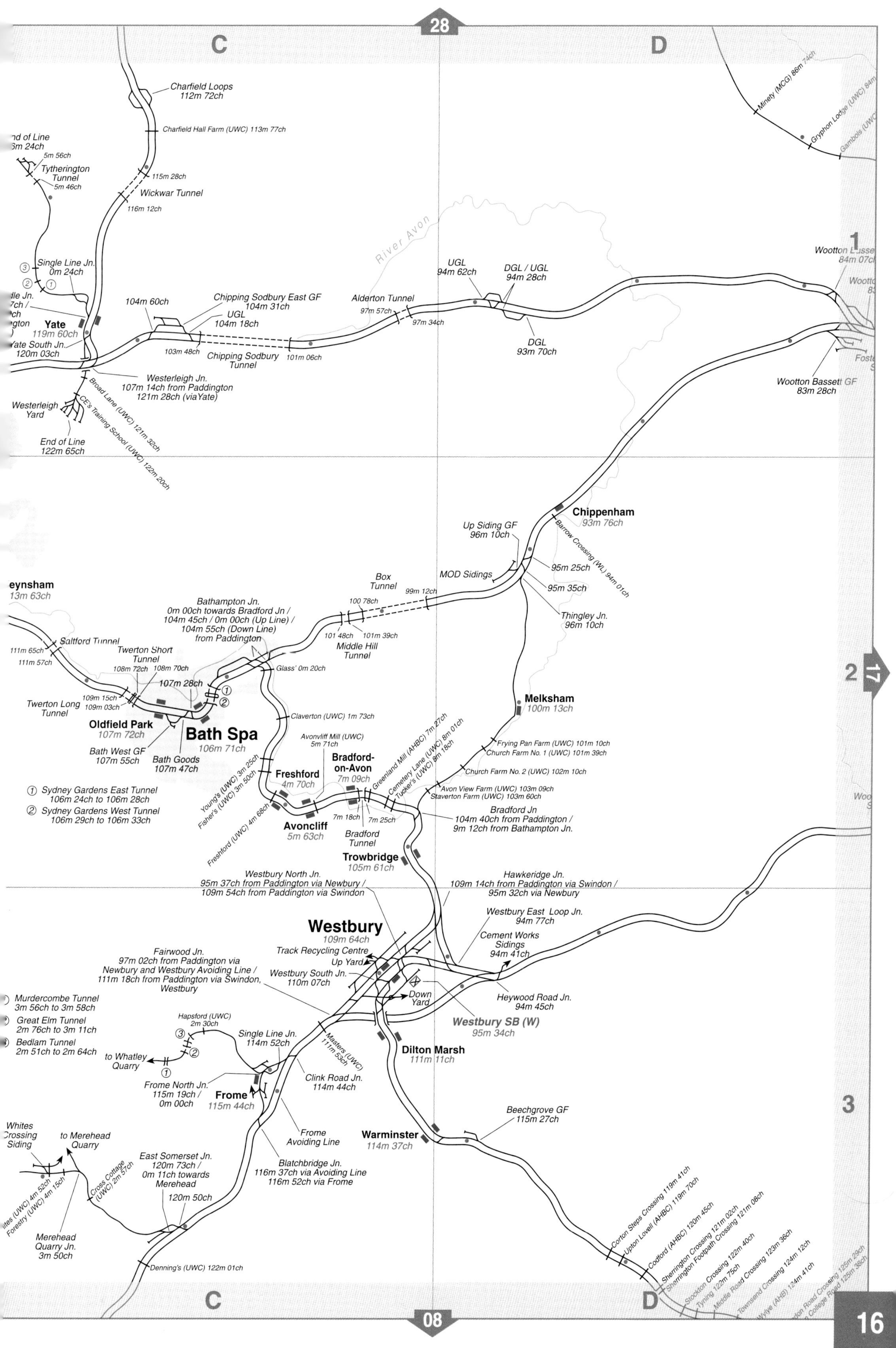
C
D
Charfield Loops
112m 72ch
Charfield Hall Farm (UWC) 113m 77ch
Tytherington Tunnel
115m 28ch
Wickwar Tunnel
116m 12ch
Single Line Jn.
0m 24ch
Yate
119m 60ch
Yate South Jn.
120m 03ch
104m 60ch
Chipping Sodbury East GF
104m 31ch
UGL
104m 18ch
103m 48ch
Chipping Sodbury Tunnel
101m 06ch
Alderton Tunnel
97m 57ch
97m 34ch
UGL
94m 62ch
DGL / UGL
94m 28ch
DGL
93m 70ch
River Avon
Minety (MCG) 86m 74ch
Wootton Bassett GF
83m 28ch
Westerleigh Jn.
107m 14ch from Paddington
121m 28ch (via Yate)
Westerleigh Yard
End of Line
122m 65ch
Broad Lane (UWC) 121m 32ch
CE's Training School (UWC) 122m 20ch
Chippenham
93m 76ch
Barrow Crossing (WL) 94m 01ch
Up Siding GF
96m 10ch
MOD Sidings
95m 25ch
95m 35ch
Thingley Jn.
96m 10ch
Box Tunnel
99m 12ch
100 78ch
101 48ch
101m 39ch
Middle Hill Tunnel
Bathampton Jn.
0m 00ch towards Bradford Jn /
104m 45ch / 0m 00ch (Up Line) /
104m 55ch (Down Line)
from Paddington
Glass' 0m 20ch
Saltford Tunnel
111m 65ch
111m 57ch
Twerton Short Tunnel
108m 72ch
108m 70ch
107m 28ch
109m 15ch
109m 03ch
Twerton Long Tunnel
Oldfield Park
107m 72ch
Bath Spa
106m 71ch
Bath West GF
107m 55ch
Bath Goods
107m 47ch
① Sydney Gardens East Tunnel
106m 24ch to 106m 28ch
② Sydney Gardens West Tunnel
106m 29ch to 106m 33ch
Claverton (UWC) 1m 73ch
Melksham
100m 13ch
Frying Pan Farm (UWC) 101m 10ch
Church Farm No. 1 (UWC) 101m 39ch
Church Farm No. 2 (UWC) 102m 10ch
Avon View Farm (UWC) 103m 09ch
Staverton Farm (UWC) 103m 60ch
Avonvliff Mill (UWC)
5m 71ch
Freshford
4m 70ch
Bradford-on-Avon
7m 09ch
Greenland Mill (AHBC) 7m 27ch
Cemetery Lane (UWC) 8m 01ch
Tucker's (UWC) 8m 18ch
Young's (UWC) 3m 25ch
Fisher's (UWC) 3m 50ch
Freshford (UWC) 4m 68ch
Avoncliff
5m 63ch
7m 18ch
7m 25ch
Bradford Tunnel
Bradford Jn
104m 40ch from Paddington /
9m 12ch from Bathampton Jn.
Trowbridge
105m 61ch
Westbury North Jn.
95m 37ch from Paddington via Newbury /
109m 54ch from Paddington via Swindon
Hawkeridge Jn.
109m 14ch from Paddington via Swindon /
95m 32ch via Newbury
Westbury
109m 64ch
Westbury East Loop Jn.
94m 77ch
Cement Works Sidings
94m 41ch
Track Recycling Centre
Up Yard
Westbury South Jn.
110m 07ch
Down Yard
Fairwood Jn.
97m 02ch from Paddington via
Newbury and Westbury Avoiding Line /
111m 18ch from Paddington via Swindon,
Westbury
Heywood Road Jn.
94m 45ch
Westbury SB (W)
95m 34ch
Murdercombe Tunnel
3m 56ch to 3m 58ch
Great Elm Tunnel
2m 76ch to 3m 11ch
Bedlam Tunnel
2m 51ch to 2m 64ch
Hapsford (UWC)
2m 30ch
Single Line Jn.
114m 52ch
Masters (UWC)
111m 53ch
Dilton Marsh
111m 11ch
to Whatley Quarry
Frome North Jn.
115m 19ch /
0m 00ch
Frome
115m 44ch
Clink Road Jn.
114m 44ch
Frome Avoiding Line
Warminster
114m 37ch
Beechgrove GF
115m 27ch
Whites Crossing Siding
to Merehead Quarry
Cross Cottage (UWC) 2m 57ch
Forestry (UWC) 4m 15ch
East Somerset Jn.
120m 73ch /
0m 11ch towards
Merehead
120m 50ch
Blatchbridge Jn.
116m 37ch via Avoiding Line
116m 52ch via Frome
Merehead Quarry Jn.
3m 50ch
Denning's (UWC) 122m 01ch
Corton Steps Crossing 119m 41ch
Upton Lovell (AHBC) 119m 70ch
Codford (AHBC) 120m 45ch
Sherrington Crossing 121m 02ch
Sherrington Footpath Crossing 121m 08ch
Stockton Crossing 122m 40ch
Tyning 122m 75ch
Middle Road Crossing 123m 36ch
Townsend Crossing 124m 12ch
Wylye (AHB) 124m 41ch
1
2
3
17
C
D

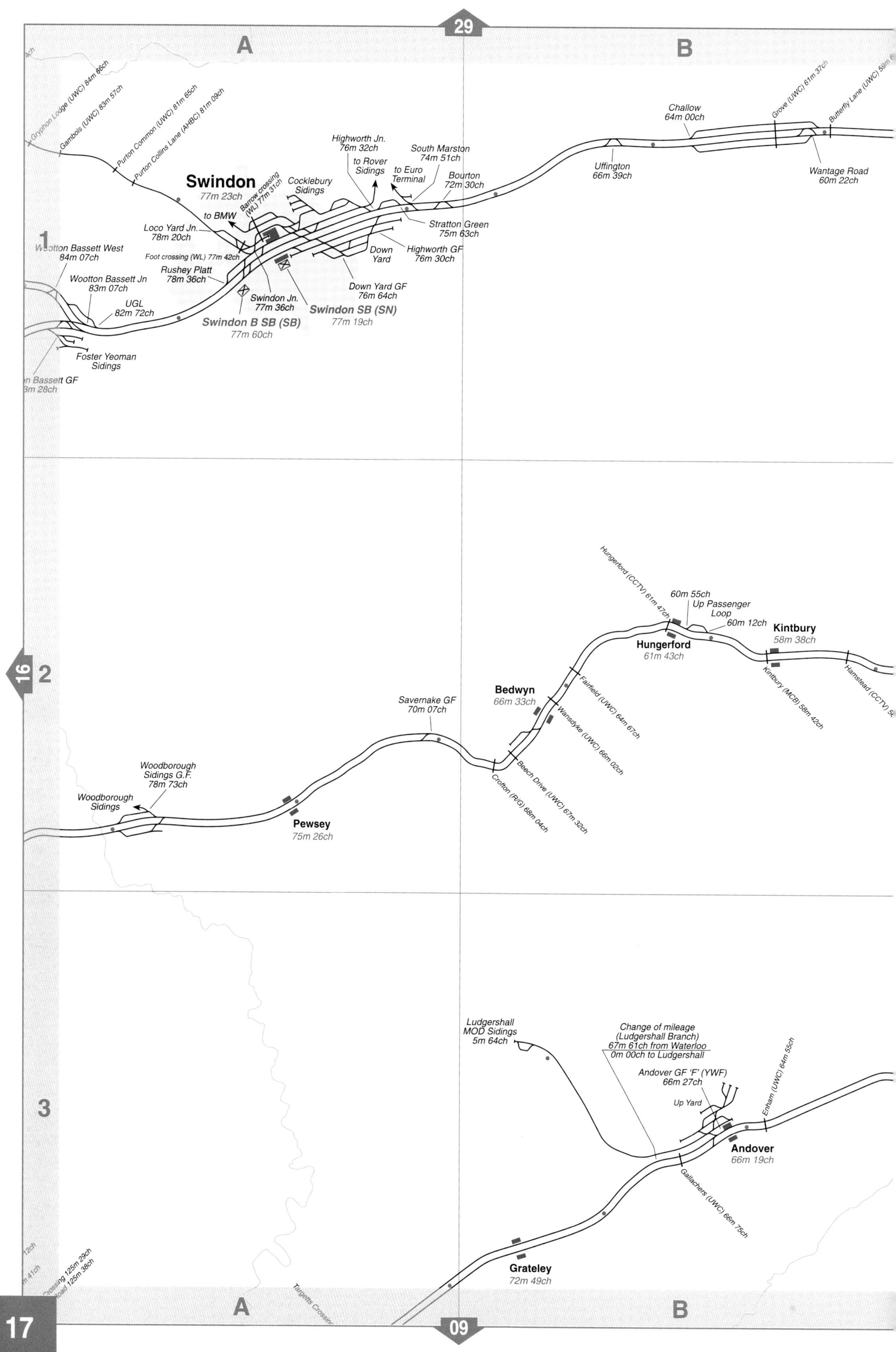
A
B
1
2
3
16
Gryphon Lodge (UWC) 84m 66ch
Gambols (UWC) 83m 57ch
Purton Common (UWC) 81m 65ch
Purton Collins Lane (AHBC) 81m 09ch
Swindon
77m 23ch
Barrow crossing (WL) 77m 31ch
Cocklebury Sidings
Highworth Jn.
76m 32ch
to Rover Sidings
to Euro Terminal
South Marston
74m 51ch
Bourton
72m 30ch
Challow
64m 00ch
Grove (UWC) 61m 37ch
Butterfly Lane (UWC) 59m
Uffington
66m 39ch
Wantage Road
60m 22ch
to BMW
Loco Yard Jn.
78m 20ch
Stratton Green
75m 63ch
Wootton Bassett West
84m 07ch
Down Yard
Highworth GF
76m 30ch
Foot crossing (WL) 77m 42ch
Rushey Platt
78m 36ch
Wootton Bassett Jn
83m 07ch
Down Yard GF
76m 64ch
UGL
82m 72ch
Swindon Jn.
77m 36ch
Swindon SB (SN)
77m 19ch
Swindon B SB (SB)
77m 60ch
Foster Yeoman Sidings
n Bassett GF
3m 28ch
Hungerford (CCTV) 61m 47ch
60m 55ch
Up Passenger Loop
60m 12ch
Kintbury
58m 38ch
Hungerford
61m 43ch
Kintbury (MCB) 58m 42ch
Hamstead (CCTV) 5
Bedwyn
66m 33ch
Fairfield (UWC) 64m 67ch
Savernake GF
70m 07ch
Wansdyke (UWC) 66m 02ch
Woodborough Sidings G.F.
78m 73ch
Woodborough Sidings
Beech Drive (UWC) 67m 32ch
Crofton (R/G) 68m 04ch
Pewsey
75m 26ch
Ludgershall MOD Sidings
5m 64ch
Change of mileage (Ludgershall Branch)
67m 61ch from Waterloo
0m 00ch to Ludgershall
Andover GF 'F' (YWF)
66m 27ch
Enham (UWC) 64m 55ch
Up Yard
Andover
66m 19ch
Gallachers (UWC) 66m 75ch
Grateley
72m 49ch
Targetts Crossing
09

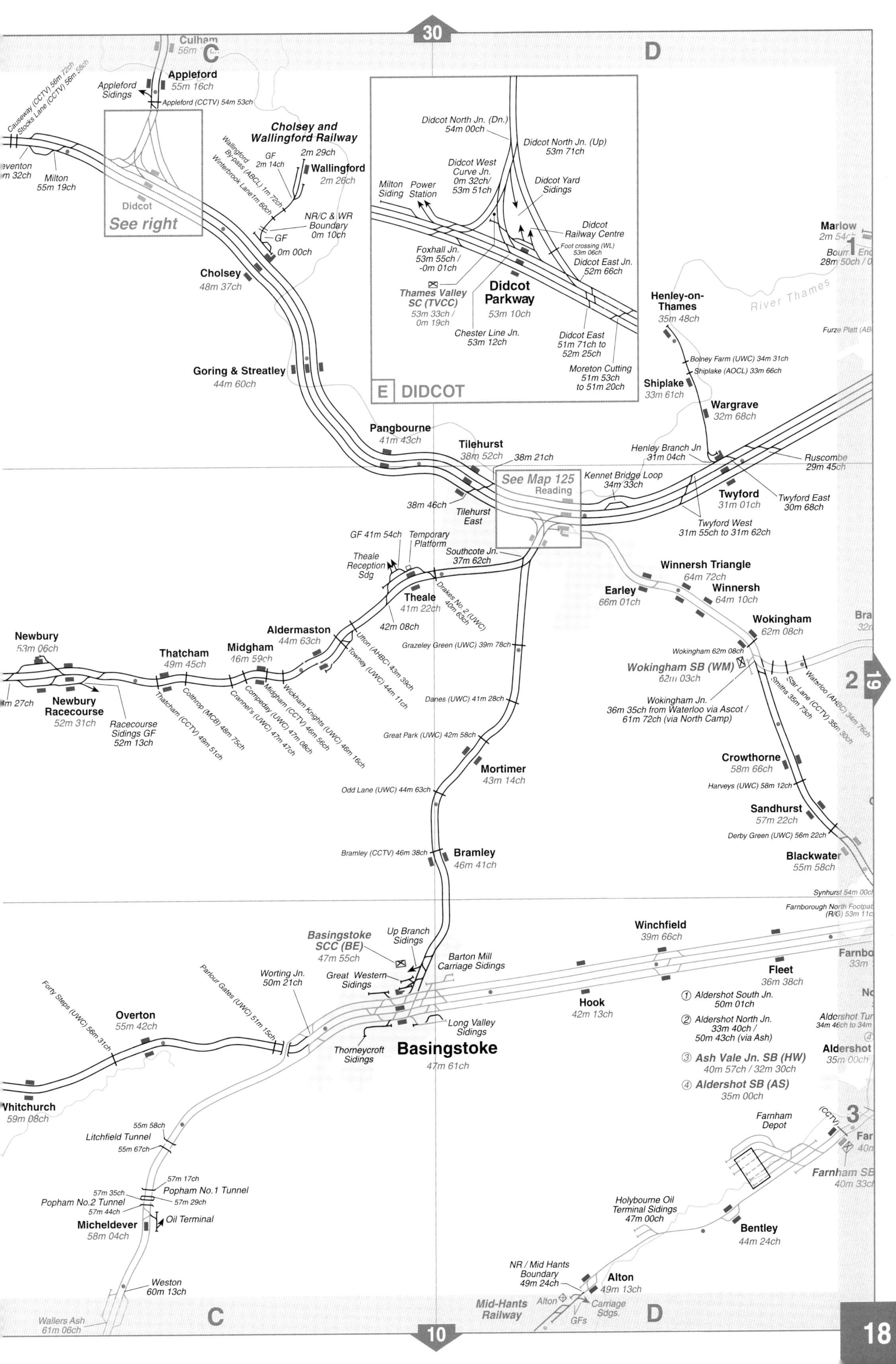

30
C
D
Culham
56m
Appleford
55m 16ch
Appleford Sidings
Appleford (CCTV) 54m 53ch
Causeway (CCTV) 56m 72ch
Stocks Lane (CCTV) 56m 58ch
Milton
55m 19ch
Didcot
See right
Cholsey and Wallingford Railway
Wallingford By-pass (ABCL) 1m 72ch
Winterbrook Lane 1m 60ch
GF 2m 14ch
2m 29ch
Wallingford
2m 26ch
NR/C & WR Boundary 0m 10ch
GF
0m 00ch
Cholsey
48m 37ch
Goring & Streatley
44m 60ch
Didcot North Jn. (Dn.) 54m 00ch
Didcot North Jn. (Up) 53m 71ch
Didcot West Curve Jn. 0m 32ch/ 53m 51ch
Milton Siding
Power Station
Didcot Yard Sidings
Didcot Railway Centre
Foot crossing (WL) 53m 06ch
Didcot East Jn. 52m 66ch
Foxhall Jn. 53m 55ch / -0m 01ch
Thames Valley SC (TVCC)
53m 33ch / 0m 19ch
Didcot Parkway
53m 10ch
Chester Line Jn. 53m 12ch
Didcot East 51m 71ch to 52m 25ch
Moreton Cutting 51m 53ch to 51m 20ch
E DIDCOT
Marlow
2m 54ch
Bourne End
28m 50ch
1
River Thames
Henley-on-Thames
35m 48ch
Furze Platt
Bolney Farm (UWC) 34m 31ch
Shiplake (AOCL) 33m 66ch
Shiplake
33m 61ch
Wargrave
32m 68ch
Pangbourne
41m 43ch
Tilehurst
38m 52ch
38m 21ch
Henley Branch Jn 31m 04ch
Ruscombe 29m 45ch
Kennet Bridge Loop 34m 33ch
See Map 125
Reading
Twyford
31m 01ch
Twyford East 30m 68ch
38m 46ch
Tilehurst East
Twyford West 31m 55ch to 31m 62ch
GF 41m 54ch
Temporary Platform
Theale Reception Sdg
Southcote Jn. 37m 62ch
Winnersh Triangle
64m 72ch
Winnersh
64m 10ch
Earley
66m 01ch
Theale
41m 22ch
Drakes No.2 (UWC) 40m 63ch
42m 08ch
Wokingham
62m 08ch
Aldermaston
44m 63ch
Newbury
53m 06ch
Thatcham
49m 45ch
Midgham
46m 59ch
Ufton (AHBC) 43m 39ch
Towney (UWC) 44m 11ch
Grazeley Green (UWC) 39m 78ch
Wokingham 62m 08ch
Wokingham SB (WM)
62m 03ch
2
19
m 27ch
Newbury Racecourse
52m 31ch
Racecourse Sidings GF 52m 13ch
Thatcham (CCTV) 49m 51ch
Colthrop (MCB) 48m 75ch
Crannel's (UWC) 47m 47ch
Compeday (UWC) 47m 08ch
Midgham (CCTV) 46m 56ch
Wickham Knights (UWC) 46m 16ch
Danes (UWC) 41m 28ch
Wokingham Jn. 36m 35ch from Waterloo via Ascot / 61m 72ch (via North Camp)
Smiths 35m 73ch
Star Lane (CCTV) 35m 30ch
Waterloo (AHBC) 34m 76ch
Great Park (UWC) 42m 58ch
Mortimer
43m 14ch
Crowthorne
58m 66ch
Harveys (UWC) 58m 12ch
Odd Lane (UWC) 44m 63ch
Sandhurst
57m 22ch
Derby Green (UWC) 56m 22ch
Bramley (CCTV) 46m 38ch
Bramley
46m 41ch
Blackwater
55m 58ch
Synhurst 54m 00ch
Farnborough North Footpath (R/G) 53m 11ch
Winchfield
39m 66ch
Basingstoke SCC (BE)
47m 55ch
Up Branch Sidings
Barton Mill Carriage Sidings
Worting Jn. 50m 21ch
Great Western Sidings
Parlour Gates (UWC) 51m 15ch
Fleet
36m 38ch
Hook
42m 13ch
Forty Steps (UWC) 56m 31ch
Overton
55m 42ch
Long Valley Sidings
① Aldershot South Jn. 50m 01ch
② Aldershot North Jn. 33m 40ch / 50m 43ch (via Ash)
③ Ash Vale Jn. SB (HW) 40m 57ch / 32m 30ch
④ Aldershot SB (AS) 35m 00ch
Aldershot
35m 00ch
Thorneycroft Sidings
Basingstoke
47m 61ch
Whitchurch
59m 08ch
55m 58ch
Litchfield Tunnel
55m 67ch
Farnham Depot
3
(CCTV)
Farnham SB
40m 33ch
57m 17ch
Popham No.1 Tunnel
57m 35ch
Popham No.2 Tunnel
57m 29ch
57m 44ch
Oil Terminal
Micheldever
58m 04ch
Holybourne Oil Terminal Sidings 47m 00ch
Bentley
44m 24ch
Weston
60m 13ch
NR / Mid Hants Boundary 49m 24ch
Alton
49m 13ch
Mid-Hants Railway
Alton
Carriage Sdgs.
GFs
Wallers Ash 61m 06ch
C
D
10
18

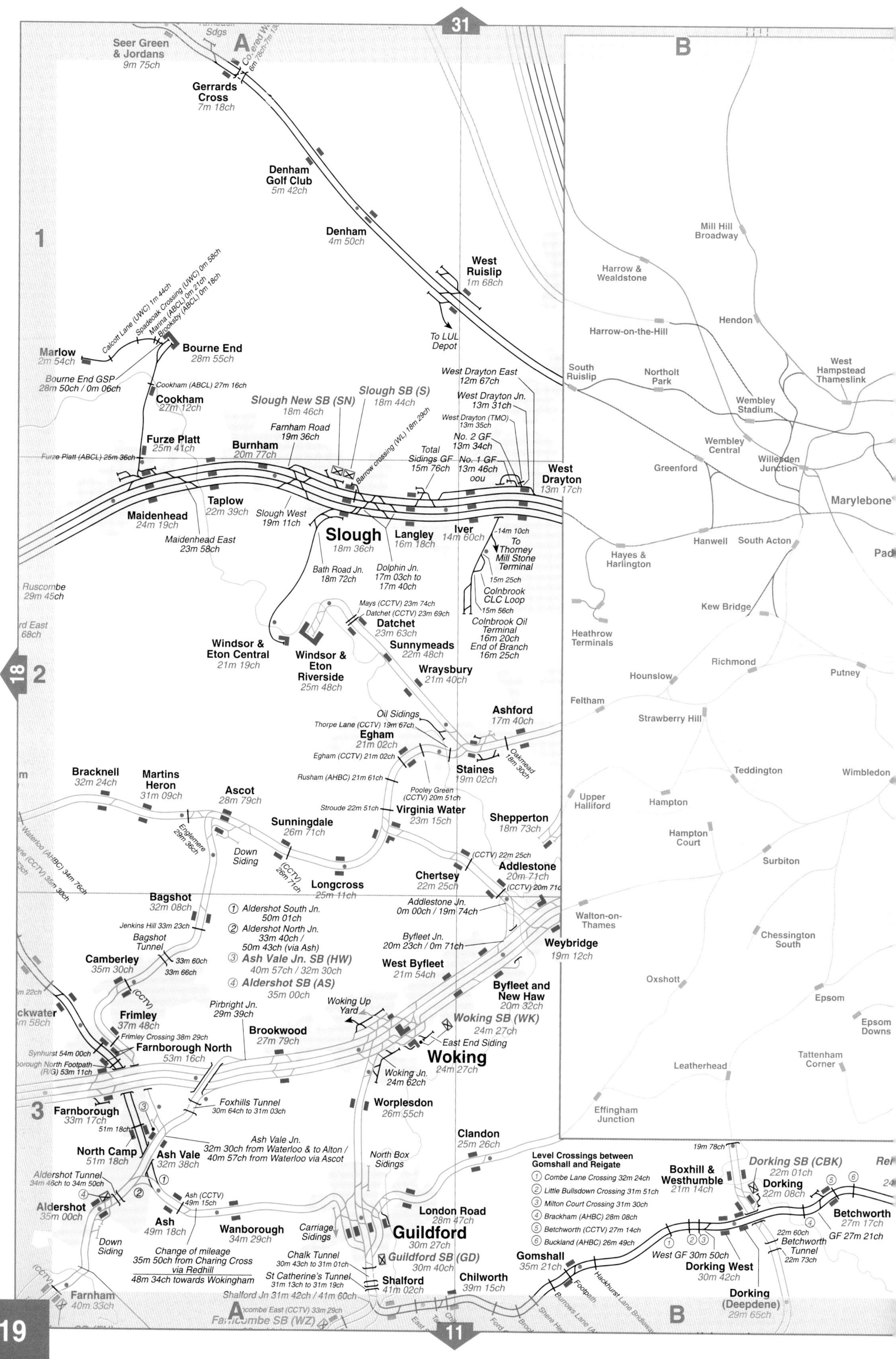
31
A
B
1
2
3
18
11
Seer Green & Jordans
9m 75ch
Gerrards Cross
7m 18ch
Denham Golf Club
5m 42ch
Denham
4m 50ch
West Ruislip
1m 68ch
To LUL Depot
Marlow
2m 54ch
Calcott Lane (UWC) 1m 44ch
Spadeoak Crossing (UWC) 0m 58ch
Marina (ABCL) 0m 21ch
Brooksby (ABCL) 0m 18ch
Bourne End
28m 55ch
Bourne End GSP
28m 50ch / 0m 06ch
Cookham (ABCL) 27m 16ch
Cookham
27m 12ch
Furze Platt
25m 41ch
Furze Platt (ABCL) 25m 36ch
Burnham
20m 77ch
Slough New SB (SN)
18m 46ch
Slough SB (S)
18m 44ch
Farnham Road
19m 36ch
Barrow crossing (WL) 18m 29ch
West Drayton East
12m 67ch
West Drayton Jn.
13m 31ch
West Drayton (TMO)
13m 35ch
No. 2 GF
13m 34ch
No. 1 GF
13m 46ch
oou
Total Sidings GF
15m 76ch
West Drayton
13m 17ch
Maidenhead
24m 19ch
Taplow
22m 39ch
Slough West
19m 11ch
Maidenhead East
23m 58ch
Slough
18m 36ch
Langley
16m 18ch
Iver
14m 60ch
14m 10ch
To Thorney Mill Stone Terminal
Bath Road Jn.
18m 72ch
Dolphin Jn.
17m 03ch to 17m 40ch
15m 25ch
Colnbrook CLC Loop
15m 56ch
Colnbrook Oil Terminal
16m 20ch
End of Branch
16m 25ch
Ruscombe
29m 45ch
Mays (CCTV) 23m 74ch
Datchet (CCTV) 23m 69ch
Datchet
23m 63ch
Windsor & Eton Central
21m 19ch
Windsor & Eton Riverside
25m 48ch
Sunnymeads
22m 48ch
Wraysbury
21m 40ch
Oil Sidings
Thorpe Lane (CCTV) 19m 67ch
Egham
21m 02ch
Egham (CCTV) 21m 02ch
Ashford
17m 40ch
Staines
19m 02ch
Oakmead 18m 30ch
Rusham (AHBC) 21m 61ch
Pooley Green (CCTV) 20m 51ch
Stroude 22m 51ch
Virginia Water
23m 15ch
Bracknell
32m 24ch
Martins Heron
31m 09ch
Ascot
28m 79ch
Sunningdale
26m 71ch
Englemere 29m 36ch
Down Siding
(CCTV) 26m 71ch
Longcross
25m 11ch
Shepperton
18m 73ch
(CCTV) 22m 25ch
Addlestone
20m 71ch
(CCTV) 20m 71ch
Chertsey
22m 25ch
Waterloo (AHBC) 34m 76ch
Bagshot
32m 08ch
Jenkins Hill 33m 23ch
Bagshot Tunnel
33m 60ch
33m 66ch
① Aldershot South Jn.
50m 01ch
② Aldershot North Jn.
33m 40ch /
50m 43ch (via Ash)
③ Ash Vale Jn. SB (HW)
40m 57ch / 32m 30ch
④ Aldershot SB (AS)
35m 00ch
Addlestone Jn.
0m 00ch / 19m 74ch
Byfleet Jn.
20m 23ch / 0m 71ch
Weybridge
19m 12ch
West Byfleet
21m 54ch
Byfleet and New Haw
20m 32ch
Camberley
35m 30ch
(CCTV)
Frimley
37m 48ch
Frimley Crossing 38m 29ch
Synhurst 54m 00ch
Farnborough North
53m 16ch
Pirbright Jn.
29m 39ch
Brookwood
27m 79ch
Woking Up Yard
Woking SB (WK)
24m 27ch
East End Siding
Woking
24m 27ch
Woking Jn.
24m 62ch
Farnborough
33m 17ch
51m 18ch
Foxhills Tunnel
30m 64ch to 31m 03ch
Worplesdon
26m 55ch
Ash Vale Jn.
32m 30ch from Waterloo & to Alton /
40m 57ch from Waterloo via Ascot
North Camp
51m 18ch
Ash Vale
32m 38ch
Clandon
25m 26ch
North Box Sidings
Aldershot Tunnel
34m 46ch to 34m 50ch
Aldershot
35m 00ch
Ash (CCTV)
49m 15ch
Ash
49m 18ch
Down Siding
Wanborough
34m 29ch
Carriage Sidings
London Road
28m 47ch
Guildford
30m 27ch
Guildford SB (GD)
30m 40ch
Change of mileage
35m 50ch from Charing Cross via Redhill
48m 34ch towards Wokingham
Chalk Tunnel
30m 43ch to 31m 01ch
St Catherine's Tunnel
31m 13ch to 31m 19ch
Shalford
41m 02ch
Shalford Jn 31m 42ch / 41m 60ch
Chilworth
39m 15ch
Farnham
40m 33ch
Gomshall
35m 21ch
Level Crossings between Gomshall and Reigate
① Combe Lane Crossing 32m 24ch
② Little Bullsdown Crossing 31m 51ch
③ Milton Court Crossing 31m 30ch
④ Brackham (AHBC) 28m 08ch
⑤ Betchworth (CCTV) 27m 14ch
⑥ Buckland (AHBC) 26m 49ch
19m 78ch
Boxhill & Westhumble
21m 14ch
Dorking SB (CBK)
22m 01ch
Dorking
22m 08ch
Betchworth
27m 17ch
GF 27m 21ch
22m 60ch
Betchworth Tunnel
22m 73ch
West GF 30m 50ch
Dorking West
30m 42ch
Dorking (Deepdene)
29m 65ch
Footpath
Hackhurst Lane Bridleway
Burrows Lane
Mill Hill Broadway
Harrow & Wealdstone
Hendon
Harrow-on-the-Hill
South Ruislip
Northolt Park
West Hampstead Thameslink
Wembley Stadium
Wembley Central
Greenford
Willesden Junction
Marylebone
Hanwell
South Acton
Hayes & Harlington
Kew Bridge
Heathrow Terminals
Richmond
Hounslow
Putney
Feltham
Strawberry Hill
Teddington
Wimbledon
Upper Halliford
Hampton
Hampton Court
Surbiton
Walton-on-Thames
Chessington South
Oxshott
Epsom
Epsom Downs
Tattenham Corner
Leatherhead
Effingham Junction

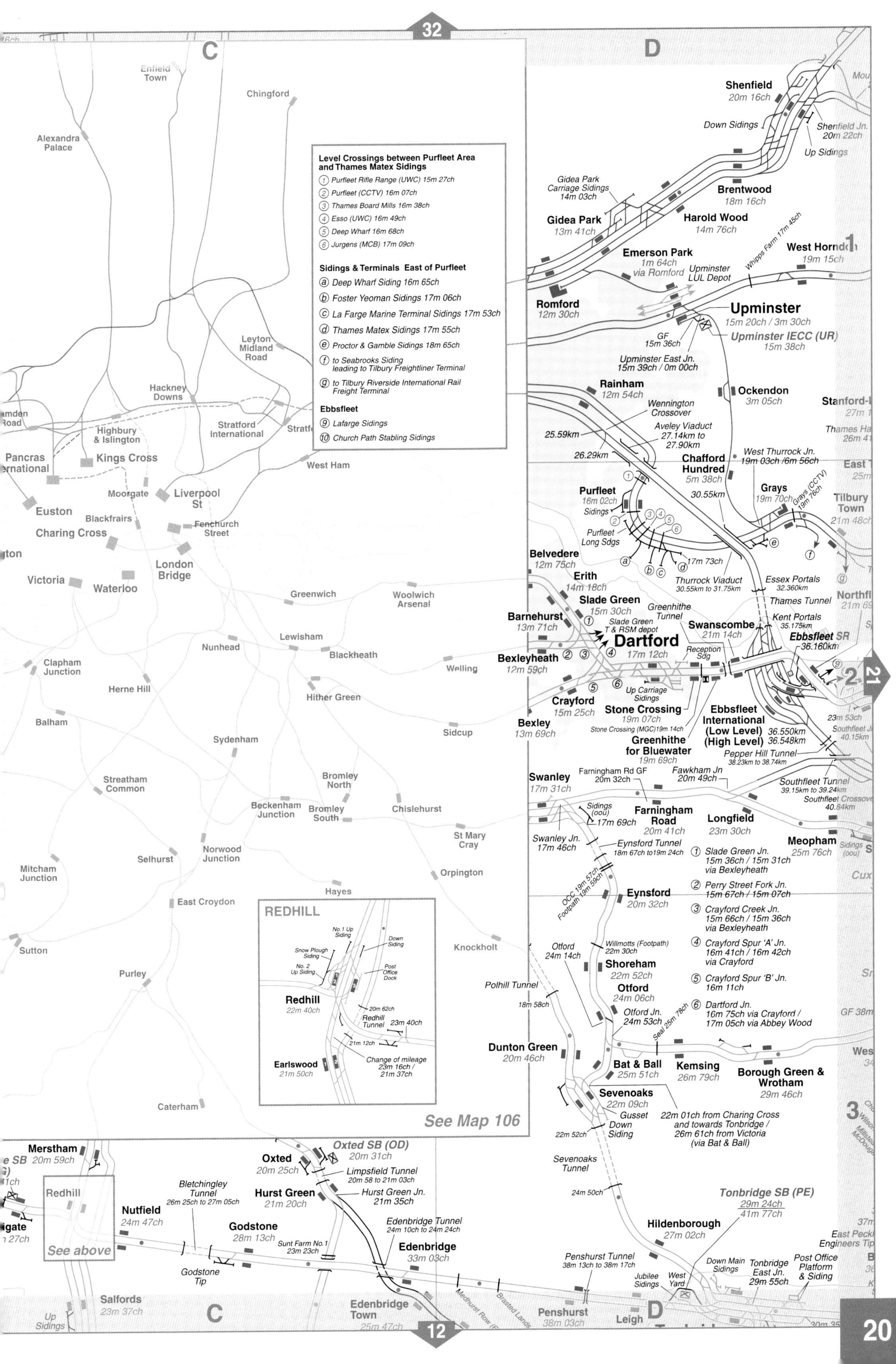
32
C
D
Enfield Town
Chingford
Alexandra Palace
Level Crossings between Purfleet Area and Thames Matex Sidings
① Purfleet Rifle Range (UWC) 15m 27ch
② Purfleet (CCTV) 16m 07ch
③ Thames Board Mills 16m 38ch
④ Esso (UWC) 16m 49ch
⑤ Deep Wharf 16m 68ch
⑥ Jurgens (MCB) 17m 09ch
Sidings & Terminals East of Purfleet
ⓐ Deep Wharf Siding 16m 65ch
ⓑ Foster Yeoman Sidings 17m 06ch
ⓒ La Farge Marine Terminal Sidings 17m 53ch
ⓓ Thames Matex Sidings 17m 55ch
ⓔ Proctor & Gamble Sidings 18m 65ch
ⓕ to Seabrooks Siding leading to Tilbury Freightliner Terminal
ⓖ to Tilbury Riverside International Rail Freight Terminal
Ebbsfleet
⑨ Lafarge Sidings
⑩ Church Path Stabling Sidings
Shenfield
20m 16ch
Down Sidings
Shenfield Jn.
20m 22ch
Up Sidings
Gidea Park Carriage Sidings
14m 03ch
Brentwood
18m 16ch
Gidea Park
13m 41ch
Harold Wood
14m 76ch
Whipps Farm 17m 45ch
West Horndon
19m 15ch
1
Emerson Park
1m 64ch via Romford
Upminster LUL Depot
Romford
12m 30ch
Upminster
15m 20ch / 3m 30ch
GF 15m 36ch
Upminster IECC (UR)
15m 38ch
Upminster East Jn.
15m 39ch / 0m 00ch
Leyton Midland Road
Hackney Downs
Rainham
12m 54ch
Ockendon
3m 05ch
Stanford-le-Hope
Camden Road
Highbury & Islington
Stratford International
Stratford
Wennington Crossover
Aveley Viaduct
27.14km to 27.90km
25.59km
26.29km
Thames Haven
St Pancras International
Kings Cross
West Ham
West Thurrock Jn.
19m 03ch /6m 56ch
Chafford Hundred
5m 38ch
East Tilbury
Moorgate
Liverpool St
Purfleet
16m 02ch
Sidings
30.55km
Grays
19m 70ch
Grays (CCTV) 19m 76ch
Tilbury Town
21m 48ch
Euston
Blackfriars
Fenchurch Street
Charing Cross
Purfleet Long Sdgs
17m 73ch
Victoria
Waterloo
London Bridge
Belvedere
12m 75ch
Erith
14m 18ch
Thurrock Viaduct
30.55km to 31.75km
Essex Portals
32.360km
Greenwich
Woolwich Arsenal
Slade Green
15m 30ch
Greenhithe Tunnel
Thames Tunnel
Northfleet
Barnehurst
13m 71ch
Slade Green T & RSM depot
Kent Portals
35.175km
Lewisham
Nunhead
Dartford
17m 12ch
Swanscombe
21m 14ch
Ebbsfleet SR
36.160km
Blackheath
Clapham Junction
Bexleyheath
12m 59ch
Reception Sdg
Welling
21
2
Herne Hill
Hither Green
Up Carriage Sidings
Crayford
15m 25ch
Stone Crossing
19m 07ch
Ebbsfleet International (Low Level) (High Level)
36.550km
36.548km
23m 53ch
Southfleet Jn.
40.15km
Balham
Bexley
13m 69ch
Sidcup
Stone Crossing (MGC)19m 14ch
Greenhithe for Bluewater
19m 69ch
Pepper Hill Tunnel
38.23km to 38.74km
Sydenham
Farningham Rd GF
20m 32ch
Fawkham Jn
20m 49ch
Streatham Common
Bromley North
Swanley
17m 31ch
Southfleet Tunnel
39.15km to 39.24km
Southfleet Crossover
40.84km
Beckenham Junction
Bromley South
Chislehurst
Sidings (oou)
17m 69ch
Farningham Road
20m 41ch
Longfield
23m 30ch
St Mary Cray
Swanley Jn.
17m 46ch
Eynsford Tunnel
18m 67ch to19m 24ch
Meopham
25m 76ch
Sidings (oou)
Mitcham Junction
Selhurst
Norwood Junction
① Slade Green Jn.
15m 36ch / 15m 31ch via Bexleyheath
Orpington
② Perry Street Fork Jn.
15m 67ch / 15m 07ch
OCC 19m 57ch
Footpath 19m 59ch
Eynsford
20m 32ch
Hayes
East Croydon
③ Crayford Creek Jn.
15m 66ch / 15m 36ch via Bexleyheath
REDHILL
No.1 Up Siding
Down Siding
④ Crayford Spur 'A' Jn.
16m 41ch / 16m 42ch via Crayford
Snow Plough Siding
No. 2 Up Siding
Post Office Dock
Knockholt
Otford
24m 14ch
Willmotts (Footpath)
22m 30ch
Sutton
Shoreham
22m 52ch
⑤ Crayford Spur 'B' Jn.
16m 11ch
Purley
Redhill
22m 40ch
Polhill Tunnel
Otford
24m 06ch
20m 62ch
Redhill Tunnel
23m 40ch
18m 58ch
Otford Jn.
24m 53ch
⑥ Dartford Jn.
16m 75ch via Crayford / 17m 05ch via Abbey Wood
Seal 25m 78ch
GF 38m
21m 12ch
Change of mileage
23m 16ch / 21m 37ch
Dunton Green
20m 46ch
Earlswood
21m 50ch
Bat & Ball
25m 51ch
Kemsing
26m 79ch
Borough Green & Wrotham
29m 46ch
Sevenoaks
22m 09ch
3
Caterham
Gusset
22m 01ch from Charing Cross and towards Tonbridge / 26m 61ch from Victoria (via Bat & Ball)
See Map 106
22m 52ch
Down Siding
Merstham
20m 59ch
Oxted SB (OD)
20m 31ch
Oxted
20m 25ch
Limpsfield Tunnel
20m 58 to 21m 03ch
Sevenoaks Tunnel
Redhill
Bletchingley Tunnel
26m 25ch to 27m 05ch
Hurst Green
21m 20ch
Hurst Green Jn.
21m 35ch
24m 50ch
Tonbridge SB (PE)
29m 24ch
41m 77ch
Nutfield
24m 47ch
Edenbridge Tunnel
24m 10ch to 24m 24ch
Hildenborough
27m 02ch
Godstone
28m 13ch
Sunt Farm No.1
23m 23ch
Edenbridge
33m 03ch
East Peckham Engineers Tip
See above
Penshurst Tunnel
38m 13ch to 38m 17ch
Down Main Sidings
Tonbridge East Jn.
29m 55ch
Post Office Platform & Siding
Godstone Tip
Jubilee Sidings
West Yard
Medhurst Row
Brasted Lands
Salfords
23m 37ch
Up Sidings
C
Edenbridge Town
25m 47ch
12
Penshurst
38m 03ch
Leigh
D

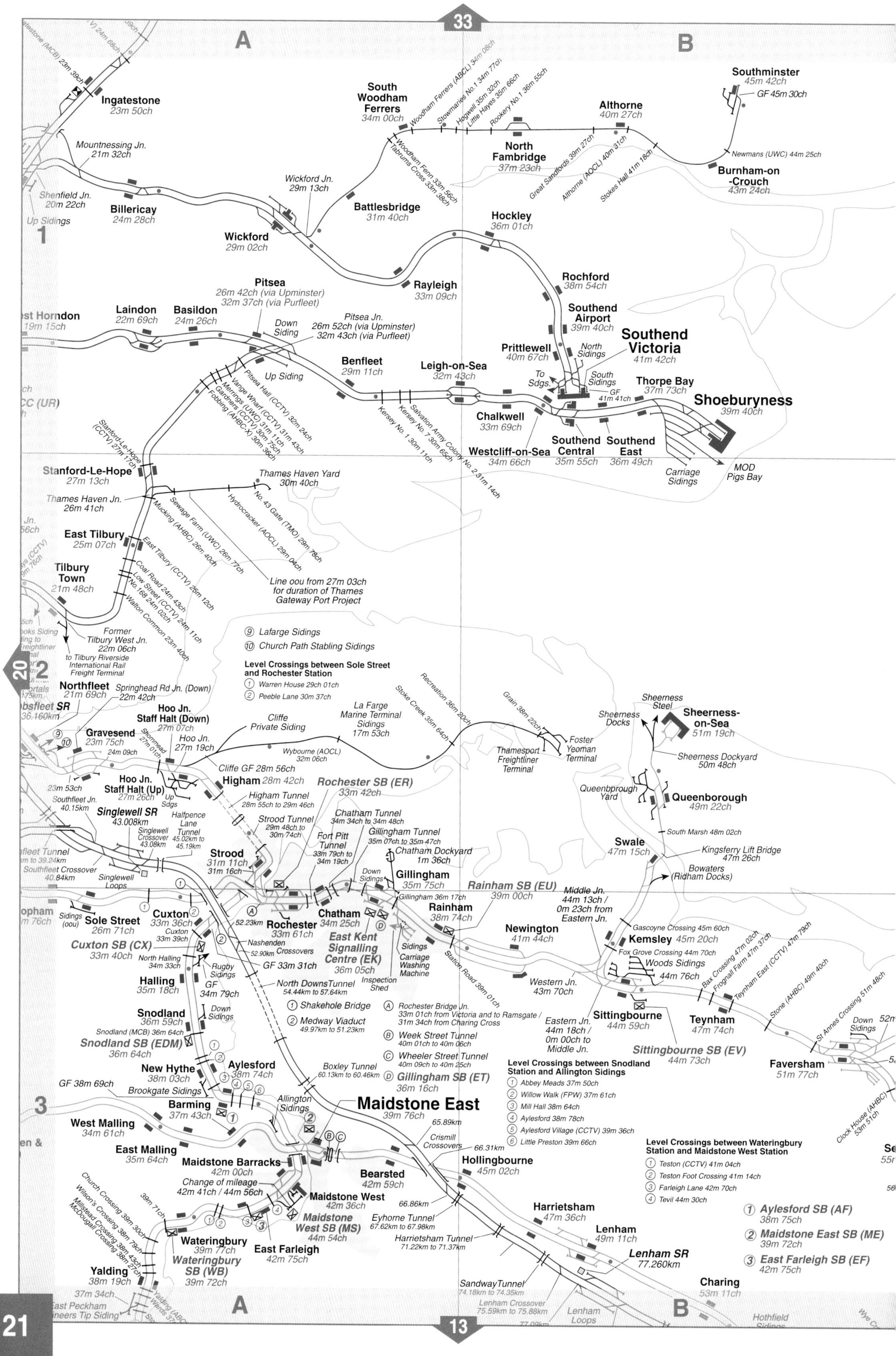

33
A
B
Ingatestone
23m 50ch
Mountnessing Jn.
21m 32ch
Shenfield Jn.
20m 22ch
Up Sidings
Billericay
24m 28ch
Wickford
29m 02ch
Wickford Jn.
29m 13ch
Battlesbridge
31m 40ch
South Woodham Ferrers
34m 00ch
Woodham Ferrers (ABCL) 34m 08ch
Stowmaries No.1 34m 77ch
Hogwell 35m 32ch
Little Hayes 35m 66ch
Rookery No.1 36m 55ch
Woodham Fenn 33m 56ch
Tabrums Cross 33m 38ch
North Fambridge
37m 23ch
Great Sandfords 39m 27ch
Althorne (AOCL) 40m 31ch
Stokes Hall 41m 18ch
Althorne
40m 27ch
Southminster
45m 42ch
GF 45m 30ch
Newmans (UWC) 44m 25ch
Burnham-on-Crouch
43m 24ch
Hockley
36m 01ch
Rayleigh
33m 09ch
Rochford
38m 54ch
Southend Airport
39m 40ch
Southend Victoria
41m 42ch
Prittlewell
40m 67ch
North Sidings
South Sidings
To Sdgs.
GF 41m 41ch
Thorpe Bay
37m 73ch
Shoeburyness
39m 40ch
Carriage Sidings
MOD Pigs Bay
Southend Central
35m 55ch
Southend East
36m 49ch
Chalkwell
33m 69ch
Westcliff-on-Sea
34m 66ch
Leigh-on-Sea
32m 43ch
Benfleet
29m 11ch
Kersey No 1 30m 11ch
Kersey No. 7 30m 65ch
Salvation Army Colony No. 2 31m 14ch
Pitsea
26m 42ch (via Upminster)
32m 37ch (via Purfleet)
Pitsea Jn.
26m 52ch (via Upminster)
32m 43ch (via Purfleet)
Down Siding
Up Siding
Laindon
22m 69ch
Basildon
24m 26ch
West Horndon
19m 15ch
Pitsea Hall (CCTV) 32m 24ch
Vange Wharf (CCTV) 31m 43ch
Merrings (UWC) 31m 11ch
Gardiners (CCTV) 30m 75ch
Fobbing (AHBC-X) 30m 36ch
Stanford-Le-Hope (CCTV) 27m 17ch
Stanford-Le-Hope
27m 13ch
Thames Haven Jn.
26m 41ch
Thames Haven Yard
30m 40ch
Sewage Farm (UWC) 26m 77ch
Mucking (AHBC) 26m 40ch
Hydrocracker (AOCL) 29m 04ch
No. 43 Gate (TMO) 29m 78ch
Line oou from 27m 03ch for duration of Thames Gateway Port Project
East Tilbury
25m 07ch
East Tilbury (CCTV) 25m 12ch
Coal Road 24m 43ch
Low Street (CCTV) 24m 11ch
No 168 24m 02ch
Walton Common 23m 40ch
Tilbury Town
21m 48ch
Former Tilbury West Jn.
22m 06ch
to Tilbury Riverside International Rail Freight Terminal
⑨ Lafarge Sidings
⑩ Church Path Stabling Sidings
Level Crossings between Sole Street and Rochester Station
① Warren House 29ch 01ch
② Peeble Lane 30m 37ch
20
2
Northfleet
21m 69ch
Springhead Rd Jn. (Down)
22m 42ch
Ebbsfleet SR
36.160km
Gravesend
23m 75ch
24m 09ch
Hoo Jn. Staff Halt (Down)
27m 07ch
Hoo Jn.
27m 19ch
Cliffe Private Siding
Cliffe GF 28m 56ch
Higham
28m 42ch
Hoo Jn. Staff Halt (Up)
27m 26ch
Up Sdgs
Higham Tunnel
28m 55ch to 29m 46ch
Strood Tunnel
29m 48ch to 30m 74ch
Southfleet Jn.
40.15km
23m 53ch
Singlewell SR
43.008km
Singlewell Crossover
43.08km
Halfpence Lane Tunnel
45.02km to 45.19km
Southfleet Crossover
40.84km
Singlewell Loops
Wybourne (AOCL) 32m 06ch
La Farge Marine Terminal Sidings
17m 53ch
Stoke Creek 35m 64ch
Recreation 36m 20ch
Grain 38m 22ch
Thamesport Freightliner Terminal
Foster Yeoman Terminal
Rochester SB (ER)
33m 42ch
Chatham Tunnel
34m 34ch to 34m 48ch
Gillingham Tunnel
35m 07ch to 35m 47ch
Fort Pitt Tunnel
33m 79ch to 34m 19ch
Chatham Dockyard
1m 36ch
Strood
31m 11ch
31m 16ch
Down Sidings
Gillingham
35m 75ch
Gillingham 36m 17ch
Rainham SB (EU)
39m 00ch
Rainham
38m 74ch
Chatham
34m 25ch
Rochester
33m 61ch
52.23km
East Kent Signalling Centre (EK)
36m 05ch
Sidings
Carriage Washing Machine
Inspection Shed
Station Road 39m 01ch
Sole Street
26m 71ch
Sidings (oou)
Cuxton
33m 36ch
Cuxton 33m 39ch
Cuxton SB (CX)
33m 40ch
North Halling
34m 33ch
Rugby Sidings
GF
34m 79ch
Nashenden Crossovers
52.90km
GF 33m 31ch
North Downs Tunnel
54.44km to 57.64km
① Shakehole Bridge
② Medway Viaduct
49.97km to 51.23km
Ⓐ Rochester Bridge Jn.
33m 01ch from Victoria and to Ramsgate / 31m 34ch from Charing Cross
Ⓑ Week Street Tunnel
40m 01ch to 40m 06ch
Ⓒ Wheeler Street Tunnel
40m 09ch to 40m 25ch
Ⓓ Gillingham SB (ET)
36m 16ch
Halling
35m 18ch
Snodland
36m 59ch
Down Sidings
Snodland (MCB) 36m 64ch
Snodland SB (EDM)
36m 64ch
New Hythe
38m 03ch
Aylesford
38m 74ch
GF 38m 69ch
Brookgate Sidings
Allington Sidings
Boxley Tunnel
60.13km to 60.46km
Sheerness Docks
Sheerness Steel
Sheerness-on-Sea
51m 19ch
Sheerness Dockyard
50m 48ch
Queenbrough Yard
Queenborough
49m 22ch
South Marsh 48m 02ch
Swale
47m 15ch
Kingsferry Lift Bridge
47m 26ch
Bowaters (Ridham Docks)
Middle Jn.
44m 13ch / 0m 23ch from Eastern Jn.
Newington
41m 44ch
Gascoyne Crossing 45m 60ch
Kemsley
45m 20ch
Fox Grove Crossing 44m 70ch
Woods Sidings
44m 76ch
Western Jn.
43m 70ch
Eastern Jn.
44m 18ch / 0m 00ch to Middle Jn.
Sittingbourne
44m 59ch
Sittingbourne SB (EV)
44m 73ch
Bax Crossing 47m 02ch
Frognall Farm 47m 37ch
Teynham East (CCTV) 47m 79ch
Stone (AHBC) 49m 40ch
St Annes Crossing 51m 48ch
Teynham
47m 74ch
Down Sidings
Faversham
51m 77ch
Clock House (AHBC) 53m 51ch
Level Crossings between Snodland Station and Allington Sidings
① Abbey Meads 37m 50ch
② Willow Walk (FPW) 37m 61ch
③ Mill Hall 38m 64ch
④ Aylesford 38m 78ch
⑤ Aylesford Village (CCTV) 39m 36ch
⑥ Little Preston 39m 66ch
3
Barming
37m 43ch
West Malling
34m 61ch
East Malling
35m 64ch
Maidstone East
39m 76ch
65.89km
Crismill Crossovers
66.31km
Maidstone Barracks
42m 00ch
Change of mileage
42m 41ch / 44m 56ch
Maidstone West
42m 36ch
Maidstone West SB (MS)
44m 54ch
Bearsted
42m 59ch
Hollingbourne
45m 02ch
66.86km
Eyhorne Tunnel
67.62km to 67.98km
Harrietsham Tunnel
71.22km to 71.37km
Harrietsham
47m 36ch
Lenham
49m 11ch
Lenham SR
77.260km
Level Crossings between Wateringbury Station and Maidstone West Station
① Teston (CCTV) 41m 04ch
② Teston Foot Crossing 41m 14ch
③ Farleigh Lane 42m 70ch
④ Tevil 44m 30ch
① Aylesford SB (AF)
38m 75ch
② Maidstone East SB (ME)
39m 72ch
③ East Farleigh SB (EF)
42m 75ch
Church Crossing 39m 30ch
Wilson's Crossing 38m 79ch
Millstead Crossing 38m 43ch
McDougal Crossing 38m 27ch
39m 71ch
Wateringbury
39m 77ch
Wateringbury SB (WB)
39m 72ch
East Farleigh
42m 75ch
Yalding
38m 19ch
37m 34ch
Sandway Tunnel
74.18km to 74.35km
Lenham Crossover
75.59km to 75.88km
Lenham Loops
Charing
53m 11ch
Hothfield Sidings
13

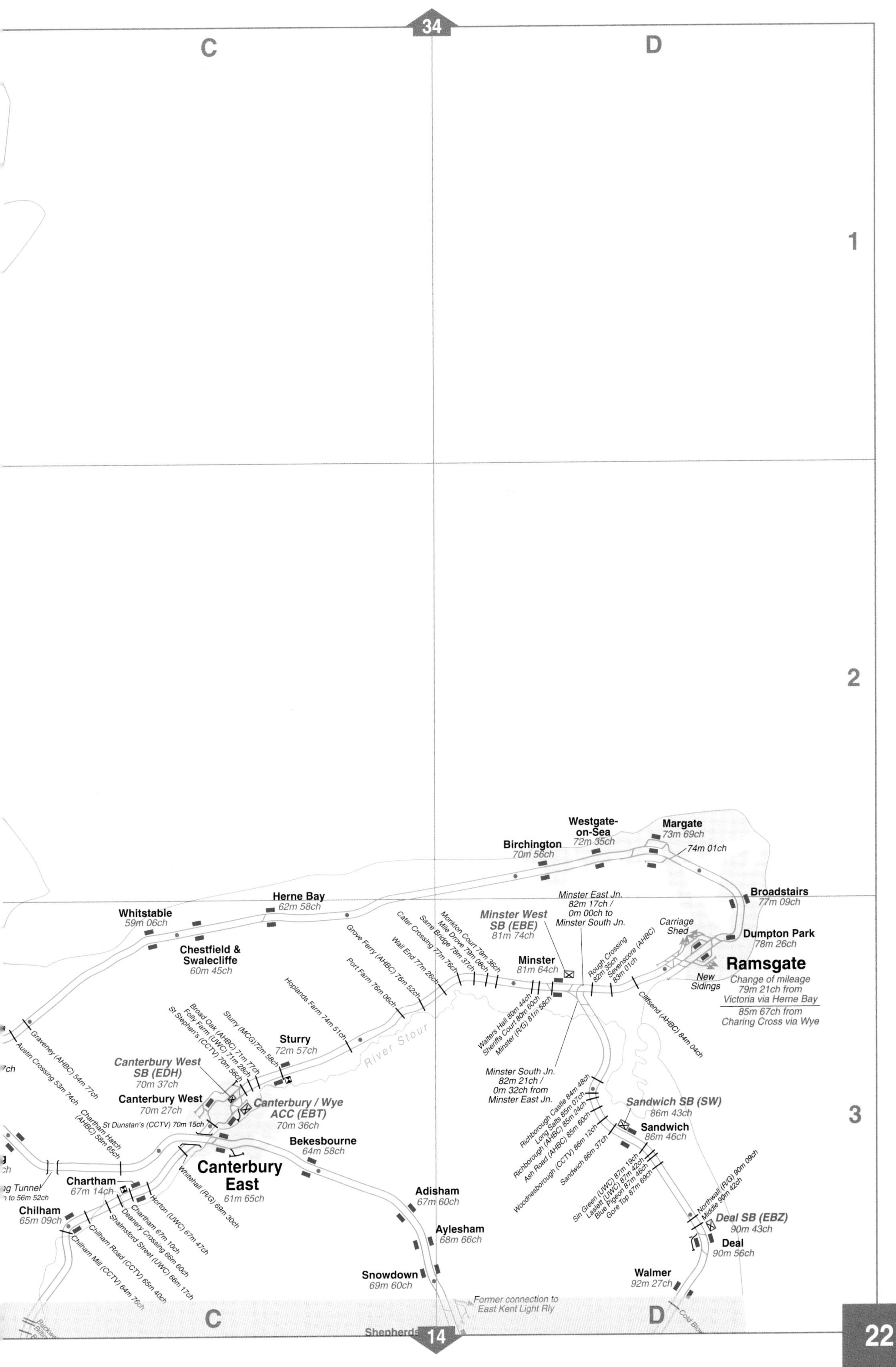
C
D
1
2
3
Margate
73m 69ch
Westgate-on-Sea
72m 35ch
Birchington
70m 56ch
74m 01ch
Broadstairs
77m 09ch
Dumpton Park
78m 26ch
Carriage Shed
Ramsgate
Change of mileage
79m 21ch from Victoria via Herne Bay
85m 67ch from Charing Cross via Wye
New Sidings
Herne Bay
62m 58ch
Whitstable
59m 06ch
Chestfield & Swalecliffe
60m 45ch
Minster East Jn.
82m 17ch /
0m 00ch to
Minster South Jn.
Minster West SB (EBE)
81m 74ch
Minster
81m 64ch
Monkton Court 79m 36ch
Mile Drove 79m 08ch
Sarre Bridge 78m 37ch
Cater Crossing 77m 76ch
Wall End 77m 26ch
Grove Ferry (AHBC) 76m 52ch
Port Farm 76m 06ch
Hoplands Farm 74m 51ch
Rough Crossing 82m 35ch
Sevenscore (AHBC) 83m 01ch
Cliffsend (AHBC) 84m 04ch
Walters Hall 80m 44ch
Sheriffs Court 80m 60ch
Minster (R/G) 81m 58ch
River Stour
Minster South Jn.
82m 21ch /
0m 32ch from
Minster East Jn.
Sturry
72m 57ch
Sturry (MCG) 72m 58ch
Broad Oak (AHBC) 71m 77ch
Folly Farm (UWC) 71m 28ch
St Stephen's (CCTV) 70m 56ch
Graveney (AHBC) 54m 77ch
Austin Crossing 53m 74ch
Canterbury West SB (EDH)
70m 37ch
Canterbury West
70m 27ch
Canterbury / Wye ACC (EBT)
70m 36ch
St Dunstan's (CCTV) 70m 15ch
Chartham Hatch (AHBC) 58m 65ch
Canterbury East
61m 65ch
Whitehall (R/G) 69m 30ch
Bekesbourne
64m 58ch
Chartham
67m 14ch
Horton (UWC) 67m 47ch
Chartham 67m 10ch
Deanery Crossing 66m 60ch
Shalmsford Street (UWC) 66m 17ch
Chilham
65m 09ch
Chilham Road (CCTV) 65m 40ch
Chilham Mill (CCTV) 64m 76ch
Adisham
67m 60ch
Aylesham
68m 66ch
Snowdown
69m 60ch
Former connection to East Kent Light Rly
Shepherds
Richborough Castle 84m 48ch
Long Salts 85m 07ch
Richborough (AHBC) 85m 24ch
Ash Road (AHBC) 85m 60ch
Woodnesborough (CCTV) 86m 12ch
Sandwich 86m 37ch
Sandwich SB (SW)
86m 43ch
Sandwich
86m 46ch
Sin Green (UWC) 87m 19ch
Laslett (UWC) 87m 42ch
Blue Pigeon 87m 46ch
Gore Top 87m 69ch
Northwall (R/G) 90m 09ch
Middle 90m 42ch
Deal SB (EBZ)
90m 43ch
Deal
90m 56ch
Walmer
92m 27ch
C
D

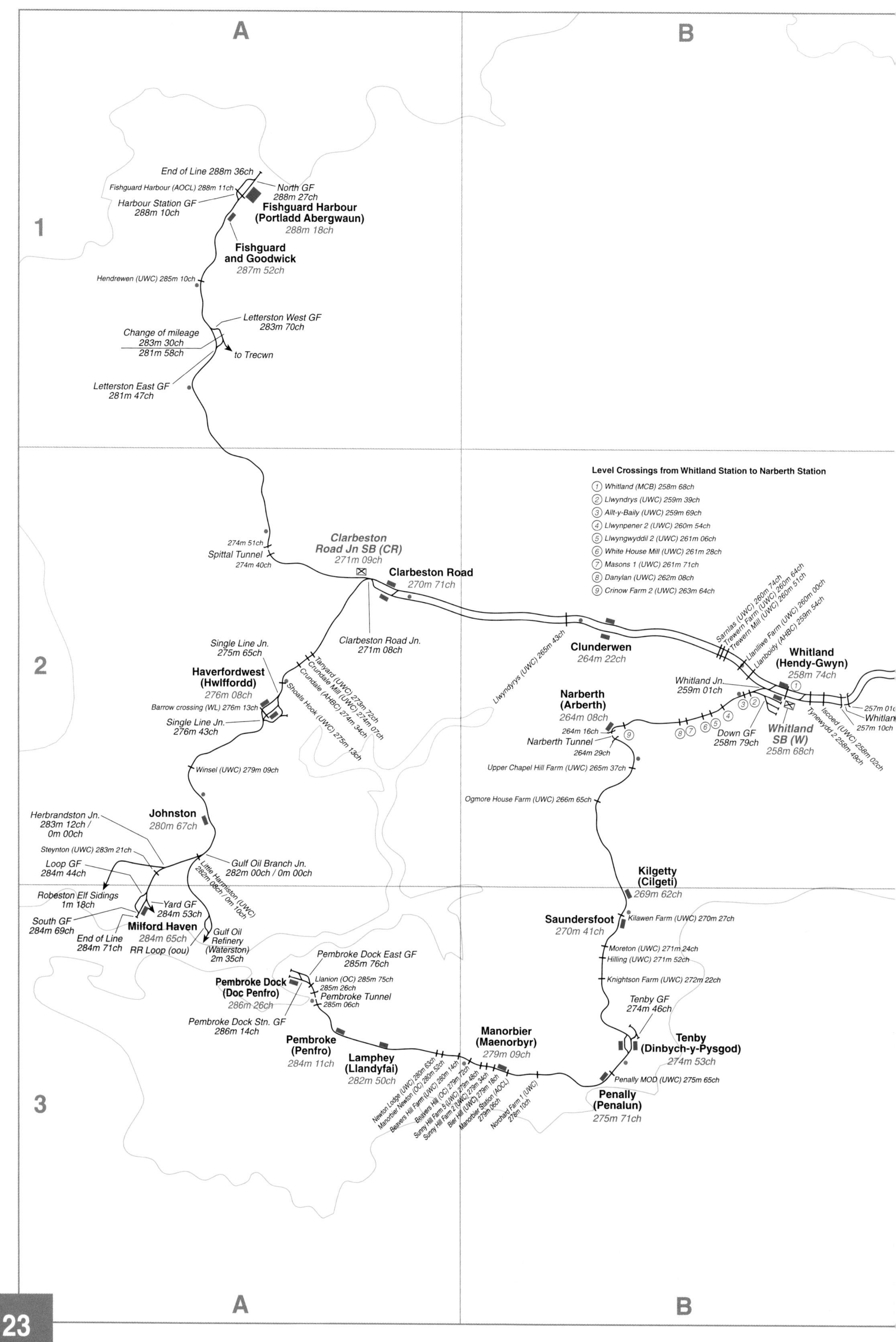
A
B
1
2
3
End of Line 288m 36ch
Fishguard Harbour (AOCL) 288m 11ch
North GF
288m 27ch
Harbour Station GF
288m 10ch
Fishguard Harbour
(Portladd Abergwaun)
288m 18ch
Fishguard
and Goodwick
287m 52ch
Hendrewen (UWC) 285m 10ch
Letterston West GF
283m 70ch
Change of mileage
283m 30ch
281m 58ch
to Trecwn
Letterston East GF
281m 47ch
274m 51ch
Spittal Tunnel
274m 40ch
Clarbeston
Road Jn SB (CR)
271m 09ch
Clarbeston Road
270m 71ch
Clarbeston Road Jn.
271m 08ch
Level Crossings from Whitland Station to Narberth Station
1 Whitland (MCB) 258m 68ch
2 Llwyndrys (UWC) 259m 39ch
3 Allt-y-Baily (UWC) 259m 69ch
4 Llwynpener 2 (UWC) 260m 54ch
5 Llwyngwyddil 2 (UWC) 261m 06ch
6 White House Mill (UWC) 261m 28ch
7 Masons 1 (UWC) 261m 71ch
8 Danylan (UWC) 262m 08ch
9 Crinow Farm 2 (UWC) 263m 64ch
Sarnlas (UWC) 260m 74ch
Trewern Farm (UWC) 260m 64ch
Trewern Mill (UWC) 260m 51ch
Llanlliwe Farm (UWC) 260m 00ch
Llanboidy (AHBC) 259m 54ch
Clunderwen
264m 22ch
Llwyndyrys (UWC) 265m 43ch
Whitland
(Hendy-Gwyn)
258m 74ch
Whitland Jn.
259m 01ch
Down GF
258m 79ch
Whitland
SB (W)
258m 68ch
Iscoed (UWC) 258m 02ch
Tynewydd 2 258m 49ch
257m 01ch
Whitland
257m 10ch
Single Line Jn.
275m 65ch
Haverfordwest
(Hwlffordd)
276m 08ch
Barrow crossing (WL) 276m 13ch
Single Line Jn.
276m 43ch
Tanyard (UWC) 273m 72ch
Crundale Mill (UWC) 274m 07ch
Crundale (AHBC) 274m 34ch
Shoals Hook (UWC) 275m 13ch
Narberth
(Arberth)
264m 08ch
264m 16ch
Narberth Tunnel
264m 29ch
Upper Chapel Hill Farm (UWC) 265m 37ch
Ogmore House Farm (UWC) 266m 65ch
Winsel (UWC) 279m 09ch
Herbrandston Jn.
283m 12ch /
0m 00ch
Johnston
280m 67ch
Steynton (UWC) 283m 21ch
Loop GF
284m 44ch
Gulf Oil Branch Jn.
282m 00ch / 0m 00ch
Little Harmiston (UWC)
282m 08ch / 0m 10ch
Robeston Elf Sidings
1m 18ch
Yard GF
284m 53ch
South GF
284m 69ch
Milford Haven
284m 65ch
End of Line
284m 71ch
RR Loop (oou)
Gulf Oil
Refinery
(Waterston)
2m 35ch
Kilgetty
(Cilgeti)
269m 62ch
Saundersfoot
270m 41ch
Kilawen Farm (UWC) 270m 27ch
Moreton (UWC) 271m 24ch
Hilling (UWC) 271m 52ch
Knightson Farm (UWC) 272m 22ch
Pembroke Dock East GF
285m 76ch
Llanion (OC) 285m 75ch
285m 26ch
Pembroke Dock
(Doc Penfro)
286m 26ch
Pembroke Tunnel
285m 06ch
Pembroke Dock Stn. GF
286m 14ch
Tenby GF
274m 46ch
Tenby
(Dinbych-y-Pysgod)
274m 53ch
Pembroke
(Penfro)
284m 11ch
Lamphey
(Llandyfai)
282m 50ch
Manorbier
(Maenorbyr)
279m 09ch
Penally MOD (UWC) 275m 65ch
Penally
(Penalun)
275m 71ch
Newton Lodge (UWC) 280m 63ch
Manorbier Newton (OC) 280m 52ch
Beavers Hill Farm (UWC) 280m 14ch
Beavers Hill (OC) 279m 72ch
Sunny Hill Farm 5 (UWC) 279m 48ch
Sunny Hill Farm 2 (UWC) 279m 34ch
Bier Hill (UWC) 279m 18ch
Manorbier Station (AOCL)
279m 06ch
Norchard Farm 1 (UWC)
278m 10ch
A
B

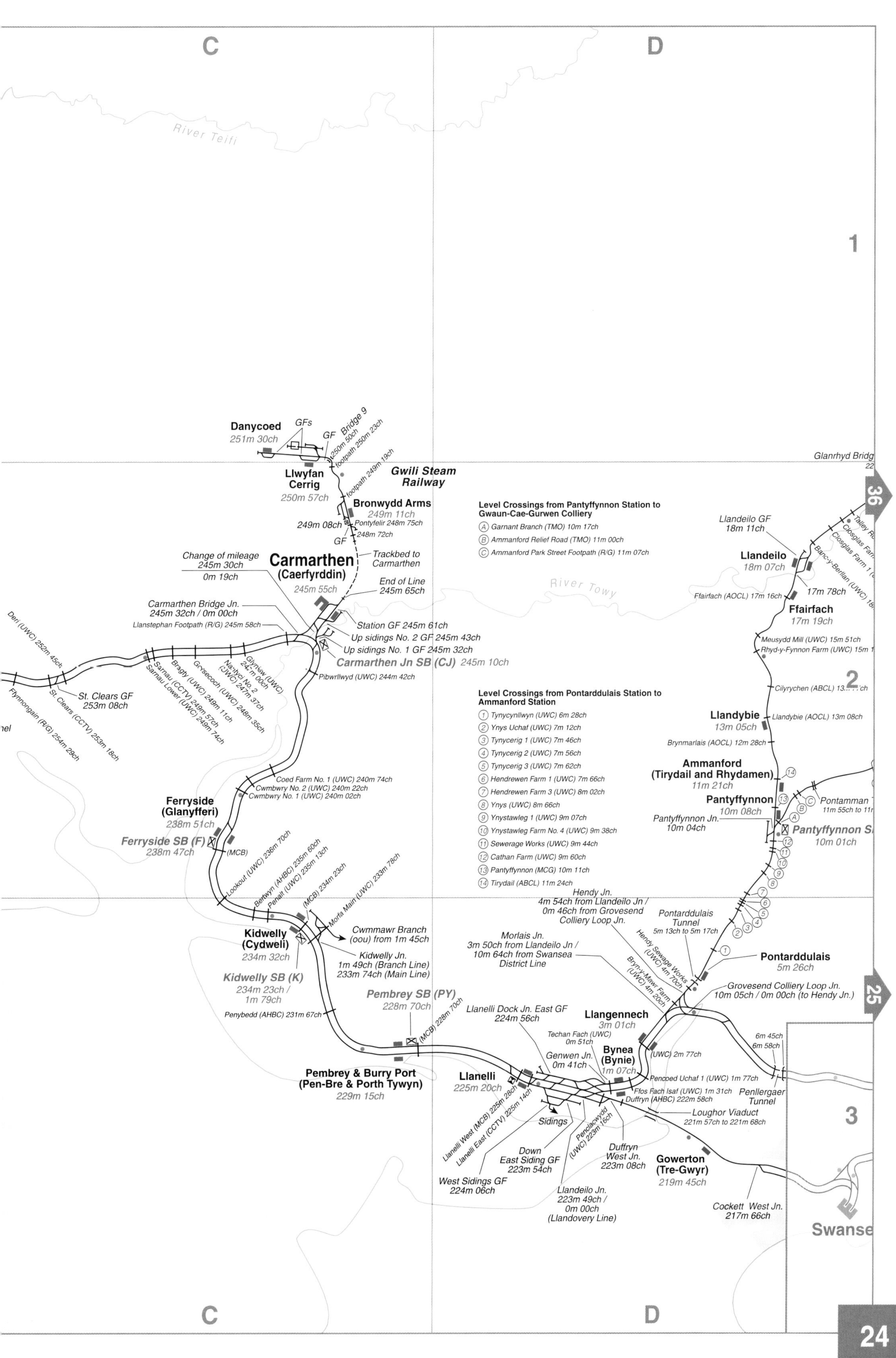
C
D
River Teifi
1
Danycoed
251m 30ch
GFs
GF
Bridge 9
250m 50ch
footpath 250m 23ch
footpath 249m 19ch
Llwyfan Cerrig
250m 57ch
Gwili Steam Railway
Bronwydd Arms
249m 11ch
249m 08ch
Pontyfelir 248m 75ch
248m 72ch
GF
Trackbed to Carmarthen
Change of mileage
245m 30ch
0m 19ch
Carmarthen
(Caerfyrddin)
245m 55ch
End of Line
245m 65ch
Carmarthen Bridge Jn.
245m 32ch / 0m 00ch
Llanstephan Footpath (R/G) 245m 58ch
Station GF 245m 61ch
Up sidings No. 2 GF 245m 43ch
Up sidings No. 1 GF 245m 32ch
Carmarthen Jn SB (CJ) 245m 10ch
Pibwrllwyd (UWC) 244m 42ch
Deri (UWC) 252m 45ch
St. Clears GF
253m 08ch
St. Clears (CCTV) 253m 18ch
Ffynnongain (R/G) 254m 29ch
Sarnau (CCTV) 249m 57ch
Sarnau Lower (UWC) 249m 74ch
Bragty (UWC) 249m 11ch
Grysecoch (UWC) 248m 35ch
Nantyci No. 2 (UWC) 247m 37ch
Glyniaw (UWC) 247m 00ch
Level Crossings from Pantyffynnon Station to Gwaun-Cae-Gurwen Colliery
(A) Garnant Branch (TMO) 10m 17ch
(B) Ammanford Relief Road (TMO) 11m 00ch
(C) Ammanford Park Street Footpath (R/G) 11m 07ch
Level Crossings from Pontarddulais Station to Ammanford Station
(1) Tynycynllwyn (UWC) 6m 28ch
(2) Ynys Uchaf (UWC) 7m 12ch
(3) Tynycerig 1 (UWC) 7m 46ch
(4) Tynycerig 2 (UWC) 7m 56ch
(5) Tynycerig 3 (UWC) 7m 62ch
(6) Hendrewen Farm 1 (UWC) 7m 66ch
(7) Hendrewen Farm 3 (UWC) 8m 02ch
(8) Ynys (UWC) 8m 66ch
(9) Ynystawleg 1 (UWC) 9m 07ch
(10) Ynystawleg Farm No. 4 (UWC) 9m 38ch
(11) Sewerage Works (UWC) 9m 44ch
(12) Cathan Farm (UWC) 9m 60ch
(13) Pantyffynnon (MCG) 10m 11ch
(14) Tirydail (ABCL) 11m 24ch
River Towy
Glanrhyd Bridg
36
Llandeilo GF
18m 11ch
Llandeilo
18m 07ch
Talley Ro
Closglas Farm
Closglas Farm 1
Banc-y-Berllan (UWC) 18
17m 78ch
Ffairfach (AOCL) 17m 16ch
Ffairfach
17m 19ch
Meusydd Mill (UWC) 15m 51ch
Rhyd-y-Fynnon Farm (UWC) 15m
Cilyrychen (ABCL) 13
2
Llandybie
13m 05ch
Llandybie (AOCL) 13m 08ch
Brynmarlais (AOCL) 12m 28ch
Ammanford
(Tirydail and Rhydamen)
11m 21ch
Pantyffynnon
10m 08ch
Pantyffynnon Jn.
10m 04ch
Pantyffynnon S
10m 01ch
Pontamman
11m 55ch to 11
Coed Farm No. 1 (UWC) 240m 74ch
Cwmbwry No. 2 (UWC) 240m 22ch
Cwmbwry No. 1 (UWC) 240m 02ch
Ferryside
(Glanyfferi)
238m 51ch
Ferryside SB (F)
238m 47ch
(MCB)
Lookout (UWC) 236m 70ch
Berwyn (AHBC) 235m 60ch
Penallt (UWC) 235m 13ch
(MCB) 234m 23ch
Morfa Main (UWC) 233m 78ch
Kidwelly
(Cydweli)
234m 32ch
Cwmmawr Branch
(oou) from 1m 45ch
Kidwelly Jn.
1m 49ch (Branch Line)
233m 74ch (Main Line)
Kidwelly SB (K)
234m 23ch /
1m 79ch
Penybedd (AHBC) 231m 67ch
Pembrey SB (PY)
228m 70ch
(MCB) 228m 70ch
Pembrey & Burry Port
(Pen-Bre & Porth Tywyn)
229m 15ch
Hendy Jn.
4m 54ch from Llandeilo Jn /
0m 46ch from Grovesend
Colliery Loop Jn.
Pontarddulais Tunnel
5m 13ch to 5m 17ch
Hendy Sewage Works (UWC) 4m 70ch
Bryn-y-Mawr Farm (UWC) 4m 20ch
Morlais Jn.
3m 50ch from Llandeilo Jn /
10m 64ch from Swansea
District Line
Pontarddulais
5m 26ch
Grovesend Colliery Loop Jn.
10m 05ch / 0m 00ch (to Hendy Jn.)
25
Llanelli Dock Jn. East GF
224m 56ch
Llangennech
3m 01ch
Techan Fach (UWC)
0m 51ch
Genwen Jn.
0m 41ch
Bynea
(Bynie)
1m 07ch
(UWC) 2m 77ch
6m 45ch
6m 58ch
Llanelli
225m 20ch
Pencoed Uchaf 1 (UWC) 1m 77ch
Ffos Fach Isaf (UWC) 1m 31ch
Duffryn (AHBC) 222m 58ch
Penllergaer Tunnel
Loughor Viaduct
221m 57ch to 221m 68ch
3
Llanelli West (MCB) 225m 28ch
Llanelli East (CCTV) 225m 14ch
Sidings
Down
East Siding GF
223m 54ch
Penclacwydd (UWC) 223m 16ch
Duffryn
West Jn.
223m 08ch
Gowerton
(Tre-Gwyr)
219m 45ch
West Sidings GF
224m 06ch
Llandeilo Jn.
223m 49ch /
0m 00ch
(Llandovery Line)
Cockett West Jn.
217m 66ch
Swanse
C
D

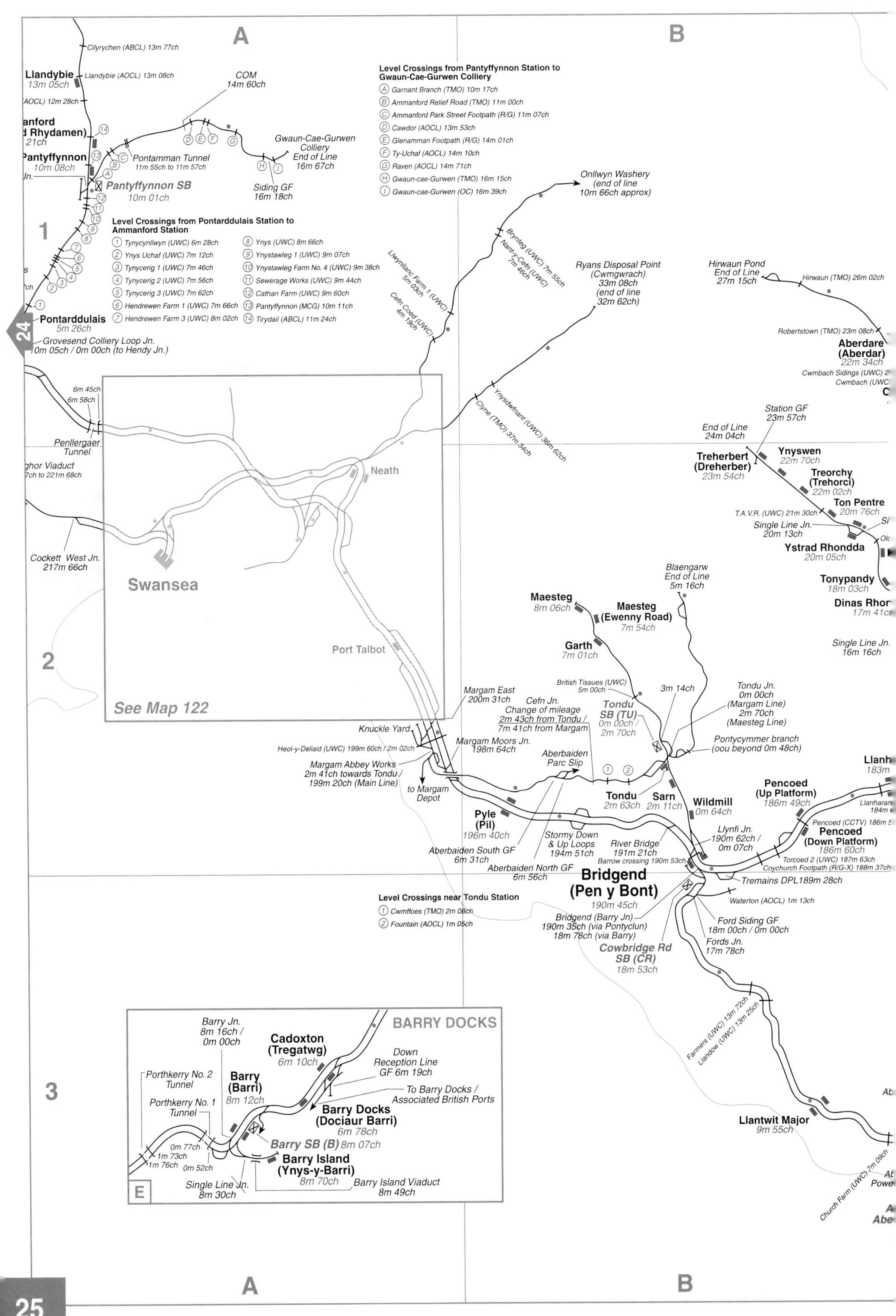

A
B
Cilyrychen (ABCL) 13m 77ch
Llandybie
13m 05ch
Llandybie (AOCL) 13m 08ch
(AOCL) 12m 28ch
anford
d Rhydamen)
21ch
Pantyffynnon
10m 08ch
Jn.
Pantyffynnon SB
10m 01ch
COM
14m 60ch
Pontamman Tunnel
11m 55ch to 11m 57ch
Gwaun-Cae-Gurwen
Colliery
End of Line
16m 67ch
Siding GF
16m 18ch
Level Crossings from Pantyffynnon Station to Gwaun-Cae-Gurwen Colliery
Ⓐ Garnant Branch (TMO) 10m 17ch
Ⓑ Ammanford Relief Road (TMO) 11m 00ch
Ⓒ Ammanford Park Street Footpath (R/G) 11m 07ch
Ⓓ Cawdor (AOCL) 13m 53ch
Ⓔ Glenamman Footpath (R/G) 14m 01ch
Ⓕ Ty-Uchaf (AOCL) 14m 10ch
Ⓖ Raven (AOCL) 14m 71ch
Ⓗ Gwaun-cae-Gurwen (TMO) 16m 15ch
Ⓘ Gwaun-cae-Gurwen (OC) 16m 39ch
Level Crossings from Pontarddulais Station to Ammanford Station
① Tynycynllwyn (UWC) 6m 28ch
② Ynys Uchaf (UWC) 7m 12ch
③ Tynycerig 1 (UWC) 7m 46ch
④ Tynycerig 2 (UWC) 7m 56ch
⑤ Tynycerig 3 (UWC) 7m 62ch
⑥ Hendrewen Farm 1 (UWC) 7m 66ch
⑦ Hendrewen Farm 3 (UWC) 8m 02ch
⑧ Ynys (UWC) 8m 66ch
⑨ Ynystawleg 1 (UWC) 9m 07ch
⑩ Ynystawleg Farm No. 4 (UWC) 9m 38ch
⑪ Sewerage Works (UWC) 9m 44ch
⑫ Cathan Farm (UWC) 9m 60ch
⑬ Pantyffynnon (MCG) 10m 11ch
⑭ Tirydail (ABCL) 11m 24ch
1
Onllwyn Washery
(end of line
10m 66ch approx)
Brynteg (UWC) 7m 55ch
Nant-y-Cefn (UWC)
7m 46ch
Llwynllanc Farm 1 (UWC)
5m 03ch
Cefn Coed (UWC)
4m 19ch
Ryans Disposal Point
(Cwmgwrach)
33m 08ch
(end of line
32m 62ch)
Hirwaun Pond
End of Line
27m 15ch
Hirwaun (TMO) 26m 02ch
Robertstown (TMO) 23m 08ch
Aberdare
(Aberdar)
22m 34ch
Cwmbach Sidings (UWC)
Cwmbach (UWC
24
Pontarddulais
5m 26ch
Grovesend Colliery Loop Jn.
10m 05ch / 0m 00ch (to Hendy Jn.)
6m 45ch
6m 58ch
Penllergaer
Tunnel
ghor Viaduct
7ch to 221m 68ch
Ynysdwfnant (UWC) 36m 62ch
Clyne (TMO) 37m 34ch
Station GF
23m 57ch
End of Line
24m 04ch
Treherbert
(Dreherber)
23m 54ch
Ynyswen
22m 70ch
Treorchy
(Trehorci)
22m 02ch
Ton Pentre
20m 76ch
T.A.V.R. (UWC) 21m 30ch
Single Line Jn.
20m 13ch
Ystrad Rhondda
20m 05ch
Tonypandy
18m 03ch
Dinas Rhor
17m 41c
Neath
Cockett West Jn.
217m 66ch
Swansea
2
Port Talbot
See Map 122
Blaengarw
End of Line
5m 16ch
Maesteg
8m 06ch
Maesteg
(Ewenny Road)
7m 54ch
Garth
7m 01ch
Single Line Jn.
16m 16ch
British Tissues (UWC)
5m 00ch
Tondu
SB (TU)
0m 00ch /
2m 70ch
3m 14ch
Tondu Jn.
0m 00ch
(Margam Line)
2m 70ch
(Maesteg Line)
Pontycymmer branch
(oou beyond 0m 48ch)
Margam East
200m 31ch
Cefn Jn.
Change of mileage
2m 43ch from Tondu /
7m 41ch from Margam
Knuckle Yard
Margam Moors Jn.
198m 64ch
Heol-y-Deliaid (UWC) 199m 60ch / 2m 02ch
Margam Abbey Works
2m 41ch towards Tondu /
199m 20ch (Main Line)
to Margam
Depot
Aberbaiden
Parc Slip
Tondu
2m 63ch
Sarn
2m 11ch
Wildmill
0m 64ch
Llanh
183m
Pencoed
(Up Platform)
186m 49ch
Llanharan
184m
Pencoed (CCTV) 186m
Pencoed
(Down Platform)
186m 60ch
Pyle
(Pil)
196m 40ch
Stormy Down
& Up Loops
194m 51ch
River Bridge
191m 21ch
Llynfi Jn.
190m 62ch /
0m 07ch
Aberbaiden South GF
6m 31ch
Aberbaiden North GF
6m 56ch
Barrow crossing 190m 53ch
Torcoed 2 (UWC) 187m 63ch
Coychurch Footpath (R/G-X) 188m 37ch
Bridgend
(Pen y Bont)
190m 45ch
Tremains DPL189m 28ch
Waterton (AOCL) 1m 13ch
Level Crossings near Tondu Station
① Cwmffoes (TMO) 2m 08ch
② Fountain (AOCL) 1m 05ch
Bridgend (Barry Jn)
190m 35ch (via Pontyclun)
18m 78ch (via Barry)
Ford Siding GF
18m 00ch / 0m 00ch
Fords Jn.
17m 78ch
Cowbridge Rd
SB (CR)
18m 53ch
Farmers (UWC) 13m 72ch
Llandow (UWC) 13m 25ch
BARRY DOCKS
Barry Jn.
8m 16ch /
0m 00ch
Cadoxton
(Tregatwg)
6m 10ch
Down
Reception Line
GF 6m 19ch
Porthkerry No. 2
Tunnel
Barry
(Barri)
8m 12ch
To Barry Docks /
Associated British Ports
Porthkerry No. 1
Tunnel
Barry Docks
(Dociaur Barri)
6m 78ch
Barry SB (B) 8m 07ch
0m 77ch
1m 73ch
1m 76ch
0m 52ch
Barry Island
(Ynys-y-Barri)
8m 70ch
Single Line Jn.
8m 30ch
Barry Island Viaduct
8m 49ch
E
3
Llantwit Major
9m 55ch
Church Farm (UWC) 7m 09ch
A
B

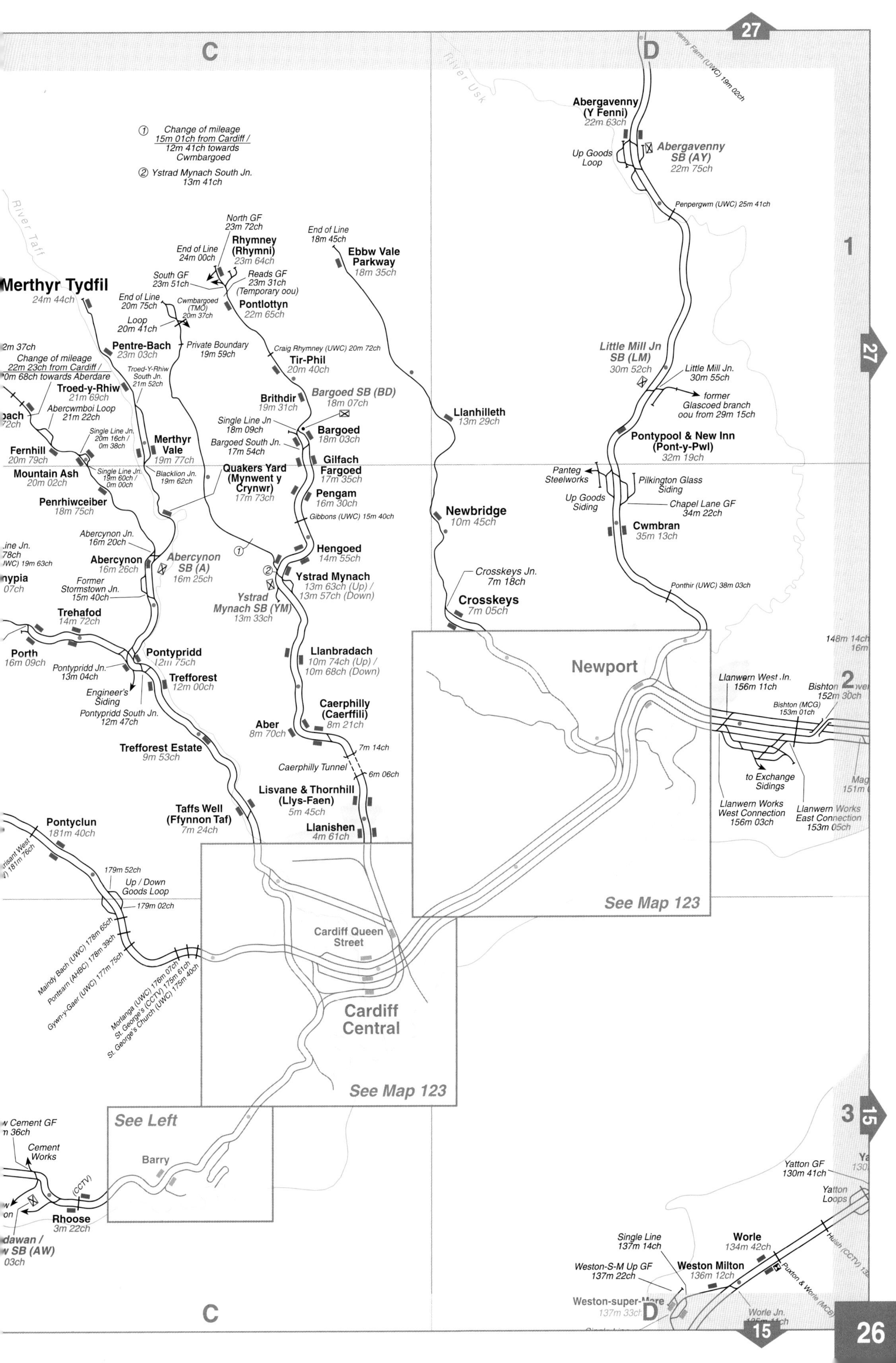

27
C
D
River Usk
① Change of mileage 15m 01ch from Cardiff / 12m 41ch towards Cwmbargoed
② Ystrad Mynach South Jn. 13m 41ch
River Taff
North GF 23m 72ch
Rhymney (Rhymni) 23m 64ch
End of Line 24m 00ch
South GF 23m 51ch
Reads GF 23m 31ch (Temporary oou)
End of Line 18m 45ch
Ebbw Vale Parkway 18m 35ch
Merthyr Tydfil 24m 44ch
End of Line 20m 75ch
Cwmbargoed (TMO) 20m 37ch
Pontlottyn 22m 65ch
Loop 20m 41ch
Pentre-Bach 23m 03ch
Private Boundary 19m 59ch
Craig Rhymney (UWC) 20m 72ch
Tir-Phil 20m 40ch
Change of mileage 22m 23ch from Cardiff / 0m 68ch towards Aberdare
Troed-Y-Rhiw South Jn. 21m 52ch
Troed-y-Rhiw 21m 69ch
Abercwmboi Loop 21m 22ch
Brithdir 19m 31ch
Bargoed SB (BD) 18m 07ch
Llanhilleth 13m 29ch
Single Line Jn 18m 09ch
Bargoed 18m 03ch
Single Line Jn. 20m 16ch / 0m 38ch
Merthyr Vale 19m 77ch
Bargoed South Jn. 17m 54ch
Fernhill 20m 79ch
Gilfach Fargoed 17m 35ch
Mountain Ash 20m 02ch
Single Line Jn. 19m 60ch / 0m 00ch
Blacklion Jn. 19m 62ch
Quakers Yard (Mynwent y Crynwr) 17m 73ch
Pengam 16m 30ch
Penrhiwceiber 18m 75ch
Gibbons (UWC) 15m 40ch
Newbridge 10m 45ch
Abercynon Jn. 16m 20ch
Hengoed 14m 55ch
Abercynon 16m 26ch
Abercynon SB (A) 16m 25ch
Ystrad Mynach 13m 63ch (Up) / 13m 57ch (Down)
Crosskeys Jn. 7m 18ch
Former Stormstown Jn. 15m 40ch
Ystrad Mynach SB (YM) 13m 33ch
Crosskeys 7m 05ch
Trehafod 14m 72ch
Porth 16m 09ch
Pontypridd 12m 75ch
Llanbradach 10m 74ch (Up) / 10m 68ch (Down)
Pontypridd Jn. 13m 04ch
Trefforest 12m 00ch
Engineer's Siding
Pontypridd South Jn. 12m 47ch
Caerphilly (Caerffili) 8m 21ch
Aber 8m 70ch
Trefforest Estate 9m 53ch
7m 14ch
Caerphilly Tunnel
6m 06ch
Lisvane & Thornhill (Llys-Faen) 5m 45ch
Taffs Well (Ffynnon Taf) 7m 24ch
Pontyclun 181m 40ch
Llanishen 4m 61ch
179m 52ch
Up / Down Goods Loop
179m 02ch
Maindy Bach (UWC) 178m 65ch
Pontsarn (AHBC) 178m 39ch
Gwyn-y-Gaer (UWC) 177m 75ch
Morlanga (UWC) 176m 07ch
St. George's (CCTV) 175m 61ch
St. George's Church (UWC) 175m 40ch
Cardiff Queen Street
Cardiff Central
See Map 123
See Left
Barry
Cement Works
(CCTV)
Rhoose 3m 22ch
Abergavenny (Y Fenni) 22m 63ch
Up Goods Loop
Abergavenny SB (AY) 22m 75ch
Penpergwm (UWC) 25m 41ch
Little Mill Jn SB (LM) 30m 52ch
Little Mill Jn. 30m 55ch
former Glascoed branch oou from 29m 15ch
Pontypool & New Inn (Pont-y-Pwl) 32m 19ch
Panteg Steelworks
Pilkington Glass Siding
Up Goods Siding
Chapel Lane GF 34m 22ch
Cwmbran 35m 13ch
Ponthir (UWC) 38m 03ch
Newport
Llanwern West Jn. 156m 11ch
Bishton (MCG) 153m 01ch
to Exchange Sidings
Llanwern Works West Connection 156m 03ch
Llanwern Works East Connection 153m 05ch
1
2
3
27
15
Yatton GF 130m 41ch
Yatton Loops
Single Line 137m 14ch
Worle 134m 42ch
Weston-S-M Up GF 137m 22ch
Weston Milton 136m 12ch
Puxton & Worle (MCB)
Worle Jn.
15

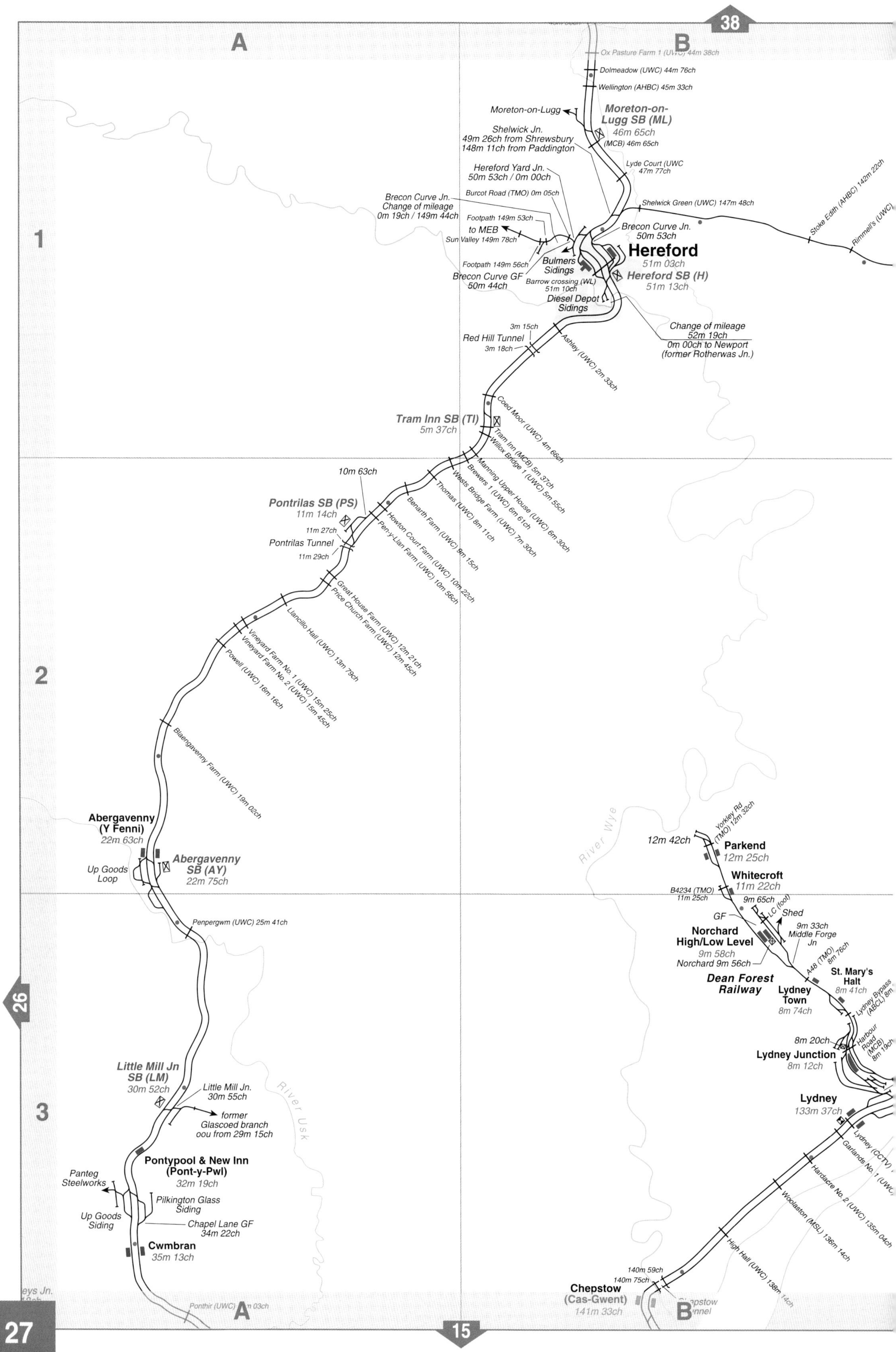
A
B
1
2
3
Ox Pasture Farm 1 (UWC) 44m 38ch
Dolmeadow (UWC) 44m 76ch
Wellington (AHBC) 45m 33ch
Moreton-on-Lugg
Moreton-on-Lugg SB (ML)
46m 65ch
Shelwick Jn.
49m 26ch from Shrewsbury
148m 11ch from Paddington
(MCB) 46m 65ch
Lyde Court (UWC
47m 77ch
Hereford Yard Jn.
50m 53ch / 0m 00ch
Burcot Road (TMO) 0m 05ch
Brecon Curve Jn.
Change of mileage
0m 19ch / 149m 44ch
Shelwick Green (UWC) 147m 48ch
Stoke Edith (AHBC) 142m 22ch
Rimmell's (UWC)
Footpath 149m 53ch
to MEB
Sun Valley 149m 78ch
Brecon Curve Jn.
50m 53ch
Hereford
51m 03ch
Bulmers Sidings
Footpath 149m 56ch
Brecon Curve GF
50m 44ch
Barrow crossing (WL)
51m 10ch
Hereford SB (H)
51m 13ch
Diesel Depot Sidings
Change of mileage
52m 19ch
0m 00ch to Newport
(former Rotherwas Jn.)
3m 15ch
Red Hill Tunnel
3m 18ch
Ashley (UWC) 2m 33ch
Coed Moor (UWC) 4m 66ch
Tram Inn SB (TI)
5m 37ch
Tram Inn (MCB) 5m 37ch
Willox Bridge 1 (UWC) 5m 55ch
Manning Upper House (UWC) 6m 30ch
Brewers 1 (UWC) 6m 61ch
Wests Bridge Farm (UWC) 7m 30ch
Thomas (UWC) 8m 11ch
Benarth Farm (UWC) 9m 15ch
Howton Court Farm (UWC) 10m 22ch
Pen-y-Lian Farm (UWC) 10m 56ch
10m 63ch
Pontrilas SB (PS)
11m 14ch
11m 27ch
Pontrilas Tunnel
11m 29ch
Great House Farm (UWC) 12m 21ch
Price Church Farm (UWC) 12m 45ch
Llancillo Hall (UWC) 13m 79ch
Vineyard Farm No. 1 (UWC) 15m 25ch
Vineyard Farm No. 2 (UWC) 15m 45ch
Powell (UWC) 16m 16ch
Blaengavenny Farm (UWC) 19m 02ch
Abergavenny (Y Fenni)
22m 63ch
Abergavenny SB (AY)
22m 75ch
Up Goods Loop
Penpergwm (UWC) 25m 41ch
River Wye
River Usk
Little Mill Jn SB (LM)
30m 52ch
Little Mill Jn.
30m 55ch
former Glascoed branch
oou from 29m 15ch
Pontypool & New Inn (Pont-y-Pwl)
32m 19ch
Panteg Steelworks
Pilkington Glass Siding
Up Goods Siding
Chapel Lane GF
34m 22ch
Cwmbran
35m 13ch
Ponthir (UWC)
12m 42ch
Yorkley Rd (TMO) 12m 32ch
Parkend
12m 25ch
Whitecroft
11m 22ch
B4234 (TMO)
11m 25ch
9m 65ch
LC (foot)
Shed
GF
9m 33ch
Middle Forge Jn
Norchard High/Low Level
9m 58ch
Norchard 9m 56ch
A48 (TMO) 8m 76ch
St. Mary's Halt
8m 41ch
Dean Forest Railway
Lydney Town
8m 74ch
Lydney Bypass (ABCL)
8m 20ch
Harbour Road (MCB)
8m 19ch
Lydney Junction
8m 12ch
Lydney
133m 37ch
Lydney (CCTV)
Garlands No. 1 (UWC)
Hardacre No. 2 (UWC) 135m 04ch
Woolaston (MSL) 136m 14ch
High Hall (UWC) 138m 14ch
140m 59ch
140m 75ch
Chepstow (Cas-Gwent)
141m 33ch
Chepstow Tunnel

26

15

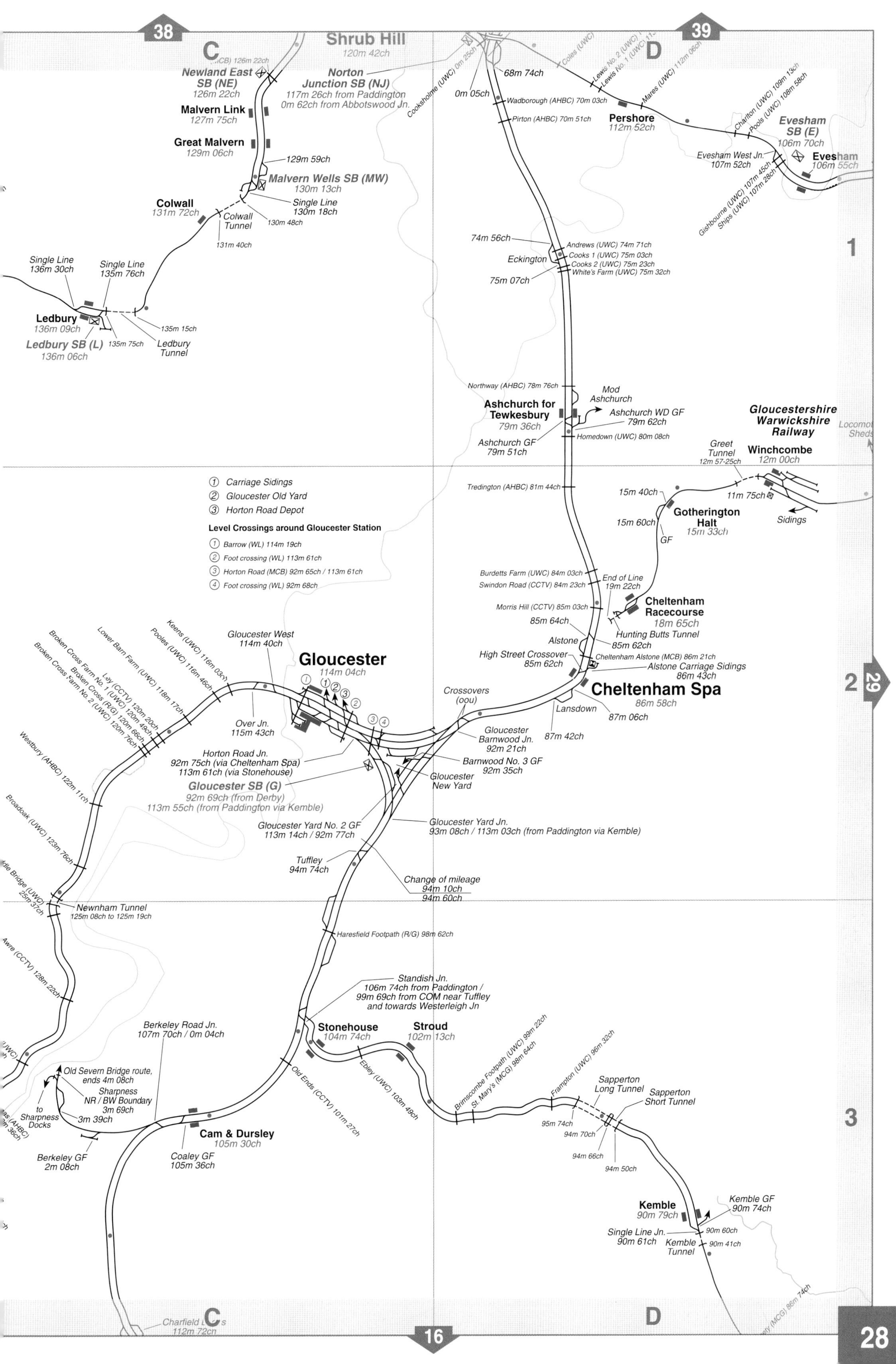

Shrub Hill
120m 42ch
Newland East SB (NE)
126m 22ch
Norton Junction SB (NJ)
117m 26ch from Paddington
0m 62ch from Abbotswood Jn.
Malvern Link
127m 75ch
Great Malvern
129m 06ch
129m 59ch
Malvern Wells SB (MW)
130m 13ch
Single Line
130m 18ch
Colwall
131m 72ch
Colwall Tunnel
130m 48ch
131m 40ch
Single Line
136m 30ch
Single Line
135m 76ch
Ledbury
136m 09ch
Ledbury SB (L)
136m 06ch
135m 75ch
135m 15ch
Ledbury Tunnel
68m 74ch
0m 05ch
Wadborough (AHBC) 70m 03ch
Pirton (AHBC) 70m 51ch
Pershore
112m 52ch
Evesham SB (E)
106m 70ch
Evesham
106m 55ch
Evesham West Jn.
107m 52ch
74m 56ch
Eckington
75m 07ch
Andrews (UWC) 74m 71ch
Cooks 1 (UWC) 75m 03ch
Cooks 2 (UWC) 75m 23ch
White's Farm (UWC) 75m 32ch
Northway (AHBC) 78m 76ch
Mod Ashchurch
Ashchurch for Tewkesbury
79m 36ch
Ashchurch WD GF
79m 62ch
Ashchurch GF
79m 51ch
Homedown (UWC) 80m 08ch
Gloucestershire Warwickshire Railway
Greet Tunnel
12m 57-25ch
Winchcombe
12m 00ch
11m 75ch
Sidings
Tredington (AHBC) 81m 44ch
15m 40ch
15m 60ch
Gotherington Halt
15m 33ch
GF
① Carriage Sidings
② Gloucester Old Yard
③ Horton Road Depot
Level Crossings around Gloucester Station
① Barrow (WL) 114m 19ch
② Foot crossing (WL) 113m 61ch
③ Horton Road (MCB) 92m 65ch / 113m 61ch
④ Foot crossing (WL) 92m 68ch
Burdetts Farm (UWC) 84m 03ch
Swindon Road (CCTV) 84m 23ch
End of Line
19m 22ch
Morris Hill (CCTV) 85m 03ch
Cheltenham Racecourse
18m 65ch
85m 64ch
Alstone
Hunting Butts Tunnel
85m 62ch
High Street Crossover
85m 62ch
Cheltenham Alstone (MCB) 86m 21ch
Alstone Carriage Sidings
86m 43ch
Cheltenham Spa
86m 58ch
Lansdown
87m 06ch
87m 42ch
Gloucester West
114m 40ch
Gloucester
114m 04ch
Over Jn.
115m 43ch
Crossovers (oou)
Gloucester Barnwood Jn.
92m 21ch
Barnwood No. 3 GF
92m 35ch
Horton Road Jn.
92m 75ch (via Cheltenham Spa)
113m 61ch (via Stonehouse)
Gloucester SB (G)
92m 69ch (from Derby)
113m 55ch (from Paddington via Kemble)
Gloucester New Yard
Gloucester Yard No. 2 GF
113m 14ch / 92m 77ch
Gloucester Yard Jn.
93m 08ch / 113m 03ch (from Paddington via Kemble)
Tuffley
94m 74ch
Change of mileage
94m 10ch
94m 60ch
Keens (UWC) 116m 03ch
Pooles (UWC) 116m 46ch
Lower Barn Farm (UWC) 118m 17ch
Ley (CCTV) 120m 20ch
Broken Cross Farm No. 1 (UWC) 120m 49ch
Broken Cross (R/G) 120m 66ch
Broken Cross Farm No. 2 (UWC) 120m 76ch
Westbury (AHBC) 122m 11ch
Broadoak (UWC) 123m 76ch
Newnham Tunnel
125m 08ch to 125m 19ch
Awre (CCTV) 128m 22ch
Haresfield Footpath (R/G) 98m 62ch
Standish Jn.
106m 74ch from Paddington /
99m 69ch from COM near Tuffley
and towards Westerleigh Jn
Berkeley Road Jn.
107m 70ch / 0m 04ch
Stonehouse
104m 74ch
Stroud
102m 13ch
Old Severn Bridge route, ends 4m 08ch
Sharpness
NR / BW Boundary
3m 69ch
to Sharpness Docks
3m 39ch
Old Ends (CCTV) 101m 27ch
Ebley (UWC) 103m 49ch
Brimscombe Footpath (UWC) 99m 22ch
St. Mary's (MCG) 98m 64ch
Frampton (UWC) 96m 32ch
Sapperton Long Tunnel
Sapperton Short Tunnel
95m 74ch
94m 70ch
94m 66ch
94m 50ch
Cam & Dursley
105m 30ch
Coaley GF
105m 36ch
Berkeley GF
2m 08ch
Kemble
90m 79ch
Kemble GF
90m 74ch
Single Line Jn.
90m 61ch
90m 60ch
Kemble Tunnel
90m 41ch
Charfield
112m 72ch
C
D
1
2
3

38 39 29 16

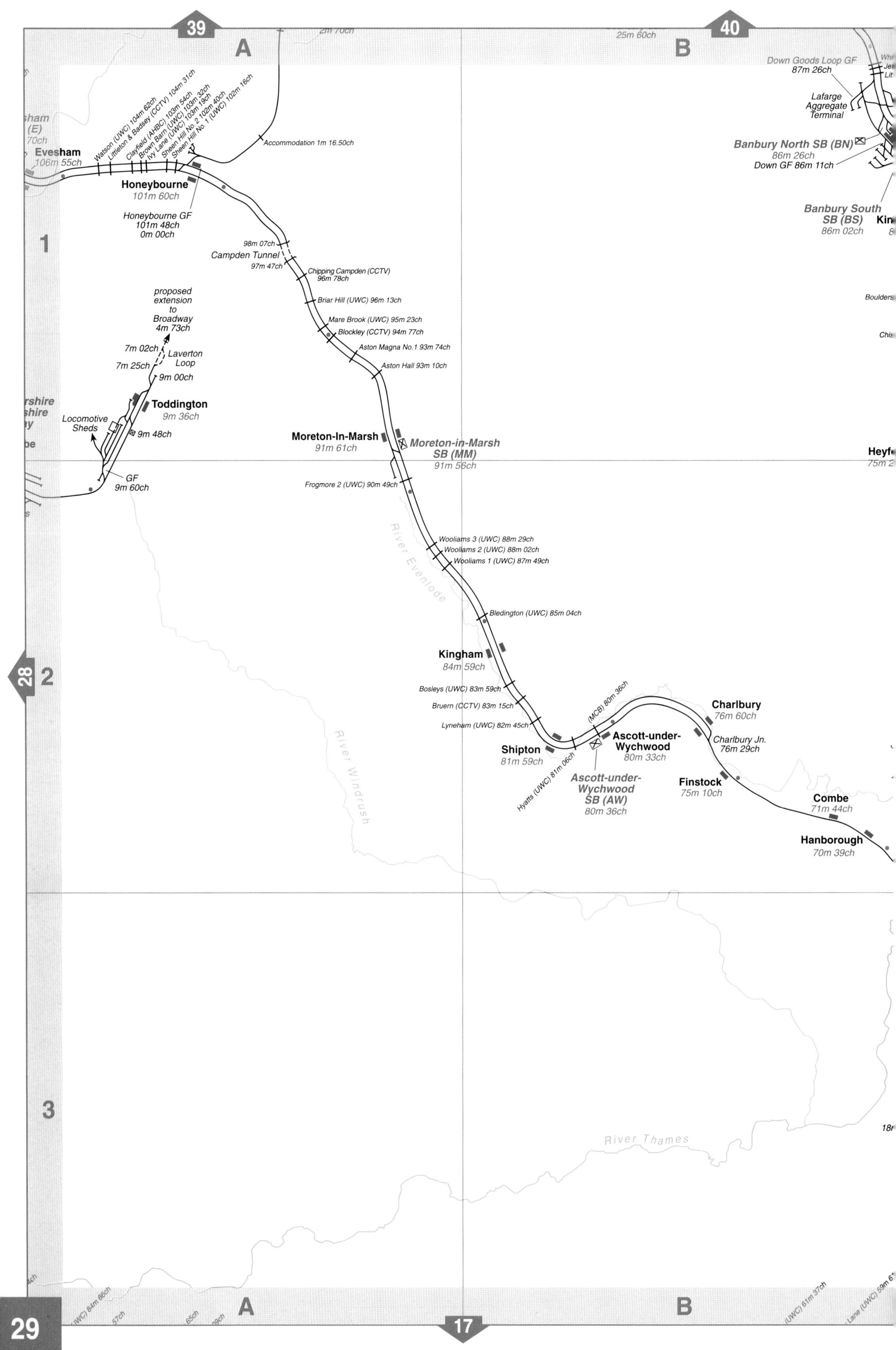
39
40
A
B
2m 70ch
25m 60ch
Evesham
106m 55ch
Watson (UWC) 104m 62ch
Littleton & Badsey (CCTV) 104m 31ch
Clayfield (AHBC) 103m 54ch
Brown Barn (UWC) 103m 32ch
Ivy Lane (UWC) 103m 19ch
Sheen Hill No.2 102m 40ch
Sheen Hill No.1 (UWC) 102m 16ch
Accommodation 1m 16.50ch
Honeybourne
101m 60ch
Honeybourne GF
101m 48ch
0m 00ch
1
98m 07ch
Campden Tunnel
97m 47ch
Chipping Campden (CCTV)
96m 78ch
Briar Hill (UWC) 96m 13ch
Mare Brook (UWC) 95m 23ch
Blockley (CCTV) 94m 77ch
Aston Magna No.1 93m 74ch
Aston Hall 93m 10ch
proposed
extension
to
Broadway
4m 73ch
7m 02ch
7m 25ch
Laverton
Loop
9m 00ch
Toddington
9m 36ch
Locomotive
Sheds
9m 48ch
GF
9m 60ch
Moreton-In-Marsh
91m 61ch
Moreton-in-Marsh
SB (MM)
91m 56ch
Frogmore 2 (UWC) 90m 49ch
River Evenlode
Wooliams 3 (UWC) 88m 29ch
Wooliams 2 (UWC) 88m 02ch
Wooliams 1 (UWC) 87m 49ch
Bledington (UWC) 85m 04ch
Kingham
84m 59ch
Bosleys (UWC) 83m 59ch
Bruern (CCTV) 83m 15ch
Lyneham (UWC) 82m 45ch
Shipton
81m 59ch
Hyatts (UWC) 81m 06ch
(MCB) 80m 36ch
Ascott-under-
Wychwood
80m 33ch
Ascott-under-
Wychwood
SB (AW)
80m 36ch
Charlbury
76m 60ch
Charlbury Jn.
76m 29ch
Finstock
75m 10ch
Combe
71m 44ch
Hanborough
70m 39ch
River Windrush
Down Goods Loop GF
87m 26ch
Lafarge
Aggregate
Terminal
Banbury North SB (BN)
86m 26ch
Down GF 86m 11ch
Banbury South
SB (BS)
86m 02ch
2
28
3
River Thames
17

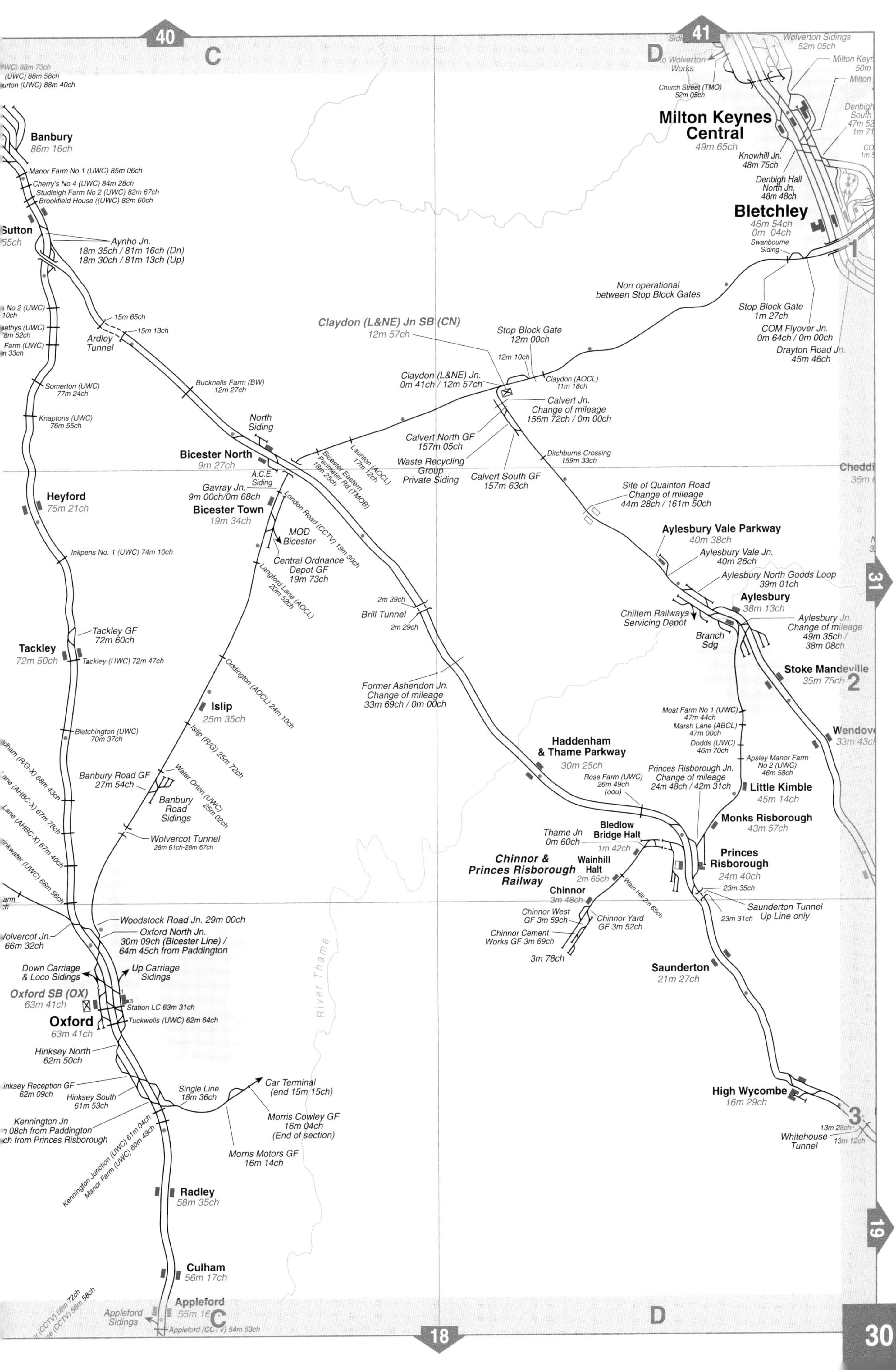

40
C
41
D
Wolverton Sidings
52m 05ch
to Wolverton Works
Church Street (TMO)
52m 05ch
Milton Keynes
50m
Milton Keynes Central
49m 65ch
Knowhill Jn.
48m 75ch
Denbigh Hall North Jn.
48m 48ch
Bletchley
46m 54ch
0m 04ch
Swanbourne Siding
1
Non operational between Stop Block Gates
Stop Block Gate
1m 27ch
COM Flyover Jn.
0m 64ch / 0m 00ch
Drayton Road Jn.
45m 46ch
Banbury
86m 16ch
Manor Farm No 1 (UWC) 85m 06ch
Cherry's No 4 (UWC) 84m 28ch
Studleigh Farm No 2 (UWC) 82m 67ch
Brookfield House ((UWC) 82m 60ch
Aynho Jn.
18m 35ch / 81m 16ch (Dn)
18m 30ch / 81m 13ch (Up)
15m 65ch
15m 13ch
Ardley Tunnel
Somerton (UWC)
77m 24ch
Knaptons (UWC)
76m 55ch
Bucknells Farm (BW)
12m 27ch
Claydon (L&NE) Jn SB (CN)
12m 57ch
Stop Block Gate
12m 00ch
12m 10ch
Claydon (AOCL)
11m 18ch
Claydon (L&NE) Jn.
0m 41ch / 12m 57ch
Calvert Jn.
Change of mileage
156m 72ch / 0m 00ch
Calvert North GF
157m 05ch
Waste Recycling Group Private Siding
Calvert South GF
157m 63ch
Ditchburns Crossing
159m 33ch
North Siding
Bicester North
9m 27ch
A.C.E. Siding
Gavray Jn.
9m 00ch/0m 68ch
Bicester Town
19m 34ch
Bicester Eastern Perimeter Rd (TMOB)
19m 25ch
Launton (AOCL)
17m 12ch
London Road (CCTV) 19m 30ch
MOD Bicester
Central Ordnance Depot GF
19m 73ch
Langford Lane (AOCL)
20m 52ch
Heyford
75m 21ch
Inkpens No. 1 (UWC) 74m 10ch
Site of Quainton Road
Change of mileage
44m 28ch / 161m 50ch
Aylesbury Vale Parkway
40m 38ch
Aylesbury Vale Jn.
40m 26ch
Aylesbury North Goods Loop
39m 01ch
Aylesbury
38m 13ch
Chiltern Railways Servicing Depot
Branch Sdg
Aylesbury Jn.
Change of mileage
49m 35ch / 38m 08ch
2m 39ch
Brill Tunnel
2m 29ch
Tackley GF
72m 60ch
Tackley
72m 50ch
Tackley (UWC) 72m 47ch
Oddington (AOCL) 24m 10ch
Islip
25m 35ch
Islip (R/G) 25m 72ch
Former Ashendon Jn.
Change of mileage
33m 69ch / 0m 00ch
Stoke Mandeville
35m 75ch
2
Wendover
33m 43ch
Moat Farm No 1 (UWC)
47m 44ch
Marsh Lane (ABCL)
47m 00ch
Dodds (UWC)
46m 70ch
Apsley Manor Farm No 2 (UWC)
46m 58ch
Bletchington (UWC)
70m 37ch
Haddenham & Thame Parkway
30m 25ch
Rose Farm (UWC)
26m 49ch
(oou)
Princes Risborough Jn.
Change of mileage
24m 48ch / 42m 31ch
Little Kimble
45m 14ch
Monks Risborough
43m 57ch
Banbury Road GF
27m 54ch
Water Orton (UWC)
25m 02ch
Banbury Road Sidings
Wolvercot Tunnel
28m 61ch-28m 67ch
Thame Jn
0m 60ch
Bledlow Bridge Halt
1m 42ch
Chinnor & Princes Risborough Railway
Wainhill Halt
2m 65ch
Wain Hill 2m 65ch
Chinnor
3m 48ch
Chinnor West GF 3m 59ch
Chinnor Yard GF 3m 52ch
Chinnor Cement Works GF 3m 69ch
3m 78ch
Princes Risborough
24m 40ch
23m 35ch
Saunderton Tunnel Up Line only
23m 31ch
Woodstock Road Jn. 29m 00ch
Oxford North Jn.
30m 09ch (Bicester Line) /
64m 45ch from Paddington
Wolvercot Jn.
66m 32ch
Down Carriage & Loco Sidings
Up Carriage Sidings
Oxford SB (OX)
63m 41ch
Station LC 63m 31ch
Oxford
63m 41ch
Tuckwells (UWC) 62m 64ch
River Thame
Saunderton
21m 27ch
Hinksey North
62m 50ch
Hinksey Reception GF
62m 09ch
Hinksey South
61m 53ch
Single Line
18m 36ch
Car Terminal
(end 15m 15ch)
Morris Cowley GF
16m 04ch
(End of section)
Kennington Jn
08ch from Paddington
ch from Princes Risborough
Kennington Junction (UWC) 61m 04ch
Manor Farm (UWC) 60m 49ch
Morris Motors GF
16m 14ch
High Wycombe
16m 29ch
3
13m 28ch
Whitehouse Tunnel
13m 12ch
Radley
58m 35ch
Culham
56m 17ch
Appleford
55m 16
Appleford Sidings
Appleford (CCTV) 54m 53ch
(CCTV) 56m 72ch
(CCTV) 56m 58ch
C
D
18
31
19

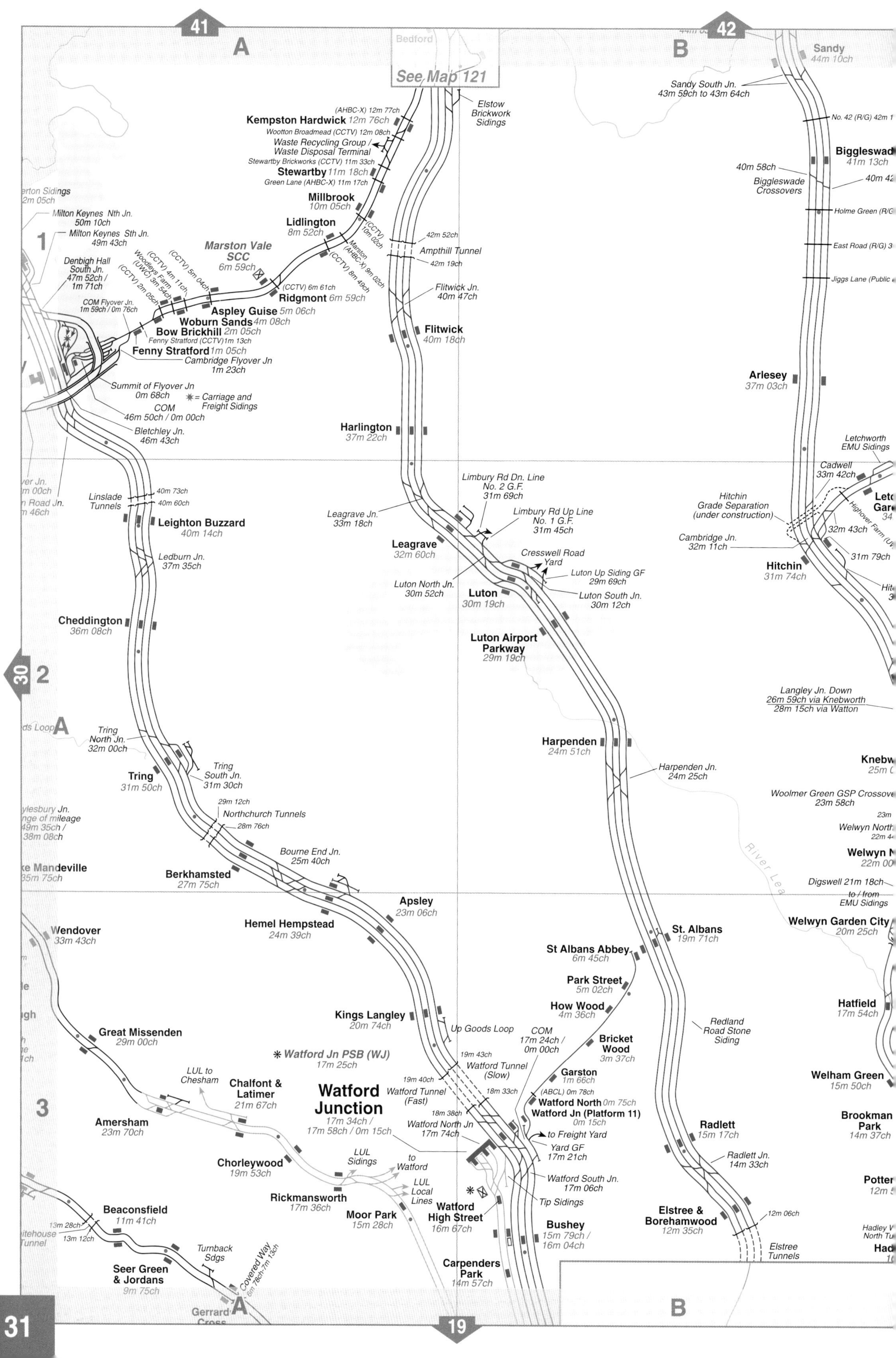

41
42
A
B
Bedford
See Map 121
Sandy
44m 10ch
Sandy South Jn.
43m 59ch to 43m 64ch
Elstow
Brickwork
Sidings
Kempston Hardwick 12m 76ch
(AHBC-X) 12m 77ch
Wootton Broadmead (CCTV) 12m 08ch
Waste Recycling Group /
Waste Disposal Terminal
Stewartby Brickworks (CCTV) 11m 33ch
Stewartby 11m 18ch
Green Lane (AHBC-X) 11m 17ch
Millbrook
10m 05ch
Lidlington
8m 52ch
Marston Vale
SCC
6m 59ch
Milton Keynes Nth Jn.
50m 10ch
Milton Keynes Sth Jn.
49m 43ch
Denbigh Hall
South Jn.
47m 52ch /
1m 71ch
COM Flyover Jn.
1m 59ch / 0m 76ch
Ridgmont 6m 59ch
(CCTV) 6m 61ch
Aspley Guise 5m 06ch
Woburn Sands 4m 08ch
Bow Brickhill 2m 05ch
Fenny Stratford (CCTV) 1m 13ch
Fenny Stratford 1m 05ch
Cambridge Flyover Jn
1m 23ch
Summit of Flyover Jn
0m 68ch
* = Carriage and
Freight Sidings
COM
46m 50ch / 0m 00ch
Bletchley Jn.
46m 43ch
42m 52ch
Ampthill Tunnel
42m 19ch
Flitwick Jn.
40m 47ch
Flitwick
40m 18ch
Harlington
37m 22ch
No. 42 (R/G) 42m 1
Biggleswade
41m 13ch
40m 58ch
Biggleswade
Crossovers
Holme Green (R/G
East Road (R/G)
Jiggs Lane (Public
Arlesey
37m 03ch
Linslade
Tunnels
40m 73ch
40m 60ch
Leighton Buzzard
40m 14ch
Ledburn Jn.
37m 35ch
Cheddington
36m 08ch
Leagrave Jn.
33m 18ch
Limbury Rd Dn. Line
No. 2 G.F.
31m 69ch
Limbury Rd Up Line
No. 1 G.F.
31m 45ch
Leagrave
32m 60ch
Cresswell Road
Yard
Luton Up Siding GF
29m 69ch
Luton North Jn.
30m 52ch
Luton
30m 19ch
Luton South Jn.
30m 12ch
Luton Airport
Parkway
29m 19ch
Letchworth
EMU Sidings
Cadwell
33m 42ch
Hitchin
Grade Separation
(under construction)
Cambridge Jn.
32m 11ch
32m 43ch
31m 79ch
Hitchin
31m 74ch
Langley Jn. Down
26m 59ch via Knebworth
28m 15ch via Watton
Tring
North Jn.
32m 00ch
Tring
South Jn.
31m 30ch
Tring
31m 50ch
29m 12ch
Northchurch Tunnels
28m 76ch
Harpenden
24m 51ch
Harpenden Jn.
24m 25ch
Woolmer Green GSP Crossove
23m 58ch
Bourne End Jn.
25m 40ch
Berkhamsted
27m 75ch
Apsley
23m 06ch
Hemel Hempstead
24m 39ch
Wendover
33m 43ch
River Lea
Digswell 21m 18ch
to / from
EMU Sidings
Welwyn Garden City
20m 25ch
St. Albans
19m 71ch
St Albans Abbey
6m 45ch
Park Street
5m 02ch
How Wood
4m 36ch
Kings Langley
20m 74ch
Up Goods Loop
Redland
Road Stone
Siding
Hatfield
17m 54ch
Great Missenden
29m 00ch
Watford Jn PSB (WJ)
17m 25ch
COM
17m 24ch /
0m 00ch
Bricket
Wood
3m 37ch
19m 43ch
Watford Tunnel
(Slow)
19m 40ch
Watford Tunnel
(Fast)
18m 33ch
Garston
1m 66ch
(ABCL) 0m 78ch
Watford North 0m 75ch
Watford Jn (Platform 11)
0m 15ch
Welham Green
15m 50ch
LUL to
Chesham
Chalfont &
Latimer
21m 67ch
Watford
Junction
17m 34ch /
17m 58ch / 0m 15ch
18m 38ch
Watford North Jn
17m 74ch
to Freight Yard
Yard GF
17m 21ch
Radlett
15m 17ch
Brookman
Park
14m 37ch
Amersham
23m 70ch
Chorleywood
19m 53ch
LUL
Sidings
to
Watford
LUL
Local
Lines
Watford South Jn.
17m 06ch
Tip Sidings
Radlett Jn.
14m 33ch
Rickmansworth
17m 36ch
Moor Park
15m 28ch
Watford
High Street
16m 67ch
Bushey
15m 79ch /
16m 04ch
Elstree &
Borehamwood
12m 35ch
12m 06ch
Elstree
Tunnels
Beaconsfield
11m 41ch
13m 28ch
13m 12ch
Turnback
Sdgs
Covered Way
Seer Green
& Jordans
9m 75ch
Carpenders
Park
14m 57ch
Gerrard
Cross
19
30
1
2
3

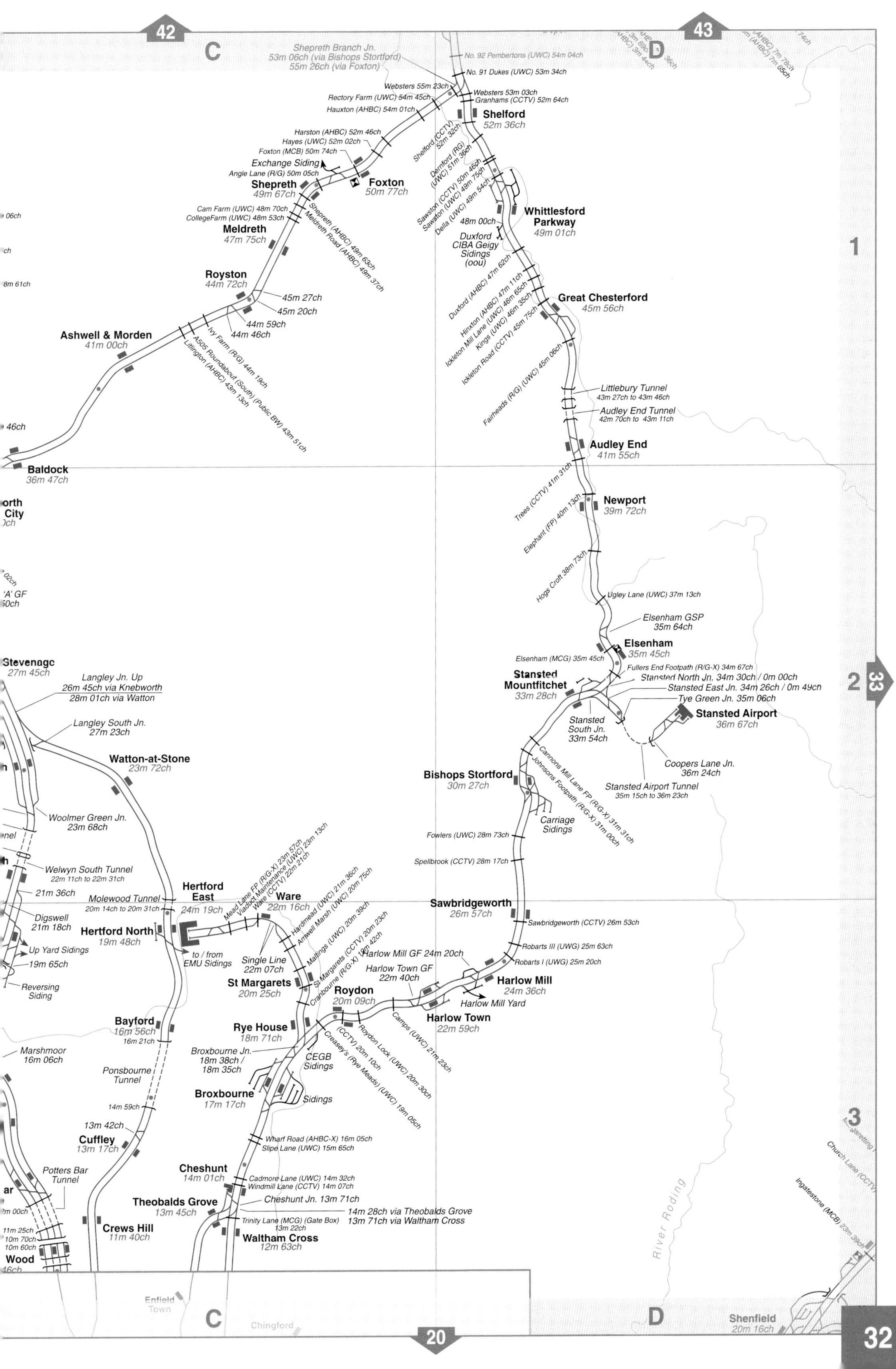
42
43
C
D
Shepreth Branch Jn.
53m 06ch (via Bishops Stortford)
55m 26ch (via Foxton)
No. 92 Pembertons (UWC) 54m 04ch
No. 91 Dukes (UWC) 53m 34ch
Websters 55m 23ch
Rectory Farm (UWC) 54m 45ch
Hauxton (AHBC) 54m 01ch
Websters 53m 03ch
Granhams (CCTV) 52m 64ch
Shelford
52m 36ch
Harston (AHBC) 52m 46ch
Hayes (UWC) 52m 02ch
Foxton (MCB) 50m 74ch
Exchange Siding
Angle Lane (R/G) 50m 05ch
Shepreth
49m 67ch
Foxton
50m 77ch
Cam Farm (UWC) 48m 70ch
CollegeFarm (UWC) 48m 53ch
Meldreth
47m 75ch
Whittlesford Parkway
49m 01ch
48m 00ch
Duxford CIBA Geigy Sidings (oou)
Royston
44m 72ch
45m 27ch
45m 20ch
44m 59ch
44m 46ch
Great Chesterford
45m 56ch
Ashwell & Morden
41m 00ch
Littlebury Tunnel
43m 27ch to 43m 46ch
Audley End Tunnel
42m 70ch to 43m 11ch
Audley End
41m 55ch
Baldock
36m 47ch
Newport
39m 72ch
Ugley Lane (UWC) 37m 13ch
Elsenham GSP
35m 64ch
Elsenham
35m 45ch
Elsenham (MCG) 35m 45ch
Fullers End Footpath (R/G-X) 34m 67ch
Stansted Mountfitchet
33m 28ch
Stansted North Jn. 34m 30ch / 0m 00ch
Stansted East Jn. 34m 26ch / 0m 49ch
Tye Green Jn. 35m 06ch
Stansted Airport
36m 67ch
Stansted South Jn.
33m 54ch
Coopers Lane Jn.
36m 24ch
Stansted Airport Tunnel
35m 15ch to 36m 23ch
Stevenage
27m 45ch
Langley Jn. Up
26m 45ch via Knebworth
28m 01ch via Watton
Langley South Jn.
27m 23ch
Watton-at-Stone
23m 72ch
Bishops Stortford
30m 27ch
Carriage Sidings
Fowlers (UWC) 28m 73ch
Spellbrook (CCTV) 28m 17ch
Woolmer Green Jn.
23m 68ch
Welwyn South Tunnel
22m 11ch to 22m 31ch
21m 36ch
Molewood Tunnel
20m 14ch to 20m 31ch
Hertford East
24m 19ch
Ware
22m 16ch
Sawbridgeworth
26m 57ch
Sawbridgeworth (CCTV) 26m 53ch
Digswell
21m 18ch
Hertford North
19m 48ch
Up Yard Sidings
19m 65ch
to / from EMU Sidings
Single Line
22m 07ch
Robarts III (UWG) 25m 63ch
Robarts I (UWG) 25m 20ch
Harlow Mill GF 24m 20ch
Harlow Town GF
22m 40ch
Harlow Mill
24m 36ch
Harlow Mill Yard
Reversing Siding
St Margarets
20m 25ch
Roydon
20m 09ch
Harlow Town
22m 59ch
Bayford
16m 56ch
16m 21ch
Rye House
18m 71ch
Marshmoor
16m 06ch
Broxbourne Jn.
18m 38ch /
18m 35ch
CEGB Sidings
Ponsbourne Tunnel
Broxbourne
17m 17ch
Sidings
14m 59ch
13m 42ch
Cuffley
13m 17ch
Wharf Road (AHBC-X) 16m 05ch
Slipe Lane (UWC) 15m 65ch
Potters Bar Tunnel
Cheshunt
14m 01ch
Cadmore Lane (UWC) 14m 32ch
Windmill Lane (CCTV) 14m 07ch
Theobalds Grove
13m 45ch
Cheshunt Jn. 13m 71ch
14m 28ch via Theobalds Grove
13m 71ch via Waltham Cross
Crews Hill
11m 40ch
Trinity Lane (MCG) (Gate Box)
13m 22ch
Waltham Cross
12m 63ch
River Roding
Enfield Town
Chingford
Shenfield
20m 16ch
1
2
3
33
20

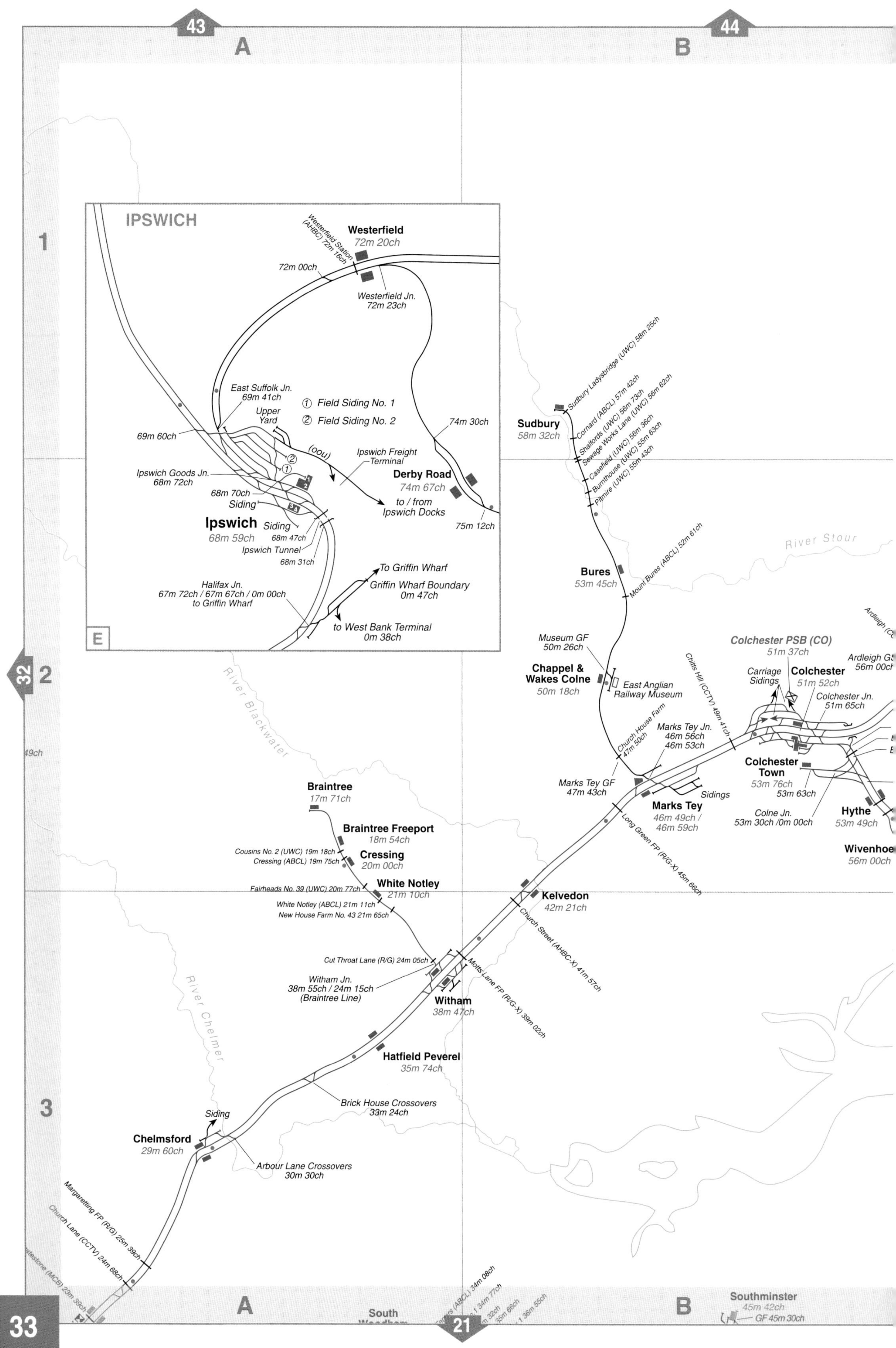
IPSWICH
Westerfield
72m 20ch
Westerfield Station (AHBC) 72m 16ch
72m 00ch
Westerfield Jn.
72m 23ch
East Suffolk Jn.
69m 41ch
Upper Yard
① Field Siding No. 1
② Field Siding No. 2
69m 60ch
(oou)
Ipswich Freight Terminal
74m 30ch
Ipswich Goods Jn.
68m 72ch
Derby Road
74m 67ch
68m 70ch
Siding
to / from Ipswich Docks
Ipswich
68m 59ch
Siding
68m 47ch
Ipswich Tunnel
68m 31ch
75m 12ch
To Griffin Wharf
Halifax Jn.
67m 72ch / 67m 67ch / 0m 00ch
to Griffin Wharf
Griffin Wharf Boundary
0m 47ch
to West Bank Terminal
0m 38ch
E
Sudbury
58m 32ch
Sudbury Ladysbridge (UWC) 58m 25ch
Cornard (ABCL) 57m 42ch
Shalfords (UWC) 56m 73ch
Sewage Works Lane (UWC) 56m 62ch
Casefield (UWC) 56m 36ch
Burnthouse (UWC) 55m 63ch
Pitmire (UWC) 55m 43ch
River Stour
Bures
53m 45ch
Mount Bures (ABCL) 52m 61ch
Museum GF
50m 26ch
Chappel & Wakes Colne
50m 18ch
East Anglian Railway Museum
Colchester PSB (CO)
51m 37ch
Ardleigh GS
56m 00ch
Carriage Sidings
Colchester
51m 52ch
Colchester Jn.
51m 65ch
Chitts Hill (CCTV) 49m 41ch
Church House Farm 47m 50ch
Marks Tey Jn.
46m 56ch
46m 53ch
Marks Tey GF
47m 43ch
Sidings
Marks Tey
46m 49ch / 46m 59ch
Colchester Town
53m 76ch
53m 63ch
Colne Jn.
53m 30ch /0m 00ch
Hythe
53m 49ch
Wivenhoe
56m 00ch
River Blackwater
49ch
Braintree
17m 71ch
Braintree Freeport
18m 54ch
Cousins No. 2 (UWC) 19m 18ch
Cressing (ABCL) 19m 75ch
Cressing
20m 00ch
Fairheads No. 39 (UWC) 20m 77ch
White Notley
21m 10ch
White Notley (ABCL) 21m 11ch
New House Farm No. 43 21m 65ch
Long Green FP (R/G-X) 45m 66ch
Kelvedon
42m 21ch
Church Street (AHBC-X) 41m 57ch
Cut Throat Lane (R/G) 24m 05ch
Witham Jn.
38m 55ch / 24m 15ch
(Braintree Line)
Motts Lane FP (R/G-X) 39m 02ch
Witham
38m 47ch
River Chelmer
Hatfield Peverel
35m 74ch
Brick House Crossovers
33m 24ch
Siding
Chelmsford
29m 60ch
Arbour Lane Crossovers
30m 30ch
Margaretting FP (R/G) 25m 39ch
Church Lane (CCTV) 24m 68ch
Southminster
45m 42ch
GF 45m 30ch
South Woodham

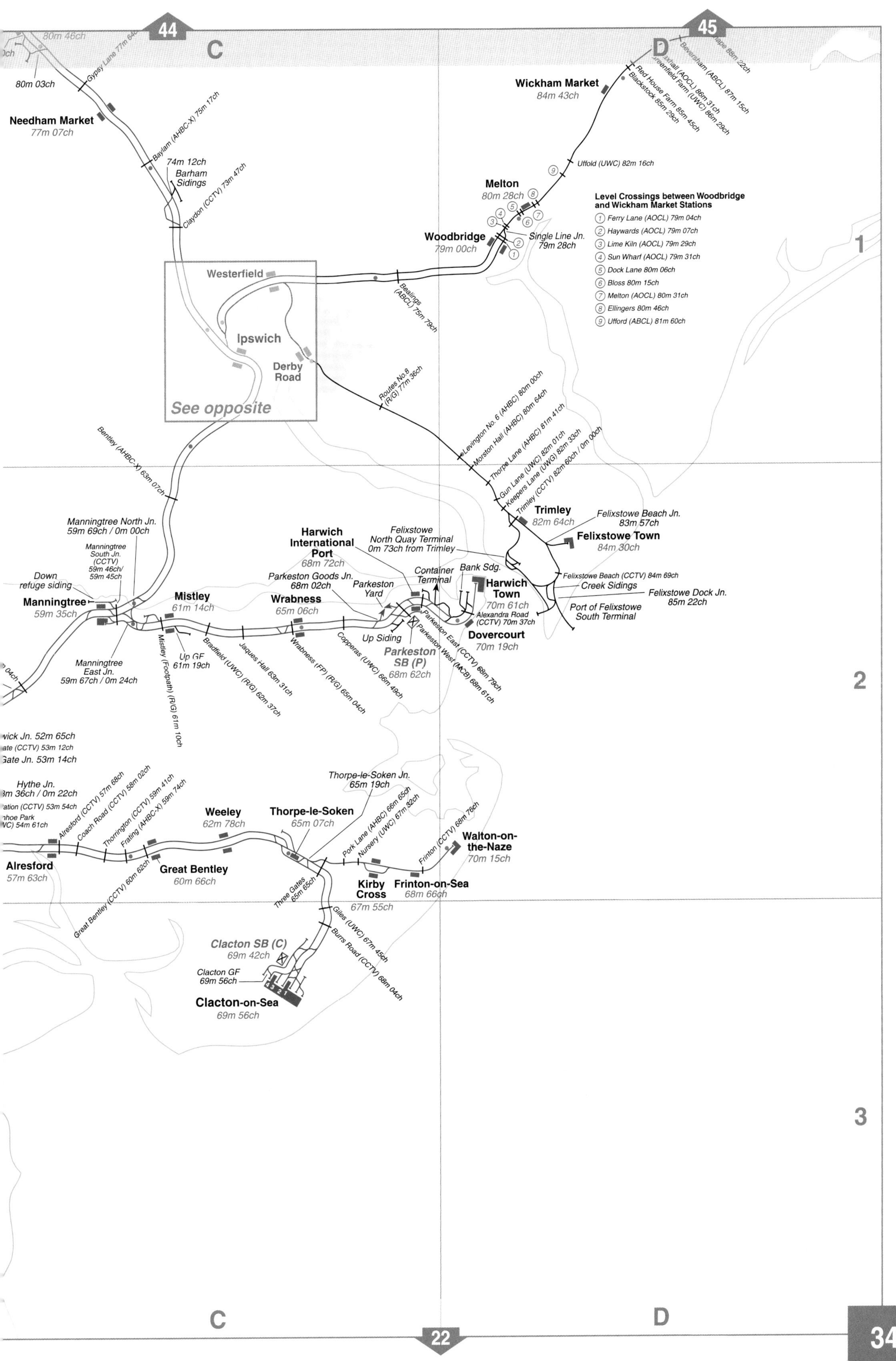

44
45
C
D
80m 46ch
80m 03ch
Gypsy Lane 77m 64ch
Needham Market
77m 07ch
Baylam (AHBC-X) 75m 17ch
74m 12ch
Barham
Sidings
Claydon (CCTV) 73m 47ch
Wickham Market
84m 43ch
Beversham (ABCL) 87m 15ch
Red House Farm 85m 45ch
Blackstock 85m 29ch
Greenfield Farm (UWC) 86m 29ch
Uffold (UWC) 82m 16ch
Melton
80m 28ch
Level Crossings between Woodbridge and Wickham Market Stations
1 Ferry Lane (AOCL) 79m 04ch
2 Haywards (AOCL) 79m 07ch
3 Lime Kiln (AOCL) 79m 29ch
4 Sun Wharf (AOCL) 79m 31ch
5 Dock Lane 80m 06ch
6 Bloss 80m 15ch
7 Melton (AOCL) 80m 31ch
8 Ellingers 80m 46ch
9 Ufford (ABCL) 81m 60ch
Woodbridge
79m 00ch
Single Line Jn.
79m 28ch
Westerfield
Bealings (ABCL) 75m 79ch
Ipswich
Derby Road
See opposite
Routes No.8 (R/G) 77m 36ch
Levington No. 6 (AHBC) 80m 00ch
Morston Hall (AHBC) 80m 64ch
Thorpe Lane (AHBC) 81m 41ch
Gun Lane (UWC) 82m 01ch
Keepers Lane (UWG) 82m 33ch
Trimley (CCTV) 82m 60ch / 0m 00ch
Bentley (AHBC-X) 63m 07ch
Trimley
82m 64ch
Felixstowe Beach Jn.
83m 57ch
Felixstowe Town
84m 30ch
Manningtree North Jn.
59m 69ch / 0m 00ch
Manningtree South Jn. (CCTV) 59m 46ch/ 59m 45ch
Down refuge siding
Manningtree
59m 35ch
Manningtree East Jn.
59m 67ch / 0m 24ch
Harwich International Port
68m 72ch
Felixstowe North Quay Terminal
0m 73ch from Trimley
Parkeston Goods Jn.
68m 02ch
Container Terminal
Bank Sdg.
Parkeston Yard
Harwich Town
70m 61ch
Alexandra Road (CCTV) 70m 37ch
Dovercourt
70m 19ch
Felixstowe Beach (CCTV) 84m 69ch
Creek Sidings
Felixstowe Dock Jn.
85m 22ch
Port of Felixstowe South Terminal
Mistley
61m 14ch
Wrabness
65m 06ch
Up Siding
Parkeston SB (P)
68m 62ch
Parkeston East (CCTV) 68m 79ch
Parkeston West (MCB) 68m 61ch
Up GF
61m 19ch
Mistley (Footpath) (R/G) 61m 10ch
Bradfield (UWC) (R/G) 62m 37ch
Jaques Hall 63m 31ch
Wrabness (FP) (R/G) 65m 04ch
Copperas (UWC) 66m 49ch
1
2
3
Thorpe-le-Soken Jn.
65m 19ch
Hythe Jn.
Alresford (CCTV) 57m 68ch
Coach Road (CCTV) 58m 02ch
Thorrington (CCTV) 59m 41ch
Frating (AHBC-X) 59m 74ch
Weeley
62m 78ch
Thorpe-le-Soken
65m 07ch
Pork Lane (AHBC) 66m 65ch
Nursery (UWC) 67m 32ch
Frinton (CCTV) 68m 76ch
Walton-on-the-Naze
70m 15ch
Alresford
57m 63ch
Great Bentley
60m 66ch
Great Bentley (CCTV) 60m 62ch
Three Gates 65m 65ch
Kirby Cross
67m 55ch
Frinton-on-Sea
68m 66ch
Giles (UWC) 67m 45ch
Burrs Road (CCTV) 68m 04ch
Clacton SB (C)
69m 42ch
Clacton GF
69m 56ch
Clacton-on-Sea
69m 56ch
22

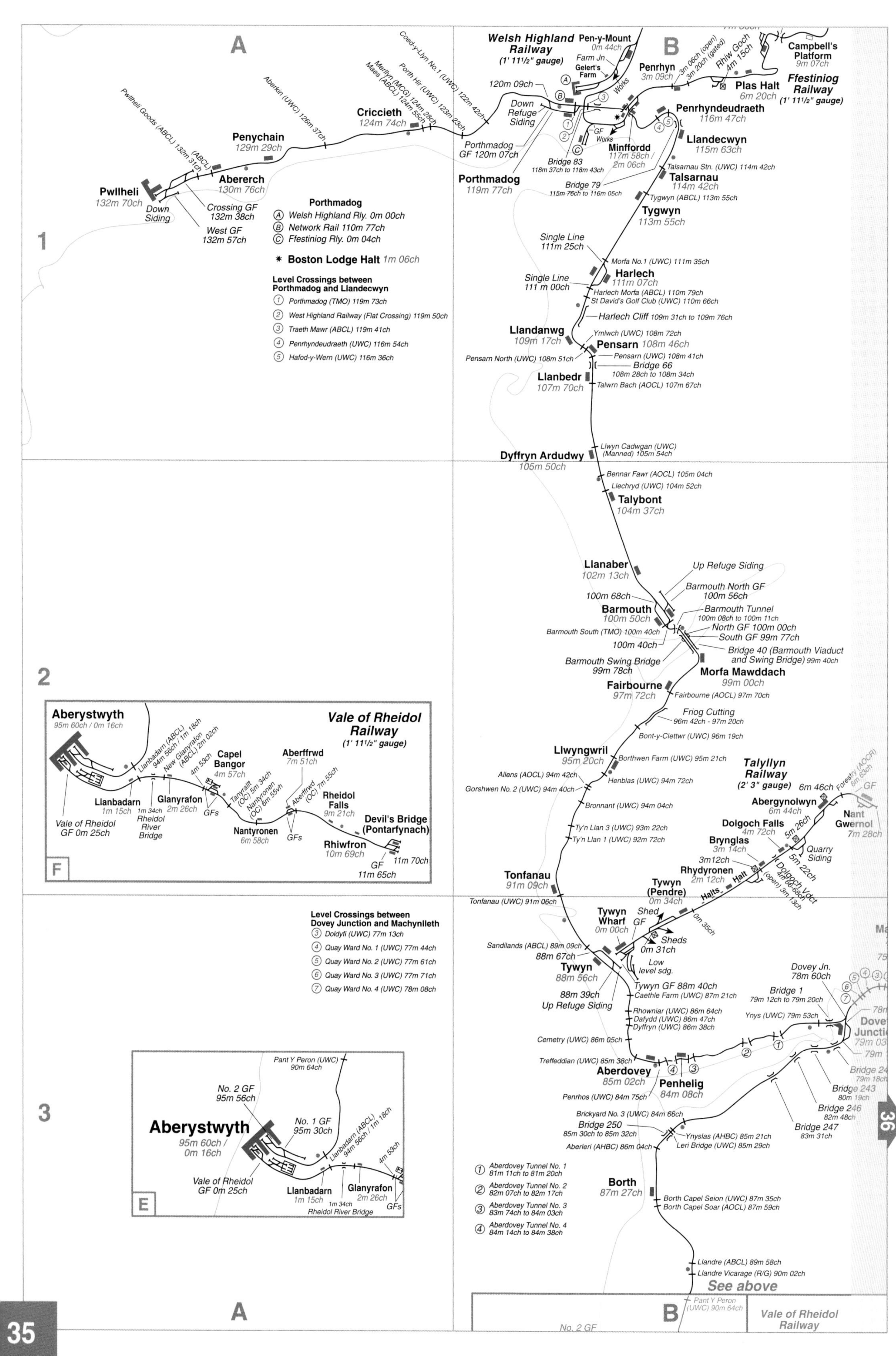

A
B
1
2
3
Welsh Highland Railway
(1' 11½" gauge)
Pen-y-Mount
0m 44ch
Farm Jn
Gelert's Farm
Works
Penrhyn
3m 09ch
3m 06ch (open)
3m 20ch (gated)
Rhiw Goch
4m 15ch
Campbell's Platform
9m 07ch
Plas Halt
6m 20ch
Ffestiniog Railway
(1' 11½" gauge)
Penrhyndeudraeth
116m 47ch
Llandecwyn
115m 63ch
Minffordd
117m 58ch / 2m 06ch
GF
Works
Coed-y-Llyn No.1 (UWC) 122m 42ch
Porth Hir (UWC) 123m 23ch
Merllyn (MCG) 124m 28ch
Maes (ABCL) 124m 55ch
Aberkin (UWC) 126m 37ch
Pwllheli Goods (ABCL) 132m 31ch
(ABCL)
120m 09ch
Down Refuge Siding
Criccieth
124m 74ch
Penychain
129m 29ch
Abererch
130m 76ch
Pwllheli
132m 70ch
Down Siding
Crossing GF
132m 38ch
West GF
132m 57ch
Porthmadog GF 120m 07ch
Porthmadog
119m 77ch
Bridge 83
118m 37ch to 118m 43ch
Bridge 79
115m 76ch to 116m 05ch
Talsarnau Stn. (UWC) 114m 42ch
Talsarnau
114m 42ch
Tygwyn (ABCL) 113m 55ch
Tygwyn
113m 55ch
Porthmadog
Ⓐ Welsh Highland Rly. 0m 00ch
Ⓑ Network Rail 110m 77ch
Ⓒ Ffestiniog Rly. 0m 04ch
✱ Boston Lodge Halt 1m 06ch
Level Crossings between Porthmadog and Llandecwyn
① Porthmadog (TMO) 119m 73ch
② West Highland Railway (Flat Crossing) 119m 50ch
③ Traeth Mawr (ABCL) 119m 41ch
④ Penrhyndeudraeth (UWC) 116m 54ch
⑤ Hafod-y-Wern (UWC) 116m 36ch
Single Line
111m 25ch
Morfa No.1 (UWC) 111m 35ch
Harlech
111m 07ch
Single Line
111 m 00ch
Harlech Morfa (ABCL) 110m 79ch
St David's Golf Club (UWC) 110m 66ch
Harlech Cliff 109m 31ch to 109m 76ch
Llandanwg
109m 17ch
Ymlwch (UWC) 108m 72ch
Pensarn 108m 46ch
Pensarn North (UWC) 108m 51ch
Pensarn (UWC) 108m 41ch
Bridge 66
108m 28ch to 108m 34ch
Llanbedr
107m 70ch
Talwrn Bach (AOCL) 107m 67ch
Llwyn Cadwgan (UWC) (Manned) 105m 54ch
Dyffryn Ardudwy
105m 50ch
Bennar Fawr (AOCL) 105m 04ch
Llechryd (UWC) 104m 52ch
Talybont
104m 37ch
Llanaber
102m 13ch
Up Refuge Siding
Barmouth North GF
100m 56ch
100m 68ch
Barmouth
100m 50ch
Barmouth Tunnel
100m 08ch to 100m 11ch
North GF 100m 00ch
South GF 99m 77ch
Barmouth South (TMO) 100m 40ch
100m 40ch
Bridge 40 (Barmouth Viaduct and Swing Bridge) 99m 40ch
Barmouth Swing Bridge
99m 78ch
Morfa Mawddach
99m 00ch
Fairbourne
97m 72ch
Fairbourne (AOCL) 97m 70ch
Friog Cutting
96m 42ch - 97m 20ch
Bont-y-Clettwr (UWC) 96m 19ch
Llwyngwril
95m 20ch
Borthwen Farm (UWC) 95m 21ch
Allens (AOCL) 94m 42ch
Henblas (UWC) 94m 72ch
Gorshwen No. 2 (UWC) 94m 40ch
Bronnant (UWC) 94m 04ch
Ty'n Llan 3 (UWC) 93m 22ch
Ty'n Llan 1 (UWC) 92m 72ch
Aberystwyth
95m 60ch / 0m 16ch
Vale of Rheidol Railway
(1' 11½" gauge)
Llanbadarn (ABCL) 94m 56ch / 1m 18ch
New Glanyrafon (ABCL) 2m 02ch
4m 53ch
Capel Bangor
4m 57ch
Aberffrwd
7m 51ch
Tanyrallt (OC) 5m 34ch
Nantyronen (OC) 6m 55ch
Aberffrwd (OC) 7m 55ch
Llanbadarn
1m 15ch
1m 34ch
Rheidol River Bridge
Glanyrafon
2m 26ch
GFs
Nantyronen
6m 58ch
GFs
Rheidol Falls
9m 21ch
Devil's Bridge (Pontarfynach)
Rhiwfron
10m 69ch
11m 70ch
GF
11m 65ch
Vale of Rheidol GF 0m 25ch
F
Talyllyn Railway
(2' 3" gauge)
Abergynolwyn
6m 44ch
6m 46ch
Forestry (AOCR) 6m 63ch
GF
Nant Gwernol
7m 28ch
Dolgoch Falls
4m 72ch
5m 26ch
Quarry Siding
5m 22ch
Brynglas
3m 14ch
3m12ch
Rhydyronen
2m 12ch
Dolgoch Vdct 4m 68ch
(open) 3m 13ch
Halt
Halts
Tywyn (Pendre)
0m 34ch
0m 35ch
Tonfanau
91m 09ch
Tonfanau (UWC) 91m 06ch
Tywyn Wharf
0m 00ch
Shed
GF
Sheds
0m 31ch
Low level sdg.
Sandilands (ABCL) 89m 09ch
88m 67ch
Tywyn
88m 56ch
88m 39ch
Up Refuge Siding
Tywyn GF 88m 40ch
Caethle Farm (UWC) 87m 21ch
Rhowniar (UWC) 86m 64ch
Dafydd (UWC) 86m 47ch
Dyffryn (UWC) 86m 38ch
Cemetry (UWC) 86m 05ch
Treffeddian (UWC) 85m 38ch
Aberdovey
85m 02ch
Penhelig
84m 08ch
Penrhos (UWC) 84m 75ch
Brickyard No. 3 (UWC) 84m 66ch
Bridge 250
85m 30ch to 85m 32ch
Aberleri (AHBC) 86m 04ch
Ynyslas (AHBC) 85m 21ch
Leri Bridge (UWC) 85m 29ch
Borth
87m 27ch
Borth Capel Seion (UWC) 87m 35ch
Borth Capel Soar (AOCL) 87m 59ch
Llandre (ABCL) 89m 58ch
Llandre Vicarage (R/G) 90m 02ch
See above
Pant Y Peron (UWC) 90m 64ch
No. 2 GF
Vale of Rheidol Railway
Dovey Jn.
78m 60ch
Bridge 1
79m 12ch to 79m 20ch
Ynys (UWC) 79m 53ch
Dovey Junction
79m 03
Bridge 243
80m 19ch
Bridge 246
82m 48ch
Bridge 247
83m 31ch
Level Crossings between Dovey Junction and Machynlleth
③ Doldyfi (UWC) 77m 13ch
④ Quay Ward No. 1 (UWC) 77m 44ch
⑤ Quay Ward No. 2 (UWC) 77m 61ch
⑥ Quay Ward No. 3 (UWC) 77m 71ch
⑦ Quay Ward No. 4 (UWC) 78m 08ch
① Aberdovey Tunnel No. 1
81m 11ch to 81m 20ch
② Aberdovey Tunnel No. 2
82m 07ch to 82m 17ch
③ Aberdovey Tunnel No. 3
83m 74ch to 84m 03ch
④ Aberdovey Tunnel No. 4
84m 14ch to 84m 38ch
Pant Y Peron (UWC)
90m 64ch
No. 2 GF
95m 56ch
No. 1 GF
95m 30ch
Aberystwyth
95m 60ch / 0m 16ch
Llanbadarn (ABCL) 94m 56ch / 1m 18ch
4m 53ch
Vale of Rheidol GF 0m 25ch
Llanbadarn
1m 15ch
1m 34ch
Rheidol River Bridge
Glanyrafon
2m 26ch
GFs
E
36

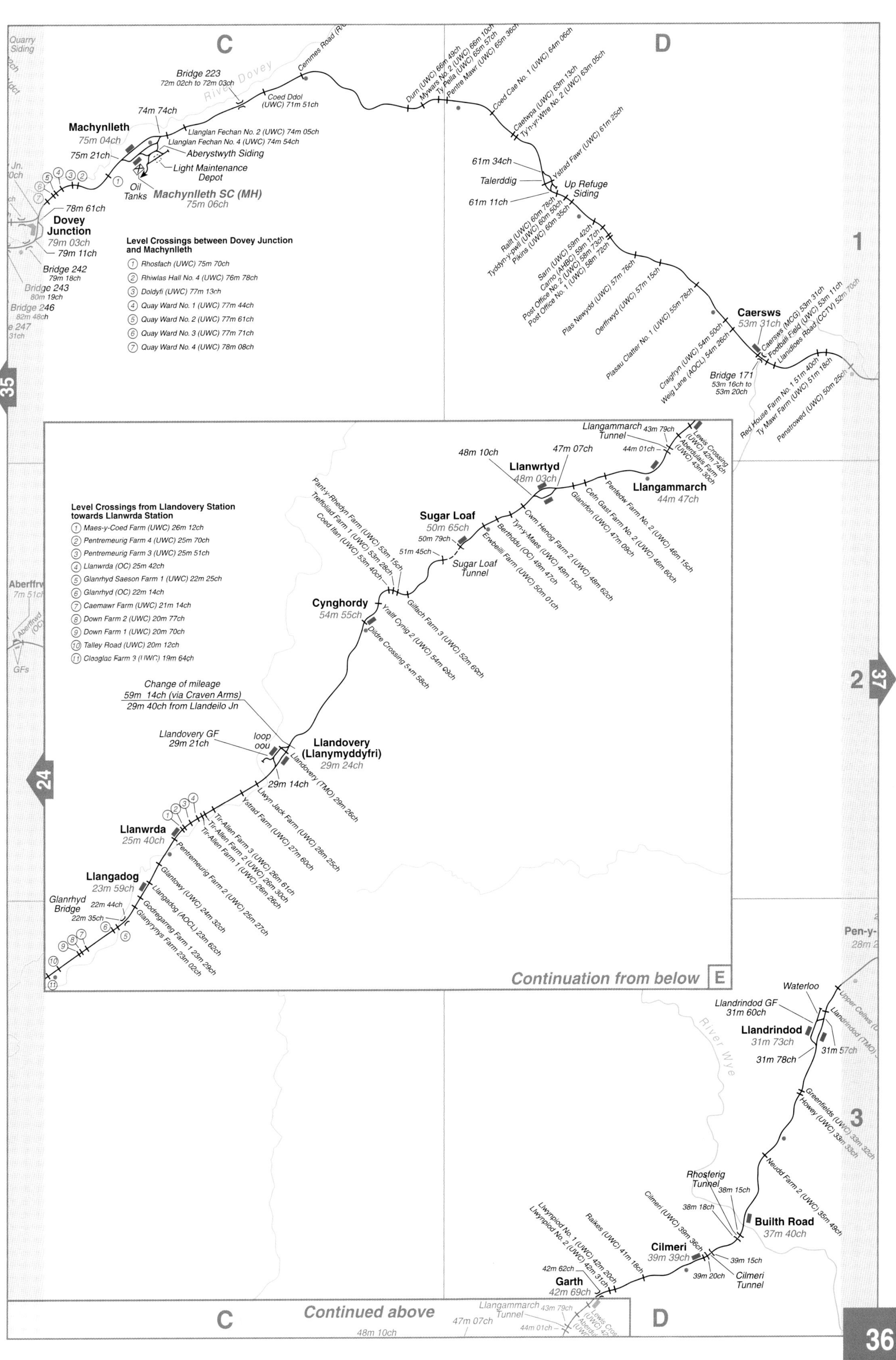

C
D
Quarry Siding
Bridge 223
72m 02ch to 72m 03ch
River Dovey
Cemmes Road
Coed Ddol (UWC) 71m 51ch
74m 74ch
Machynlleth
75m 04ch
75m 21ch
Llanglan Fechan No. 2 (UWC) 74m 05ch
Llanglan Fechan No. 4 (UWC) 74m 54ch
Aberystwyth Siding
Light Maintenance Depot
Oil Tanks
Machynlleth SC (MH)
75m 06ch
78m 61ch
Dovey Junction
79m 03ch
79m 11ch
Bridge 242
79m 18ch
Bridge 243
80m 19ch
Bridge 246
82m 48ch
Level Crossings between Dovey Junction and Machynlleth
1 Rhosfach (UWC) 75m 70ch
2 Rhiwlas Hall No. 4 (UWC) 76m 78ch
3 Doldyfi (UWC) 77m 13ch
4 Quay Ward No. 1 (UWC) 77m 44ch
5 Quay Ward No. 2 (UWC) 77m 61ch
6 Quay Ward No. 3 (UWC) 77m 71ch
7 Quay Ward No. 4 (UWC) 78m 08ch
Durn (UWC) 66m 49ch
Mywars No. 2 (UWC) 66m 10ch
Ty Pella (UWC) 65m 57ch
Pentre Mawr (UWC) 65m 36ch
Coed Cae No. 1 (UWC) 64m 06ch
Caetwpa (UWC) 63m 13ch
Ty'n-yr-Wtre No. 2 (UWC) 63m 05ch
Ystrad Fawr (UWC) 61m 25ch
61m 34ch
Talerddig
Up Refuge Siding
61m 11ch
Rallt (UWC) 60m 78ch
Tyddyn-y-pwll (UWC) 60m 50ch
Pikins (UWC) 60m 35ch
Sarn (UWC) 59m 42ch
Carno (AHBC) 59m 17ch
Post Office No. 2 (UWC) 58m 73ch
Post Office No. 1 (UWC) 58m 72ch
Plas Newydd (UWC) 57m 76ch
Oerffrwyd (UWC) 57m 15ch
Plasau Clatter No. 1 (UWC) 55m 78ch
Craigfryn (UWC) 54m 50ch
Weig Lane (AOCL) 54m 26ch
Caersws
53m 31ch
Caersws (MCG) 53m 31ch
Football Field (UWC) 53m 11ch
Llanidloes Road (CCTV) 52m 70ch
Bridge 171
53m 16ch to 53m 20ch
Red House Farm No. 1 51m 40ch
Ty Mawr Farm (UWC) 51m 18ch
Penstrowed (UWC) 50m 25ch
1
2
3
35
37
24
Llangammarch Tunnel
43m 79ch
44m 01ch
Lewis Crossing (UWC) 42m 74ch
Aberdulais Farm (UWC) 43m 30ch
48m 10ch
47m 07ch
Llanwrtyd
48m 03ch
Llangammarch
44m 47ch
Level Crossings from Llandovery Station towards Llanwrda Station
1 Maes-y-Coed Farm (UWC) 26m 12ch
2 Pentremeurig Farm 4 (UWC) 25m 70ch
3 Pentremeurig Farm 3 (UWC) 25m 51ch
4 Llanwrda (OC) 25m 42ch
5 Glanrhyd Saeson Farm 1 (UWC) 22m 25ch
6 Glanrhyd (OC) 22m 14ch
7 Caemawr Farm (UWC) 21m 14ch
8 Down Farm 2 (UWC) 20m 77ch
9 Down Farm 1 (UWC) 20m 70ch
10 Talley Road (UWC) 20m 12ch
11 Cloeglas Farm 3 (UWC) 19m 64ch
Pant-y-Rhedyn Farm (UWC) 53m 15ch
Treffoilad Farm 1 (UWC) 53m 28ch
Coed Ifan (UWC) 53m 40ch
Sugar Loaf
50m 65ch
50m 79ch
51m 45ch
Sugar Loaf Tunnel
Cefn Gast Farm No. 2 (UWC) 46m 60ch
Penfedw Farm No. 2 (UWC) 46m 15ch
Glanirfon (UWC) 47m 09ch
Cwm Henog Farm 2 (UWC) 48m 62ch
Tyn-y-Maes (UWC) 49m 15ch
Berthddu (OC) 49m 47ch
Erwbeilli Farm (UWC) 50m 01ch
Aberffrwd
7m 51ch
GFs
Cynghordy
54m 55ch
Gilfach Farm 3 (UWC) 52m 69ch
Yrallt Cynig 2 (UWC) 54m 03ch
Dildre Crossing 54m 58ch
Change of mileage
59m 14ch (via Craven Arms)
29m 40ch from Llandeilo Jn
Llandovery GF
29m 21ch
loop
Llandovery (Llanymyddyfri)
29m 24ch
Llandovery (TMO) 29m 26ch
29m 14ch
Llwyn Jack Farm (UWC) 28m 25ch
Ystrad Farm (UWC) 27m 60ch
Tir-Allen Farm 3 (UWC) 26m 61ch
Tir-Allen Farm 2 (UWC) 26m 30ch
Tir-Allen Farm 1 (UWC) 26m 26ch
Llanwrda
25m 40ch
Pentremeurig Farm 2 (UWC) 25m 27ch
Glantowy (UWC) 24m 32ch
Llangadog (AOCL) 23m 62ch
Godregarreg Farm 1 23m 29ch
Glanyrynys Farm 23m 02ch
Llangadog
23m 59ch
Glanrhyd Bridge
22m 44ch
22m 35ch
Continuation from below
E
Pen-y-
28m
Waterloo
Llandrindod GF
31m 60ch
Llandrindod
31m 73ch
31m 57ch
31m 78ch
Upper Ceilws
Llandrindod (TMO)
River Wye
Greenfields (UWC) 33m 32ch
Howey (UWC) 33m 33ch
Neudd Farm 2 (UWC) 35m 49ch
Rhosferig Tunnel
38m 15ch
38m 18ch
Builth Road
37m 40ch
Cilmeri (UWC) 39m 36ch
Raikes (UWC) 41m 18ch
Llwynpiod No. 1 (UWC) 42m 20ch
Llwynpiod No. 2 (UWC) 42m 31ch
42m 62ch
Cilmeri
39m 39ch
39m 15ch
39m 20ch
Cilmeri Tunnel
Garth
42m 69ch
Continued above

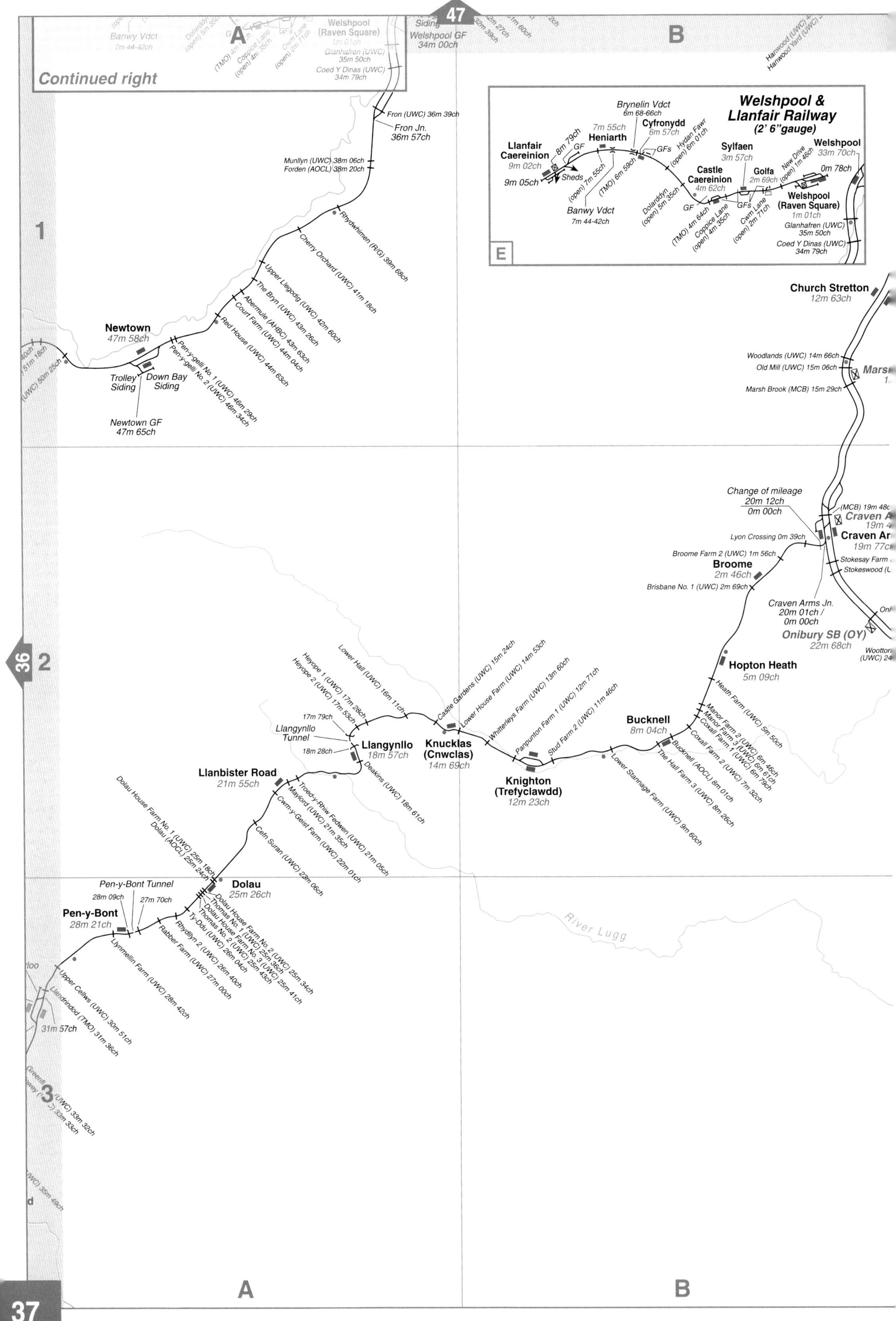
47
A
B
Continued right
Banwy Vdct
7m 44-42ch
Welshpool (Raven Square)
1m 01ch
Glanhafren (UWC)
35m 50ch
Coed Y Dinas (UWC)
34m 79ch
Welshpool GF
34m 00ch
Fron (UWC) 36m 39ch
Fron Jn.
36m 57ch
Munllyn (UWC) 38m 06ch
Forden (AOCL) 38m 20ch
Rhydwhimen (R/G) 39m 68ch
Cherry Orchard (UWC) 41m 18ch
Upper Llegodig (UWC) 42m 60ch
The Bryn (UWC) 43m 26ch
Abermule (AHBC) 43m 63ch
Court Farm (UWC) 44m 04ch
Red House (UWC) 44m 63ch
Newtown
47m 58ch
Trolley Siding
Down Bay Siding
Newtown GF
47m 65ch
Pen-y-gelli No. 1 (UWC) 46m 29ch
Pen-y-gelli No. 2 (UWC) 46m 34ch
(UWC) 50m 25ch
1
Welshpool & Llanfair Railway
(2' 6" gauge)
Llanfair Caereinion
9m 02ch
9m 05ch
8m 79ch
GF
Sheds
Heniarth
7m 55ch
(open) 7m 55ch
Banwy Vdct
7m 44-42ch
Brynelin Vdct
6m 68-66ch
Cyfronydd
6m 57ch
(TMO) 6m 59ch
GFs
Hydan Fawr (open) 6m 01ch
Dolarddyn (open) 5m 35ch
Castle Caereinion
4m 62ch
GF
(TMO) 4m 64ch
Coppice Lane (open) 4m 35ch
Sylfaen
3m 57ch
GFs
Cwm Lane (open) 2m 71ch
Golfa
2m 69ch
New Drive (open) 1m 46ch
Welshpool
33m 70ch
0m 78ch
Welshpool (Raven Square)
1m 01ch
Glanhafren (UWC)
35m 50ch
Coed Y Dinas (UWC)
34m 79ch
E
Church Stretton
12m 63ch
Woodlands (UWC) 14m 66ch
Old Mill (UWC) 15m 06ch
Marsh Brook (MCB) 15m 29ch
Change of mileage
20m 12ch
0m 00ch
(MCB) 19m 48ch
Craven Arms
19m 77ch
Lyon Crossing 0m 39ch
Broome Farm 2 (UWC) 1m 56ch
Broome
2m 46ch
Stokesay Farm
Stokeswood
Brisbane No. 1 (UWC) 2m 69ch
Craven Arms Jn.
20m 01ch /
0m 00ch
Onibury SB (OY)
22m 68ch
36
2
Hopton Heath
5m 09ch
Heath Farm (UWC) 5m 50ch
Manor Farm 2 (UWC) 6m 46ch
Manor Farm 3 (UWC) 6m 61ch
Coxall Farm 1 (UWC) 6m 79ch
Coxall Farm 2 (UWC) 7m 32ch
Bucknell
8m 04ch
Bucknell (AOCL) 8m 01ch
The Hall Farm 3 (UWC) 8m 26ch
Lower Stannage Farm (UWC) 9m 60ch
Stud Farm 2 (UWC) 11m 46ch
Panpunton Farm 1 (UWC) 12m 71ch
Knighton (Trefyclawdd)
12m 23ch
Whitterleys Farm (UWC) 13m 60ch
Lower House Farm (UWC) 14m 53ch
Castle Gardens (UWC) 15m 24ch
Knucklas (Cnwclas)
14m 69ch
Lower Hall (UWC) 16m 11ch
Heyope 1 (UWC) 17m 28ch
Heyope 2 (UWC) 17m 53ch
17m 79ch
Llangynllo Tunnel
18m 28ch
Llangynllo
18m 57ch
Deakins (UWC) 18m 61ch
Llanbister Road
21m 55ch
Troed-y-Rhiw Fedwen (UWC) 21m 05ch
Mayiord (UWC) 21m 35ch
Cwm-y-Geist Farm (UWC) 22m 01ch
Cefn Suran (UWC) 23m 06ch
Dolau House Farm No. 1 (UWC) 25m 18ch
Dolau (AOCL) 25m 24ch
Dolau
25m 26ch
Dolau House Farm No. 2 (UWC) 25m 34ch
Thomas No. 1 (UWC) 25m 36ch
Dolau House Farm No. 3 (UWC) 25m 41ch
Thomas No. 2 (UWC) 25m 43ch
Ty-Ddu (UWC) 26m 04ch
Rhydllyn 2 (UWC) 26m 40ch
Rabber Farm (UWC) 27m 00ch
Pen-y-Bont Tunnel
28m 09ch
27m 70ch
Pen-y-Bont
28m 21ch
Llynmellin Farm (UWC) 28m 42ch
Upper Cellws (UWC) 30m 51ch
Llandrindod (TMO) 31m 36ch
31m 57ch
(UWC) 33m 32ch
33m 33ch
(UWC) 35m 49ch
3
River Lugg

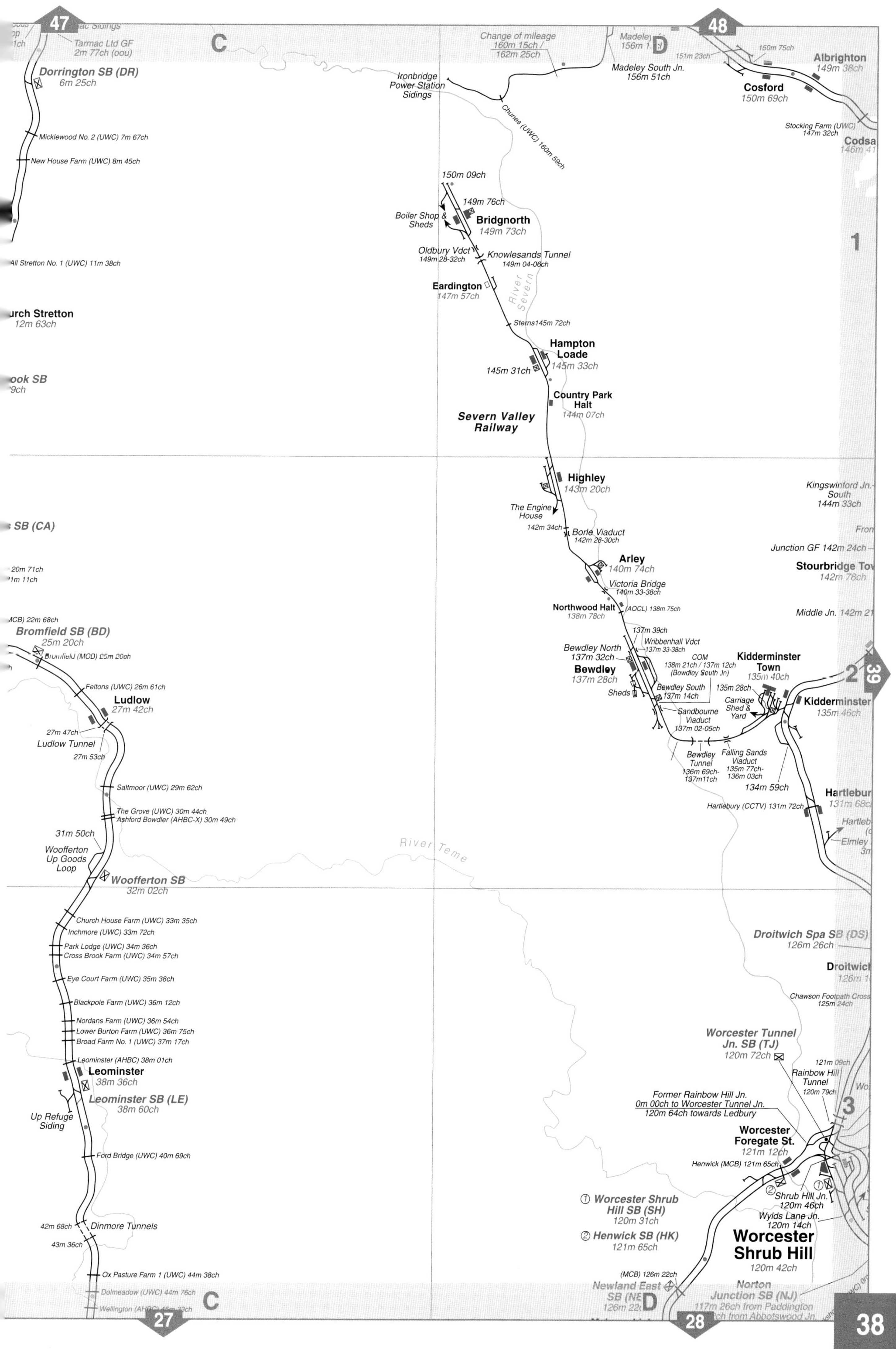
C
D
Tarmac Ltd GF
2m 77ch (oou)
Dorrington SB (DR)
6m 25ch
Micklewood No. 2 (UWC) 7m 67ch
New House Farm (UWC) 8m 45ch
All Stretton No. 1 (UWC) 11m 38ch
urch Stretton
12m 63ch
ook SB
SB (CA)
Bromfield SB (BD)
25m 20ch
Bromfield (MOD) 25m 20ch
Feltons (UWC) 26m 61ch
Ludlow
27m 42ch
27m 47ch
Ludlow Tunnel
27m 53ch
Saltmoor (UWC) 29m 62ch
The Grove (UWC) 30m 44ch
Ashford Bowdler (AHBC-X) 30m 49ch
31m 50ch
Woofferton
Up Goods
Loop
Woofferton SB
32m 02ch
Church House Farm (UWC) 33m 35ch
Inchmore (UWC) 33m 72ch
Park Lodge (UWC) 34m 36ch
Cross Brook Farm (UWC) 34m 57ch
Eye Court Farm (UWC) 35m 38ch
Blackpole Farm (UWC) 36m 12ch
Nordans Farm (UWC) 36m 54ch
Lower Burton Farm (UWC) 36m 75ch
Broad Farm No. 1 (UWC) 37m 17ch
Leominster (AHBC) 38m 01ch
Leominster
38m 36ch
Leominster SB (LE)
38m 60ch
Up Refuge
Siding
Ford Bridge (UWC) 40m 69ch
42m 68ch
Dinmore Tunnels
43m 36ch
Ox Pasture Farm 1 (UWC) 44m 38ch
Dolmeadow (UWC) 44m 76ch
Change of mileage
160m 15ch /
162m 25ch
Madeley South Jn.
156m 51ch
Ironbridge
Power Station
Sidings
Chunes (UWC) 160m 59ch
151m 23ch
150m 75ch
Albrighton
149m 38ch
Cosford
150m 69ch
Stocking Farm (UWC)
147m 32ch
150m 09ch
149m 76ch
Boiler Shop &
Sheds
Bridgnorth
149m 73ch
Oldbury Vdct
149m 28-32ch
Knowlesands Tunnel
149m 04-06ch
Eardington
147m 57ch
River Severn
Sterns 145m 72ch
Hampton
Loade
145m 33ch
145m 31ch
Country Park
Halt
144m 07ch
Severn Valley
Railway
1
Highley
143m 20ch
The Engine
House
142m 34ch
Borle Viaduct
142m 28-30ch
Arley
140m 74ch
Victoria Bridge
140m 33-38ch
Northwood Halt
138m 78ch
(AOCL) 138m 75ch
137m 39ch
Wribbenhall Vdct
137m 33-38ch
Bewdley North
137m 32ch
Bewdley
137m 28ch
COM
138m 21ch / 137m 12ch
(Bewdley South Jn)
Kidderminster
Town
135m 40ch
Sheds
Bewdley South
137m 14ch
135m 28ch
Carriage
Shed &
Yard
Kidderminster
135m 46ch
Sandbourne
Viaduct
137m 02-05ch
Bewdley
Tunnel
136m 69ch-
137m11ch
Falling Sands
Viaduct
135m 77ch-
136m 03ch
134m 59ch
Hartlebury (CCTV) 131m 72ch
Kingswinford Jn.
South
144m 33ch
Junction GF 142m 24ch
Middle Jn.
2
River Teme
Droitwich Spa SB (DS)
126m 26ch
Chawson Footpath Cross
125m 24ch
Worcester Tunnel
Jn. SB (TJ)
120m 72ch
121m 09ch
Rainbow Hill
Tunnel
120m 79ch
3
Former Rainbow Hill Jn.
0m 00ch to Worcester Tunnel Jn.
120m 64ch towards Ledbury
Worcester
Foregate St.
121m 12ch
Henwick (MCB) 121m 65ch
① Worcester Shrub
Hill SB (SH)
120m 31ch
② Henwick SB (HK)
121m 65ch
Shrub Hill Jn.
120m 46ch
Wylds Lane Jn.
120m 14ch
Worcester
Shrub Hill
120m 42ch
(MCB) 126m 22ch
Newland East
Norton
Junction SB (NJ)
117m 26ch from Paddington

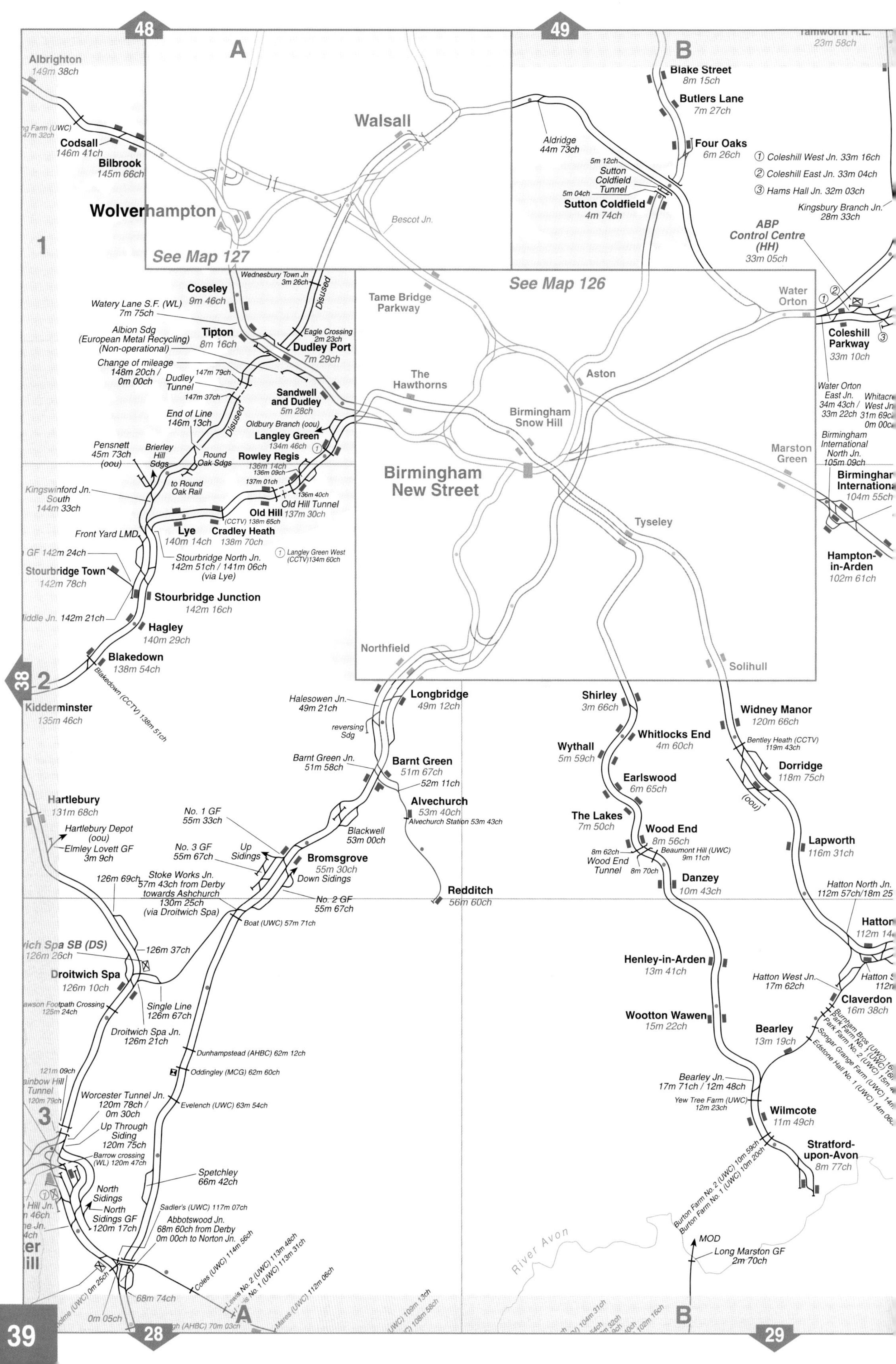
48
49
A
B
Albrighton
149m 38ch
Codsall
146m 41ch
Bilbrook
145m 66ch
Wolverhampton
Walsall
Bescot Jn.
See Map 127
Aldridge
44m 73ch
Blake Street
8m 15ch
Butlers Lane
7m 27ch
Four Oaks
6m 26ch
5m 12ch
Sutton Coldfield Tunnel
5m 04ch
Sutton Coldfield
4m 74ch
① Coleshill West Jn. 33m 16ch
② Coleshill East Jn. 33m 04ch
③ Hams Hall Jn. 32m 03ch
Kingsbury Branch Jn.
28m 33ch
ABP Control Centre (HH)
33m 05ch
Tamworth H.L.
23m 58ch
1
Coseley
9m 46ch
Wednesbury Town Jn
3m 26ch
Disused
Watery Lane S.F. (WL)
7m 75ch
Albion Sdg
(European Metal Recycling)
(Non-operational)
Tipton
8m 16ch
Eagle Crossing
2m 23ch
Dudley Port
7m 29ch
Change of mileage
148m 20ch /
0m 00ch
Dudley Tunnel
147m 79ch
147m 37ch
Sandwell and Dudley
5m 28ch
End of Line
146m 13ch
Disused
Oldbury Branch (oou)
Langley Green
134m 46ch
Pensnett
45m 73ch
(oou)
Brierley Hill Sdgs
Round Oak Sdgs
to Round Oak Rail
Rowley Regis
136m 14ch
136m 09ch
137m 01ch
136m 40ch
Old Hill Tunnel
Old Hill
137m 30ch
Kingswinford Jn. South
144m 33ch
Lye
140m 14ch
(CCTV) 138m 65ch
Cradley Heath
138m 70ch
Front Yard LMD
GF 142m 24ch
Stourbridge Town
142m 78ch
Stourbridge North Jn.
142m 51ch / 141m 06ch
(via Lye)
① Langley Green West (CCTV) 134m 60ch
Stourbridge Junction
142m 16ch
Middle Jn. 142m 21ch
Hagley
140m 29ch
Blakedown
138m 54ch
Blakedown (CCTV) 138m 51ch
38
2
Kidderminster
135m 46ch
See Map 126
Tame Bridge Parkway
The Hawthorns
Aston
Birmingham Snow Hill
Birmingham New Street
Water Orton
Coleshill Parkway
33m 10ch
Water Orton East Jn.
34m 43ch /
33m 22ch
Whitacre West Jn.
31m 69ch
0m 00ch
Birmingham International North Jn.
105m 09ch
Marston Green
Birmingham International
104m 55ch
Tyseley
Hampton-in-Arden
102m 61ch
Northfield
Solihull
Longbridge
49m 12ch
Halesowen Jn.
49m 21ch
reversing Sdg
Barnt Green Jn.
51m 58ch
Barnt Green
51m 67ch
52m 11ch
Alvechurch
53m 40ch
Alvechurch Station 53m 43ch
Redditch
56m 60ch
Shirley
3m 66ch
Whitlocks End
4m 60ch
Wythall
5m 59ch
Earlswood
6m 65ch
The Lakes
7m 50ch
Wood End
8m 56ch
8m 62ch
Wood End Tunnel
8m 70ch
Beaumont Hill (UWC)
9m 11ch
Danzey
10m 43ch
Widney Manor
120m 66ch
Bentley Heath (CCTV)
119m 43ch
Dorridge
118m 75ch
(oou)
Lapworth
116m 31ch
Hatton North Jn.
112m 57ch/18m 25
Hatton
112m 14
Hartlebury
131m 68ch
Hartlebury Depot
(oou)
Elmley Lovett GF
3m 9ch
No. 1 GF
55m 33ch
No. 3 GF
55m 67ch
Up Sidings
Blackwell
53m 00ch
Bromsgrove
55m 30ch
Down Sidings
No. 2 GF
55m 67ch
126m 69ch
Stoke Works Jn.
57m 43ch from Derby
towards Ashchurch
130m 25ch
(via Droitwich Spa)
Boat (UWC) 57m 71ch
Droitwich Spa SB (DS)
126m 26ch
126m 37ch
Droitwich Spa
126m 10ch
Single Line
126m 67ch
Droitwich Spa Jn.
126m 21ch
Dunhampstead (AHBC) 62m 12ch
Oddingley (MCG) 62m 60ch
Evelench (UWC) 63m 54ch
121m 09ch
Rainbow Hill Tunnel
120m 79ch
Worcester Tunnel Jn.
120m 78ch /
0m 30ch
Up Through Siding
120m 75ch
Barrow crossing (WL) 120m 47ch
3
North Sidings
North Sidings GF
120m 17ch
Spetchley
66m 42ch
Sadler's (UWC) 117m 07ch
Abbotswood Jn.
68m 60ch from Derby
0m 00ch to Norton Jn.
Coles (UWC) 114m 56ch
Lewis No. 2 (UWC) 113m 48ch
Lewis No. 1 (UWC) 113m 31ch
Mares (UWC) 112m 06ch
68m 74ch
0m 05ch
Henley-in-Arden
13m 41ch
Wootton Wawen
15m 22ch
Hatton West Jn.
17m 62ch
Claverdon
16m 38ch
Bearley
13m 19ch
Bearley Jn.
17m 71ch / 12m 48ch
Yew Tree Farm (UWC)
12m 23ch
Wilmcote
11m 49ch
Stratford-upon-Avon
8m 77ch
Burton Farm No. 2 (UWC) 10m 59ch
Burton Farm No. 1 (UWC) 10m 20ch
River Avon
MOD
Long Marston GF
2m 70ch
28
29

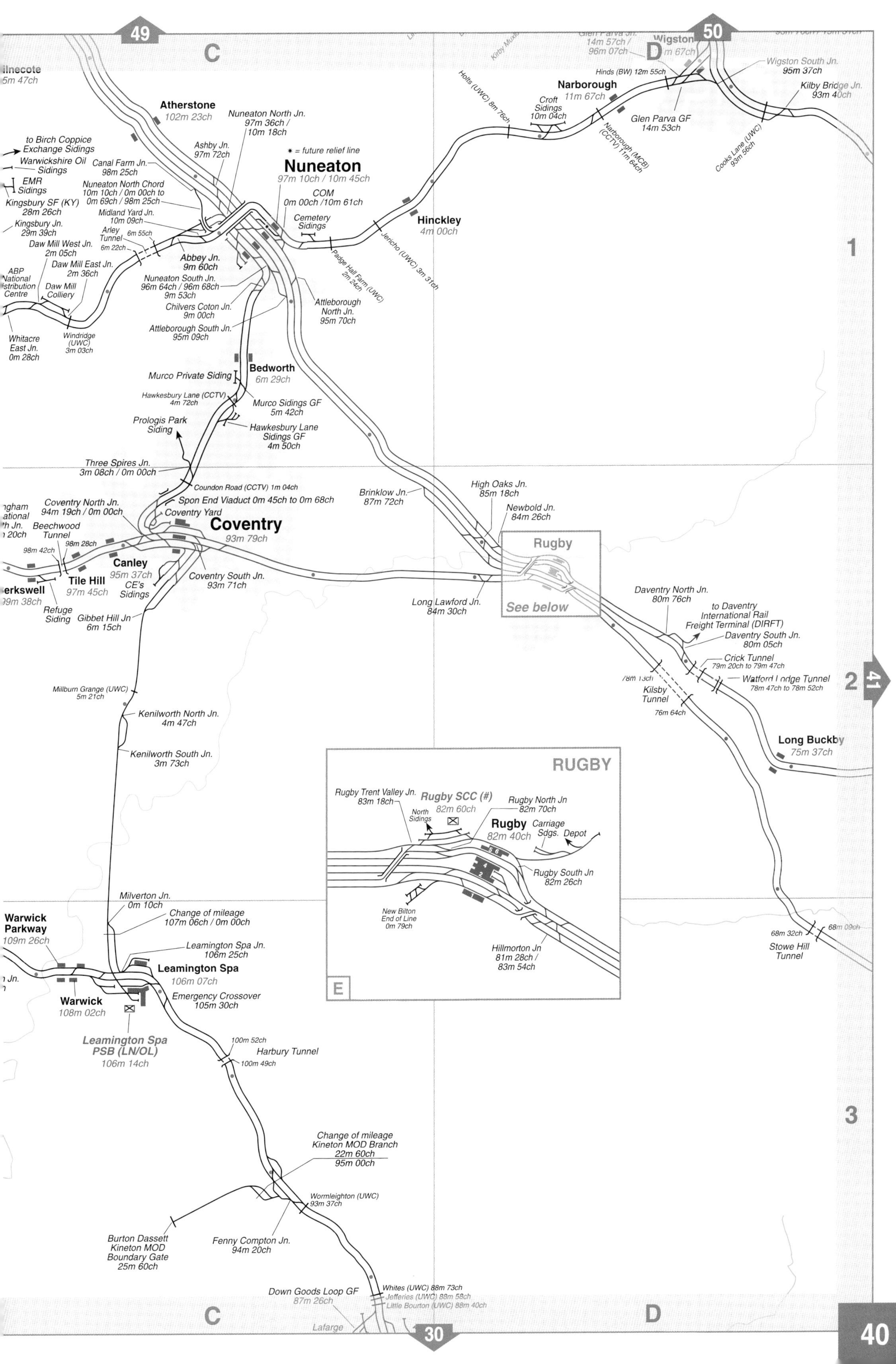
C
D
Atherstone
102m 23ch
Nuneaton North Jn.
97m 36ch /
10m 18ch
Ashby Jn.
97m 72ch
* = future relief line
Nuneaton
97m 10ch / 10m 45ch
COM
0m 00ch /10m 61ch
to Birch Coppice
Exchange Sidings
Warwickshire Oil
Sidings
Canal Farm Jn.
98m 25ch
EMR
Sidings
Nuneaton North Chord
10m 10ch / 0m 00ch to
0m 69ch / 98m 25ch
Kingsbury SF (KY)
28m 26ch
Midland Yard Jn.
10m 09ch
Kingsbury Jn.
29m 39ch
Arley
Tunnel
6m 55ch
6m 22ch
Daw Mill West Jn.
2m 05ch
Abbey Jn.
9m 60ch
Cemetery
Sidings
Daw Mill East Jn.
2m 36ch
Nuneaton South Jn.
96m 64ch / 96m 68ch
9m 53ch
Daw Mill
Colliery
Chilvers Coton Jn.
9m 00ch
Attleborough South Jn.
95m 09ch
Attleborough
North Jn.
95m 70ch
Padge Hall Farm (UWC)
2m 24ch
Whitacre
East Jn.
0m 28ch
Windridge
(UWC)
3m 03ch
Hinckley
4m 00ch
Jericho (UWC) 3m 31ch
Narborough
11m 67ch
Croft
Sidings
10m 04ch
Holts (UWC) 8m 76ch
Narborough (MCB)
(CCTV) 11m 64ch
Hinds (BW) 12m 55ch
Glen Parva GF
14m 53ch
Glen Parva Jn.
14m 57ch /
96m 07ch
Wigston
Wigston South Jn.
95m 37ch
Kilby Bridge Jn.
93m 40ch
Cooks Lane (UWC)
93m 56ch
1
Bedworth
6m 29ch
Murco Private Siding
Hawkesbury Lane (CCTV)
4m 72ch
Murco Sidings GF
5m 42ch
Prologis Park
Siding
Hawkesbury Lane
Sidings GF
4m 50ch
Three Spires Jn.
3m 08ch / 0m 00ch
Coundon Road (CCTV) 1m 04ch
Spon End Viaduct 0m 45ch to 0m 68ch
Coventry North Jn.
94m 19ch / 0m 00ch
Coventry Yard
Coventry
93m 79ch
Beechwood
Tunnel
98m 28ch
98m 42ch
Tile Hill
97m 45ch
Canley
95m 37ch
CE's
Sidings
Coventry South Jn.
93m 71ch
Refuge
Siding
Gibbet Hill Jn
6m 15ch
Brinklow Jn.
87m 72ch
High Oaks Jn.
85m 18ch
Newbold Jn.
84m 26ch
Rugby
See below
Long Lawford Jn.
84m 30ch
Daventry North Jn.
80m 76ch
to Daventry
International Rail
Freight Terminal (DIRFT)
Daventry South Jn.
80m 05ch
Crick Tunnel
79m 20ch to 79m 47ch
Watford Lodge Tunnel
78m 47ch to 78m 52ch
Kilsby
Tunnel
76m 64ch
2
Long Buckby
75m 37ch
Millburn Grange (UWC)
5m 21ch
Kenilworth North Jn.
4m 47ch
Kenilworth South Jn.
3m 73ch
RUGBY
Rugby Trent Valley Jn.
83m 18ch
Rugby SCC (#)
82m 60ch
North
Sidings
Rugby North Jn
82m 70ch
Rugby
82m 40ch
Carriage
Sdgs.
Depot
Rugby South Jn
82m 26ch
New Bilton
End of Line
0m 79ch
Hillmorton Jn
81m 28ch /
83m 54ch
E
68m 32ch
68m 09ch
Stowe Hill
Tunnel
Warwick
Parkway
109m 26ch
Milverton Jn.
0m 10ch
Change of mileage
107m 06ch / 0m 00ch
Leamington Spa Jn.
106m 25ch
Leamington Spa
106m 07ch
Warwick
108m 02ch
Emergency Crossover
105m 30ch
Leamington Spa
PSB (LN/OL)
106m 14ch
100m 52ch
Harbury Tunnel
100m 49ch
3
Change of mileage
Kineton MOD Branch
22m 60ch
95m 00ch
Wormleighton (UWC)
93m 37ch
Burton Dassett
Kineton MOD
Boundary Gate
25m 60ch
Fenny Compton Jn.
94m 20ch
Down Goods Loop GF
87m 26ch
Whites (UWC) 88m 73ch
Jefferies (UWC) 88m 58ch
Little Bourton (UWC) 88m 40ch
Lafarge
C
D
41

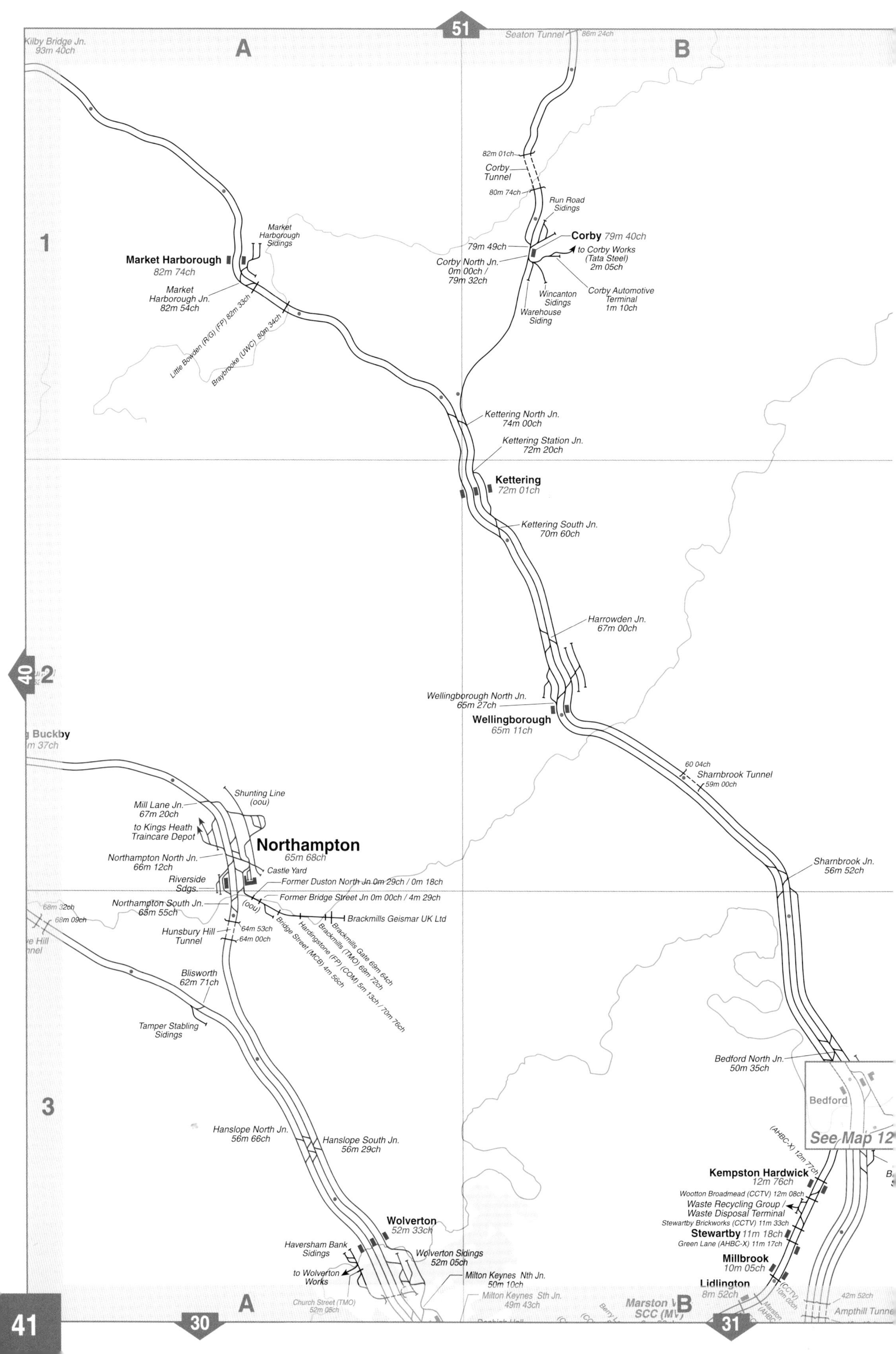

51
Kilby Bridge Jn.
93m 40ch
A
Seaton Tunnel
86m 24ch
B
1
82m 01ch
Corby
Tunnel
80m 74ch
Run Road
Sidings
Corby 79m 40ch
79m 49ch
to Corby Works
(Tata Steel)
2m 05ch
Corby North Jn.
0m 00ch /
79m 32ch
Wincanton
Sidings
Warehouse
Siding
Corby Automotive
Terminal
1m 10ch
Market
Harborough
Sidings
Market Harborough
82m 74ch
Market
Harborough Jn.
82m 54ch
Little Bowden (R/G) (FP) 82m 33ch
Braybrooke (UWC) 80m 34ch
Kettering North Jn.
74m 00ch
Kettering Station Jn.
72m 20ch
Kettering
72m 01ch
Kettering South Jn.
70m 60ch
Harrowden Jn.
67m 00ch
40
2
Wellingborough North Jn.
65m 27ch
Wellingborough
65m 11ch
Buckby
m 37ch
60 04ch
Sharnbrook Tunnel
59m 00ch
Shunting Line
(oou)
Mill Lane Jn.
67m 20ch
to Kings Heath
Traincare Depot
Northampton
65m 68ch
Northampton North Jn.
66m 12ch
Castle Yard
Riverside
Sdgs.
Former Duston North Jn 0m 29ch / 0m 18ch
Former Bridge Street Jn 0m 00ch / 4m 29ch
Northampton South Jn.
65m 55ch
(oou)
Brackmills Geismar UK Ltd
68m 32ch
68m 09ch
Hunsbury Hill
Tunnel
64m 53ch
64m 00ch
Bridge Street (MCB) 4m 56ch
Hardingstone (FP) (COM) 5m 13ch / 70m 76ch
Brackmills (TMO) 69m 72ch
Brackmills Gate 69m 64ch
Sharnbrook Jn.
56m 52ch
Blisworth
62m 71ch
Tamper Stabling
Sidings
Bedford North Jn.
50m 35ch
Bedford
See Map 12
3
Hanslope North Jn.
56m 66ch
Hanslope South Jn.
56m 29ch
(AHBC-X) 12m 77ch
Kempston Hardwick
12m 76ch
Wootton Broadmead (CCTV) 12m 08ch
Waste Recycling Group /
Waste Disposal Terminal
Stewartby Brickworks (CCTV) 11m 33ch
Stewartby 11m 18ch
Green Lane (AHBC-X) 11m 17ch
Millbrook
10m 05ch
Lidlington
8m 52ch
Wolverton
52m 33ch
Haversham Bank
Sidings
Wolverton Sidings
52m 05ch
to Wolverton
Works
Milton Keynes Nth Jn.
50m 10ch
Milton Keynes Sth Jn.
49m 43ch
Church Street (TMO)
52m 05ch
A
B
42m 52ch
Ampthill Tunnel
30
31

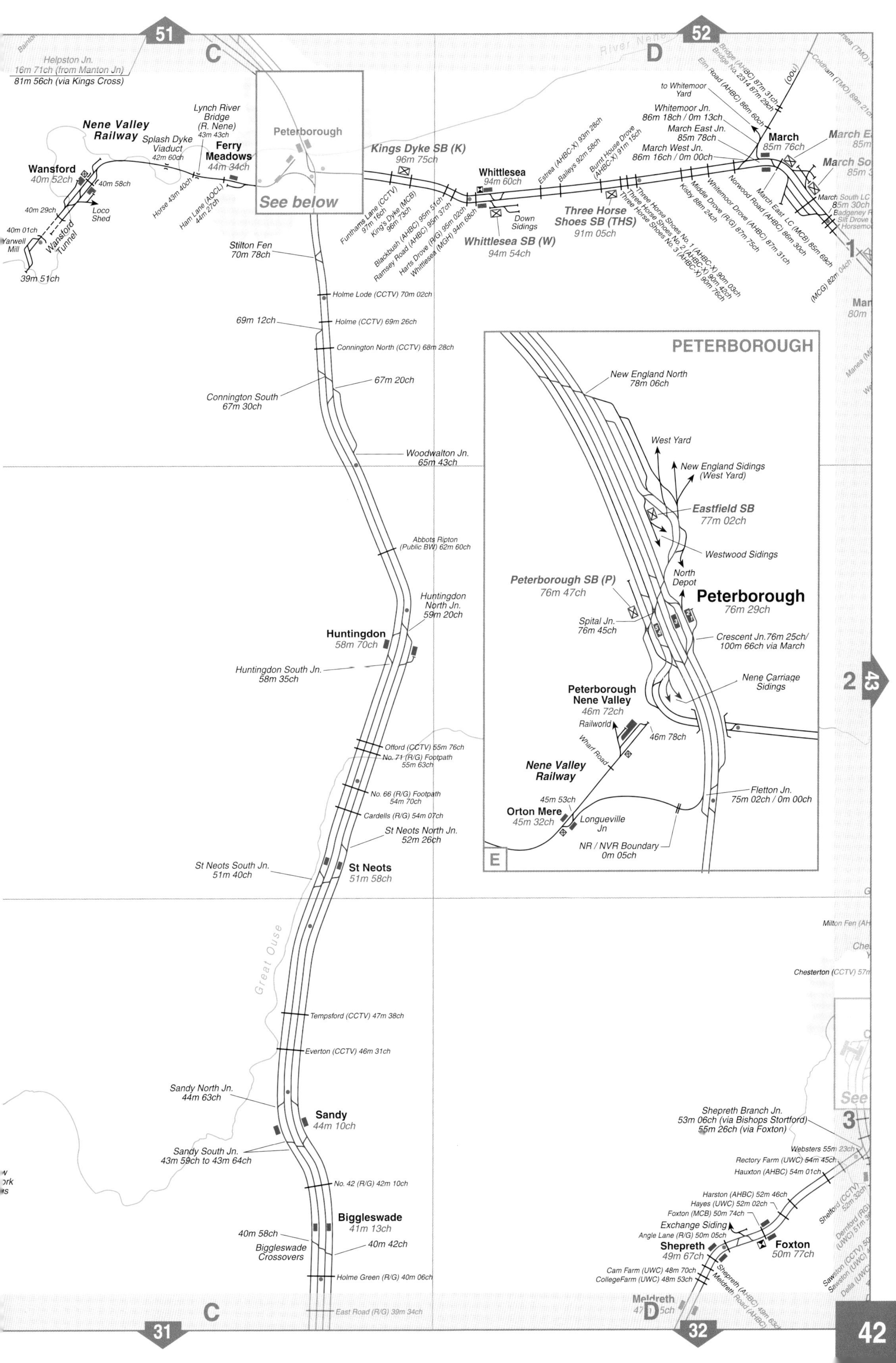
C
D
Helpston Jn.
16m 71ch (from Manton Jn)
81m 56ch (via Kings Cross)
Nene Valley Railway
Wansford
40m 52ch
Wansford Tunnel
Yarwell Mill
39m 51ch
Loco Shed
Splash Dyke Viaduct
42m 60ch
Lynch River Bridge (R. Nene)
43m 43ch
Ferry Meadows
44m 34ch
Peterborough
See below
Kings Dyke SB (K)
96m 75ch
Whittlesea
94m 60ch
Whittlesea SB (W)
94m 54ch
Down Sidings
Three Horse Shoes SB (THS)
91m 05ch
to Whitemoor Yard
Whitemoor Jn.
86m 18ch / 0m 13ch
March East Jn.
85m 78ch
March West Jn.
86m 16ch / 0m 00ch
March
85m 76ch
Stilton Fen
70m 78ch
Holme Lode (CCTV) 70m 02ch
69m 12ch
Holme (CCTV) 69m 26ch
Connington North (CCTV) 68m 28ch
67m 20ch
Connington South
67m 30ch
Woodwalton Jn.
65m 43ch
Abbots Ripton (Public BW) 62m 60ch
Huntingdon North Jn.
59m 20ch
Huntingdon
58m 70ch
Huntingdon South Jn.
58m 35ch
Offord (CCTV) 55m 76ch
No. 71 (R/G) Footpath 55m 63ch
No. 66 (R/G) Footpath 54m 70ch
Cardells (R/G) 54m 07ch
St Neots North Jn.
52m 26ch
St Neots South Jn.
51m 40ch
St Neots
51m 58ch
Great Ouse
Tempsford (CCTV) 47m 38ch
Everton (CCTV) 46m 31ch
Sandy North Jn.
44m 63ch
Sandy
44m 10ch
Sandy South Jn.
43m 59ch to 43m 64ch
No. 42 (R/G) 42m 10ch
Biggleswade
41m 13ch
40m 58ch
Biggleswade Crossovers
40m 42ch
Holme Green (R/G) 40m 06ch
East Road (R/G) 39m 34ch
PETERBOROUGH
New England North
78m 06ch
West Yard
New England Sidings (West Yard)
Eastfield SB
77m 02ch
Westwood Sidings
North Depot
Peterborough SB (P)
76m 47ch
Peterborough
76m 29ch
Spital Jn.
76m 45ch
Crescent Jn. 76m 25ch/ 100m 66ch via March
Nene Carriage Sidings
Peterborough Nene Valley
46m 72ch
Railworld
46m 78ch
Wharf Road
Nene Valley Railway
45m 53ch
Orton Mere
45m 32ch
Longueville Jn
NR / NVR Boundary
0m 05ch
Fletton Jn.
75m 02ch / 0m 00ch
E
Shepreth Branch Jn.
53m 06ch (via Bishops Stortford)
55m 26ch (via Foxton)
Websters 55m 23ch
Rectory Farm (UWC) 54m 45ch
Hauxton (AHBC) 54m 01ch
Harston (AHBC) 52m 46ch
Hayes (UWC) 52m 02ch
Foxton (MCB) 50m 74ch
Exchange Siding
Angle Lane (R/G) 50m 05ch
Shepreth
49m 67ch
Foxton
50m 77ch
Cam Farm (UWC) 48m 70ch
CollegeFarm (UWC) 48m 53ch
Meldreth
1
2
3
43

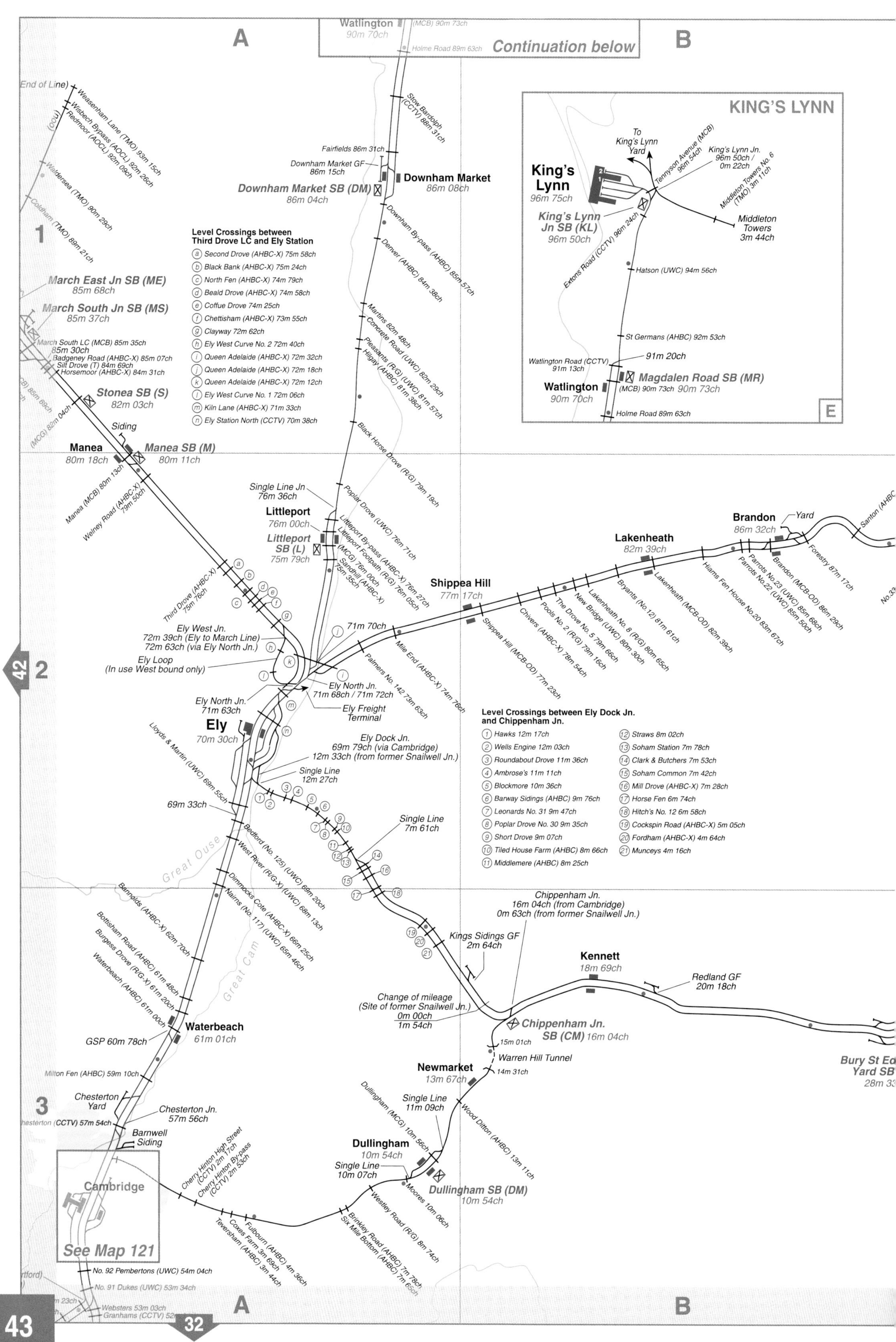

Watlington
90m 70ch
(MCB) 90m 73ch
Holme Road 89m 63ch
Continuation below
A
B
End of Line)
Weasenham Lane (TMO) 93m 15ch
Wisbech Bypass (AOCL) 92m 26ch
Redmoor (AOCL) 92m 09ch
(OOU)
Waldersea (TMO) 90m 29ch
Coldham (TMO) 89m 21ch
1
March East Jn SB (ME)
85m 68ch
March South Jn SB (MS)
85m 37ch
March South LC (MCB) 85m 35ch
85m 30ch
Badgeney Road (AHBC-X) 85m 07ch
Silt Drove (T) 84m 69ch
Horsemoor (AHBC-X) 84m 31ch
Stonea SB (S)
82m 03ch
(MCG) 82m 04ch
Siding
Manea
80m 18ch
Manea SB (M)
80m 11ch
Manea (MCB) 80m 13ch
Welney Road (AHBC-X) 79m 50ch
Stow Bardolph (CCTV) 88m 31ch
Fairfields 86m 31ch
Downham Market GF
86m 15ch
Downham Market
86m 08ch
Downham Market SB (DM)
86m 04ch
Downham By-pass (AHBC) 85m 57ch
Denver (AHBC) 84m 38ch
Level Crossings between Third Drove LC and Ely Station
a Second Drove (AHBC-X) 75m 58ch
b Black Bank (AHBC-X) 75m 24ch
c North Fen (AHBC-X) 74m 79ch
d Beald Drove (AHBC-X) 74m 58ch
e Coffue Drove 74m 25ch
f Chettisham (AHBC-X) 73m 55ch
g Clayway 72m 62ch
h Ely West Curve No. 2 72m 40ch
i Queen Adelaide (AHBC-X) 72m 32ch
j Queen Adelaide (AHBC-X) 72m 18ch
k Queen Adelaide (AHBC-X) 72m 12ch
l Ely West Curve No. 1 72m 06ch
m Kiln Lane (AHBC-X) 71m 33ch
n Ely Station North (CCTV) 70m 38ch
Martins 82m 48ch
Concrete Road (UWC) 82m 29ch
Pleasants (R/G) (UWC) 81m 57ch
Hilgay (AHBC) 81m 38ch
Black Horse Drove (R/G) 79m 19ch
Single Line Jn
76m 36ch
Littleport
76m 00ch
Littleport SB (L)
75m 79ch
Poplar Drove (UWC) 76m 71ch
Littleport By-pass (AHBC-X) 76m 27ch
Littleport Footpath (R/G) 76m 05ch
(MCG) 76m 00ch
Sandhill (AHBC-X) 75m 35ch
Third Drove (AHBC-X) 75m 76ch
71m 70ch
Ely West Jn.
72m 39ch (Ely to March Line)
72m 63ch (via Ely North Jn.)
Ely Loop
(In use West bound only)
2
Ely North Jn.
71m 68ch / 71m 72ch
Ely Freight Terminal
Ely North Jn.
71m 63ch
Ely
70m 30ch
Ely Dock Jn.
69m 79ch (via Cambridge)
12m 33ch (from former Snailwell Jn.)
Single Line
12m 27ch
Lloyds & Martin (UWC) 69m 55ch
69m 33ch
Bedford (No. 125) (UWC) 69m 20ch
West River (R/G-X) (UWC) 68m 13ch
Dimmocks Cote (AHBC-X) 66m 25ch
Nairns (No. 117) (UWC) 65m 46ch
Great Ouse
Great Cam
Bannolds (AHBC-X) 62m 70ch
Bottisham Road (AHBC) 61m 48ch
Burgess Drove (R/G-X) 61m 20ch
Waterbeach (AHBC) 61m 00ch
Waterbeach
61m 01ch
GSP 60m 78ch
Milton Fen (AHBC) 59m 10ch
3
Chesterton Yard
Chesterton Jn.
57m 56ch
Chesterton (CCTV) 57m 54ch
Barnwell Siding
Cambridge
See Map 121
No. 92 Pembertons (UWC) 54m 04ch
No. 91 Dukes (UWC) 53m 34ch
Websters 53m 03ch
Granhams (CCTV)
Cherry Hinton High Street (CCTV) 2m 17ch
Cherry Hinton By-pass (CCTV) 2m 53ch
Fulbourn (AHBC) 4m 36ch
Coxes Farm 3m 69ch
Teversham (AHBC) 3m 44ch
Brinkley Road (AHBC) 7m 78ch
Six Mile Bottom (AHBC) 7m 65ch
Westley Road (R/G) 8m 74ch
Moores 10m 06ch
Dullingham
10m 54ch
Single Line
10m 07ch
Dullingham (MCG) 10m 56ch
Dullingham SB (DM)
10m 54ch
Single Line
11m 09ch
Wood Ditton (AHBC) 13m 11ch
Newmarket
13m 67ch
14m 31ch
Warren Hill Tunnel
15m 01ch
Chippenham Jn.
SB (CM) 16m 04ch
Change of mileage
(Site of former Snailwell Jn.)
0m 00ch
1m 54ch
Chippenham Jn.
16m 04ch (from Cambridge)
0m 63ch (from former Snailwell Jn.)
Kings Sidings GF
2m 64ch
Kennett
18m 69ch
Redland GF
20m 18ch
Bury St Ed
Yard SB
28m 33
Single Line
7m 61ch
Level Crossings between Ely Dock Jn. and Chippenham Jn.
1 Hawks 12m 17ch
2 Wells Engine 12m 03ch
3 Roundabout Drove 11m 36ch
4 Ambrose's 11m 11ch
5 Blockmore 10m 36ch
6 Barway Sidings (AHBC) 9m 76ch
7 Leonards No. 31 9m 47ch
8 Poplar Drove No. 30 9m 35ch
9 Short Drove 9m 07ch
10 Tiled House Farm (AHBC) 8m 66ch
11 Middlemere (AHBC) 8m 25ch
12 Straws 8m 02ch
13 Soham Station 7m 78ch
14 Clark & Butchers 7m 53ch
15 Soham Common 7m 42ch
16 Mill Drove (AHBC-X) 7m 28ch
17 Horse Fen 6m 74ch
18 Hitch's No. 12 6m 58ch
19 Cockspin Road (AHBC-X) 5m 05ch
20 Fordham (AHBC-X) 4m 64ch
21 Munceys 4m 16ch
Shippea Hill
77m 17ch
Mile End (AHBC-X) 74m 76ch
Palmers No. 142 73m 63ch
Shippea Hill (MCB-OD) 77m 23ch
Chivers (AHBC-X) 78m 54ch
Pools No. 2 (R/G) 79m 16ch
The Drove No. 5 79m 66ch
New Bridge (UWC) 80m 30ch
Lakenheath No. 8 (R/G) 80m 65ch
Bryants (No. 12) 81m 61ch
Lakenheath
82m 39ch
Lakenheath (MCB-OD) 82m 39ch
Hiams Fen House No. 20 83m 67ch
Parrots No. 22 (UWC) 85m 50ch
Parrots No. 23 (UWC) 85m 68ch
Brandon
86m 32ch
Yard
Brandon (MCB-OD) 86m 29ch
Forestry 87m 17ch
Santon (AHBC
KING'S LYNN
To King's Lynn Yard
King's Lynn
96m 75ch
King's Lynn Jn SB (KL)
96m 50ch
Tennyson Avenue (MCB) 96m 54ch
King's Lynn Jn.
96m 50ch /
0m 22ch
Middleton Towers No. 6 (TMO) 3m 11ch
Middleton Towers
3m 44ch
Extons Road (CCTV) 96m 24ch
Hatson (UWC) 94m 56ch
St Germans (AHBC) 92m 53ch
91m 20ch
Watlington Road (CCTV)
91m 13ch
Watlington
90m 70ch
Magdalen Road SB (MR)
(MCB) 90m 73ch 90m 73ch
Holme Road 89m 63ch
E

42
32

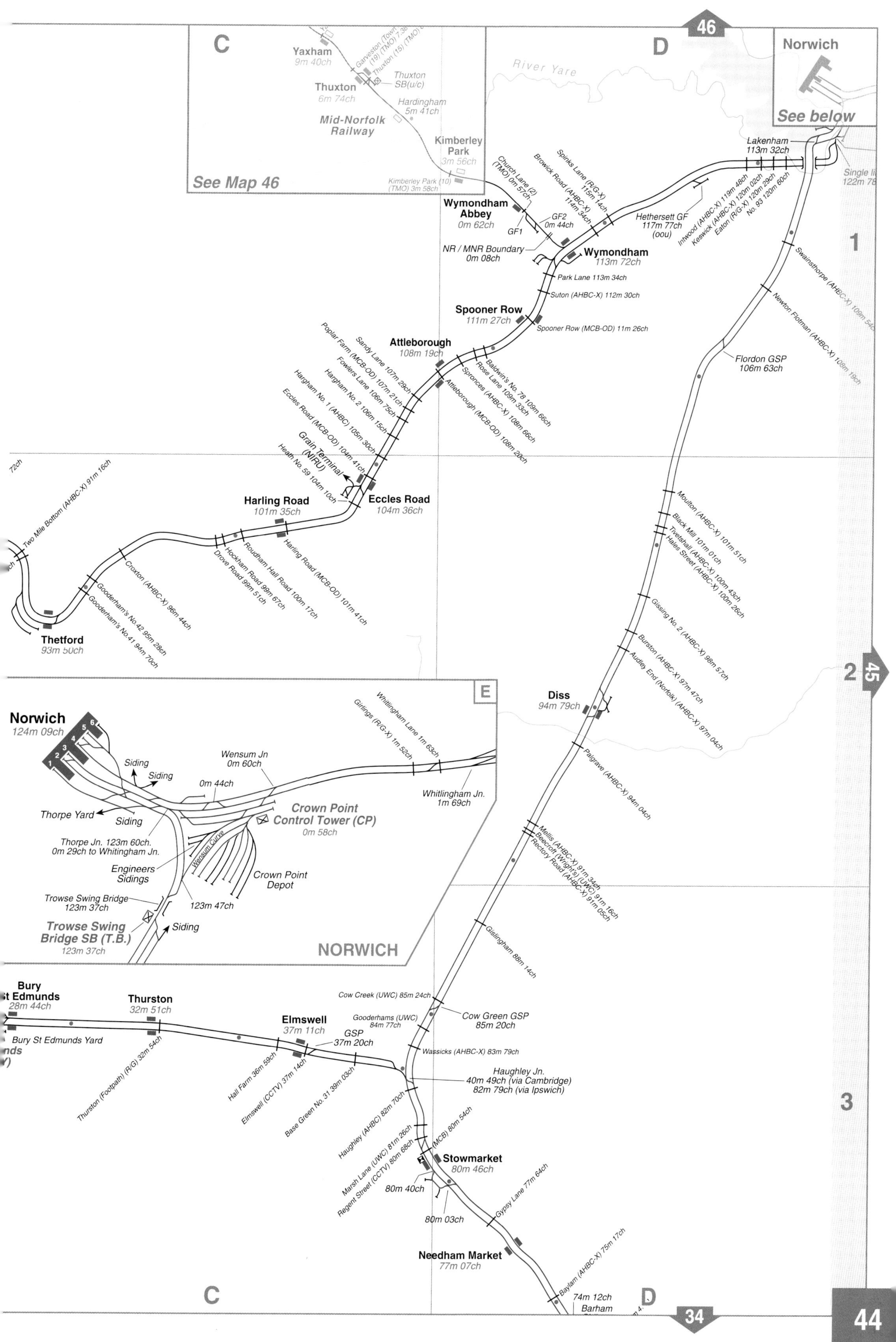

C
D
Norwich
See below
Yaxham
9m 40ch
Thuxton
6m 74ch
Thuxton SB(u/c)
Hardingham
5m 41ch
Mid-Norfolk Railway
Kimberley Park
3m 56ch
Kimberley Park (10) (TMO) 3m 58ch
See Map 46
River Yare
Church Lane (2) (TMO) 0m 57ch
Browick Road (AHBC-X) 114m 34ch
Spinks Lane (R/G-X) 115m 14ch
Wymondham Abbey
0m 62ch
GF1
GF2
0m 44ch
NR / MNR Boundary
0m 08ch
Wymondham
113m 72ch
Hethersett GF
117m 77ch
(oou)
Intwood (AHBC-X) 119m 48ch
Keswick (AHBC-X) 120m 02ch
Eaton (R/G-X) 120m 29ch
No. 93 120m 60ch
Lakenham
113m 32ch
Single li
122m 78
Park Lane 113m 34ch
Suton (AHBC-X) 112m 30ch
Spooner Row
111m 27ch
Spooner Row (MCB-OD) 11m 26ch
Swainsthorpe (AHBC-X) 109m 54ch
Newton Flotman (AHBC-X) 108m 19ch
Flordon GSP
106m 63ch
1
Attleborough
108m 19ch
Sandy Lane 107m 29ch
Poplar Farm (MCB-OD) 107m 21ch
Fowlers Lane 106m 75ch
Hargham No. 2 106m 15ch
Hargham No. 1 (AHBC) 105m 30ch
Eccles Road (MCB-OD) 104m 41ch
Grain Terminal (NRU)
Heath No. 59 104m 10ch
Baldwin's No. 78 109m 66ch
Rose Lane 109m 33ch
Spronces (AHBC-X) 108m 66ch
Attleborough (MCB-OD) 108m 20ch
Eccles Road
104m 36ch
Harling Road
101m 35ch
Two Mile Bottom (AHBC-X) 91m 16ch
Croxton (AHBC-X) 96m 44ch
Hockham Road 99m 67ch
Drove Road 99m 51ch
Roudham Hall Road 100m 17ch
Harling Road (MCB-OD) 101m 41ch
Gooderham's No.42 95m 28ch
Gooderham's No.41 94m 70ch
Thetford
93m 50ch
Moulton (AHBC-X) 101m 51ch
Black Mill 101m 01ch
Tivetshall (AHBC-X) 100m 43ch
Hales Street (AHBC-X) 100m 26ch
Gissing No. 2 (AHBC-X) 98m 57ch
Burston (AHBC-X) 97m 47ch
Audley End (Norfolk) (AHBC-X) 97m 04ch
Diss
94m 79ch
2
Palgrave (AHBC-X) 94m 04ch
Mellis (AHBC-X) 91m 34ch
Beecroft (Wrights) (UWC) 91m 16ch
Rectory Road (AHBC-X) 91m 05ch
Gislingham 88m 14ch
E
Norwich
124m 09ch
Whitlingham Lane 1m 63ch
Girlings (R/G-X) 1m 52ch
Wensum Jn
0m 60ch
0m 44ch
Siding
Siding
Siding
Thorpe Yard
Whitlingham Jn.
1m 69ch
Crown Point Control Tower (CP)
0m 58ch
Thorpe Jn. 123m 60ch.
0m 29ch to Whitingham Jn.
Wensum Curve
Engineers Sidings
Crown Point Depot
Trowse Swing Bridge
123m 37ch
123m 47ch
Trowse Swing Bridge SB (T.B.)
123m 37ch
Siding
NORWICH
Bury St Edmunds
28m 44ch
Bury St Edmunds Yard
Thurston
32m 51ch
Thurston (Footpath) (R/G) 32m 54ch
Elmswell
37m 11ch
GSP
37m 20ch
Hall Farm 36m 59ch
Elmswell (CCTV) 37m 14ch
Base Green No. 31 39m 03ch
Cow Creek (UWC) 85m 24ch
Cow Green GSP
85m 20ch
Gooderhams (UWC)
84m 77ch
Wassicks (AHBC-X) 83m 79ch
Haughley Jn.
40m 49ch (via Cambridge)
82m 79ch (via Ipswich)
Haughley (AHBC) 82m 70ch
(MCB) 80m 54ch
Marsh Lane (UWC) 81m 26ch
Regent Street (CCTV) 80m 68ch
Stowmarket
80m 46ch
80m 40ch
80m 03ch
Gypsy Lane 77m 64ch
Needham Market
77m 07ch
Baylam (AHBC-X) 75m 17ch
74m 12ch
Barham
3
C
D

46
45
34

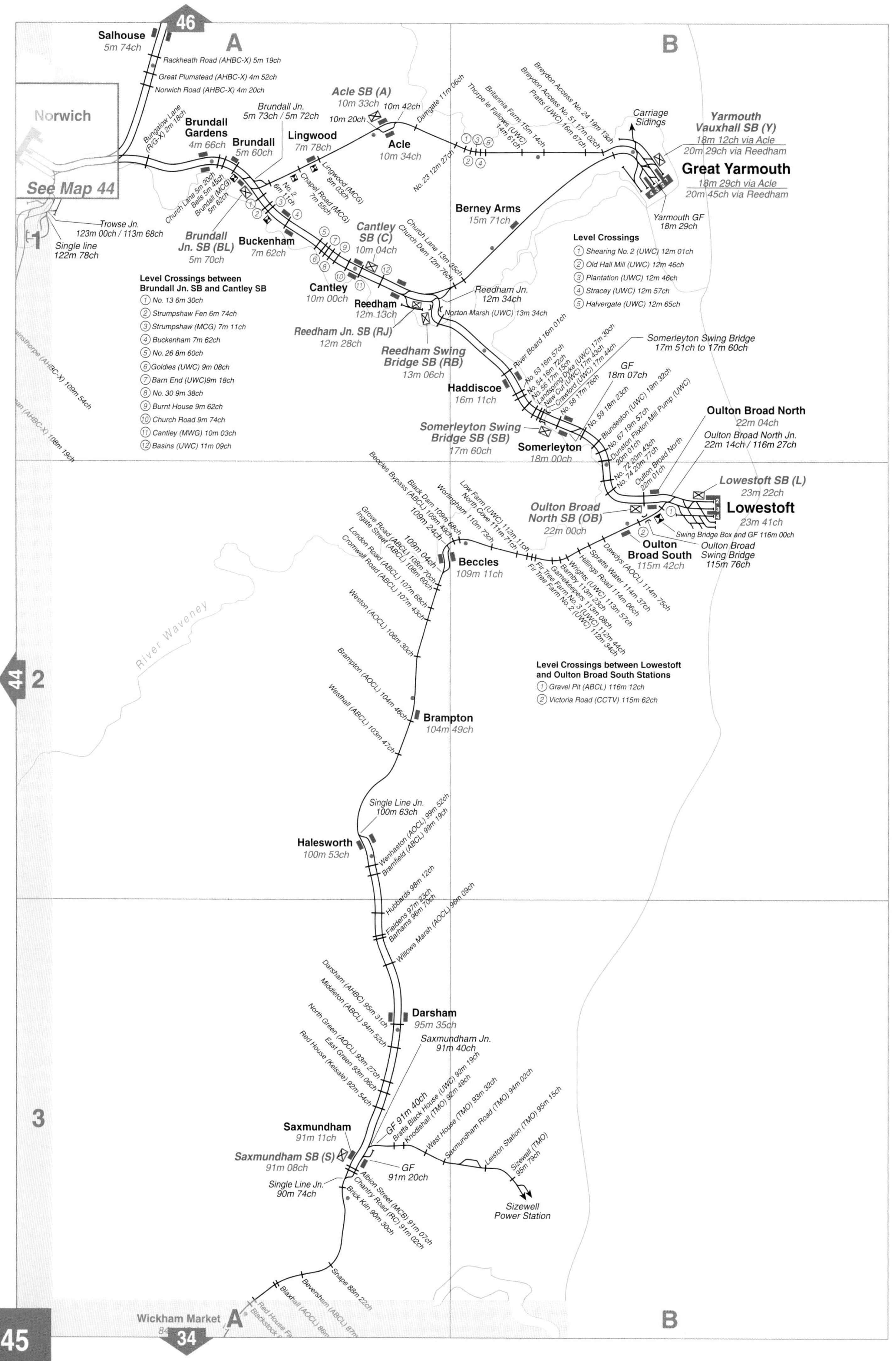

46
A
B
Salhouse
5m 74ch
Rackheath Road (AHBC-X) 5m 19ch
Great Plumstead (AHBC-X) 4m 52ch
Norwich Road (AHBC-X) 4m 20ch
Norwich
See Map 44
Bungalow Lane (R/G-X) 2m 18ch
Brundall Gardens
4m 66ch
Brundall
5m 60ch
Brundall Jn.
5m 73ch / 5m 72ch
Lingwood
7m 78ch
Acle SB (A)
10m 33ch
10m 20ch
10m 42ch
Acle
10m 34ch
Damgate 11m 06ch
No. 23 12m 27ch
Britannia Farm 15m 14ch
Thorpe le Fallows (UWC) 14m 61ch
Breydon Access No. 24 19m 13ch
Breydon Access No. 51 17m 02ch
Pratts (UWC) 16m 67ch
Carriage Sidings
Yarmouth Vauxhall SB (Y)
18m 12ch via Acle
20m 29ch via Reedham
Great Yarmouth
18m 29ch via Acle
20m 45ch via Reedham
Yarmouth GF
18m 29ch
Church Lane 5m 20ch
Bells 5m 45ch
Brundall (MCG) 5m 62ch
No. 2 6m 11ch
Chapel Road (MCG) 7m 55ch
Lingwood (MCG) 8m 03ch
Trowse Jn.
123m 00ch / 113m 68ch
Single line
122m 78ch
Brundall Jn. SB (BL)
5m 70ch
Buckenham
7m 62ch
Cantley SB (C)
10m 04ch
Church Lane 13m 35ch
Church Dam 12m 76ch
Berney Arms
15m 71ch
Level Crossings
1 Shearing No. 2 (UWC) 12m 01ch
2 Old Hall Mill (UWC) 12m 46ch
3 Plantation (UWC) 12m 46ch
4 Stracey (UWC) 12m 57ch
5 Halvergate (UWC) 12m 65ch
Level Crossings between Brundall Jn. SB and Cantley SB
1 No. 13 6m 30ch
2 Strumpshaw Fen 6m 74ch
3 Strumpshaw (MCG) 7m 11ch
4 Buckenham 7m 62ch
5 No. 26 8m 60ch
6 Goldies (UWC) 9m 08ch
7 Barn End (UWC) 9m 18ch
8 No. 30 9m 38ch
9 Burnt House 9m 62ch
10 Church Road 9m 74ch
11 Cantley (MWG) 10m 03ch
12 Basins (UWC) 11m 09ch
Cantley
10m 00ch
Reedham
12m 13ch
Reedham Jn.
12m 34ch
Norton Marsh (UWC) 13m 34ch
Reedham Jn. SB (RJ)
12m 28ch
Reedham Swing Bridge SB (RB)
13m 06ch
Somerleyton Swing Bridge
17m 51ch to 17m 60ch
River Board 16m 01ch
No. 53 16m 57ch
No. 54 16m 72ch
No. 56 17m 15ch
Landspring Dyke (UWC) 17m 30ch
New Cut (UWC) 17m 43ch
Crawford (UWC) 17m 44ch
No. 58 17m 76ch
GF
18m 07ch
No. 59 18m 23ch
Blundeston (UWC) 19m 32ch
No. 67 19m 57ch
Dunston Flixton Mill Pump (UWC)
20m 01ch
No. 72 20m 43ch
No. 74 20m 77ch
Oulton Broad North
22m 01ch
Haddiscoe
16m 11ch
Somerleyton Swing Bridge SB (SB)
17m 60ch
Somerleyton
18m 00ch
Oulton Broad North
22m 04ch
Oulton Broad North Jn.
22m 14ch / 116m 27ch
Lowestoft SB (L)
23m 22ch
Lowestoft
23m 41ch
Swing Bridge Box and GF 116m 00ch
Oulton Broad North SB (OB)
22m 00ch
Oulton Broad South
115m 42ch
Oulton Broad Swing Bridge
115m 76ch
Beccles Bypass (ABCL) 109m 68ch
Black Dam 109m 49ch
Worlingham 110m 73ch
Low Farm (UWC) 112m 11ch
North Cove 111m 71ch
109m 24ch
109m 04ch
Beccles
109m 11ch
Grove Road (ABCL) 108m 70ch
Ingate Street (ABCL) 108m 60ch
London Road (ABCL) 107m 68ch
Cromwell Road (ABCL) 107m 43ch
Weston (AOCL) 106m 30ch
Fir Tree Farm No. 3 (UWC) 112m 44ch
Fir Tree Farm No. 2 (UWC) 112m 34ch
Gamekeepers 113m 08ch
Barnby 113m 23ch
Wrights (UWC) 113m 57ch
Hillings Road 114m 06ch
Spratts Water 114m 37ch
Dawdys (AOCL) 114m 75ch
Level Crossings between Lowestoft and Oulton Broad South Stations
1 Gravel Pit (ABCL) 116m 12ch
2 Victoria Road (CCTV) 115m 62ch
River Waveney
44
1
2
3
Brampton (AOCL) 104m 46ch
Brampton
104m 49ch
Westhall (ABCL) 103m 47ch
Single Line Jn.
100m 63ch
Halesworth
100m 53ch
Wenhaston (AOCL) 99m 52ch
Bramfield (ABCL) 99m 19ch
Hubbards 98m 12ch
Fieldens 97m 23ch
Barhams 96m 70ch
Willows Marsh (AOCL) 96m 09ch
Darsham (AHBC) 95m 31ch
Middleton (ABCL) 94m 52ch
North Green (AOCL) 93m 27ch
East Green 93m 06ch
Red House (Kelsale) 92m 54ch
Darsham
95m 35ch
Saxmundham Jn.
91m 40ch
GF 91m 40ch
Bratts Black House (UWC) 92m 19ch
Knodishall (TMO) 92m 49ch
West House (TMO) 93m 32ch
Saxmundham Road (TMO) 94m 02ch
Leiston Station (TMO) 95m 15ch
Sizewell (TMO) 95m 79ch
Saxmundham
91m 11ch
Saxmundham SB (S)
91m 08ch
GF
91m 20ch
Single Line Jn.
90m 74ch
Albion Street (MCB) 91m 07ch
Chantry Road (RC) 91m 02ch
Brick Kiln 90m 30ch
Sizewell Power Station
Snape 88m 22ch
Beversham (ABCL) 87m
Blaxhall (AOCL) 86m
Wickham Market
34

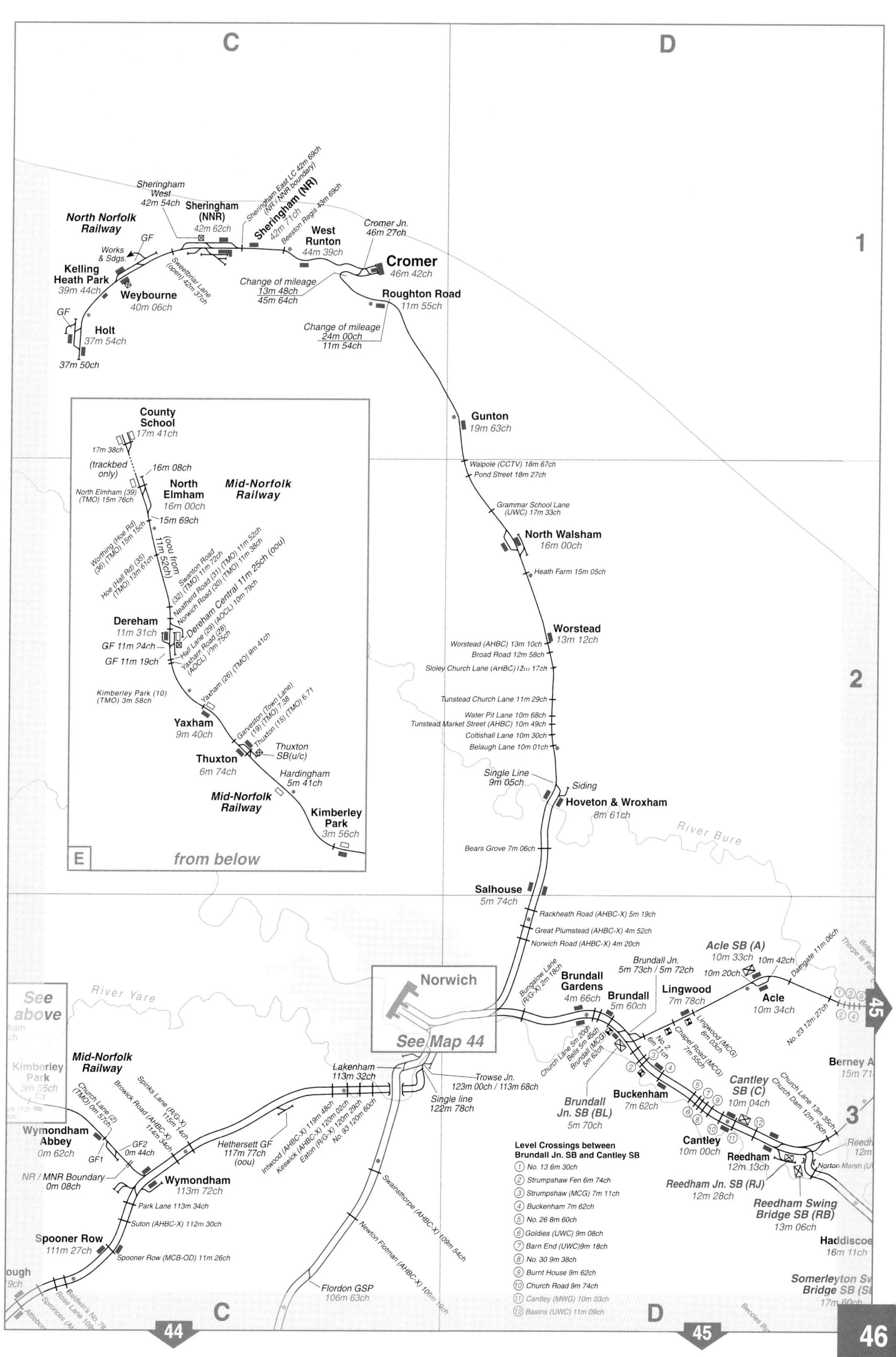
C
D
1
2
3
North Norfolk Railway
Sheringham West 42m 54ch
Sheringham (NNR) 42m 62ch
Sheringham East LC 42m 69ch (NR / NNR boundary)
Sheringham (NR) 42m 71ch
Beeston Regis 43m 69ch
West Runton 44m 39ch
Cromer Jn. 46m 27ch
Cromer 46m 42ch
GF
Works & Sdgs.
Kelling Heath Park 39m 44ch
Weybourne 40m 06ch
Sweetbriar Lane (open) 42m 37ch
Change of mileage 13m 48ch 45m 64ch
Roughton Road 11m 55ch
Change of mileage 24m 00ch 11m 54ch
Holt 37m 54ch
37m 50ch
Gunton 19m 63ch
Walpole (CCTV) 18m 67ch
Pond Street 18m 27ch
Grammar School Lane (UWC) 17m 33ch
North Walsham 16m 00ch
Heath Farm 15m 05ch
Worstead 13m 12ch
Worstead (AHBC) 13m 10ch
Broad Road 12m 58ch
Sloley Church Lane (AHBC) 12m 17ch
Tunstead Church Lane 11m 29ch
Water Pit Lane 10m 68ch
Tunstead Market Street (AHBC) 10m 49ch
Coltishall Lane 10m 30ch
Belaugh Lane 10m 01ch
Single Line 9m 05ch
Siding
Hoveton & Wroxham 8m 61ch
River Bure
Bears Grove 7m 06ch
Salhouse 5m 74ch
Rackheath Road (AHBC-X) 5m 19ch
Great Plumstead (AHBC-X) 4m 52ch
Norwich Road (AHBC-X) 4m 20ch
County School 17m 41ch
17m 38ch
(trackbed only)
16m 08ch
North Elmham (39) (TMO) 15m 76ch
North Elmham 16m 00ch
Mid-Norfolk Railway
15m 69ch
Worthing (Hoe Rd) (36) (TMO) 15m 15ch
Hoe (Hall Rd) (35) (TMO) 13m 61ch
(oou from 11m 52ch)
Swanton Road (32) (TMO) 11m 72ch
Neatherd Road (31) (TMO) 11m 52ch
Norwich Road (30) (TMO) 11m 38ch
Dereham Central 11m 25ch (oou)
Dereham 11m 31ch
GF 11m 24ch
GF 11m 19ch
Hall Lane (29) (AOCL) 10m 79ch
Yaxham Road (28) (AOCL) 10m 75ch
Kimberley Park (10) (TMO) 3m 58ch
Yaxham (26) (TMO) 9m 41ch
Yaxham 9m 40ch
Garveston (Town Lane) (19) (TMO) 7.38
Thuxton (15) (TMO) 6.71
Thuxton 6m 74ch
Thuxton SB(u/c)
Hardingham 5m 41ch
Mid-Norfolk Railway
Kimberley Park 3m 56ch
E
from below
Acle SB (A) 10m 33ch
10m 42ch
10m 20ch
Damgate 11m 06ch
Acle 10m 34ch
No. 23 12m 27ch
Brundall Jn. 5m 73ch / 5m 72ch
Bungalow Lane (R/G-X) 2m 18ch
Brundall Gardens 4m 66ch
Brundall 5m 60ch
Lingwood 7m 78ch
Lingwood (MCG) 8m 03ch
Chapel Road (MCG) 7m 55ch
No. 2 6m 11ch
Church Lane 5m 20ch
Bells 5m 45ch
Brundall (MCG) 5m 62ch
Norwich
See Map 44
River Yare
See above
Kimberley Park 3m 55ch
Mid-Norfolk Railway
Church Lane (2) (TMO) 0m 57ch
Browick Road (AHBC-X) 114m 34ch
Spinks Lane (R/G-X) 115m 14ch
Lakenham 113m 32ch
Trowse Jn. 123m 00ch / 113m 68ch
Single line 122m 78ch
Intwood (AHBC-X) 119m 48ch
Keswick (AHBC-X) 120m 02ch
Eaton (R/G-X) 120m 29ch
No. 93 120m 60ch
Hethersett GF 117m 77ch (oou)
Wymondham Abbey 0m 62ch
GF1
GF2 0m 44ch
NR / MNR Boundary 0m 08ch
Wymondham 113m 72ch
Park Lane 113m 34ch
Suton (AHBC-X) 112m 30ch
Spooner Row 111m 27ch
Spooner Row (MCB-OD) 111m 26ch
Swainsthorpe (AHBC-X) 109m 54ch
Newton Flotman (AHBC-X) 108m 19ch
Flordon GSP 106m 63ch
Brundall Jn. SB (BL) 5m 70ch
Buckenham 7m 62ch
Cantley SB (C) 10m 04ch
Church Lane 13m 35ch
Church Dam 12m 76ch
Cantley 10m 00ch
Reedham 12m 13ch
Reedham Jn. SB (RJ) 12m 28ch
Reedham Swing Bridge SB (RB) 13m 06ch
Norton Marsh
Berney A 15m 71
Haddiscoe 16m 11ch
Somerleyton Swing Bridge SB 17m 60ch
Level Crossings between Brundall Jn. SB and Cantley SB
1 No. 13 6m 30ch
2 Strumpshaw Fen 6m 74ch
3 Strumpshaw (MCG) 7m 11ch
4 Buckenham 7m 62ch
5 No. 26 8m 60ch
6 Goldies (UWC) 9m 08ch
7 Barn End (UWC) 9m 18ch
8 No. 30 9m 38ch
9 Burnt House 9m 62ch
10 Church Road 9m 74ch
11 Cantley (MWG) 10m 03ch
12 Basins (UWC) 11m 09ch
Baldwin's No. 78
Rose Lane 109
Spronces
Attleborough
Becoles
44
45

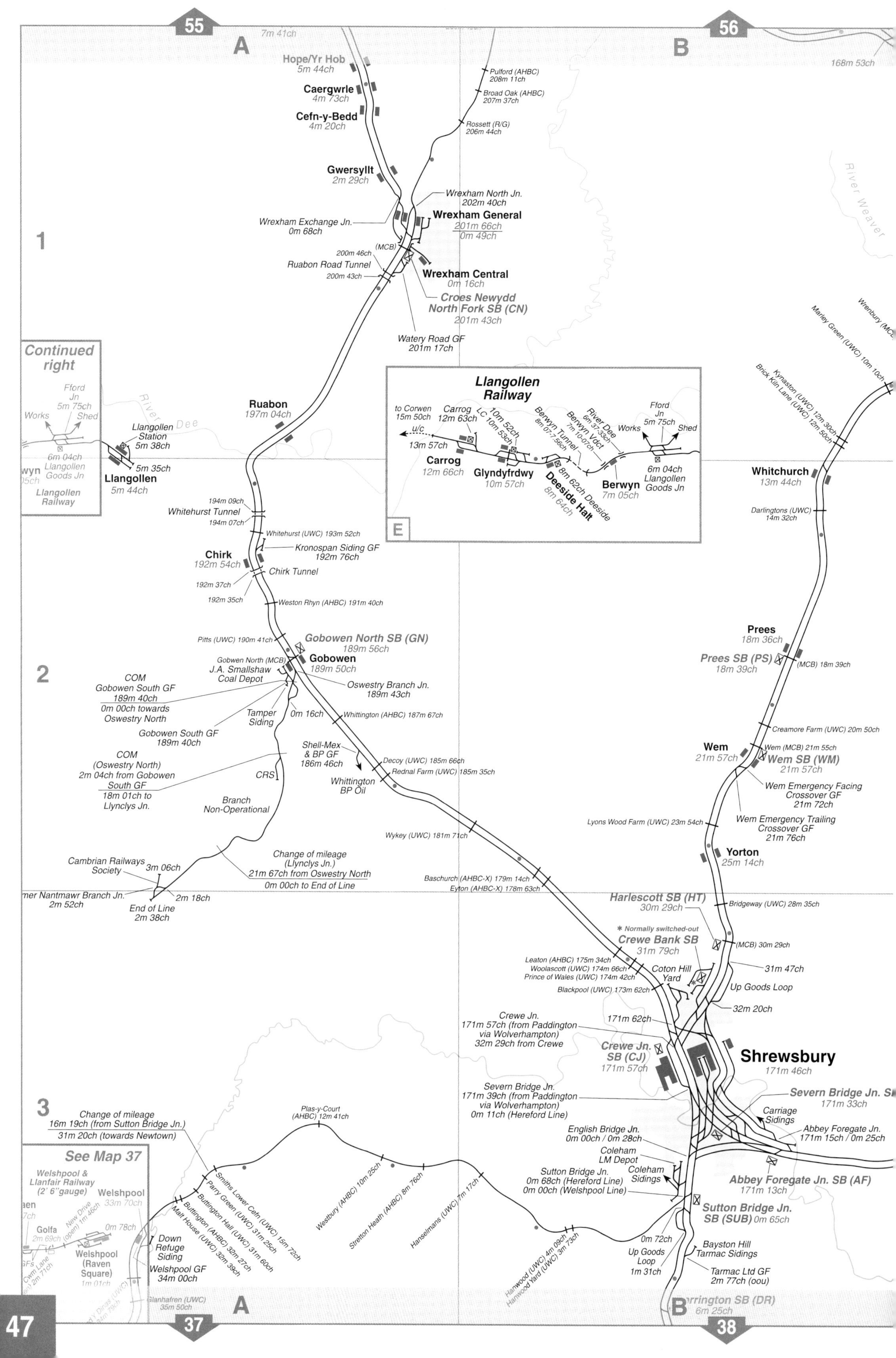

55
56
A
B
7m 41ch
168m 53ch
Hope/Yr Hob
5m 44ch
Caergwrle
4m 73ch
Cefn-y-Bedd
4m 20ch
Pulford (AHBC)
208m 11ch
Broad Oak (AHBC)
207m 37ch
Rossett (R/G)
206m 44ch
River Weaver
Gwersyllt
2m 29ch
Wrexham North Jn.
202m 40ch
Wrexham General
201m 66ch
0m 49ch
Wrexham Exchange Jn.
0m 68ch
1
(MCB)
200m 46ch
Ruabon Road Tunnel
200m 43ch
Wrexham Central
0m 16ch
Croes Newydd
North Fork SB (CN)
201m 43ch
Watery Road GF
201m 17ch
Wrenbury (MCB)
Marley Green (UWC) 10m 10ch
Kynaston (UWC) 12m 30ch
Brick Kiln Lane (UWC) 12m 50ch
Continued right
Ffford Jn
5m 75ch
Works
Shed
River Dee
6m 04ch
Llangollen Goods Jn
Llangollen Railway
Llangollen Station
5m 38ch
5m 35ch
Llangollen
5m 44ch
Ruabon
197m 04ch
Llangollen Railway
to Corwen
15m 50ch
u/c
13m 57ch
Carrog
12m 63ch
Carrog
12m 66ch
LC 10m 53ch
10m 52ch
Glyndyfrdwy
10m 57ch
Berwyn Tunnel
8m 07-7.56ch
Berwyn Vdct
7m 10-07ch
River Dee
6m 37-33ch
8m 62ch Deeside
Deeside Halt
8m 64ch
Berwyn
7m 05ch
Fford Jn
5m 75ch
Works
Shed
6m 04ch
Llangollen Goods Jn
E
Whitchurch
13m 44ch
Darlingtons (UWC)
14m 32ch
194m 09ch
Whitehurst Tunnel
194m 07ch
Whitehurst (UWC) 193m 52ch
Kronospan Siding GF
192m 76ch
Chirk
192m 54ch
Chirk Tunnel
192m 37ch
192m 35ch
Weston Rhyn (AHBC) 191m 40ch
Pitts (UWC) 190m 41ch
Gobowen North SB (GN)
189m 56ch
Gobowen North (MCB)
Gobowen
189m 50ch
J.A. Smallshaw Coal Depot
2
COM
Gobowen South GF
189m 40ch
0m 00ch towards Oswestry North
Oswestry Branch Jn.
189m 43ch
Tamper Siding
0m 16ch
Whittington (AHBC) 187m 67ch
Gobowen South GF
189m 40ch
Shell-Mex & BP GF
186m 46ch
Whittington BP Oil
Decoy (UWC) 185m 66ch
Rednal Farm (UWC) 185m 35ch
COM
(Oswestry North)
2m 04ch from Gobowen South GF
18m 01ch to Llynclys Jn.
CRS
Branch Non-Operational
Prees
18m 36ch
Prees SB (PS)
18m 39ch
(MCB) 18m 39ch
Creamore Farm (UWC) 20m 50ch
Wem
21m 57ch
Wem (MCB) 21m 55ch
Wem SB (WM)
21m 57ch
Wem Emergency Facing Crossover GF
21m 72ch
Wem Emergency Trailing Crossover GF
21m 76ch
Lyons Wood Farm (UWC) 23m 54ch
Yorton
25m 14ch
Wykey (UWC) 181m 71ch
Cambrian Railways Society
3m 06ch
Change of mileage
(Llynclys Jn.)
21m 67ch from Oswestry North
0m 00ch to End of Line
Baschurch (AHBC-X) 179m 14ch
Eyton (AHBC-X) 178m 63ch
Nantmawr Branch Jn.
2m 52ch
2m 18ch
End of Line
2m 38ch
Harlescott SB (HT)
30m 29ch
Bridgeway (UWC) 28m 35ch
* Normally switched-out
Crewe Bank SB
31m 79ch
(MCB) 30m 29ch
Leaton (AHBC) 175m 34ch
Woolascott (UWC) 174m 66ch
Prince of Wales (UWC) 174m 42ch
Coton Hill Yard
31m 47ch
Up Goods Loop
Blackpool (UWC) 173m 62ch
32m 20ch
Crewe Jn.
171m 57ch (from Paddington via Wolverhampton)
32m 29ch from Crewe
171m 62ch
Crewe Jn. SB (CJ)
171m 57ch
Shrewsbury
171m 46ch
3
Severn Bridge Jn.
171m 39ch (from Paddington via Wolverhampton)
0m 11ch (Hereford Line)
Severn Bridge Jn. SB
171m 33ch
Carriage Sidings
Abbey Foregate Jn.
171m 15ch / 0m 25ch
Change of mileage
16m 19ch (from Sutton Bridge Jn.)
31m 20ch (towards Newtown)
Plas-y-Court (AHBC) 12m 41ch
English Bridge Jn.
0m 00ch / 0m 28ch
Coleham LM Depot
Coleham Sidings
Abbey Foregate Jn. SB (AF)
171m 13ch
See Map 37
Welshpool & Llanfair Railway (2' 6" gauge)
Welshpool
33m 70ch
Golfa
2m 69ch
New Drive (open) 1m 46ch
0m 78ch
Welshpool (Raven Square)
1m 01ch
Smiths Lower Cefn (UWC) 15m 72ch
Parry Green (UWC) 31m 25ch
Buttington Hall (UWC) 31m 60ch
Buttington (AHBC) 32m 27ch
Malt House (UWC) 32m 39ch
Westbury (AHBC) 10m 25ch
Stretton Heath (AHBC) 8m 76ch
Hanselmans (UWC) 7m 17ch
Sutton Bridge Jn.
0m 68ch (Hereford Line)
0m 00ch (Welshpool Line)
Sutton Bridge Jn. SB (SUB) 0m 65ch
Down Refuge Siding
Welshpool GF
34m 00ch
Hanwood (UWC) 4m 09ch
Hanwood Yard (UWC) 3m 73ch
0m 72ch
Up Goods Loop
1m 31ch
Bayston Hill Tarmac Sidings
Tarmac Ltd GF
2m 77ch (oou)
Glanhafren (UWC)
35m 50ch
Dorrington SB (DR)
6m 25ch
47
37
38

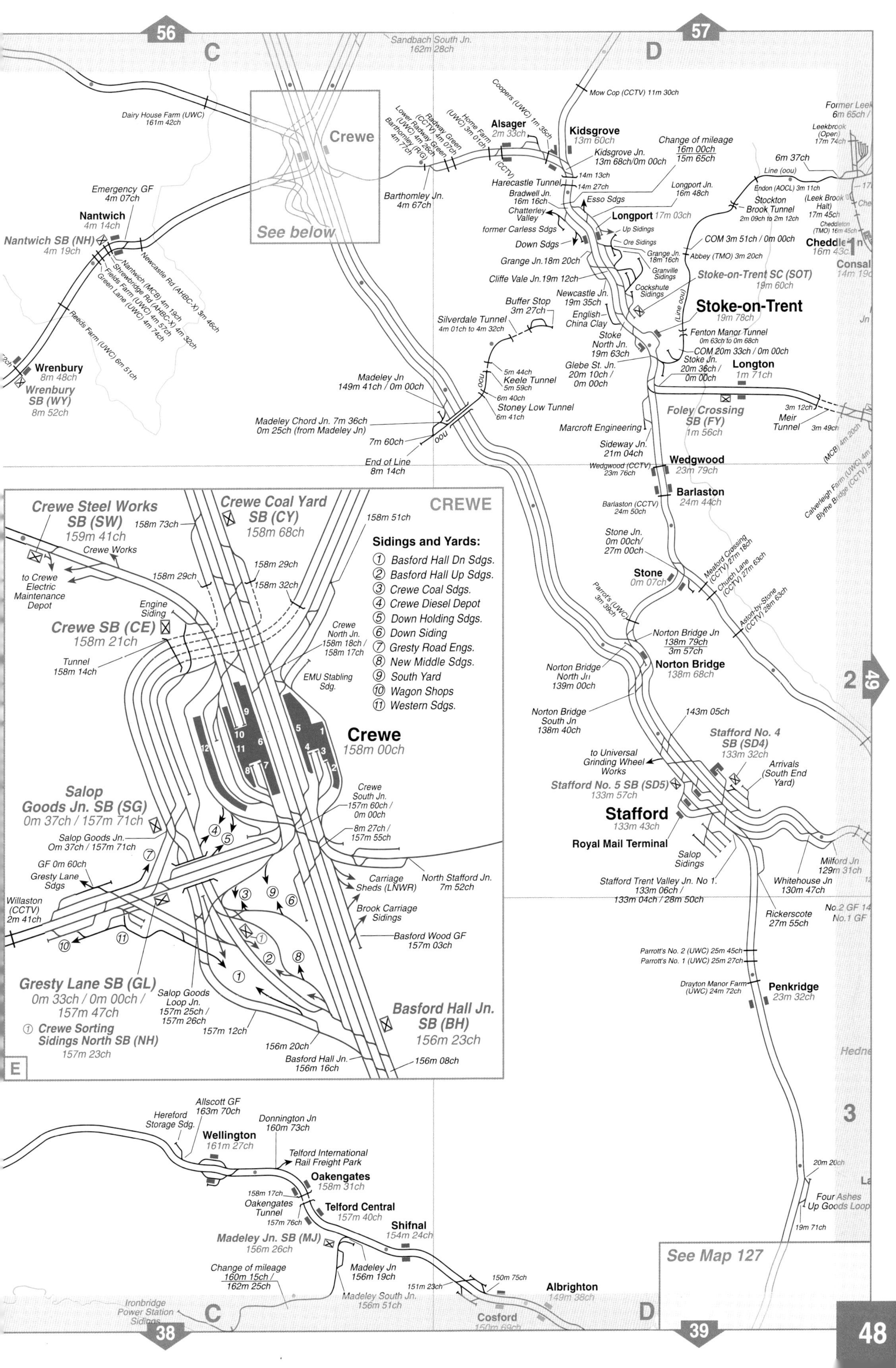

56
57
C
D
Sandbach South Jn.
162m 28ch
Dairy House Farm (UWC)
161m 42ch
Crewe
See below
Mow Cop (CCTV) 11m 30ch
Coopers (UWC) 1m 35ch
Alsager
2m 33ch
Home Farm (UWC) 3m 01ch
Radway Green (CCTV) 4m 07ch
Lower Radway Green (UWC) 4m 26ch
Barthomley (R/G) 4m 77ch
Barthomley Jn.
4m 67ch
Kidsgrove
13m 60ch
Kidsgrove Jn.
13m 68ch/0m 00ch
Change of mileage
16m 00ch
15m 65ch
Harecastle Tunnel
14m 13ch
14m 27ch
Bradwell Jn.
16m 16ch
Esso Sdgs
Longport Jn.
16m 48ch
Chatterley Valley
former Carless Sdgs
Longport 17m 03ch
Up Sidings
Ore Sidings
Down Sdgs
Grange Jn.
18m 16ch
Granville Sidings
Grange Jn.18m 20ch
Cliffe Vale Jn.19m 12ch
Cockshute Sidings
Newcastle Jn.
19m 35ch
English China Clay
Stoke North Jn.
19m 63ch
Glebe St. Jn.
20m 10ch /
0m 00ch
Stoke-on-Trent
19m 78ch
Stoke-on-Trent SC (SOT)
19m 60ch
Fenton Manor Tunnel
0m 63ch to 0m 68ch
COM 20m 33ch / 0m 00ch
Stoke Jn.
20m 36ch /
0m 00ch
Longton
1m 71ch
Foley Crossing SB (FY)
1m 56ch
3m 12ch
Meir Tunnel
3m 49ch
Line (oou)
6m 37ch
Endon (AOCL) 3m 11ch
Stockton Brook Tunnel
2m 09ch to 2m 12ch
COM 3m 51ch / 0m 00ch
Abbey (TMO) 3m 20ch
Former Leek
6m 65ch
Leekbrook (Open)
17m 74ch
(Leek Brook Halt)
17m 45ch
Cheddleton (TMO) 16m 45ch
Cheddleton
16m 43ch
Consall
14m 19ch
Buffer Stop
3m 27ch
Silverdale Tunnel
4m 01ch to 4m 32ch
5m 44ch
Keele Tunnel
5m 59ch
6m 40ch
Stoney Low Tunnel
6m 41ch
Madeley Jn
149m 41ch / 0m 00ch
Madeley Chord Jn. 7m 36ch
0m 25ch (from Madeley Jn)
7m 60ch
End of Line
8m 14ch
Marcroft Engineering
Sideway Jn.
21m 04ch
Wedgwood (CCTV)
23m 76ch
Wedgwood
23m 79ch
Barlaston (CCTV)
24m 50ch
Barlaston
24m 44ch
Stone Jn.
0m 00ch/
27m 00ch
Stone
0m 07ch
Meaford Crossing (CCTV) 27m 18ch
Church Lane (CCTV) 27m 63ch
Aston-by-Stone (CCTV) 28m 63ch
Parrot's (UWC) 3m 39ch
Norton Bridge Jn
138m 79ch
3m 57ch
Norton Bridge
138m 68ch
Norton Bridge North Jn
139m 00ch
Norton Bridge South Jn
138m 40ch
143m 05ch
Stafford No. 4 SB (SD4)
133m 32ch
Arrivals (South End Yard)
to Universal Grinding Wheel Works
Stafford No. 5 SB (SD5)
133m 57ch
Stafford
133m 43ch
Royal Mail Terminal
Salop Sidings
Stafford Trent Valley Jn. No 1.
133m 06ch /
133m 04ch / 28m 50ch
Milford Jn
129m 31ch
Whitehouse Jn
130m 47ch
Rickerscote
27m 55ch
No.2 GF
No.1 GF
Parrott's No. 2 (UWC) 25m 45ch
Parrott's No. 1 (UWC) 25m 27ch
Drayton Manor Farm (UWC) 24m 72ch
Penkridge
23m 32ch
(MCB) 4m 20ch
Calverleigh Farm (UWC) 4m
Blythe Bridge (CCTV)
2
49
3
20m 20ch
Four Ashes Up Goods Loop
19m 71ch
See Map 127
Emergency GF
4m 07ch
Nantwich
4m 14ch
Nantwich SB (NH)
4m 19ch
Newcastle Rd (AHBC-X) 3m 46ch
Nantwich (MCB) 4m 19ch
Shrewbridge Rd (AHBC-X) 4m 32ch
Fields Farm (UWC) 4m 57ch
Green Lane (UWC) 4m 74ch
Reeds Farm (UWC) 6m 51ch
Wrenbury
8m 48ch
Wrenbury SB (WY)
8m 52ch
CREWE
Crewe Steel Works SB (SW)
159m 41ch
158m 73ch
Crewe Coal Yard SB (CY)
158m 68ch
158m 51ch
Crewe Works
to Crewe Electric Maintenance Depot
158m 29ch
158m 29ch
158m 32ch
Engine Siding
Crewe SB (CE)
158m 21ch
Tunnel
158m 14ch
Crewe North Jn.
158m 18ch /
158m 17ch
EMU Stabling Sdg.
Sidings and Yards:
① Basford Hall Dn Sdgs.
② Basford Hall Up Sdgs.
③ Crewe Coal Sdgs.
④ Crewe Diesel Depot
⑤ Down Holding Sdgs.
⑥ Down Siding
⑦ Gresty Road Engs.
⑧ New Middle Sdgs.
⑨ South Yard
⑩ Wagon Shops
⑪ Western Sdgs.
Crewe
158m 00ch
Salop Goods Jn. SB (SG)
0m 37ch / 157m 71ch
Salop Goods Jn.
Om 37ch / 157m 71ch
Crewe South Jn.
157m 60ch /
0m 00ch
8m 27ch /
157m 55ch
GF 0m 60ch
Gresty Lane Sdgs
Willaston (CCTV)
2m 41ch
Carriage Sheds (LNWR)
North Stafford Jn.
7m 52ch
Brook Carriage Sidings
Basford Wood GF
157m 03ch
Gresty Lane SB (GL)
0m 33ch / 0m 00ch /
157m 47ch
① Crewe Sorting Sidings North SB (NH)
157m 23ch
Salop Goods Loop Jn.
157m 25ch /
157m 26ch
157m 12ch
156m 20ch
Basford Hall Jn.
156m 16ch
Basford Hall Jn. SB (BH)
156m 23ch
156m 08ch
E
Allscott GF
163m 70ch
Hereford Storage Sdg.
Wellington
161m 27ch
Donnington Jn
160m 73ch
Telford International Rail Freight Park
Oakengates
158m 31ch
158m 17ch
Oakengates Tunnel
157m 76ch
Telford Central
157m 40ch
Shifnal
154m 24ch
Madeley Jn. SB (MJ)
156m 26ch
Change of mileage
160m 15ch /
162m 25ch
Madeley Jn
156m 19ch
151m 23ch
150m 75ch
Albrighton
149m 38ch
Madeley South Jn.
156m 51ch
Ironbridge Power Station Sidings
Cosford
150m 69ch
38
39

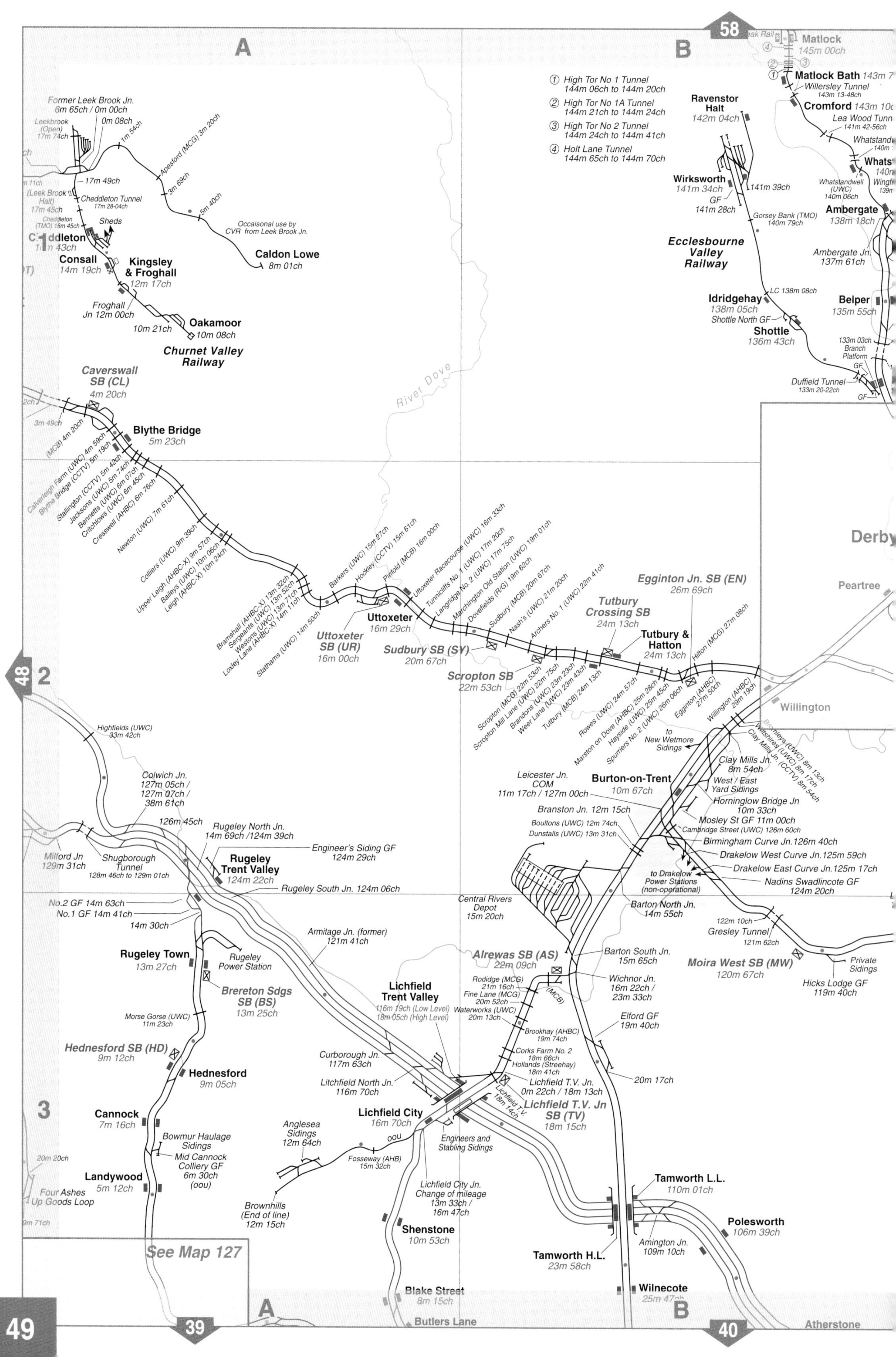

A
B
58
Matlock
145m 00ch
Matlock Bath 143m 7
Willersley Tunnel
143m 13-48ch
Cromford 143m 10
Lea Wood Tunn
141m 42-56ch
Whatstandwell (UWC) 140m 06ch
Ambergate
138m 18ch
Ambergate Jn.
137m 61ch
Belper
135m 55ch
133m 03ch Branch Platform
Duffield Tunnel
133m 20-22ch
① High Tor No 1 Tunnel
144m 06ch to 144m 20ch
② High Tor No 1A Tunnel
144m 21ch to 144m 24ch
③ High Tor No 2 Tunnel
144m 24ch to 144m 41ch
④ Holt Lane Tunnel
144m 65ch to 144m 70ch
Ravenstor Halt
142m 04ch
Wirksworth
141m 34ch
GF
141m 28ch
141m 39ch
Gorsey Bank (TMO)
140m 79ch
Ecclesbourne Valley Railway
Idridgehay
138m 05ch
LC 138m 08ch
Shottle North GF
Shottle
136m 43ch
Former Leek Brook Jn.
6m 65ch / 0m 00ch
Leekbrook (Open)
17m 74ch
0m 08ch
1m 54ch
Apesford (MCG) 3m 20ch
3m 69ch
5m 40ch
17m 49ch
(Leek Brook Halt)
17m 45ch
Cheddleton Tunnel
17m 28-04ch
Sheds
Occaisonal use by CVR from Leek Brook Jn.
Caldon Lowe
8m 01ch
Consall
14m 19ch
Kingsley & Froghall
12m 17ch
Froghall Jn 12m 00ch
10m 21ch
Oakamoor
10m 08ch
Churnet Valley Railway
Caverswall SB (CL)
4m 20ch
Blythe Bridge
5m 23ch
3m 49ch
(MCB) 4m 20ch
River Dove
Calverleigh Farm (UWC) 4m 59ch
Blythe Bridge (CCTV) 5m 19ch
Stallington (CCTV) 5m 42ch
Jacksons (UWC) 5m 74ch
Bennetts (UWC) 6m 07ch
Critchlows (UWC) 6m 45ch
Cresswell (AHBC) 6m 76ch
Newton (UWC) 7m 61ch
Colliers (UWC) 9m 39ch
Upper Leigh (AHBC-X) 9m 57ch
Baileys (UWC) 10m 06ch
Leigh (AHBC-X) 10m 24ch
Bramshall (AHBC-X) 13m 32ch
Sergeants (UWC) 13m 52ch
Westons (UWC) 13m 71ch
Loxley Lane (AHBC-X) 14m 11ch
Stathams (UWC) 14m 50ch
Barkers (UWC) 15m 27ch
Hockley (CCTV) 15m 61ch
Pinfold (MCB) 16m 00ch
Uttoxeter Racecourse (UWC) 16m 33ch
Tunnicliffs No. 1 (UWC) 17m 20ch
Langridge No. 2 (UWC) 17m 75ch
Marchington Old Station (UWC) 19m 01ch
Dovefields (R/G) 19m 62ch
Sudbury (MCB) 20m 67ch
Nash's (UWC) 21m 20ch
Archers No. 1 (UWC) 22m 41ch
Uttoxeter
16m 29ch
Uttoxeter SB (UR)
16m 00ch
Sudbury SB (SY)
20m 67ch
Scropton SB
22m 53ch
Tutbury Crossing SB
24m 13ch
Tutbury & Hatton
24m 13ch
Egginton Jn. SB (EN)
26m 69ch
Hilton (MCG) 27m 08ch
Scropton (MCG) 22m 53ch
Scropton Mill Lane (UWC) 22m 75ch
Brandons (UWC) 23m 23ch
Weer Lane (UWC) 23m 43ch
Tutbury (MCB) 24m 13ch
Rowes (UWC) 24m 57ch
Marston on Dove (AHBC) 25m 28ch
Hayside (UWC) 25m 45ch
Spurriers No. 2 (UWC) 26m 06ch
Egginton (AHBC) 27m 50ch
Willington (AHBC) 29m 19ch
Derby
Peartree
Willington
48
2
Highfields (UWC)
33m 42ch
Colwich Jn.
127m 05ch / 127m 07ch / 38m 61ch
126m 45ch
Rugeley North Jn.
14m 69ch /124m 39ch
Milford Jn
129m 31ch
Shugborough Tunnel
128m 46ch to 129m 01ch
Rugeley Trent Valley
124m 22ch
Engineer's Siding GF
124m 29ch
Rugeley South Jn. 124m 06ch
No.2 GF 14m 63ch
No.1 GF 14m 41ch
14m 30ch
Rugeley Town
13m 27ch
Rugeley Power Station
Brereton Sdgs SB (BS)
13m 25ch
Morse Gorse (UWC)
11m 23ch
Hednesford SB (HD)
9m 12ch
Hednesford
9m 05ch
3
Cannock
7m 16ch
Bowmur Haulage Sidings
Mid Cannock Colliery GF
6m 30ch (oou)
Landywood
5m 12ch
20m 20ch
Four Ashes Up Goods Loop
See Map 127
Armitage Jn. (former)
121m 41ch
Lichfield Trent Valley
116m 19ch (Low Level)
18m 05ch (High Level)
Curborough Jn.
117m 63ch
Litchfield North Jn.
116m 70ch
Lichfield City
16m 70ch
oou
Anglesea Sidings
12m 64ch
Fosseway (AHB)
15m 32ch
Brownhills (End of line)
12m 15ch
Engineers and Stabling Sidings
Lichfield City Jn. Change of mileage
13m 33ch / 16m 47ch
Shenstone
10m 53ch
Blake Street
8m 15ch
Butlers Lane
Central Rivers Depot
15m 20ch
Alrewas SB (AS)
22m 09ch
Rodidge (MCG)
21m 16ch
Fine Lane (MCG)
20m 52ch
Waterworks (UWC)
20m 13ch
(MCB)
Brookhay (AHBC)
19m 74ch
Corks Farm No. 2
18m 66ch
Hollands (Streehay)
18m 41ch
Lichfield T.V. Jn.
0m 22ch / 18m 13ch
Lichfield T.V. 18m 14ch
Lichfield T.V. Jn SB (TV)
18m 15ch
Wichnor Jn.
16m 22ch / 23m 33ch
Elford GF
19m 40ch
20m 17ch
Barton South Jn.
15m 65ch
Barton North Jn.
14m 55ch
Leicester Jn. COM
11m 17ch / 127m 00ch
Burton-on-Trent
10m 67ch
Branston Jn. 12m 15ch
Boultons (UWC) 12m 74ch
Dunstalls (UWC) 13m 31ch
to New Wetmore Sidings
Clay Mills Jn.
8m 54ch
West / East Yard Sidings
Horninglow Bridge Jn
10m 33ch
Mosley St GF 11m 00ch
Cambridge Street (UWC) 126m 60ch
Birmingham Curve Jn.126m 40ch
Drakelow West Curve Jn.125m 59ch
Drakelow East Curve Jn.125m 17ch
Nadins Swadlincote GF
124m 20ch
to Drakelow Power Stations (non-operational)
Bromleys (UWC) 8m 13ch
Wiltshires (UWC) 8m 17ch
Clay Mills Jn. (CCTV) 8m 54ch
122m 10ch
Gresley Tunnel
121m 62ch
Moira West SB (MW)
120m 67ch
Private Sidings
Hicks Lodge GF
119m 40ch
Tamworth L.L.
110m 01ch
Tamworth H.L.
23m 58ch
Amington Jn.
109m 10ch
Polesworth
106m 39ch
Wilnecote
39
40
Atherstone

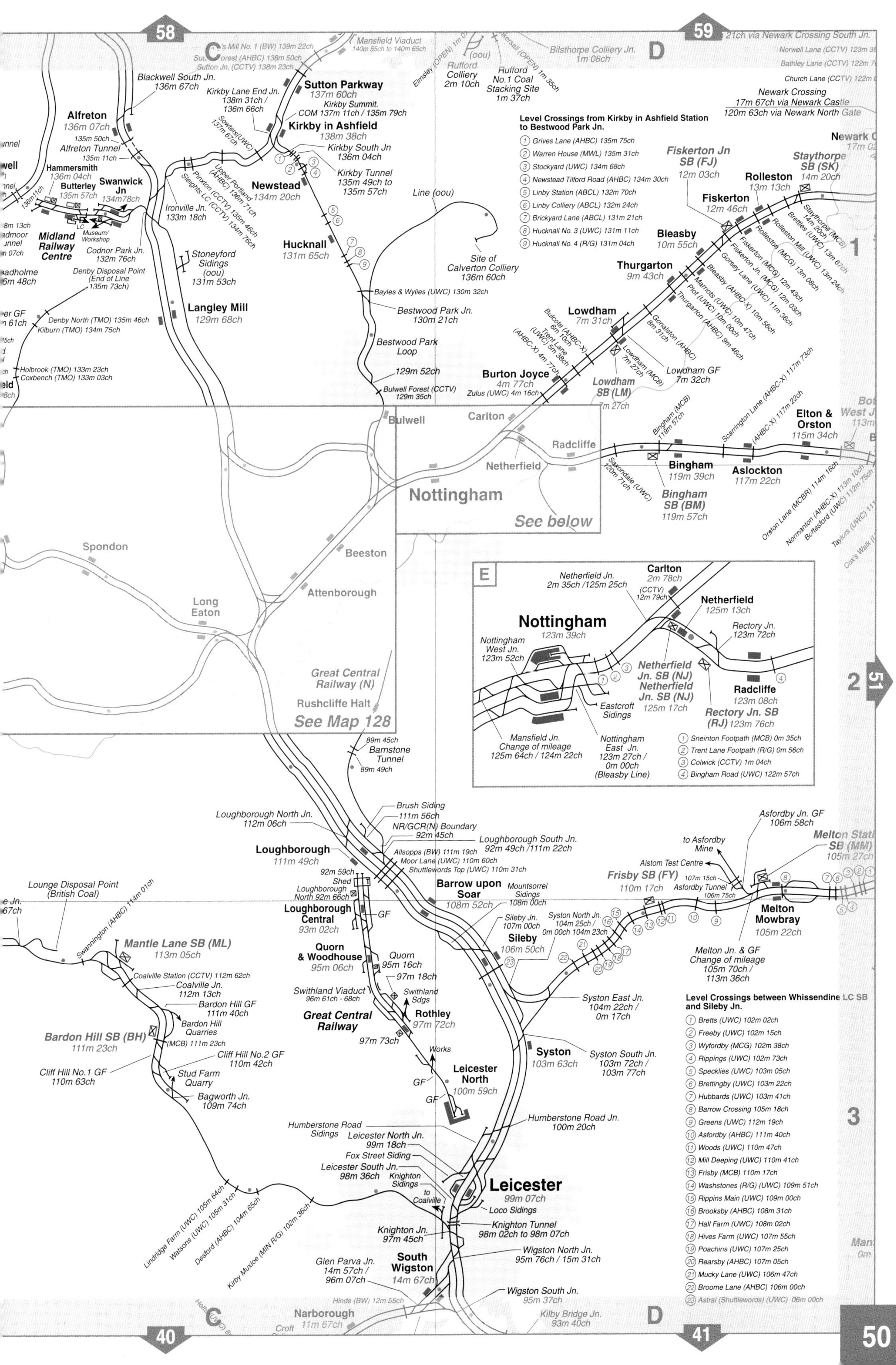

58
59
C
D
40
41
51
1
2
3
Mansfield Viaduct
140m 55ch to 140m 65ch
Sutton Jn. (CCTV) 138m 23ch
Blackwell South Jn.
136m 67ch
Sutton Parkway
137m 60ch
Kirkby Summit.
COM 137m 11ch / 135m 79ch
Kirkby Lane End Jn.
138m 31ch /
136m 66ch
Kirkby in Ashfield
138m 38ch
Kirkby South Jn
136m 04ch
Kirkby Tunnel
135m 49ch to
135m 57ch
Alfreton
136m 07ch
135m 50ch
Alfreton Tunnel
135m 11ch
Hammersmith
136m 04ch
Butterley
135m 57ch
Swanwick Jn
134m78ch
Midland Railway Centre
Museum/ Workshop
Codnor Park Jn.
132m 76ch
Ironville Jn.
133m 18ch
Newstead
134m 20ch
Hucknall
131m 65ch
Stoneyford Sidings (oou)
131m 53ch
Denby Disposal Point (End of Line
135m 73ch)
Denby North (TMO) 135m 46ch
Kilburn (TMO) 134m 75ch
Holbrook (TMO) 133m 23ch
Coxbench (TMO) 133m 03ch
Langley Mill
129m 68ch
Elmsley (OPEN)
Rufford Colliery
2m 10ch
Rufford No.1 Coal Stacking Site
1m 37ch
Bilsthorpe Colliery Jn.
1m 08ch
Line (oou)
Site of Calverton Colliery
136m 60ch
Bayles & Wylies (UWC) 130m 32ch
Bestwood Park Jn.
130m 21ch
Bestwood Park Loop
129m 52ch
Bulwell Forest (CCTV)
129m 35ch
Level Crossings from Kirkby in Ashfield Station to Bestwood Park Jn.
1 Grives Lane (AHBC) 135m 75ch
2 Warren House (MWL) 135m 31ch
3 Stockyard (UWC) 134m 68ch
4 Newstead Tilford Road (AHBC) 134m 30ch
5 Linby Station (ABCL) 132m 70ch
6 Linby Colliery (ABCL) 132m 24ch
7 Brickyard Lane (ABCL) 131m 21ch
8 Hucknall No. 3 (UWC) 131m 11ch
9 Hucknall No. 4 (R/G) 131m 04ch
21ch via Newark Crossing South Jn.
Newark Crossing
17m 67ch via Newark Castle
120m 63ch via Newark North Gate
Fiskerton Jn SB (FJ)
12m 03ch
Staythorpe SB (SK)
14m 20ch
Rolleston
13m 13ch
Fiskerton
12m 46ch
Bleasby
10m 55ch
Thurgarton
9m 43ch
Lowdham
7m 31ch
Burton Joyce
4m 77ch
Zulus (UWC) 4m 16ch
Lowdham SB (LM)
7m 27ch
Lowdham GF
7m 32ch
Carlton
Bulwell
Radcliffe
Netherfield
Nottingham
See below
Bingham
119m 39ch
Bingham SB (BM)
119m 57ch
Aslockton
117m 22ch
Elton & Orston
115m 34ch
Spondon
Beeston
Attenborough
Long Eaton
Great Central Railway (N)
Rushcliffe Halt
See Map 128
E
Netherfield Jn.
2m 35ch /125m 25ch
Carlton
2m 78ch
(CCTV)
12m 79ch
Netherfield
125m 13ch
Nottingham
123m 39ch
Nottingham West Jn.
123m 52ch
Rectory Jn.
123m 72ch
Netherfield Jn. SB (NJ)
125m 17ch
Radcliffe
123m 08ch
Rectory Jn. SB (RJ)
123m 76ch
Eastcroft Sidings
Mansfield Jn.
Change of mileage
125m 64ch / 124m 22ch
Nottingham East Jn.
123m 27ch /
0m 00ch
(Bleasby Line)
1 Sneinton Footpath (MCB) 0m 35ch
2 Trent Lane Footpath (R/G) 0m 56ch
3 Colwick (CCTV) 1m 04ch
4 Bingham Road (UWC) 122m 57ch
89m 45ch
Barnstone Tunnel
89m 49ch
Brush Siding
111m 56ch
NR/GCR(N) Boundary
92m 45ch
Loughborough North Jn.
112m 06ch
Loughborough South Jn.
92m 49ch /111m 22ch
Loughborough
111m 49ch
Allsopps (BW) 111m 19ch
Moor Lane (UWC) 110m 60ch
Shuttlewords Top (UWC) 110m 31ch
92m 59ch
Shed
Loughborough North 92m 66ch
Loughborough Central
93m 02ch
GF
Barrow upon Soar
108m 52ch
Mountsorrel Sidings
108m 00ch
Sileby Jn.
107m 00ch
Syston North Jn.
104m 25ch /
0m 00ch 104m 23ch
Sileby
106m 50ch
Quorn & Woodhouse
95m 06ch
Quorn
95m 16ch
97m 18ch
Swithland Viaduct
96m 61ch - 68ch
Swithland Sdgs
Great Central Railway
Rothley
97m 72ch
97m 73ch
Works
Leicester North
100m 59ch
GF
Lounge Disposal Point (British Coal)
Swannington (AHBC) 114m 01ch
Mantle Lane SB (ML)
113m 05ch
Coalville Station (CCTV) 112m 62ch
Coalville Jn.
112m 13ch
Bardon Hill GF
111m 40ch
Bardon Hill Quarries
Bardon Hill SB (BH)
111m 23ch
(MCB) 111m 23ch
Cliff Hill No.2 GF
110m 42ch
Cliff Hill No.1 GF
110m 63ch
Stud Farm Quarry
Bagworth Jn.
109m 74ch
Asfordby Jn. GF
106m 58ch
to Asfordby Mine
Alstom Test Centre
Frisby SB (FY)
110m 17ch
107m 15ch
Asfordby Tunnel
106m 75ch
Melton Station SB (MM)
105m 27ch
Melton Mowbray
105m 22ch
Melton Jn. & GF
Change of mileage
105m 70ch /
113m 36ch
Syston East Jn.
104m 22ch /
0m 17ch
Syston
103m 63ch
Syston South Jn.
103m 72ch /
103m 77ch
Humberstone Road Jn.
100m 20ch
Humberstone Road Sidings
Leicester North Jn.
99m 18ch
Fox Street Siding
Leicester South Jn.
98m 36ch
Knighton Sidings
to Coalville
Leicester
99m 07ch
Loco Sidings
Knighton Tunnel
98m 02ch to 98m 07ch
Knighton Jn.
97m 45ch
South Wigston
14m 67ch
Wigston North Jn.
95m 76ch / 15m 31ch
Glen Parva Jn.
14m 57ch /
96m 07ch
Wigston South Jn.
95m 37ch
Hinds (BW) 12m 55ch
Narborough
11m 67ch
Kilby Bridge Jn.
93m 40ch
Lindridge Farm (UWC) 105m 64ch
Watsons (UWC) 105m 31ch
Desford (AHBC) 104m 65ch
Kirby Muxloe (MIN R/G) 102m 36ch
Level Crossings between Whissendine LC SB and Sileby Jn.
1 Bretts (UWC) 102m 02ch
2 Freeby (UWC) 102m 15ch
3 Wyfordby (MCG) 102m 38ch
4 Rippings (UWC) 102m 73ch
5 Specklies (UWC) 103m 05ch
6 Brettingby (UWC) 103m 22ch
7 Hubbards (UWC) 103m 41ch
8 Barrow Crossing 105m 18ch
9 Greens (UWC) 112m 19ch
10 Asfordby (AHBC) 111m 40ch
11 Woods (UWC) 110m 47ch
12 Mill Deeping (UWC) 110m 41ch
13 Frisby (MCB) 110m 17ch
14 Washstones (R/G) (UWC) 109m 51ch
15 Rippins Main (UWC) 109m 00ch
16 Brooksby (AHBC) 108m 31ch
17 Hall Farm (UWC) 108m 02ch
18 Hives Farm (UWC) 107m 55ch
19 Poachins (UWC) 107m 25ch
20 Rearsby (AHBC) 107m 05ch
21 Mucky Lane (UWC) 106m 47ch
22 Broome Lane (AHBC) 106m 00ch
23 Astral (Shuttlewords) (UWC) 06m 00ch

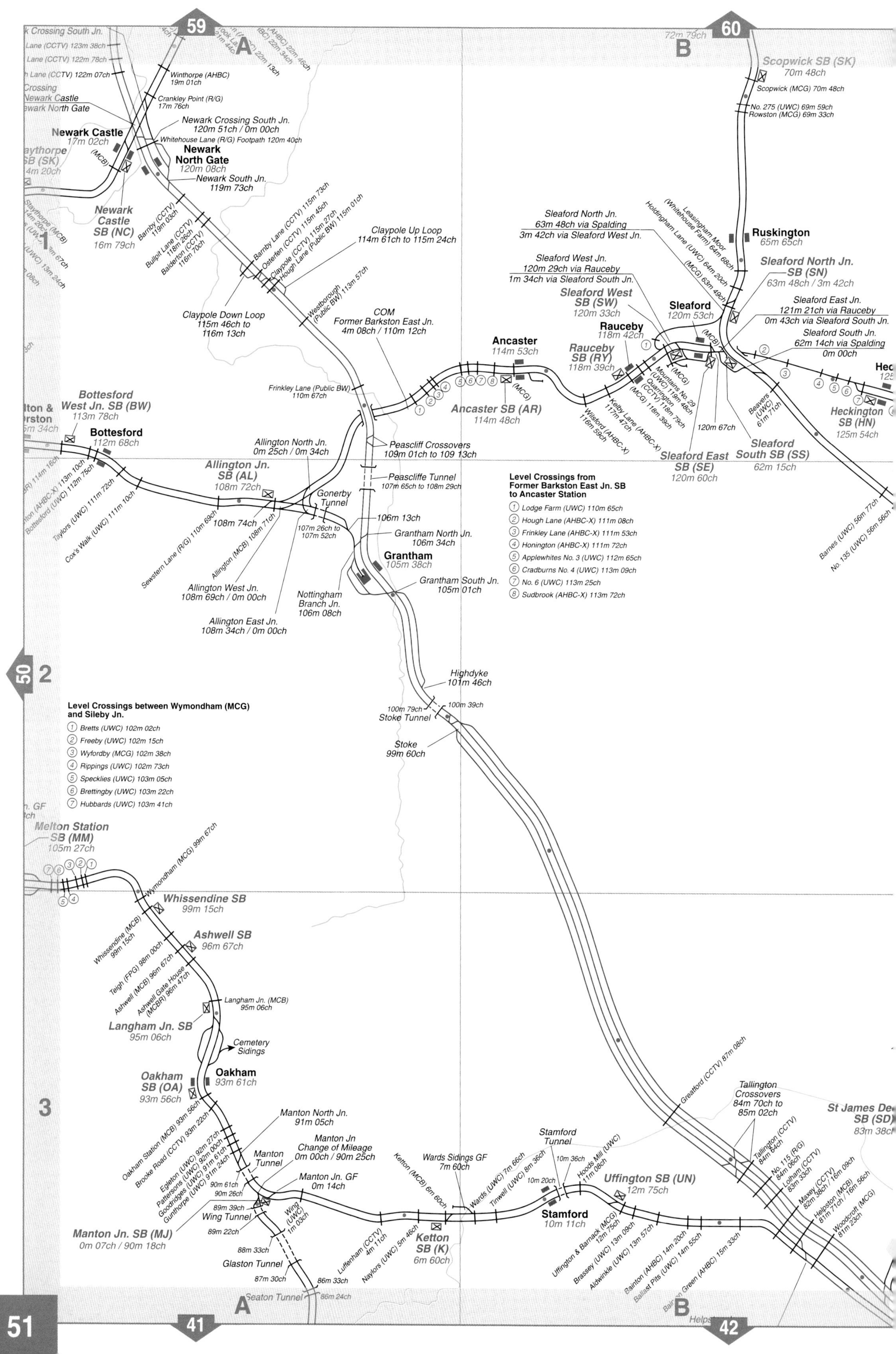

59
60
A
B
Scopwick SB (SK)
70m 48ch
Scopwick (MCG) 70m 48ch
No. 275 (UWC) 69m 59ch
Rowston (MCG) 69m 33ch
Winthorpe (AHBC)
19m 01ch
Crankley Point (R/G)
17m 76ch
Newark Crossing South Jn.
120m 51ch / 0m 00ch
Whitehouse Lane (R/G) Footpath 120m 40ch
Newark Castle
17m 02ch
Newark
North Gate
120m 08ch
Newark South Jn.
119m 73ch
Newark
Castle
SB (NC)
16m 79ch
Barnby (CCTV)
119m 03ch
Bullpit Lane (CCTV)
118m 26ch
Balderton (CCTV)
116m 70ch
Barnby Lane (CCTV) 115m 73ch
Claypole Up Loop
114m 61ch to 115m 24ch
Claypole Down Loop
115m 46ch to
116m 13ch
Westborough
(Public BW) 113m 57ch
COM
Former Barkston East Jn.
4m 08ch / 110m 12ch
Ruskington
65m 65ch
Sleaford North Jn.
63m 48ch via Spalding
3m 42ch via Sleaford West Jn.
Sleaford North Jn.
SB (SN)
63m 48ch / 3m 42ch
Sleaford West Jn.
120m 29ch via Rauceby
1m 34ch via Sleaford South Jn.
Sleaford West
SB (SW)
120m 33ch
Sleaford
120m 53ch
Sleaford East Jn.
121m 21ch via Rauceby
0m 43ch via Sleaford South Jn.
Sleaford South Jn.
62m 14ch via Spalding
0m 00ch
Rauceby
118m 42ch
Rauceby
SB (RY)
118m 39ch
Ancaster
114m 53ch
Ancaster SB (AR)
114m 48ch
Frinkley Lane (Public BW)
110m 67ch
Bottesford
West Jn. SB (BW)
113m 78ch
Bottesford
112m 68ch
Allington North Jn.
0m 25ch / 0m 34ch
Peascliff Crossovers
109m 01ch to 109 13ch
Peascliffe Tunnel
107m 65ch to 108m 29ch
Allington Jn.
SB (AL)
108m 72ch
Gonerby
Tunnel
108m 74ch
107m 26ch to
107m 52ch
106m 13ch
Grantham North Jn.
106m 34ch
Grantham
105m 38ch
Grantham South Jn.
105m 01ch
Nottingham
Branch Jn.
106m 08ch
Allington West Jn.
108m 69ch / 0m 00ch
Allington East Jn.
108m 34ch / 0m 00ch
Taylors (UWC) 111m 72ch
Cox's Walk (UWC) 111m 10ch
Sewstern Lane (R/G) 110m 69ch
Allington (MCB) 108m 71ch
Wilsford (AHBC-X)
116m 59ch
Kelby Lane (AHBC-X)
117m 47ch
120m 67ch
Sleaford East
SB (SE)
120m 60ch
Sleaford
South SB (SS)
62m 15ch
Heckington
SB (HN)
125m 54ch
Barnes (UWC) 56m 77ch
No. 135 (UWC) 56m 56ch
Level Crossings from
Former Barkston East Jn. SB
to Ancaster Station
1 Lodge Farm (UWC) 110m 65ch
2 Hough Lane (AHBC-X) 111m 08ch
3 Frinkley Lane (AHBC-X) 111m 53ch
4 Honington (AHBC-X) 111m 72ch
5 Applewhites No. 3 (UWC) 112m 65ch
6 Cradburns No. 4 (UWC) 113m 09ch
7 No. 6 (UWC) 113m 25ch
8 Sudbrook (AHBC-X) 113m 72ch
50
2
Highdyke
101m 46ch
100m 79ch
Stoke Tunnel
100m 39ch
Stoke
99m 60ch
Level Crossings between Wymondham (MCG)
and Sileby Jn.
1 Bretts (UWC) 102m 02ch
2 Freeby (UWC) 102m 15ch
3 Wyfordby (MCG) 102m 38ch
4 Rippings (UWC) 102m 73ch
5 Specklies (UWC) 103m 05ch
6 Brettingby (UWC) 103m 22ch
7 Hubbards (UWC) 103m 41ch
Melton Station
SB (MM)
105m 27ch
Wymondham (MCG) 99m 67ch
Whissendine SB
99m 15ch
Whissendine (MCB)
99m 15ch
Ashwell SB
96m 67ch
Teigh (FPG) 98m 00ch
Ashwell (MCB) 96m 67ch
Ashwell Gate House
(MCBR) 96m 47ch
Langham Jn. (MCB)
95m 06ch
Langham Jn. SB
95m 06ch
Cemetery
Sidings
Oakham
SB (OA)
93m 56ch
Oakham
93m 61ch
3
Oakham Station (MCB) 93m 56ch
Brooke Road (CCTV) 93m 22ch
Manton North Jn.
91m 05ch
Manton Jn
Change of Mileage
0m 00ch / 90m 25ch
Manton
Tunnel
Manton Jn. GF
0m 14ch
90m 61ch
90m 26ch
89m 39ch
Wing Tunnel
89m 22ch
Wing
(UWC)
1m 03ch
Manton Jn. SB (MJ)
0m 07ch / 90m 18ch
88m 33ch
Glaston Tunnel
87m 30ch
86m 33ch
Seaton Tunnel
86m 24ch
Ketton (MCB) 6m 60ch
Wards Sidings GF
7m 60ch
Ketton
SB (K)
6m 60ch
Luffenham (CCTV)
4m 11ch
Naylors (UWC) 5m 46ch
Wards (UWC) 7m 66ch
Tinwell (UWC) 8m 36ch
Stamford
Tunnel
10m 36ch
10m 20ch
Hoods Mill (UWC)
11m 08ch
Stamford
10m 11ch
Uffington SB (UN)
12m 75ch
Uffington & Barnack (MCG)
12m 75ch
Brassey (UWC) 13m 09ch
Aldwinkle (UWC) 13m 57ch
Bainton (AHBC) 14m 20ch
Ballast Pits (UWC) 14m 55ch
Bainton Green (AHBC) 15m 33ch
Greatford (CCTV) 87m 08ch
Tallington
Crossovers
84m 70ch to
85m 02ch
Tallington (CCTV)
84m 64ch
No. 115 (R/G)
84m 06ch
Lolham (CCTV)
83m 33ch
Maxey (CCTV)
82m 38ch / 16m 09ch
Helpston (MCB)
81m 71ch / 16m 56ch
Woodcroft (MCG)
81m 23ch
St James Deeping
SB (SD)
83m 38ch
A
B
41
42

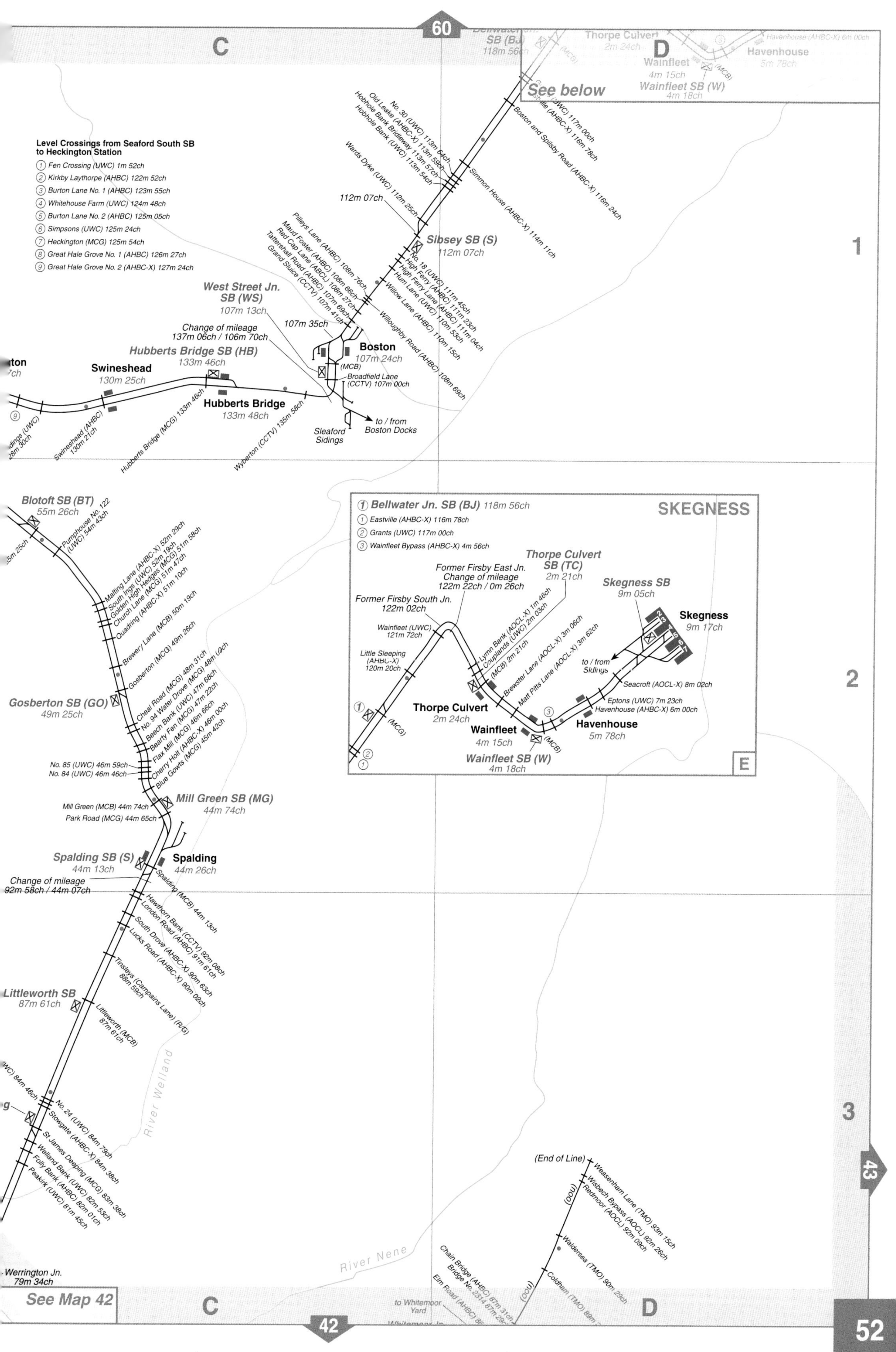

Level Crossings from Seaford South SB to Heckington Station
1 Fen Crossing (UWC) 1m 52ch
2 Kirkby Laythorpe (AHBC) 122m 52ch
3 Burton Lane No. 1 (AHBC) 123m 55ch
4 Whitehouse Farm (UWC) 124m 48ch
5 Burton Lane No. 2 (AHBC) 125m 05ch
6 Simpsons (UWC) 125m 24ch
7 Heckington (MCG) 125m 54ch
8 Great Hale Grove No. 1 (AHBC) 126m 27ch
9 Great Hale Grove No. 2 (AHBC-X) 127m 24ch
See below
Sibsey SB (S)
112m 07ch
West Street Jn. SB (WS)
107m 13ch
Change of mileage 137m 06ch / 106m 70ch
Hubberts Bridge SB (HB)
133m 46ch
Swineshead
130m 25ch
Hubberts Bridge
133m 48ch
Boston
107m 24ch
Broadfield Lane (CCTV) 107m 00ch
Sleaford Sidings
to / from Boston Docks
Blotoft SB (BT)
55m 26ch
Gosberton SB (GO)
49m 25ch
Mill Green SB (MG)
44m 74ch
Spalding SB (S)
44m 13ch
Spalding
44m 26ch
Change of mileage 92m 58ch / 44m 07ch
Littleworth SB
87m 61ch
Werrington Jn.
79m 34ch
River Welland
River Nene
SKEGNESS
1 Bellwater Jn. SB (BJ) 118m 56ch
1 Eastville (AHBC-X) 116m 78ch
2 Grants (UWC) 117m 00ch
3 Wainfleet Bypass (AHBC-X) 4m 56ch
Thorpe Culvert SB (TC)
2m 21ch
Former Firsby East Jn. Change of mileage 122m 22ch / 0m 26ch
Former Firsby South Jn. 122m 02ch
Skegness SB
9m 05ch
Skegness
9m 17ch
Thorpe Culvert
2m 24ch
Wainfleet
4m 15ch
Wainfleet SB (W)
4m 18ch
Havenhouse
5m 78ch
Seacroft (AOCL-X) 8m 02ch
Eptons (UWC) 7m 23ch
Havenhouse (AHBC-X) 6m 00ch
E
(End of Line)
See Map 42
C
D
1
2
3
60
42
43

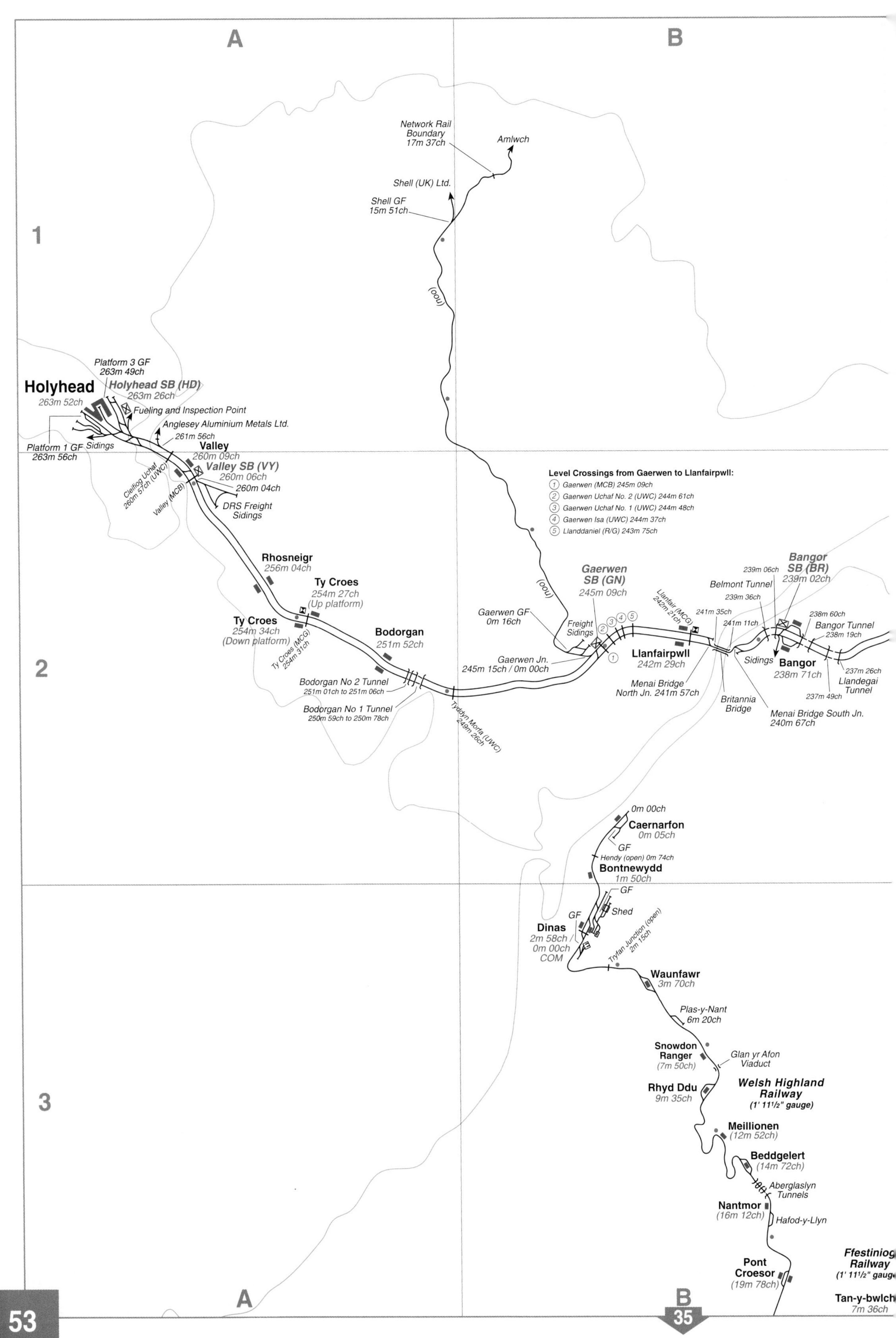
A
B
1
2
3
Network Rail Boundary 17m 37ch
Amlwch
Shell (UK) Ltd.
Shell GF 15m 51ch
(noo)
Platform 3 GF 263m 49ch
Holyhead
263m 52ch
Holyhead SB (HD)
263m 26ch
Fueling and Inspection Point
Anglesey Aluminium Metals Ltd.
261m 56ch
Platform 1 GF 263m 56ch
Sidings
Valley
260m 09ch
Valley SB (VY)
260m 06ch
260m 04ch
Cleifiog Uchaf 260m 57ch (UWC)
Valley (MCB)
DRS Freight Sidings
Level Crossings from Gaerwen to Llanfairpwll:
1 Gaerwen (MCB) 245m 09ch
2 Gaerwen Uchaf No. 2 (UWC) 244m 61ch
3 Gaerwen Uchaf No. 1 (UWC) 244m 48ch
4 Gaerwen Isa (UWC) 244m 37ch
5 Llanddaniel (R/G) 243m 75ch
Rhosneigr
256m 04ch
Ty Croes
254m 27ch
(Up platform)
Ty Croes
254m 34ch
(Down platform)
Ty Croes (MCG) 254m 31ch
Bodorgan
251m 52ch
Bodorgan No 2 Tunnel
251m 01ch to 251m 06ch
Bodorgan No 1 Tunnel
250m 59ch to 250m 78ch
Tydyn Morfa (UWC) 249m 26ch
Gaerwen SB (GN)
245m 09ch
Gaerwen GF 0m 16ch
Freight Sidings
Gaerwen Jn. 245m 15ch / 0m 00ch
Llanfair (MCG) 242m 21ch
241m 35ch
241m 11ch
Llanfairpwll
242m 29ch
Menai Bridge North Jn. 241m 57ch
Britannia Bridge
Menai Bridge South Jn. 240m 67ch
239m 06ch
Belmont Tunnel
239m 36ch
Bangor SB (BR)
239m 02ch
238m 60ch
Bangor Tunnel
238m 19ch
Sidings
Bangor
238m 71ch
237m 26ch
Llandegai Tunnel
237m 49ch
0m 00ch
Caernarfon
0m 05ch
GF
Hendy (open) 0m 74ch
Bontnewydd
1m 50ch
GF
Shed
GF
Dinas
2m 58ch /
0m 00ch
COM
Tryfan Junction (open) 2m 15ch
Waunfawr
3m 70ch
Plas-y-Nant
6m 20ch
Snowdon Ranger
(7m 50ch)
Glan yr Afon Viaduct
Welsh Highland Railway
(1' 11½" gauge)
Rhyd Ddu
9m 35ch
Meillionen
(12m 52ch)
Beddgelert
(14m 72ch)
Aberglaslyn Tunnels
Nantmor
(16m 12ch)
Hafod-y-Llyn
Pont Croesor
(19m 78ch)
Ffestiniog Railway
(1' 11½" gauge)
Tan-y-bwlch
7m 36ch
35

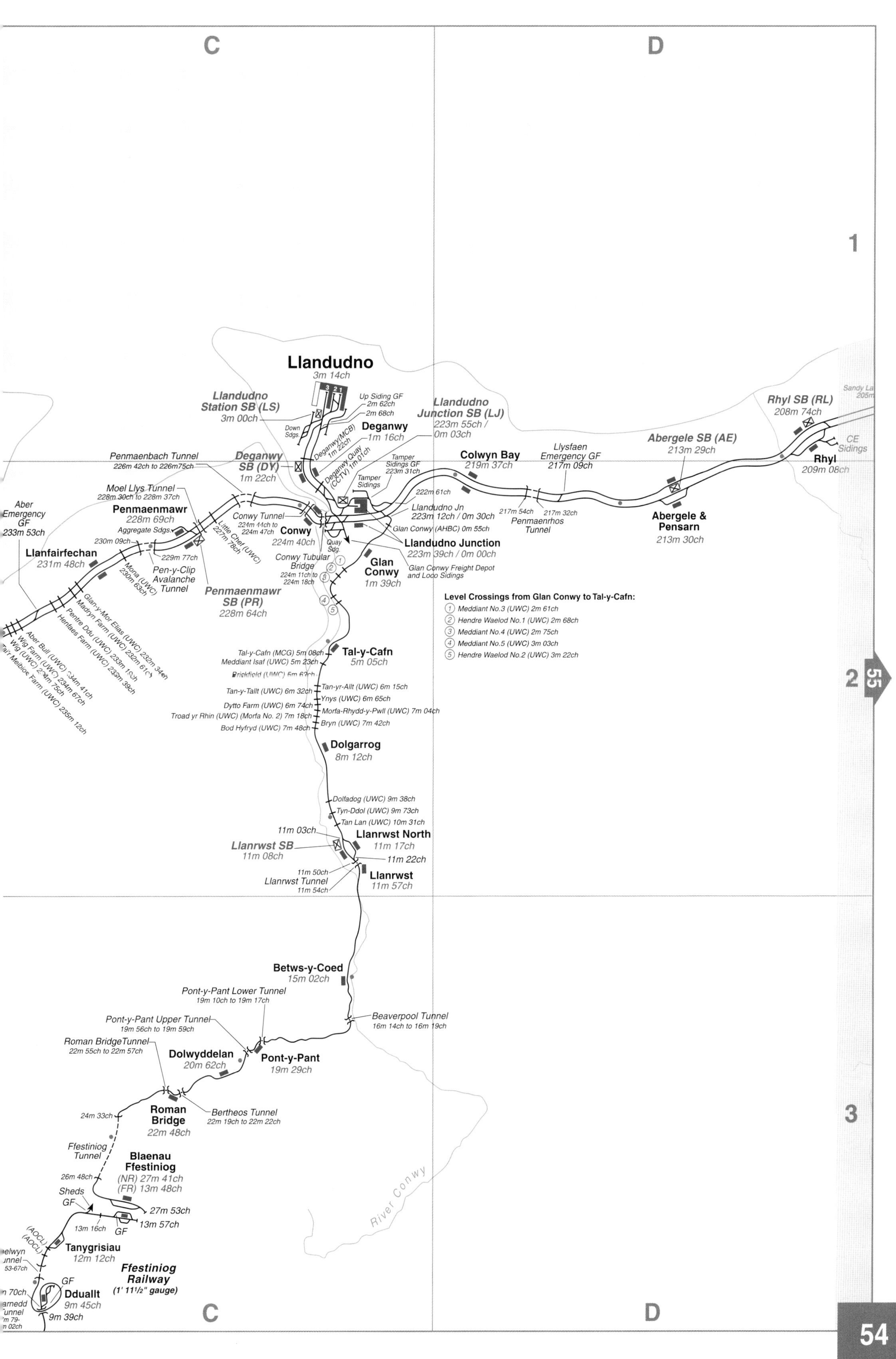
C
D
1
2
3
55
Llandudno
3m 14ch
Llandudno Station SB (LS)
3m 00ch
Up Siding GF
2m 62ch
2m 68ch
Down Sdgs.
Deganwy
1m 16ch
Deganwy(MCB) 1m 22ch
Deganwy Quay (CCTV) 1m 01ch
Llandudno Junction SB (LJ)
223m 55ch / 0m 03ch
Tamper Sidings GF 223m 31ch
Tamper Sidings
Deganwy SB (DY)
1m 22ch
Penmaenbach Tunnel
226m 42ch to 226m75ch
Moel Llys Tunnel
228m 30ch to 228m 37ch
Aber Emergency GF
233m 53ch
Penmaenmawr
228m 69ch
Aggregate Sdgs.
Conwy Tunnel
224m 44ch to 224m 47ch
Conwy
224m 40ch
Little Chef (UWC) 227m 78ch
230m 09ch
229m 77ch
Llanfairfechan
231m 48ch
Pen-y-Clip Avalanche Tunnel
Mona (UWC) 230m 63ch
Penmaenmawr SB (PR)
228m 64ch
Quay Sdg.
Conwy Tubular Bridge
224m 11ch to 224m 18ch
Glan-y-Mor Elias (UWC) 232m 34ch
Madryn Farm (UWC) 232m 61ch
Pentre Du (UWC) 233m 10ch
Henfaes Farm (UWC) 233m 39ch
Aber Bull (UWC) 234m 41ch
Wig Farm (UWC) 234m 67ch
Wig (UWC) 234m 75ch
Tai'r Meibion Farm (UWC) 235m 12ch
Colwyn Bay
219m 37ch
222m 61ch
Llandudno Jn
223m 12ch / 0m 30ch
Glan Conwy (AHBC) 0m 55ch
Llandudno Junction
223m 39ch / 0m 00ch
Glan Conwy
1m 39ch
Glan Conwy Freight Depot and Loop Sidings
Llysfaen Emergency GF
217m 09ch
217m 54ch
217m 32ch
Penmaenrhos Tunnel
Abergele SB (AE)
213m 29ch
Abergele & Pensarn
213m 30ch
Rhyl SB (RL)
208m 74ch
CE Sidings
Rhyl
209m 08ch
Level Crossings from Glan Conwy to Tal-y-Cafn:
1 Meddiant No.3 (UWC) 2m 61ch
2 Hendre Waelod No.1 (UWC) 2m 68ch
3 Meddiant No.4 (UWC) 2m 75ch
4 Meddiant No.5 (UWC) 3m 03ch
5 Hendre Waelod No.2 (UWC) 3m 22ch
Tal-y-Cafn (MCG) 5m 08ch
Meddiant Isaf (UWC) 5m 23ch
Tal-y-Cafn
5m 05ch
Tan-y-Tallt (UWC) 6m 32ch
Tan-yr-Allt (UWC) 6m 15ch
Ynys (UWC) 6m 65ch
Dytto Farm (UWC) 6m 74ch
Morfa-Rhydd-y-Pwll (UWC) 7m 04ch
Troad yr Rhin (UWC) (Morfa No. 2) 7m 18ch
Bryn (UWC) 7m 42ch
Bod Hyfryd (UWC) 7m 48ch
Dolgarrog
8m 12ch
Dolfadog (UWC) 9m 38ch
Tyn-Ddol (UWC) 9m 73ch
Tan Lan (UWC) 10m 31ch
11m 03ch
Llanrwst North
11m 17ch
Llanrwst SB
11m 08ch
11m 22ch
11m 50ch
Llanrwst Tunnel
11m 54ch
Llanrwst
11m 57ch
Betws-y-Coed
15m 02ch
Pont-y-Pant Lower Tunnel
19m 10ch to 19m 17ch
Pont-y-Pant Upper Tunnel
19m 56ch to 19m 59ch
Beaverpool Tunnel
16m 14ch to 16m 19ch
Roman BridgeTunnel
22m 55ch to 22m 57ch
Dolwyddelan
20m 62ch
Pont-y-Pant
19m 29ch
Roman Bridge
22m 48ch
Bertheos Tunnel
22m 19ch to 22m 22ch
24m 33ch
Ffestiniog Tunnel
26m 48ch
Blaenau Ffestiniog
(NR) 27m 41ch
(FR) 13m 48ch
Sheds GF
27m 53ch
13m 16ch
GF
13m 57ch
River Conwy
(AOCL)
(AOCL)
Tanygrisiau
12m 12ch
Ffestiniog Railway
(1' 11½" gauge)
GF
Dduallt
9m 45ch
9m 39ch

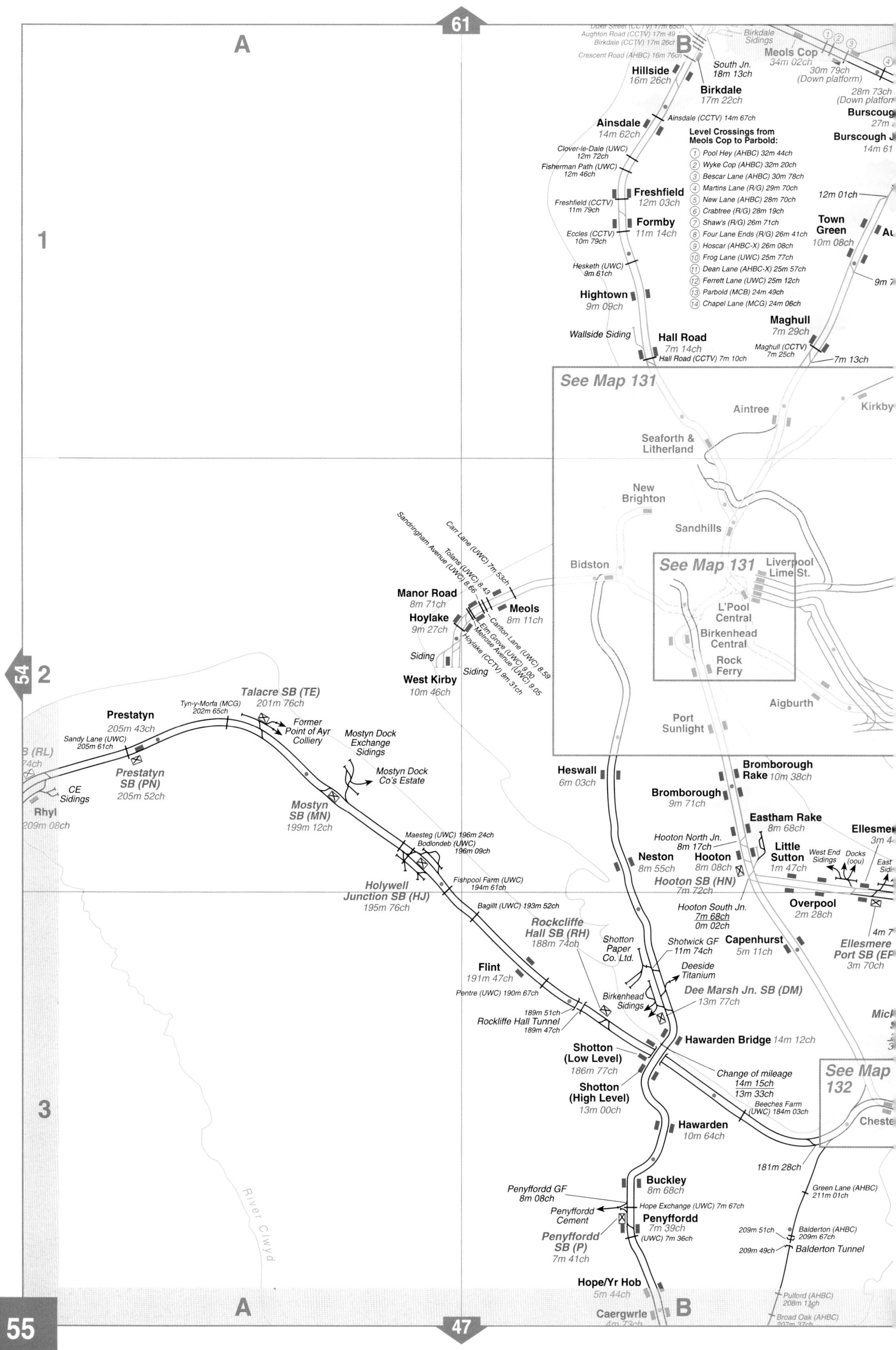

61
A
B
Aughton Road (CCTV) 17m 49
Birkdale (CCTV) 17m 26ch
Crescent Road (AHBC) 16m 76ch
Birkdale Sidings
Meols Cop
34m 02ch
South Jn.
18m 13ch
30m 79ch
(Down platform)
Hillside
16m 26ch
Birkdale
17m 22ch
28m 73ch
Ainsdale
14m 62ch
Ainsdale (CCTV) 14m 67ch
Level Crossings from Meols Cop to Parbold:
1 Pool Hey (AHBC) 32m 44ch
2 Wyke Cop (AHBC) 32m 20ch
3 Bescar Lane (AHBC) 30m 78ch
4 Martins Lane (R/G) 29m 70ch
5 New Lane (AHBC) 28m 70ch
6 Crabtree (R/G) 28m 19ch
7 Shaw's (R/G) 26m 71ch
8 Four Lane Ends (R/G) 26m 41ch
9 Hoscar (AHBC-X) 26m 08ch
10 Frog Lane (UWC) 25m 77ch
11 Dean Lane (AHBC-X) 25m 57ch
12 Ferrett Lane (UWC) 25m 12ch
13 Parbold (MCB) 24m 49ch
14 Chapel Lane (MCG) 24m 06ch
Clover-le-Dale (UWC)
12m 72ch
Fisherman Path (UWC)
12m 46ch
Freshfield
12m 03ch
Freshfield (CCTV)
11m 79ch
Formby
11m 14ch
Eccles (CCTV)
10m 79ch
12m 01ch
Town Green
10m 08ch
1
Hesketh (UWC)
9m 61ch
Hightown
9m 09ch
Maghull
7m 29ch
Wallside Siding
Hall Road
7m 14ch
Hall Road (CCTV) 7m 10ch
Maghull (CCTV)
7m 25ch
7m 13ch
See Map 131
Aintree
Seaforth & Litherland
New Brighton
Sandhills
Bidston
See Map 131
Liverpool Lime St.
L'Pool Central
Birkenhead Central
Rock Ferry
Aigburth
Port Sunlight
Sandringham Avenue (UWC) 8.66
Carr Lane (UWC) 7m 53ch
Tolans (UWC) 8.43
Manor Road
8m 71ch
Meols
8m 11ch
Hoylake
9m 27ch
Carlton Lane (UWC) 8.59
Elm Grove (UWC) 9.00
Melrose Avenue (UWC) 9.05
Hoylake (CCTV) 9m 31ch
Siding
Siding
West Kirby
10m 46ch
54
2
Prestatyn
205m 43ch
Tyn-y-Morfa (MCG)
202m 65ch
Talacre SB (TE)
201m 76ch
Former Point of Ayr Colliery
Mostyn Dock Exchange Sidings
Mostyn Dock Co's Estate
Sandy Lane (UWC)
205m 61ch
Prestatyn SB (PN)
205m 52ch
CE Sidings
Rhyl
209m 08ch
Mostyn SB (MN)
199m 12ch
Heswall
6m 03ch
Bromborough Rake 10m 38ch
Bromborough
9m 71ch
Eastham Rake
8m 68ch
Maesteg (UWC) 196m 24ch
Bodlondeb (UWC)
196m 09ch
Hooton North Jn.
8m 17ch
Neston
8m 55ch
Hooton
8m 08ch
Little Sutton
1m 47ch
West End Sidings
Docks (oou)
Fishpool Farm (UWC)
194m 61ch
Holywell Junction SB (HJ)
195m 76ch
Hooton SB (HN)
7m 72ch
Hooton South Jn.
7m 68ch
0m 02ch
Overpool
2m 28ch
Bagillt (UWC) 193m 52ch
Rockcliffe Hall SB (RH)
188m 74ch
Shotton Paper Co. Ltd.
Shotwick GF
11m 74ch
Capenhurst
5m 11ch
Ellesmere Port SB (EP)
3m 70ch
Flint
191m 47ch
Deeside Titanium
Pentre (UWC) 190m 67ch
Dee Marsh Jn. SB (DM)
13m 77ch
Birkenhead Sidings
189m 51ch
Rockliffe Hall Tunnel
189m 47ch
Hawarden Bridge 14m 12ch
Shotton (Low Level)
186m 77ch
Change of mileage
14m 15ch
13m 33ch
See Map 132
Shotton (High Level)
13m 00ch
Beeches Farm (UWC) 184m 03ch
3
Hawarden
10m 64ch
181m 28ch
Buckley
8m 68ch
Green Lane (AHBC)
211m 01ch
Penyffordd GF
8m 08ch
Hope Exchange (UWC) 7m 67ch
Penyffordd Cement
Penyffordd
7m 39ch
(UWC) 7m 36ch
209m 51ch
Balderton (AHBC)
209m 67ch
Penyffordd SB (P)
7m 41ch
209m 49ch
Balderton Tunnel
River Clwyd
Hope/Yr Hob
5m 44ch
Pulford (AHBC)
208m 11ch
A
B
Caergwrle
Broad Oak (AHBC)
47

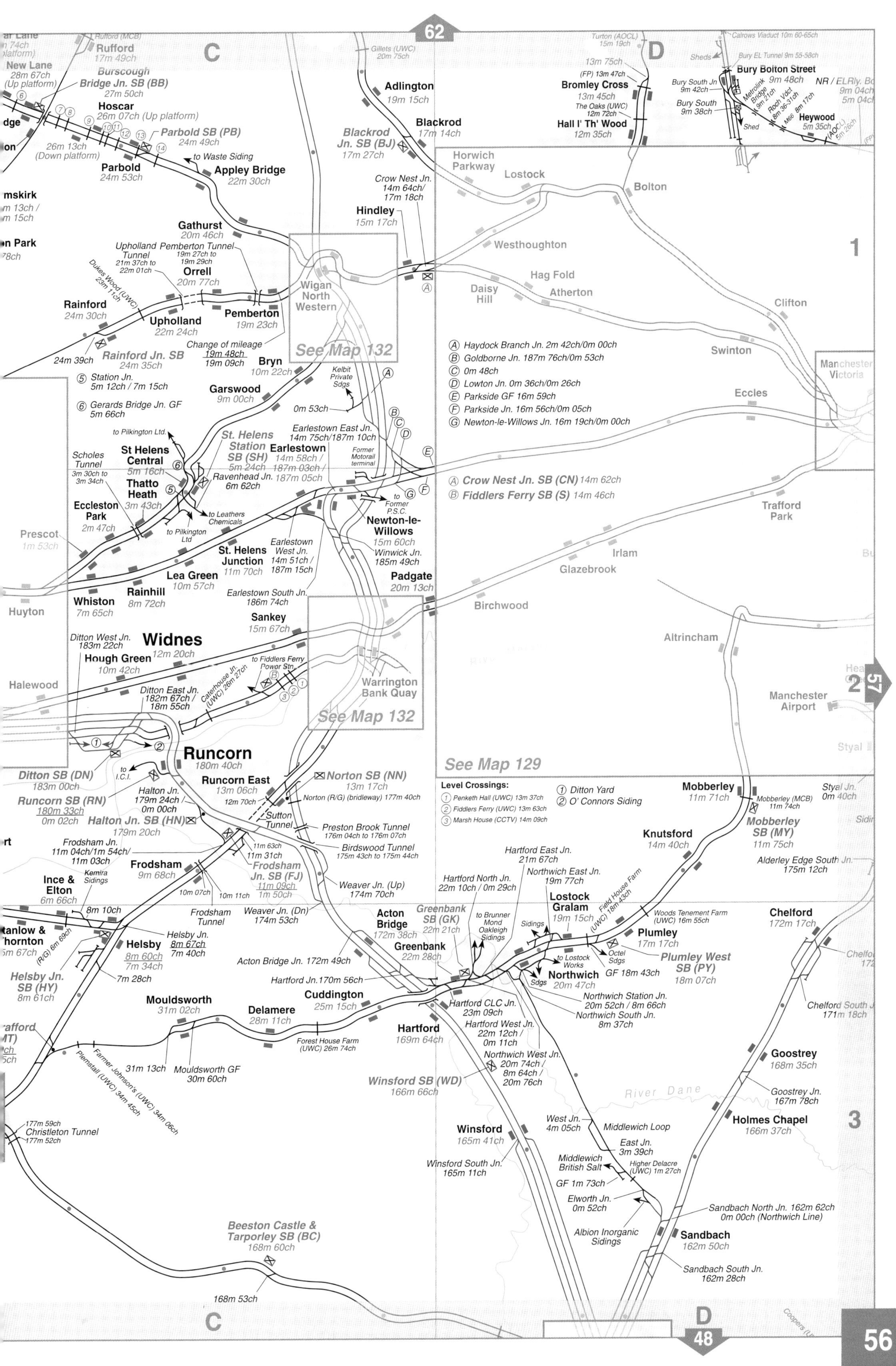
62
C
D
Rufford (MCB)
Rufford
17m 49ch
Burscough
Bridge Jn. SB (BB)
27m 50ch
New Lane
28m 67ch
(Up platform)
Hoscar
26m 07ch (Up platform)
26m 13ch
(Down platform)
Parbold SB (PB)
24m 49ch
Parbold
24m 53ch
to Waste Siding
Appley Bridge
22m 30ch
Gathurst
20m 46ch
Upholland Tunnel
21m 37ch to
22m 01ch
Pemberton Tunnel
19m 27ch to
19m 29ch
Orrell
20m 77ch
Dukes Wood (UWC)
23m 11ch
Rainford
24m 30ch
Upholland
22m 24ch
Pemberton
19m 23ch
Change of mileage
19m 48ch
19m 09ch
24m 39ch
Rainford Jn. SB
24m 35ch
Bryn
10m 22ch
Wigan
North
Western
See Map 132
Gillets (UWC)
20m 75ch
Adlington
19m 15ch
Blackrod
17m 14ch
Blackrod
Jn. SB (BJ)
17m 27ch
Crow Nest Jn.
14m 64ch/
17m 18ch
Hindley
15m 17ch
Horwich
Parkway
Lostock
Bolton
Westhoughton
Hag Fold
Daisy
Hill
Atherton
Turton (AOCL)
15m 19ch
13m 75ch
(FP) 13m 47ch
Bromley Cross
13m 45ch
The Oaks (UWC)
12m 72ch
Hall I' Th' Wood
12m 35ch
Calrows Viaduct 10m 60-65ch
Sheds
Bury EL Tunnel 9m 55-58ch
Bury Bolton Street
Bury South Jn
9m 42ch
Metrolink
Bridge
9m 21ch
9m 48ch
NR / ELRly. Bdy
9m 04ch
5m 04ch
Bury South
9m 38ch
Roch Vdct
8m 36-31ch
8m 17ch
Heywood
5m 35ch
M66
Shed
1
Clifton
Swinton
Manchester
Victoria
Eccles
Trafford
Park
Ⓐ Haydock Branch Jn. 2m 42ch/0m 00ch
Ⓑ Goldborne Jn. 187m 76ch/0m 53ch
Ⓒ 0m 48ch
Ⓓ Lowton Jn. 0m 36ch/0m 26ch
Ⓔ Parkside GF 16m 59ch
Ⓕ Parkside Jn. 16m 56ch/0m 05ch
Ⓖ Newton-le-Willows Jn. 16m 19ch/0m 00ch
Ⓐ Crow Nest Jn. SB (CN) 14m 62ch
Ⓑ Fiddlers Ferry SB (S) 14m 46ch
Kelbit
Private
Sdgs
0m 53ch
Garswood
9m 00ch
⑤ Station Jn.
5m 12ch / 7m 15ch
⑥ Gerards Bridge Jn. GF
5m 66ch
to Pilkington Ltd.
St. Helens
Station
SB (SH)
5m 24ch
Earlestown
14m 58ch /
187m 03ch /
187m 05ch
Earlestown East Jn.
14m 75ch/187m 10ch
Former
Motorail
terminal
Scholes
Tunnel
3m 30ch to
3m 34ch
St Helens
Central
5m 16ch
Ravenhead Jn.
6m 62ch
Thatto
Heath
3m 43ch
Eccleston
Park
2m 47ch
to Leathers
Chemicals
to Pilkington
Ltd
to
Former
P.S.C.
Newton-le-
Willows
15m 60ch
Winwick Jn.
185m 49ch
Prescot
1m 53ch
Earlestown
West Jn.
14m 51ch /
187m 15ch
St. Helens
Junction
11m 70ch
Lea Green
10m 57ch
Rainhill
8m 72ch
Whiston
7m 65ch
Huyton
Earlestown South Jn.
186m 74ch
Padgate
20m 13ch
Irlam
Glazebrook
Birchwood
Widnes
Ditton West Jn.
183m 22ch
Hough Green
10m 42ch
12m 20ch
Sankey
15m 67ch
Halewood
to Fiddlers Ferry
Power Stn.
Caterhouse Jn.
(UWC) 26m 27ch
Ditton East Jn.
182m 67ch /
18m 55ch
Warrington
Bank Quay
See Map 132
Altrincham
Manchester
Airport
2
57
Styal
Runcorn
180m 40ch
Ditton SB (DN)
183m 00ch
to
I.C.I.
Runcorn East
13m 06ch
12m 70ch
Norton SB (NN)
13m 17ch
Norton (R/G) (bridleway) 177m 40ch
Runcorn SB (RN)
180m 33ch
0m 02ch
Halton Jn.
179m 24ch /
0m 00ch
Halton Jn. SB (HN)
179m 20ch
Sutton
Tunnel
Preston Brook Tunnel
176m 04ch to 176m 07ch
Birdswood Tunnel
175m 43ch to 175m 44ch
See Map 129
Level Crossings:
① Penketh Hall (UWC) 13m 37ch
② Fiddlers Ferry (UWC) 13m 63ch
③ Marsh House (CCTV) 14m 09ch
① Ditton Yard
② O' Connors Siding
Mobberley
11m 71ch
Mobberley (MCB)
11m 74ch
Mobberley
SB (MY)
11m 75ch
Styal Jn.
0m 40ch
Frodsham Jn.
11m 04ch/1m 54ch/
11m 03ch
11m 63ch
11m 31ch
Frodsham
Jn. SB (FJ)
11m 09ch
1m 50ch
Kemira
Sidings
Ince &
Elton
6m 66ch
Frodsham
9m 68ch
10m 07ch
10m 11ch
Weaver Jn. (Up)
174m 70ch
Hartford East Jn.
21m 67ch
Northwich East Jn.
19m 77ch
Hartford North Jn.
22m 10ch / 0m 29ch
Knutsford
14m 40ch
Alderley Edge South Jn.
175m 12ch
8m 10ch
Stanlow &
Thornton
6m 67ch
(R/G) 6m 69ch
Helsby Jn.
8m 67ch
7m 40ch
Helsby
8m 60ch
7m 34ch
Frodsham
Tunnel
Weaver Jn. (Dn)
174m 53ch
Acton
Bridge
172m 38ch
Greenbank
SB (GK)
22m 21ch
to Brunner
Mond
Oakleigh
Sidings
Lostock
Gralam
19m 15ch
Field House Farm
(UWC) 18m 43ch
Woods Tenement Farm
(UWC) 16m 55ch
Plumley
17m 17ch
Chelford
172m 17ch
Sidings
Octel
Sdgs
Plumley West
SB (PY)
18m 07ch
Helsby Jn.
SB (HY)
8m 61ch
7m 28ch
Acton Bridge Jn. 172m 49ch
Greenbank
22m 28ch
to Lostock
Works
Northwich
20m 47ch
GF 18m 43ch
Sdgs
Hartford Jn.170m 56ch
Mouldsworth
31m 02ch
Cuddington
25m 15ch
Delamere
28m 11ch
Hartford CLC Jn.
23m 09ch
Northwich Station Jn.
20m 52ch / 8m 66ch
Northwich South Jn.
8m 37ch
Chelford South Jn.
171m 18ch
Hartford
169m 64ch
Hartford West Jn.
22m 12ch /
0m 11ch
Forest House Farm
(UWC) 26m 74ch
Farmer Johnson's (UWC) 34m 06ch
Plemstall (UWC) 34m 45ch
31m 13ch
Mouldsworth GF
30m 60ch
Northwich West Jn.
20m 74ch /
8m 64ch /
20m 76ch
Winsford SB (WD)
166m 66ch
Goostrey
168m 35ch
River Dane
Goostrey Jn.
167m 78ch
177m 59ch
Christleton Tunnel
177m 52ch
Winsford
165m 41ch
West Jn.
4m 05ch
Middlewich Loop
Holmes Chapel
166m 37ch
3
East Jn.
3m 39ch
Middlewich
British Salt
Higher Delacre
(UWC) 1m 27ch
Winsford South Jn.
165m 11ch
GF 1m 73ch
Elworth Jn.
0m 52ch
Sandbach North Jn. 162m 62ch
0m 00ch (Northwich Line)
Beeston Castle &
Tarporley SB (BC)
168m 60ch
Albion Inorganic
Sidings
Sandbach
162m 50ch
Sandbach South Jn.
162m 28ch
168m 53ch
C
D
48

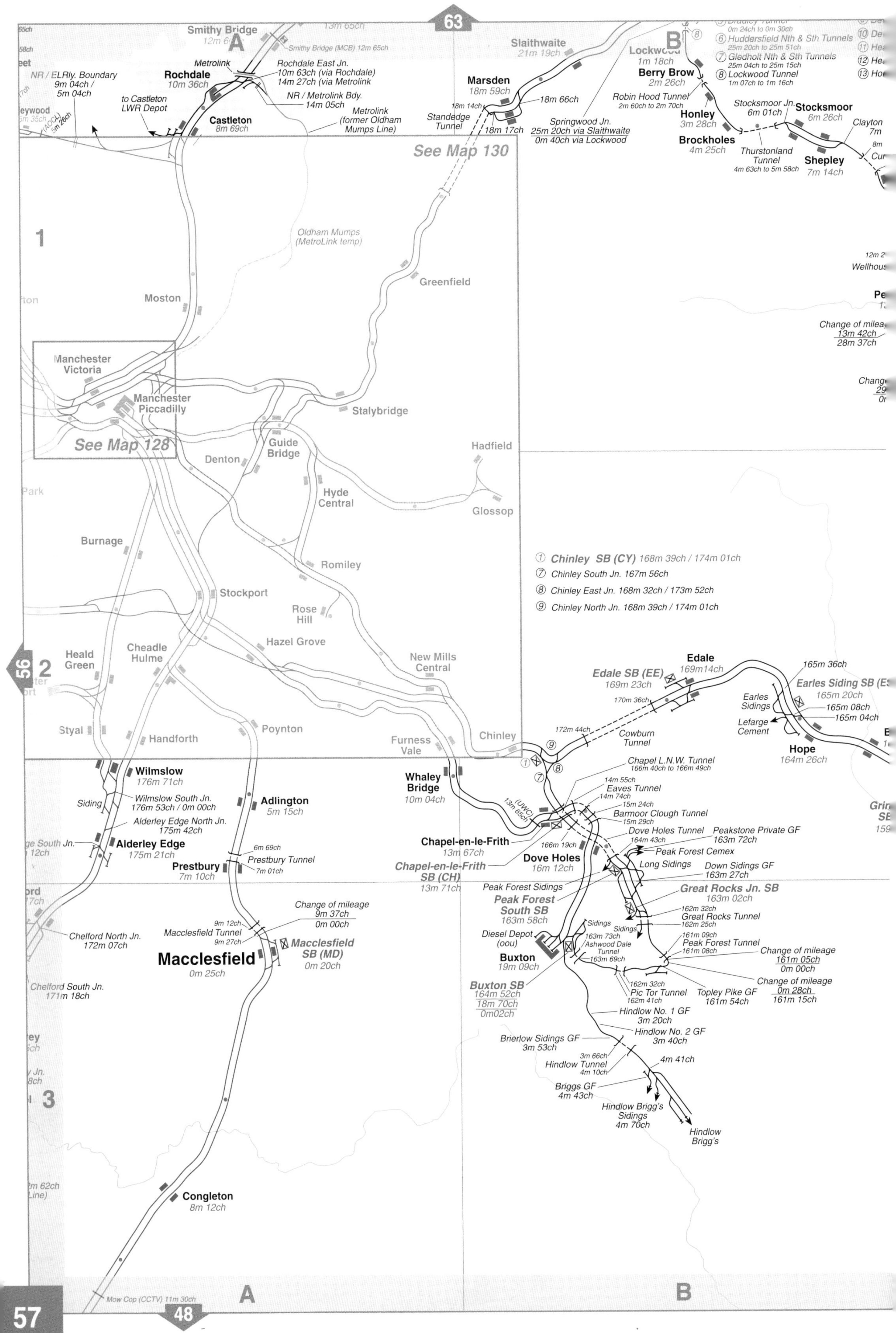

63
Smithy Bridge
Smithy Bridge (MCB) 12m 65ch
Slaithwaite
21m 19ch
Lockwood
1m 18ch
Berry Brow
2m 26ch
Honley
3m 28ch
Brockholes
4m 25ch
Robin Hood Tunnel
2m 60ch to 2m 70ch
Stocksmoor Jn.
6m 01ch
Stocksmoor
6m 26ch
Thurstonland Tunnel
4m 63ch to 5m 58ch
Shepley
7m 14ch
Clayton
6 Huddersfield Nth & Sth Tunnels
25m 20ch to 25m 51ch
7 Gledholt Nth & Sth Tunnels
25m 04ch to 25m 15ch
8 Lockwood Tunnel
1m 07ch to 1m 16ch
NR / ELRly. Boundary
9m 04ch /
5m 04ch
Metrolink
Rochdale
10m 36ch
Rochdale East Jn.
10m 63ch (via Rochdale)
14m 27ch (via Metrolink
NR / Metrolink Bdy.
14m 05ch
Metrolink
(former Oldham
Mumps Line)
to Castleton
LWR Depot
Castleton
8m 69ch
Marsden
18m 59ch
18m 14ch
18m 66ch
18m 17ch
Standedge
Tunnel
Springwood Jn.
25m 20ch via Slaithwaite
0m 40ch via Lockwood
See Map 130
1
Oldham Mumps
(MetroLink temp)
Greenfield
Moston
Change of mileage
13m 42ch
28m 37ch
Manchester
Victoria
Manchester
Piccadilly
See Map 128
Stalybridge
Guide
Bridge
Denton
Hadfield
Hyde
Central
Glossop
Burnage
Romiley
Stockport
Rose
Hill
Hazel Grove
1 Chinley SB (CY) 168m 39ch / 174m 01ch
7 Chinley South Jn. 167m 56ch
8 Chinley East Jn. 168m 32ch / 173m 52ch
9 Chinley North Jn. 168m 39ch / 174m 01ch
56
2
Heald
Green
Cheadle
Hulme
New Mills
Central
Edale
169m14ch
Edale SB (EE)
169m 23ch
170m 36ch
165m 36ch
Earles Siding SB
165m 20ch
Earles
Sidings
165m 08ch
165m 04ch
Lefarge
Cement
Hope
164m 26ch
Styal
Handforth
Poynton
Furness
Vale
Chinley
172m 44ch
Cowburn
Tunnel
Chapel L.N.W. Tunnel
166m 40ch to 166m 49ch
Wilmslow
176m 71ch
Siding
Wilmslow South Jn.
176m 53ch / 0m 00ch
Alderley Edge North Jn.
175m 42ch
Alderley Edge
175m 21ch
Adlington
5m 15ch
6m 69ch
Prestbury Tunnel
7m 01ch
Prestbury
7m 10ch
Whaley
Bridge
10m 04ch
(UWC)
13m 65ch
14m 55ch
Eaves Tunnel
14m 74ch
15m 24ch
Barmoor Clough Tunnel
15m 29ch
Dove Holes Tunnel
164m 43ch
Peakstone Private GF
163m 72ch
Peak Forest Cemex
Chapel-en-le-Frith
13m 67ch
166m 19ch
Dove Holes
16m 12ch
Chapel-en-le-Frith
SB (CH)
13m 71ch
Long Sidings
Down Sidings GF
163m 27ch
Peak Forest Sidings
Peak Forest
South SB
163m 58ch
Great Rocks Jn. SB
163m 02ch
162m 32ch
Great Rocks Tunnel
162m 25ch
Change of mileage
9m 37ch
0m 00ch
9m 12ch
Macclesfield Tunnel
9m 27ch
Macclesfield
SB (MD)
0m 20ch
Macclesfield
0m 25ch
Chelford North Jn.
172m 07ch
Chelford South Jn.
171m 18ch
Diesel Depot
(oou)
Sidings
Sidings
163m 73ch
Ashwood Dale
Tunnel
163m 69ch
161m 09ch
Peak Forest Tunnel
161m 08ch
Change of mileage
161m 05ch
0m 00ch
Buxton
19m 09ch
Buxton SB
164m 52ch
18m 70ch
0m02ch
162m 32ch
Pic Tor Tunnel
162m 41ch
Topley Pike GF
161m 54ch
Change of mileage
0m 28ch
161m 15ch
Hindlow No. 1 GF
3m 20ch
Hindlow No. 2 GF
3m 40ch
Brierlow Sidings GF
3m 53ch
3m 66ch
Hindlow Tunnel
4m 10ch
4m 41ch
Briggs GF
4m 43ch
Hindlow Brigg's
Sidings
4m 70ch
Hindlow
Brigg's
3
Congleton
8m 12ch
Mow Cop (CCTV) 11m 30ch
A
B
48

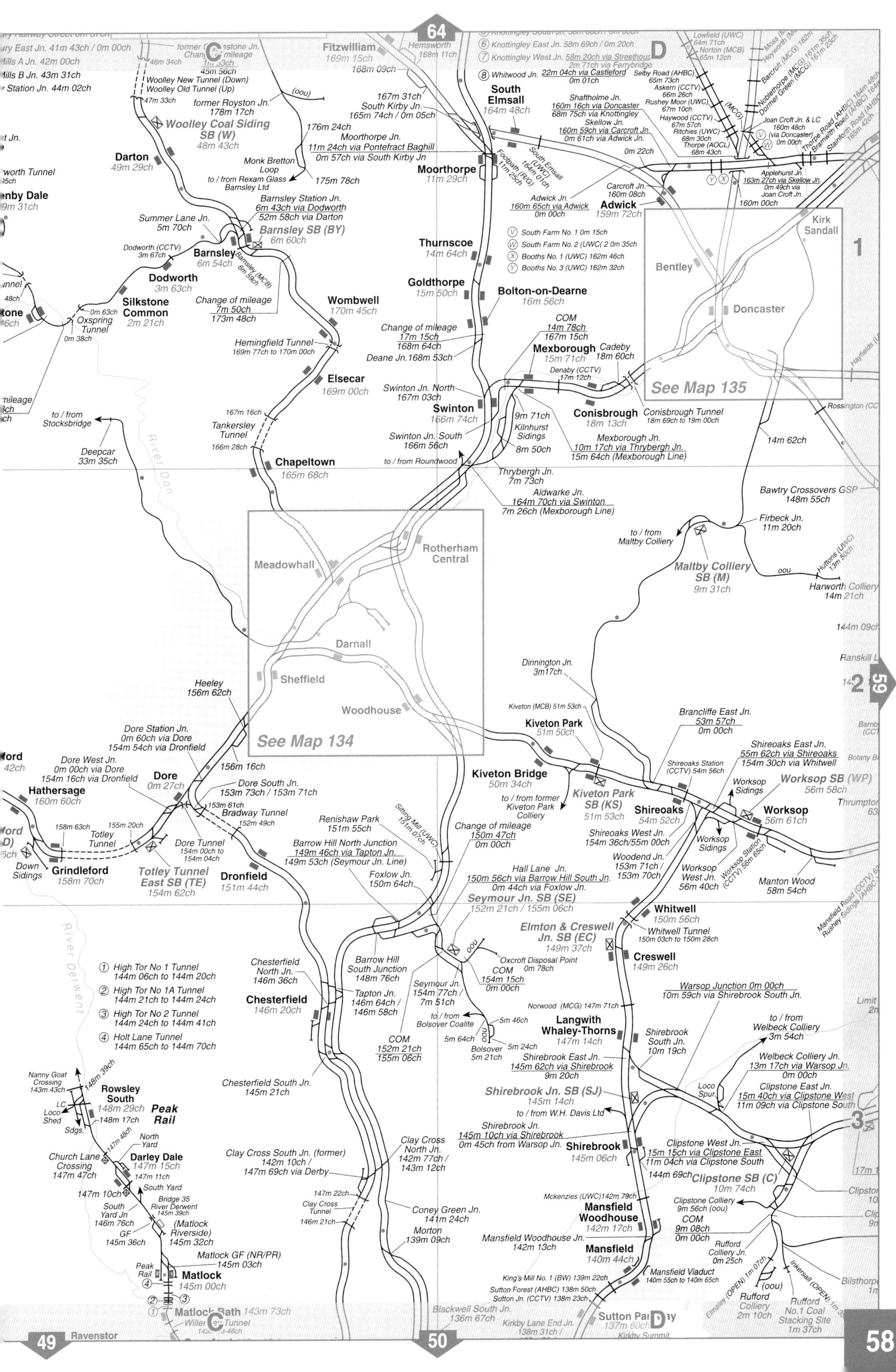

64
Woolley Coal Siding SB (W)
48m 43ch
Woolley New Tunnel (Down)
Woolley Old Tunnel (Up)
former Royston Jn.
178m 17ch
Darton
49m 29ch
Monk Bretton Loop
to / from Rexam Glass Barnsley Ltd
176m 24ch
175m 78ch
Fitzwilliam
169m 15ch
168m 09ch
Hemsworth
168m 11ch
167m 31ch
South Kirby Jn.
165m 74ch / 0m 05ch
Moorthorpe Jn.
11m 24ch via Pontefract Baghill
0m 57ch via South Kirby Jn
South Elmsall
164m 48ch
Moorthorpe
11m 29ch
Knottingley East Jn. 58m 69ch / 0m 20ch
Knottingley West Jn. 58m 20ch via Streethouse
2m 71ch via Ferrybridge
Whitwood Jn. 22m 04ch via Castleford
0m 01ch
Shaftholme Jn.
160m 16ch via Doncaster
68m 75ch via Knottingley
Skellow Jn.
160m 59ch via Carcroft Jn.
0m 61ch via Adwick Jn.
Selby Road (AHBC)
65m 73ch
Askern (CCTV)
66m 26ch
Rushey Moor (UWC)
67m 10ch
Haywood (CCTV)
67m 57ch
Ritchies (UWC)
68m 30ch
Thorpe (AOCL)
68m 43ch
Lowfield (UWC)
64m 71ch
Norton (MCB)
65m 12ch
Joan Croft Jn. & LC
160m 48ch
(via Doncaster)
0m 00ch
Applehurst Jn.
163m 27ch via Skellow Jn.
0m 49ch via Joan Croft Jn.
160m 00ch
0m 22ch
Carcroft Jn.
160m 08ch
Adwick Jn.
160m 65ch via Adwick
0m 00ch
Adwick
159m 72ch
Kirk Sandall
Bentley
Doncaster
See Map 135
Barnsley Station Jn.
6m 43ch via Dodworth
52m 58ch via Darton
Summer Lane Jn.
5m 70ch
Barnsley SB (BY)
6m 60ch
Barnsley
6m 54ch
Dodworth (CCTV)
3m 67ch
Dodworth
3m 63ch
Silkstone Common
2m 21ch
Oxspring Tunnel
Change of mileage
7m 50ch
173m 48ch
Wombwell
170m 45ch
Hemingfield Tunnel
169m 77ch to 170m 00ch
Elsecar
169m 00ch
Thurnscoe
14m 64ch
Goldthorpe
15m 50ch
Bolton-on-Dearne
16m 56ch
South Farm No. 1 0m 15ch
South Farm No. 2 (UWC(2 0m 35ch
Booths No. 1 (UWC) 162m 46ch
Booths No. 3 (UWC) 162m 32ch
Change of mileage
17m 15ch
168m 64ch
Deane Jn.168m 53ch
COM
14m 78ch
167m 15ch
Mexborough
15m 71ch
Cadeby
18m 60ch
Denaby (CCTV)
17m 12ch
Conisbrough
18m 13ch
Conisbrough Tunnel
18m 69ch to 19m 00ch
Swinton Jn. North
167m 03ch
Swinton
166m 74ch
Swinton Jn. South
166m 56ch
Kilnhurst Sidings
9m 71ch
8m 50ch
Mexborough Jn.
10m 17ch via Thrybergh Jn.
15m 64ch (Mexborough Line)
Thrybergh Jn.
7m 73ch
Aldwarke Jn.
164m 70ch via Swinton
7m 26ch (Mexborough Line)
to / from Roundwood
14m 62ch
to / from Stocksbridge
Deepcar
33m 35ch
River Don
Tankersley Tunnel
167m 16ch
166m 28ch
Chapeltown
165m 68ch
Meadowhall
Rotherham Central
Darnall
Sheffield
Woodhouse
See Map 134
Bawtry Crossovers GSP
148m 55ch
Firbeck Jn.
11m 20ch
to / from Maltby Colliery
Maltby Colliery SB (M)
9m 31ch
Harworth Colliery
14m 21ch
Dinnington Jn.
3m17ch
Heeley
156m 62ch
Dore Station Jn.
0m 60ch via Dore
154m 54ch via Dronfield
Dore West Jn.
0m 00ch via Dore
154m 16ch via Dronfield
Dore
0m 27ch
Dore South Jn.
153m 73ch / 153m 71ch
156m 16ch
Hathersage
160m 60ch
Totley Tunnel
Dore Tunnel
154m 00ch to 154m 04ch
Bradway Tunnel
Grindleford
158m 70ch
Totley Tunnel East SB (TE)
154m 62ch
Dronfield
151m 44ch
Down Sidings
Kiveton (MCB) 51m 53ch
Kiveton Park
51m 50ch
Kiveton Bridge
50m 34ch
to / from former Kiveton Park Colliery
Kiveton Park SB (KS)
51m 53ch
Brancliffe East Jn.
53m 57ch
0m 00ch
Shireoaks East Jn.
55m 62ch via Shireoaks
154m 30ch via Whitwell
Shireoaks Station (CCTV) 54m 56ch
Shireoaks
54m 52ch
Worksop SB (WP)
56m 58ch
Worksop
56m 61ch
Worksop Sidings
Shireoaks West Jn.
154m 36ch/55m 00ch
Woodend Jn.
153m 71ch / 153m 70ch
Worksop West Jn.
56m 40ch
Manton Wood
58m 54ch
Renishaw Park
151m 55ch
Barrow Hill North Junction
149m 46ch via Tapton Jn.
149m 53ch (Seymour Jn. Line)
Change of mileage
150m 47ch
0m 00ch
Foxlow Jn.
150m 64ch
Hall Lane Jn.
150m 56ch via Barrow Hill South Jn.
0m 44ch via Foxlow Jn.
Seymour Jn. SB (SE)
152m 21ch / 155m 06ch
Whitwell
150m 56ch
Whitwell Tunnel
150m 03ch to 150m 28ch
Elmton & Creswell Jn. SB (EC)
149m 37ch
Creswell
149m 26ch
Oxcroft Disposal Point
0m 78ch
COM
154m 15ch
0m 00ch
Barrow Hill South Junction
148m 76ch
Chesterfield North Jn.
146m 36ch
Chesterfield
146m 20ch
Tapton Jn.
146m 64ch / 146m 58ch
Seymour Jn.
154m 77ch / 7m 51ch
to / from Bolsover Coalite
Bolsover
5m 21ch
COM
152m 21ch
155m 06ch
Warsop Junction 0m 00ch
10m 59ch via Shirebrook South Jn.
Norwood (MCG) 147m 71ch
Langwith Whaley-Thorns
147m 14ch
Shirebrook South Jn.
10m 19ch
to / from Welbeck Colliery
3m 54ch
Welbeck Colliery Jn.
13m 17ch via Warsop Jn.
0m 00ch
Shirebrook East Jn.
145m 62ch via Shirebrook
9m 20ch
Shirebrook Jn. SB (SJ)
145m 14ch
to / from W.H. Davis Ltd
Clipstone East Jn.
15m 40ch via Clipstone West
11m 09ch via Clipstone South
Shirebrook Jn.
145m 10ch via Shirebrook
0m 45ch from Warsop Jn.
Shirebrook
145m 06ch
Clipstone West Jn.
15m 15ch via Clipstone East
11m 04ch via Clipstone South
Clipstone SB (C)
10m 74ch
Clipstone Colliery
9m 56ch (oou)
COM
9m 08ch
0m 00ch
Rufford Colliery Jn.
0m 25ch
Chesterfield South Jn.
145m 21ch
High Tor No 1 Tunnel
144m 06ch to 144m 20ch
High Tor No 1A Tunnel
144m 21ch to 144m 24ch
High Tor No 2 Tunnel
144m 24ch to 144m 41ch
Holt Lane Tunnel
144m 65ch to 144m 70ch
River Derwent
Clay Cross South Jn. (former)
142m 10ch /
147m 69ch via Derby
Clay Cross North Jn.
142m 77ch /
143m 12ch
Clay Cross Tunnel
147m 22ch
146m 21ch
Coney Green Jn.
141m 24ch
Morton
139m 09ch
Mckenzies (UWC)142m 79ch
Mansfield Woodhouse
142m 17ch
Mansfield Woodhouse Jn.
142m 13ch
Mansfield
140m 44ch
Mansfield Viaduct
140m 55ch to 140m 65ch
King's Mill No. 1 (BW) 139m 22ch
Sutton Forest (AHBC) 138m 50ch
Sutton Jn. (CCTV) 138m 23ch
Blackwell South Jn.
136m 67ch
Kirkby Lane End Jn.
138m 31ch /
Rufford Colliery
2m 10ch
Rufford No.1 Coal Stacking Site
1m 37ch
Nanny Goat Crossing
143m 43ch
Rowsley South
148m 29ch
Peak Rail
Loco Shed
Sdgs.
North Yard
Church Lane Crossing
147m 47ch
Darley Dale
147m 15ch
147m 11ch
South Yard
147m 10ch
Bridge 35 River Derwent
145m 39ch
South Yard Jn
146m 76ch
(Matlock Riverside)
145m 32ch
GF
145m 36ch
Matlock GF (NR/PR)
145m 03ch
Peak Rail
Matlock
145m 00ch
Matlock Bath 143m 73ch
Ravenstor
49
50
59

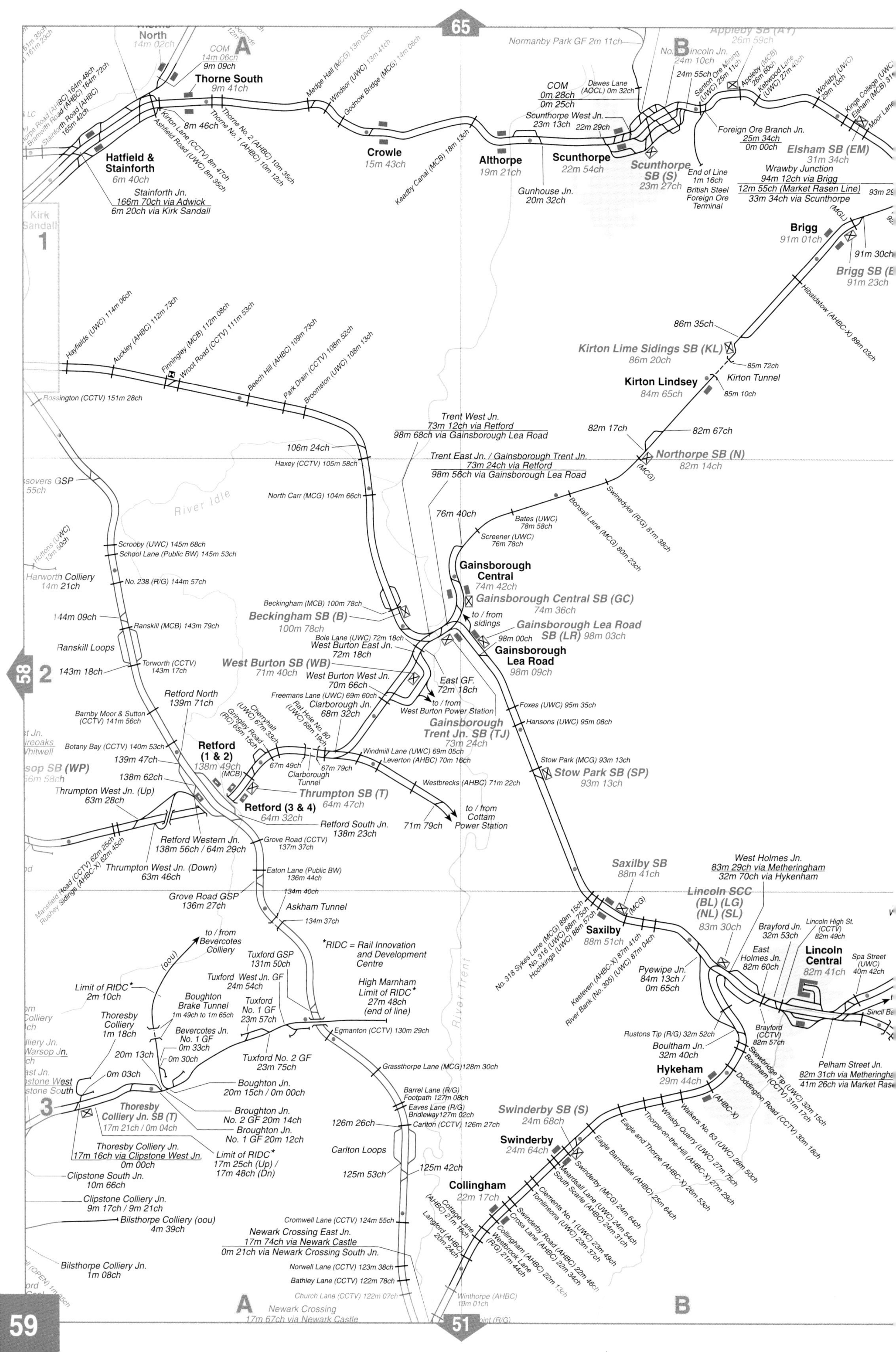
65
A
B
Thorne North
14m 02ch
COM
14m 06ch
9m 09ch
Thorne South
9m 41ch
8m 46ch
Hatfield & Stainforth
6m 40ch
Stainforth Jn.
166m 70ch via Adwick
6m 20ch via Kirk Sandall
Kirk Sandall
1
Stainforth Road (AHBC) 165m 42ch
Bramwith Road (AHBC) 164m 72ch
Kirton Lane (CCTV) 8m 47ch
Ashfield Road (UWC) 8m 35ch
Thorne No. 2 (AHBC) 10m 35ch
Thorne No. 1 (AHBC) 10m 12ch
Medge Hall (MCG) 13m 02ch
Windsor (UWC) 13m 41ch
Godnow Bridge (MCG) 14m 08ch
Crowle
15m 43ch
Keadby Canal (MCB) 18m 13ch
Althorpe
19m 21ch
Normanby Park GF 2m 11ch
COM
0m 28ch
0m 25ch
Dawes Lane (AOCL) 0m 32ch
Scunthorpe West Jn.
23m 13ch
22m 29ch
Scunthorpe
22m 54ch
Gunhouse Jn.
20m 32ch
Scunthorpe SB (S)
23m 27ch
No. Lincoln Jn.
24m 10ch
24m 55ch
Santon Ore Mining (UWC) 25m 11ch
Appleby SB (AY)
26m 59ch
Appleby (MCB) 26m 60ch
Kelwood Lane (UWC) 27m 40ch
Worlaby (UWC) 29m 10ch
Kings College (UWC)
Elsham (MCB)
Moor Lane
Foreign Ore Branch Jn.
25m 34ch
0m 00ch
End of Line
1m 16ch
British Steel Foreign Ore Terminal
Elsham SB (EM)
31m 34ch
Wrawby Junction
94m 12ch via Brigg
12m 55ch (Market Rasen Line)
33m 34ch via Scunthorpe
(MGL)
Brigg
91m 01ch
91m 30ch
Brigg SB
91m 23ch
Hibaldstow (AHBC-X) 89m 03ch
86m 35ch
Kirton Lime Sidings SB (KL)
86m 20ch
85m 72ch
Kirton Tunnel
85m 10ch
Kirton Lindsey
84m 65ch
82m 17ch
82m 67ch
Northorpe SB (N)
82m 14ch
(MCG)
Swinedyke (R/G) 81m 38ch
Bonsall Lane (MCG) 80m 23ch
Bates (UWC)
78m 58ch
Screener (UWC)
76m 78ch
76m 40ch
Hayfields (UWC) 114m 06ch
Auckley (AHBC) 112m 73ch
Finningley (MCB) 112m 08ch
Wroot Road (CCTV) 111m 53ch
Beech Hill (AHBC) 109m 73ch
Park Drain (CCTV) 108m 52ch
Broomston (UWC) 108m 13ch
Rossington (CCTV) 151m 28ch
Trent West Jn.
73m 12ch via Retford
98m 68ch via Gainsborough Lea Road
106m 24ch
Trent East Jn. / Gainsborough Trent Jn.
73m 24ch via Retford
98m 56ch via Gainsborough Lea Road
Haxey (CCTV) 105m 58ch
North Carr (MCG) 104m 66ch
River Idle
Scrooby (UWC) 145m 68ch
School Lane (Public BW) 145m 53ch
Harworth Colliery
14m 21ch
No. 238 (R/G) 144m 57ch
144m 09ch
Ranskill (MCB) 143m 79ch
Ranskill Loops
58
2
143m 18ch
Torworth (CCTV) 143m 17ch
Gainsborough Central
74m 42ch
Gainsborough Central SB (GC)
74m 36ch
to / from sidings
Gainsborough Lea Road SB (LR) 98m 03ch
98m 00ch
Gainsborough Lea Road
98m 09ch
Beckingham (MCB) 100m 78ch
Beckingham SB (B)
100m 78ch
Bole Lane (UWC) 72m 18ch
West Burton East Jn.
72m 18ch
West Burton SB (WB)
71m 40ch
West Burton West Jn.
70m 66ch
East GF.
72m 18ch
to / from West Burton Power Station
Freemans Lane (UWC) 69m 60ch
Clarborough Jn.
68m 32ch
Gainsborough Trent Jn. SB (TJ)
73m 24ch
Foxes (UWC) 95m 35ch
Hansons (UWC) 95m 08ch
Stow Park (MCG) 93m 13ch
Stow Park SB (SP)
93m 13ch
Retford North
139m 71ch
Barnby Moor & Sutton (CCTV) 141m 56ch
Botany Bay (CCTV) 140m 53ch
139m 47ch
138m 62ch
Retford (1 & 2)
138m 49ch
(MCB)
Cherryholt (UWC) 67m 33ch
Gringley Road (RC) 65m 15ch
Rat Hole No. 80 (UWC) 68m 19ch
67m 49ch
67m 79ch
Clarborough Tunnel
Windmill Lane (UWC) 69m 05ch
Leverton (AHBC) 70m 16ch
Westbrecks (AHBC) 71m 22ch
Thrumpton SB (T)
64m 47ch
Retford (3 & 4)
64m 32ch
Retford South Jn.
138m 23ch
71m 79ch
to / from Cottam Power Station
Worksop SB (WP)
Thrumpton West Jn. (Up)
63m 28ch
Retford Western Jn.
138m 56ch / 64m 29ch
Thrumpton West Jn. (Down)
63m 46ch
Grove Road (CCTV) 137m 37ch
Eaton Lane (Public BW) 136m 44ch
Mansfield Road (CCTV) 62m 25ch
Rushey Sidings (AHBC-X) 62m 45ch
Grove Road GSP
136m 27ch
134m 40ch
Askham Tunnel
134m 37ch
Saxilby SB
88m 41ch
(MCG)
Saxilby
88m 51ch
No. 318 Sykes Lane (MCG) 89m 15ch
No. 316 (UWC) 88m 75ch
Hockings (UWC) 88m 57ch
Kesteven (AHBC-X) 87m 41ch
River Bank (No. 305) (UWC) 87m 04ch
West Holmes Jn.
83m 29ch via Metheringham
32m 70ch via Hykenham
Lincoln SCC
(BL) (LG)
(NL) (SL)
83m 30ch
Pyewipe Jn.
84m 13ch /
0m 65ch
Brayford Jn.
32m 53ch
Lincoln High St. (CCTV)
82m 49ch
East Holmes Jn.
82m 60ch
Lincoln Central
82m 41ch
Spa Street (UWC)
40m 42ch
Brayford (CCTV)
82m 57ch
Rustons Tip (R/G) 32m 52ch
Boultham Jn.
32m 40ch
Pelham Street Jn.
82m 31ch via Metheringham
41m 26ch via Market Rasen
Hykeham
29m 44ch
(AHBC-X)
Skewbridge Tip (UWC) 32m 15ch
Boultham (CCTV) 31m 17ch
Doddington Road (CCTV) 30m 18ch
Walkers No. 63 (UWC) 28m 50ch
Whisby Quarry (UWC) 27m 75ch
Thorpe-on-the-Hill (AHBC-X) 27m 29ch
Eagle and Thorpe (AHBC-X) 26m 53ch
Eagle Barnsdale (AHBC) 25m 64ch
Swinderby SB (S)
24m 68ch
Swinderby
24m 64ch
Swinderby (MCG) 24m 64ch
Meadsall Lane (UWC) 24m 54ch
South Scarle (AHBC) 24m 31ch
Clements No. 1 (UWC) 23m 49ch
Tomlinsons (UWC) 23m 37ch
Swinderby Road (AHBC) 22m 46ch
Cross Lane (AHBC) 22m 34ch
Collingham (AHBC) 22m 13ch
Collingham
22m 17ch
Cottage Lane (AHBC) 21m 18ch
Westbrook Lane (R/G) 21m 44ch
Langford (AHBC) 20m 24ch
Winthorpe (AHBC)
19m 01ch
*RIDC = Rail Innovation and Development Centre
Tuxford GSP
131m 50ch
Tuxford West Jn. GF
24m 54ch
High Marnham Limit of RIDC*
27m 48ch
(end of line)
Tuxford No. 1 GF
23m 57ch
Egmanton (CCTV) 130m 29ch
Grassthorpe Lane (MCG) 128m 30ch
Barrel Lane (R/G) Footpath 127m 08ch
Eaves Lane (R/G) Bridleway 127m 02ch
Carlton (CCTV) 126m 27ch
126m 26ch
Carlton Loops
125m 42ch
125m 53ch
Cromwell Lane (CCTV) 124m 55ch
Newark Crossing East Jn.
17m 74ch via Newark Castle
0m 21ch via Newark Crossing South Jn.
Norwell Lane (CCTV) 123m 38ch
Bathley Lane (CCTV) 122m 78ch
Church Lane (CCTV) 122m 07ch
River Trent
to / from Bevercotes Colliery
(oou)
Limit of RIDC*
2m 10ch
Boughton Brake Tunnel
1m 49ch to 1m 65ch
Thoresby Colliery
1m 18ch
Bevercotes Jn. No. 1 GF
0m 33ch
0m 30ch
Tuxford No. 2 GF
23m 75ch
20m 13ch
0m 03ch
Boughton Jn.
20m 15ch / 0m 00ch
Broughton Jn. No. 2 GF 20m 14ch
Broughton Jn. No. 1 GF 20m 12ch
Thoresby Colliery Jn. SB (T)
17m 21ch / 0m 04ch
3
Thoresby Colliery Jn.
17m 16ch via Clipstone West Jn.
0m 00ch
Limit of RIDC*
17m 25ch (Up) /
17m 48ch (Dn)
Clipstone South Jn.
10m 66ch
Clipstone Colliery Jn.
9m 17ch / 9m 21ch
Bilsthorpe Colliery (oou)
4m 39ch
Bilsthorpe Colliery Jn.
1m 08ch
A
Newark Crossing
17m 67ch via Newark Castle
51
B

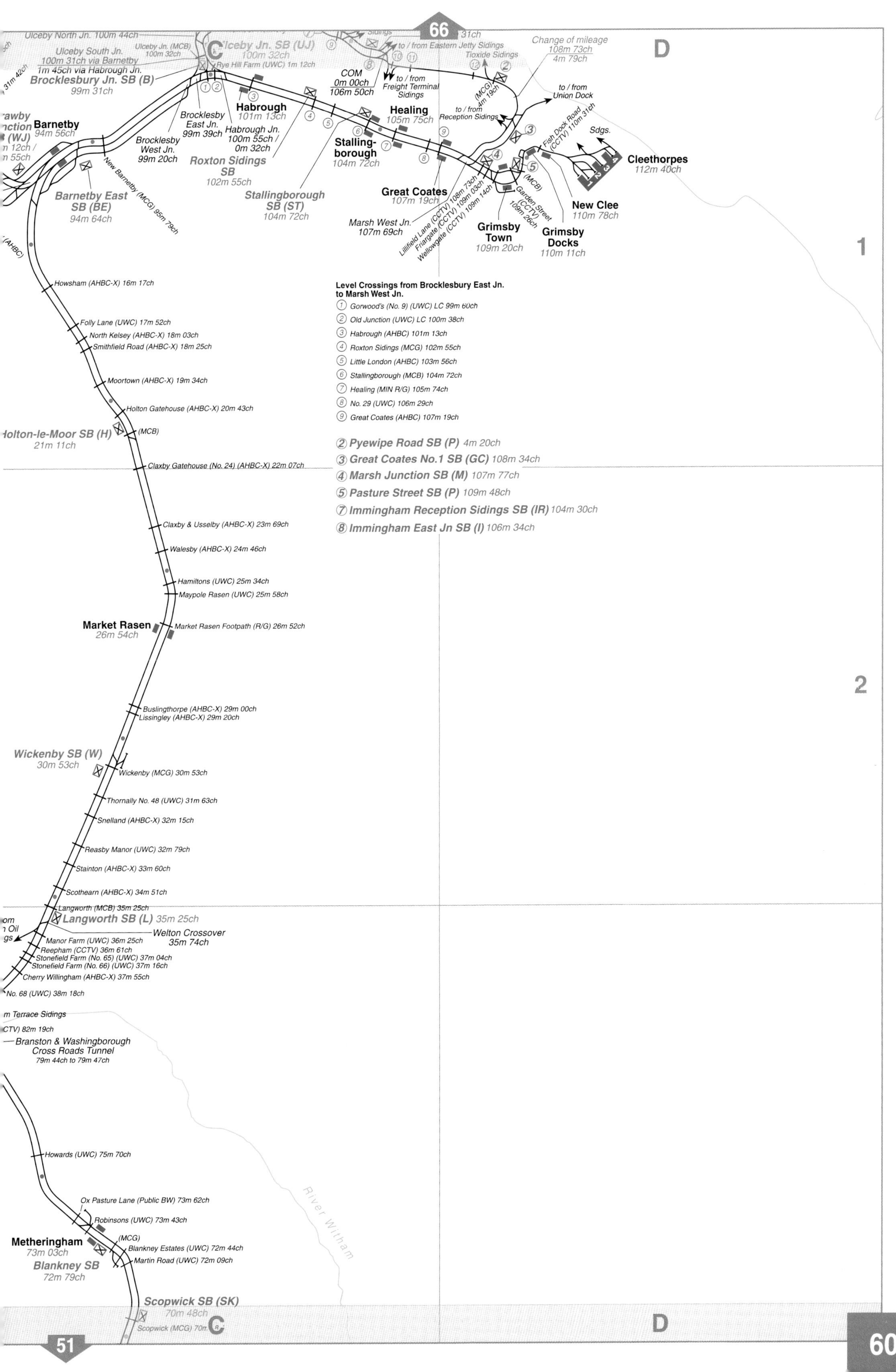

Level Crossings from Brocklesbury East Jn. to Marsh West Jn.

1. Gorwood's (No. 9) (UWC) LC 99m 60ch
2. Old Junction (UWC) LC 100m 38ch
3. Habrough (AHBC) 101m 13ch
4. Roxton Sidings (MCG) 102m 55ch
5. Little London (AHBC) 103m 56ch
6. Stallingborough (MCB) 104m 72ch
7. Healing (MIN R/G) 105m 74ch
8. No. 29 (UWC) 106m 29ch
9. Great Coates (AHBC) 107m 19ch

(2) Pyewipe Road SB (P) 4m 20ch
(3) Great Coates No.1 SB (GC) 108m 34ch
(4) Marsh Junction SB (M) 107m 77ch
(5) Pasture Street SB (P) 109m 48ch
(7) Immingham Reception Sidings SB (IR) 104m 30ch
(8) Immingham East Jn SB (I) 106m 34ch

67

A
B
1
2
3
Barrow-in-Furness SB (BF)
29m 05ch
29m 28ch
Abbey
25m 44ch
Carriage sidings
Barrow-in-Furness
28m 76ch
Roose
27m 13ch
to Port of Barrow
Salthouse Jn. GF
27m 57ch
Salthouse Jn.
27m 59ch
Morecamb
2m 10ch
Buffer stop
2m 12ch
Morecambe Jn. GF
1m 71ch /0m 00ch
Port of Heysham (UWC) 3m 69ch
Heysham Port
4m 01ch
Heysham Power Station Sidings GF
3m 53ch
End of Line
18m 08ch
Oil Sidings GF
17m 73ch
Hillhouse No.4 GF
17m 61ch
Hillhouse No.5 GF
17m 45ch
Hillhouse No.3 GF
17m 44ch
Burn Naze ICI Power Station
to Burn Naze Hillhouse VCM Sidings
Hilly Laid (TMO) 16m 43ch
Thornton (TMO) 16m 10ch
Tarn Gate (UWC) 15m 58ch
Line permanently oou from 14m 75ch
Poulton SB (PT)
14m 44ch
Poulton Jn. 14m 40ch
Poulton-le-Fylde
14m 31ch
Blackpool North PGF
16m 70ch
to Blackpool Carriage Sidings Depot
Layton
16m 32ch
Blackpool North No.2 SB (BN2)
17m 30ch
(MCB)
Carleton Crossing (CN) SB
15m 44ch
Blackpool North
17m 49ch
Kirkham S
8m 29c
8m 42ch
Kirkham Tip Sidings
Tarnbrick (UWC) 8m 54ch
Sidings
Blackpool South
20m 00ch
Blackpool Pleasure Beach
19m 18ch
Squires Gate
18m 34ch
Moss Side
11m 14ch
Moss Side (ABCL) 11m 09ch
St Annes-on-the-Sea
16m 51ch
Lytham
13m 56ch
Ansdell & Fairhaven
14m 75ch
Level Crossings from Meols Cop to Parbold:
1 Pool Hey (AHBC) 32m 44ch
2 Wyke Cop (AHBC) 32m 20ch
3 Bescar Lane (AHBC) 30m 78ch
4 Martins Lane (R/G) 29m 70ch
Southport
35m 27ch
18m 35ch
North Jn. 35m 08ch
Bradford Siding
Stabling Sidings
to former Steamport (oou)
Goods Yard GF
34m 58ch
Wallside Sidings
Carriage Sidings
Portland Street (CCTV) 18m 00ch
Duke Street (CCTV) 17m 65ch
Aughton Road (CCTV) 17m 49ch
Birkdale (CCTV) 17m 26ch
Crescent Road (AHBC) 16m 76ch
Birkdale Sidings
Meols Cop
34m 02ch
South Jn.
18m 13ch
Hillside
16m 26ch
Birkdale
30m 79ch (Down platform)
28m 73ch

55

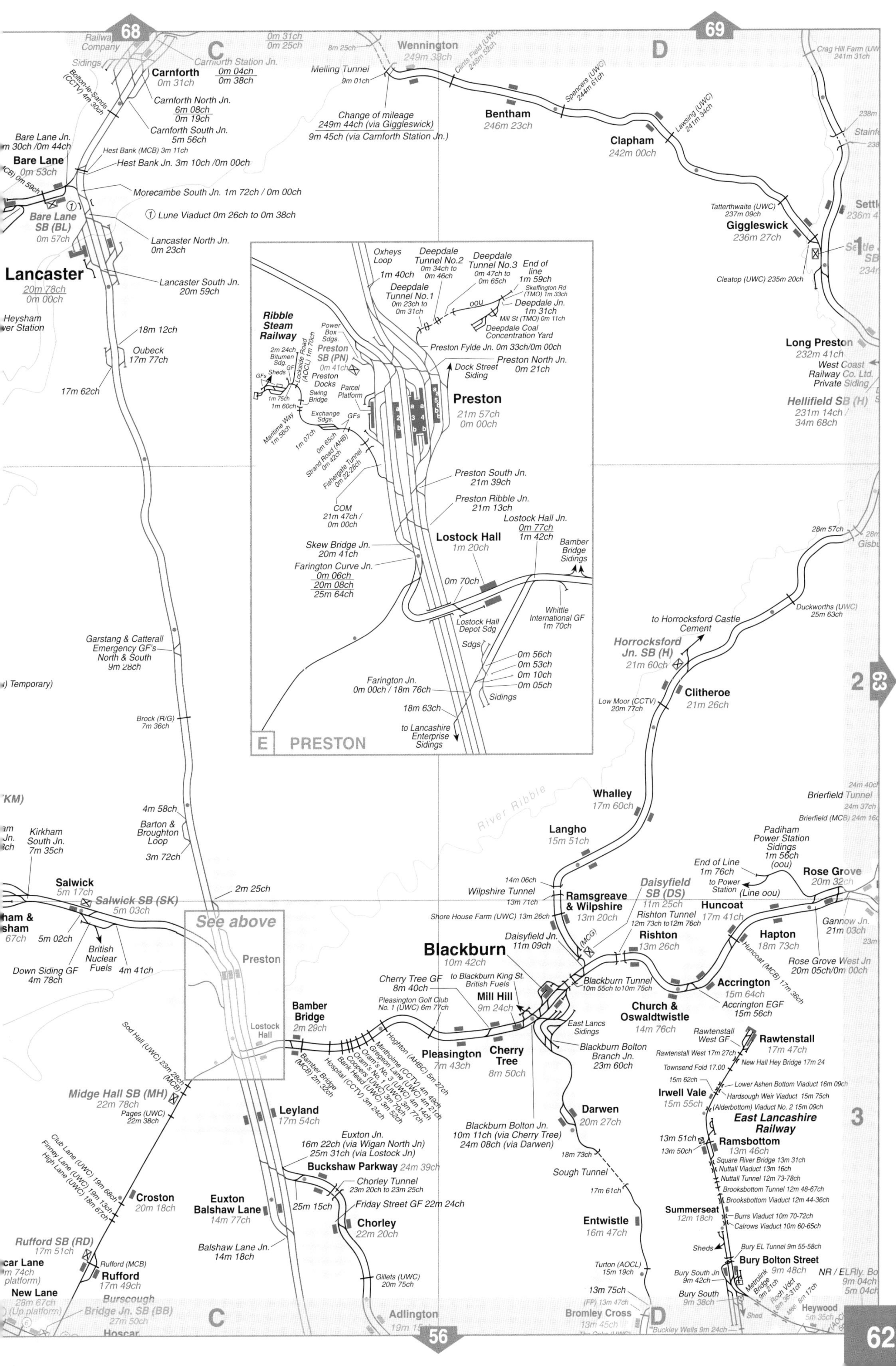
68
69
C
D
E PRESTON
See above
63
56
62

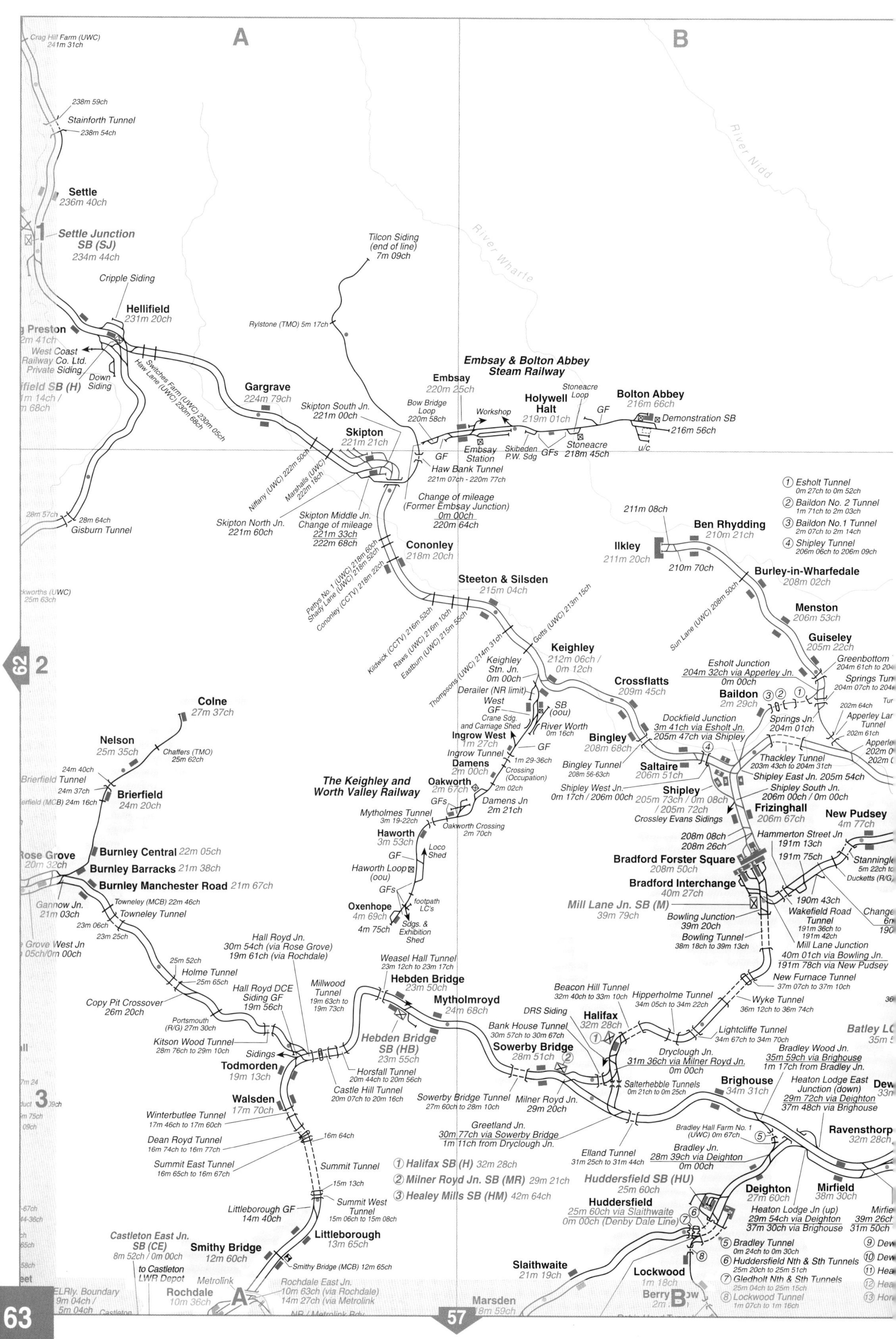
A
B
Crag Hill Farm (UWC) 241m 31ch
238m 59ch
Stainforth Tunnel
238m 54ch
River Nidd
Settle
236m 40ch
Settle Junction SB (SJ)
234m 44ch
Tilcon Siding (end of line) 7m 09ch
River Wharfe
Cripple Siding
Hellifield
231m 20ch
West Coast Railway Co. Ltd. Private Siding
Down Siding
Switches Farm (UWC) 230m 05ch
Haw Lane (UWC) 230m 68ch
Rylstone (TMO) 5m 17ch
Embsay & Bolton Abbey Steam Railway
Gargrave
224m 79ch
Embsay
220m 25ch
Stoneacre Loop
Bolton Abbey
216m 66ch
Holywell Halt
219m 01ch
Skipton South Jn.
221m 00ch
Bow Bridge Loop 220m 58ch
Workshop
GF
Demonstration SB
216m 56ch
Skipton
221m 21ch
Embsay Station
Skibeden P.W. Sdg
GFs
Stoneacre 218m 45ch
u/c
Niffany (UWC) 222m 50ch
Marshalls (UWC) 222m 18ch
Haw Bank Tunnel 221m 07ch - 220m 77ch
Change of mileage (Former Embsay Junction) 0m 00ch 220m 64ch
Skipton North Jn. 221m 60ch
Skipton Middle Jn. Change of mileage 221m 33ch 222m 68ch
28m 57ch
28m 64ch
Gisburn Tunnel
1 Esholt Tunnel 0m 27ch to 0m 52ch
2 Baildon No. 2 Tunnel 1m 71ch to 2m 03ch
3 Baildon No.1 Tunnel 2m 07ch to 2m 14ch
4 Shipley Tunnel 206m 06ch to 206m 09ch
211m 08ch
Ben Rhydding
210m 21ch
Ilkley
211m 20ch
210m 70ch
Cononley
218m 20ch
Burley-in-Wharfedale
208m 02ch
Pettys No. 1 (UWC) 218m 60ch
Shady Lane (UWC) 218m 52ch
Cononley (CCTV) 218m 22ch
Steeton & Silsden
215m 04ch
Menston
206m 53ch
Sun Lane (UWC) 208m 50ch
Kildwick (CCTV) 216m 52ch
Raws (UWC) 216m 10ch
Eastburn (UWC) 215m 55ch
Gotts (UWC) 213m 15ch
Thompsons (UWC) 214m 31ch
Guiseley
205m 22ch
Keighley
212m 06ch / 0m 12ch
Esholt Junction 204m 32ch via Apperley Jn. 0m 00ch
Greenbottom
Springs Tunnel
Keighley Stn. Jn. 0m 00ch
Crossflatts
209m 45ch
Baildon
2m 29ch
Derailer (NR limit)
West GF
SB (oou)
Crane Sdg. and Carriage Shed
River Worth 0m 16ch
Dockfield Junction 3m 41ch via Esholt Jn. 205m 47ch via Shipley
Springs Jn. 204m 01ch
Apperley Lane Tunnel 202m 61ch
Colne
27m 37ch
Ingrow West
1m 27ch
GF
Bingley
208m 68ch
Ingrow Tunnel
1m 29-36ch
Thackley Tunnel 203m 43ch to 204m 31ch
Nelson
25m 35ch
Chaffers (TMO) 25m 62ch
Damens
2m 00ch
Crossing (Occupation)
Bingley Tunnel 208m 56-63ch
Saltaire
206m 51ch
Shipley East Jn. 205m 54ch
24m 40ch
Brierfield Tunnel
24m 37ch
Oakworth
2m 67ch
2m 02ch
Shipley West Jn. 0m 17ch / 206m 00ch
Shipley
205m 73ch / 0m 08ch / 205m 72ch
Shipley South Jn. 206m 00ch / 0m 00ch
Brierfield
24m 20ch
The Keighley and Worth Valley Railway
GFs
Damens Jn 2m 21ch
Crossley Evans Sidings
Frizinghall
206m 67ch
New Pudsey
4m 77ch
Mytholmes Tunnel 3m 19-22ch
Oakworth Crossing 2m 70ch
Haworth
3m 53ch
208m 08ch
208m 26ch
Hammerton Street Jn 191m 13ch
Rose Grove
20m 32ch
Burnley Central 22m 05ch
Burnley Barracks 21m 38ch
Burnley Manchester Road 21m 67ch
Loco Shed
GF
Haworth Loop (oou)
Bradford Forster Square
208m 50ch
191m 75ch
Stanningley Tunnel 5m 22ch
Ducketts (R/G)
Gannow Jn. 21m 03ch
Towneley (MCB) 22m 46ch
Towneley Tunnel
GFs
footpath LC's
Oxenhope
4m 69ch
4m 75ch
Sdgs. & Exhibition Shed
Bradford Interchange
40m 27ch
Mill Lane Jn. SB (M)
39m 79ch
190m 43ch
Wakefield Road Tunnel 191m 36ch to 191m 42ch
23m 06ch
23m 25ch
Bowling Junction 39m 20ch
Bowling Tunnel 38m 18ch to 39m 13ch
Mill Lane Junction 40m 01ch via Bowling Jn. 191m 78ch via New Pudsey
Hall Royd Jn. 30m 54ch (via Rose Grove) 19m 61ch (via Rochdale)
Weasel Hall Tunnel 23m 12ch to 23m 17ch
25m 52ch
Holme Tunnel
25m 65ch
Millwood Tunnel 19m 63ch to 19m 73ch
Hebden Bridge
23m 50ch
New Furnace Tunnel 37m 07ch to 37m 10ch
Hall Royd DCE Siding GF 19m 56ch
Mytholmroyd
24m 68ch
Beacon Hill Tunnel 32m 40ch to 33m 10ch
Hipperholme Tunnel 34m 05ch to 34m 22ch
Wyke Tunnel 36m 12ch to 36m 74ch
Copy Pit Crossover 26m 20ch
Portsmouth (R/G) 27m 30ch
DRS Siding
Halifax
32m 28ch
Kitson Wood Tunnel 28m 76ch to 29m 10ch
Hebden Bridge SB (HB)
23m 55ch
Bank House Tunnel 30m 57ch to 30m 67ch
Lightcliffe Tunnel 34m 67ch to 34m 70ch
Batley
Sidings
Todmorden
19m 13ch
Horsfall Tunnel 20m 44ch to 20m 56ch
Sowerby Bridge
28m 51ch
Dryclough Jn. 31m 36ch via Milner Royd Jn. 0m 00ch
Bradley Wood Jn. 35m 59ch via Brighouse 1m 17ch from Bradley Jn.
Castle Hill Tunnel 20m 07ch to 20m 16ch
Salterhebble Tunnels 0m 21ch to 0m 25ch
Brighouse
34m 31ch
Heaton Lodge East Junction (down) 29m 72ch via Deighton 37m 48ch via Brighouse
Walsden
17m 70ch
Sowerby Bridge Tunnel 27m 60ch to 28m 10ch
Milner Royd Jn. 29m 20ch
Winterbutlee Tunnel 17m 46ch to 17m 60ch
Greetland Jn. 30m 77ch via Sowerby Bridge 1m 11ch from Dryclough Jn.
Ravensthorpe
32m 28ch
Bradley Hall Farm No. 1 (UWC) 0m 67ch
Dean Royd Tunnel 16m 74ch to 16m 77ch
16m 64ch
Elland Tunnel 31m 25ch to 31m 44ch
Bradley Jn. 28m 39ch via Deighton 0m 00ch
Summit East Tunnel 16m 65ch to 16m 67ch
Summit Tunnel
1 Halifax SB (H) 32m 28ch
2 Milner Royd Jn. SB (MR) 29m 21ch
3 Healey Mills SB (HM) 42m 64ch
Huddersfield SB (HU) 25m 60ch
Deighton
27m 60ch
Mirfield
38m 30ch
15m 13ch
Summit West Tunnel 15m 06ch to 15m 08ch
Littleborough GF 14m 40ch
Huddersfield 25m 60ch via Slaithwaite 0m 00ch (Denby Dale Line)
Heaton Lodge Jn (up) 29m 54ch via Deighton 37m 30ch via Brighouse
31m 50ch
Castleton East Jn. SB (CE) 8m 52ch / 0m 00ch
Smithy Bridge
12m 60ch
Littleborough
13m 65ch
5 Bradley Tunnel 0m 24ch to 0m 30ch
6 Huddersfield Nth & Sth Tunnels 25m 20ch to 25m 51ch
7 Gledholt Nth & Sth Tunnels 25m 04ch to 25m 15ch
8 Lockwood Tunnel 1m 07ch to 1m 16ch
to Castleton LWR Depot
Smithy Bridge (MCB) 12m 65ch
Slaithwaite
21m 19ch
Lockwood
1m 18ch
Metrolink
Rochdale East Jn. 10m 63ch (via Rochdale) 14m 27ch (via Metrolink
Rochdale
10m 36ch
Marsden
Berry Brow
63
57
62
2
3

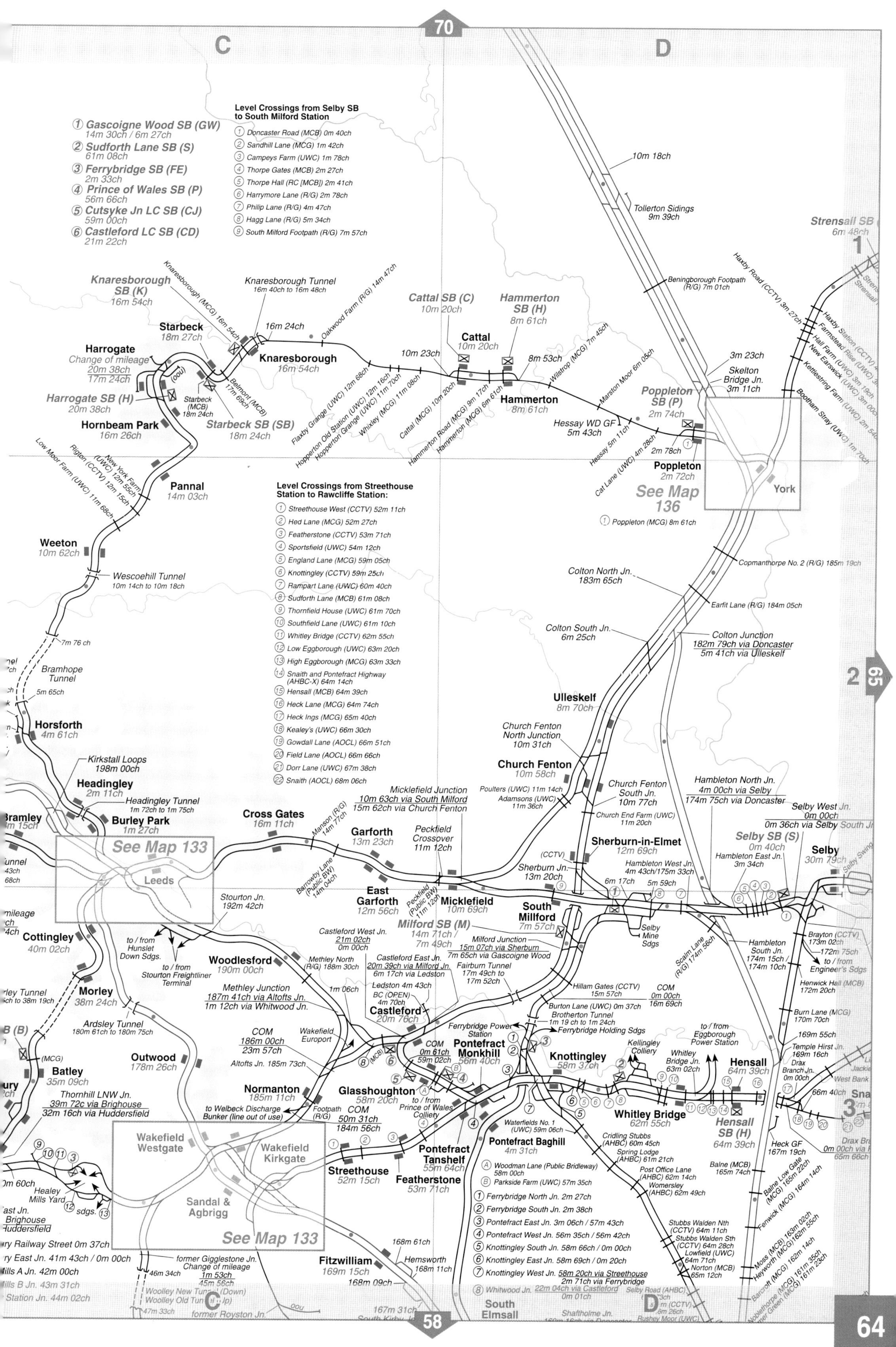

70
C
D
① Gascoigne Wood SB (GW)
14m 30ch / 6m 27ch
② Sudforth Lane SB (S)
61m 08ch
③ Ferrybridge SB (FE)
2m 33ch
④ Prince of Wales SB (P)
56m 66ch
⑤ Cutsyke Jn LC SB (CJ)
59m 00ch
⑥ Castleford LC SB (CD)
21m 22ch
Level Crossings from Selby SB to South Milford Station
① Doncaster Road (MCB) 0m 40ch
② Sandhill Lane (MCG) 1m 42ch
③ Campeys Farm (UWC) 1m 78ch
④ Thorpe Gates (MCB) 2m 27ch
⑤ Thorpe Hall (RC [MCB]) 2m 41ch
⑥ Harrymore Lane (R/G) 2m 78ch
⑦ Philip Lane (R/G) 4m 47ch
⑧ Hagg Lane (R/G) 5m 34ch
⑨ South Milford Footpath (R/G) 7m 57ch
10m 18ch
Tollerton Sidings
9m 39ch
Strensall SB
6m 48ch
1
Beningborough Footpath (R/G) 7m 01ch
Haxby Road (CCTV) 3m 27ch
Knaresborough SB (K)
16m 54ch
Knaresborough Tunnel
16m 40ch to 16m 48ch
Knaresborough (MCG) 16m 54ch
Oakwood Farm (R/G) 14m 47ch
16m 24ch
Starbeck
18m 27ch
Harrogate
Change of mileage
20m 38ch
17m 24ch
Knaresborough
16m 54ch
Cattal SB (C)
10m 20ch
Hammerton SB (H)
8m 61ch
Cattal
10m 20ch
10m 23ch
8m 53ch
Wilstrop (MCG) 7m 45ch
3m 23ch
Skelton Bridge Jn.
3m 11ch
Harrogate SB (H)
20m 38ch
Starbeck (MCB) 18m 24ch
Belmont (MCB) 17m 69ch
Starbeck SB (SB)
18m 24ch
Hornbeam Park
16m 26ch
Flaxby Grange (UWC) 12m 68ch
Hopperton Old Station (UWC) 12m 16ch
Hopperton Grange (UWC) 11m 70ch
Whixley (MCG) 11m 08ch
Cattal (MCG) 10m 20ch
Hammerton Road (MCG) 9m 17ch
Hammerton (MCG) 8m 61ch
Hammerton
8m 61ch
Marston Moor 6m 05ch
Poppleton SB (P)
2m 74ch
Hessay WD GF
5m 43ch
Hessay 5m 11ch
Cat Lane (UWC) 4m 28ch
2m 78ch
Poppleton
2m 72ch
See Map 136
York
① Poppleton (MCG) 8m 61ch
Low Moor Farm (UWC) 11m 68ch
Rigton (CCTV) 12m 15ch
New York Farm (UWC) 12m 55ch
Pannal
14m 03ch
Level Crossings from Streethouse Station to Rawcliffe Station:
① Streethouse West (CCTV) 52m 11ch
② Hed Lane (MCG) 52m 27ch
③ Featherstone (CCTV) 53m 71ch
④ Sportsfield (UWC) 54m 12ch
⑤ England Lane (MCG) 59m 05ch
⑥ Knottingley (CCTV) 59m 25ch
⑦ Rampart Lane (UWC) 60m 40ch
⑧ Sudforth Lane (MCB) 61m 08ch
⑨ Thornfield House (UWC) 61m 70ch
⑩ Southfield Lane (UWC) 61m 10ch
⑪ Whitley Bridge (CCTV) 62m 55ch
⑫ Low Eggborough (UWC) 63m 20ch
⑬ High Eggborough (MCG) 63m 33ch
⑭ Snaith and Pontefract Highway (AHBC-X) 64m 14ch
⑮ Hensall (MCB) 64m 39ch
⑯ Heck Lane (MCG) 64m 74ch
⑰ Heck Ings (MCG) 65m 40ch
⑱ Kealey's (UWC) 66m 30ch
⑲ Gowdall Lane (AOCL) 66m 51ch
⑳ Field Lane (AOCL) 66m 66ch
㉑ Dorr Lane (UWC) 67m 38ch
㉒ Snaith (AOCL) 68m 06ch
Copmanthorpe No. 2 (R/G) 185m 19ch
Weeton
10m 62ch
Wescoehill Tunnel
10m 14ch to 10m 18ch
Colton North Jn.
183m 65ch
Earfit Lane (R/G) 184m 05ch
7m 76 ch
Bramhope Tunnel
Colton South Jn.
6m 25ch
Colton Junction
182m 79ch via Doncaster
5m 41ch via Ulleskelf
5m 65ch
2
65
Ulleskelf
8m 70ch
Horsforth
4m 61ch
Church Fenton North Junction
10m 31ch
Kirkstall Loops
198m 00ch
Church Fenton
10m 58ch
Headingley
2m 11ch
Headingley Tunnel
1m 72ch to 1m 75ch
Micklefield Junction
10m 63ch via South Milford
15m 62ch via Church Fenton
Poulters (UWC) 11m 14ch
Adamsons (UWC) 11m 36ch
Church Fenton South Jn.
10m 77ch
Church End Farm (UWC) 11m 20ch
Hambleton North Jn.
4m 00ch via Selby
174m 75ch via Doncaster
Selby West Jn.
0m 36ch via Selby South Jn.
Bramley
Burley Park
1m 27ch
Cross Gates
16m 11ch
Manston (R/G) 14m 77ch
Garforth
13m 23ch
Peckfield Crossover
11m 12ch
Sherburn-in-Elmet
12m 69ch
Selby SB (S)
0m 40ch
Hambleton East Jn.
3m 34ch
Selby
30m 79ch
See Map 133
Leeds
Barrowby Lane (Public BW) 14m 04ch
East Garforth
12m 56ch
Peckfield (Public BW) 11m 12ch
Micklefield
10m 69ch
Sherburn Jn.
13m 20ch
Hambleton West Jn.
4m 43ch/175m 33ch
6m 17ch
5m 59ch
Stourton Jn.
192m 42ch
South Milford
7m 57ch
Selby Mine Sdgs
Cottingley
40m 02ch
to / from Hunslet Down Sdgs.
to / from Stourton Freightliner Terminal
Castleford West Jn.
21m 02ch
0m 00ch
Milford SB (M)
14m 71ch / 7m 49ch
Milford Junction
15m 07ch via Sherburn
7m 65ch via Gascoigne Wood
Scalm Lane (R/G) 174m 56ch
Hambleton South Jn.
174m 15ch / 174m 10ch
Brayton (CCTV) 173m 02ch
172m 75ch
to / from Engineer's Sdgs
Woodlesford
190m 00ch
Methley North (R/G) 188m 30ch
Castleford East Jn.
20m 39ch via Milford Jn.
6m 17ch via Ledston
Fairburn Tunnel
17m 49ch to 17m 52ch
Henwick Hall (MCB) 172m 20ch
Morley
38m 24ch
Methley Junction
187m 41ch via Altofts Jn.
1m 12ch via Whitwood Jn.
1m 06ch
Ledston 4m 43ch
BC (OPEN) 4m 70ch
Castleford
20m 76ch
Hillam Gates (CCTV) 15m 57ch
COM
0m 00ch
16m 69ch
Burton Lane (UWC) 0m 37ch
Brotherton Tunnel
1m 19 ch to 1m 24ch
Ferrybridge Holding Sdgs
Burn Lane (MCG) 170m 70ch
Ardsley Tunnel
180m 61ch to 180m 75ch
COM
186m 00ch
23m 57ch
Wakefield Europort
Ferrybridge Power Station
to / from Eggborough Power Station
169m 55ch
Temple Hirst Jn.
169m 16ch
Drax Branch Jn.
0m 00ch
Batley
35m 09ch
Outwood
178m 26ch
Altofts Jn. 185m 73ch
COM
0m 61ch
59m 02ch
Pontefract Monkhill
56m 40ch
Knottingley
58m 37ch
Kellingley Colliery
Whitley Bridge Jn.
63m 02ch
Hensall
64m 39ch
66m 40ch
Thornhill LNW Jn.
39m 72c via Brighouse
32m 16ch via Huddersfield
Normanton
185m 11ch
to Welbeck Discharge Bunker (line out of use)
Footpath (R/G)
Glasshoughton
58m 20ch
COM
50m 31ch
184m 56ch
to / from Prince of Wales Colliery
Whitley Bridge
62m 55ch
Hensall SB (H)
64m 39ch
Heck GF
167m 19ch
Waterfields No. 1 (UWC) 59m 06ch
Pontefract Baghill
4m 31ch
Cridling Stubbs (AHBC) 60m 45ch
Spring Lodge (AHBC) 61m 21ch
Post Office Lane (AHBC) 62m 14ch
Womersley (AHBC) 62m 49ch
Balne (MCB) 165m 74ch
Balne Low Gate (MCG) 165m 22ch
Fenwick (MCG) 164m 14ch
Wakefield Westgate
Wakefield Kirkgate
Streethouse
52m 15ch
Pontefract Tanshelf
55m 64ch
Featherstone
53m 71ch
Ⓐ Woodman Lane (Public Bridleway) 58m 00ch
Ⓑ Parkside Farm (UWC) 57m 35ch
① Ferrybridge North Jn. 2m 27ch
② Ferrybridge South Jn. 2m 38ch
③ Pontefract East Jn. 3m 06ch / 57m 43ch
④ Pontefract West Jn. 56m 35ch / 56m 42ch
⑤ Knottingley South Jn. 58m 66ch / 0m 00ch
⑥ Knottingley East Jn. 58m 69ch / 0m 20ch
⑦ Knottingley West Jn. 58m 20ch via Streethouse
2m 71ch via Ferrybridge
⑧ Whitwood Jn. 22m 04ch via Castleford
0m 01ch
Healey Mills Yard
sdgs.
Sandal & Agbrigg
See Map 133
168m 61ch
Hemsworth
168m 11ch
Fitzwilliam
169m 15ch
168m 09ch
Stubbs Walden Nth (CCTV) 64m 11ch
Stubbs Walden Sth (CCTV) 64m 28ch
Lowfield (UWC) 64m 71ch
Norton (MCB) 65m 12ch
Moss (MCB) 163m 02ch
Heyworth (MCG) 162m 55ch
Barcroft (MCG) 162m 14ch
Noblethorpe (MCG) 161m 35ch
former Gigglestone Jn.
Change of mileage
1m 53ch
45m 56ch
46m 34ch
47m 33ch
former Royston Jn.
167m 31ch
South Elmsall
Shaftholme Jn.
Selby Road (AHBC)
Rushey Moor (UWC)
58

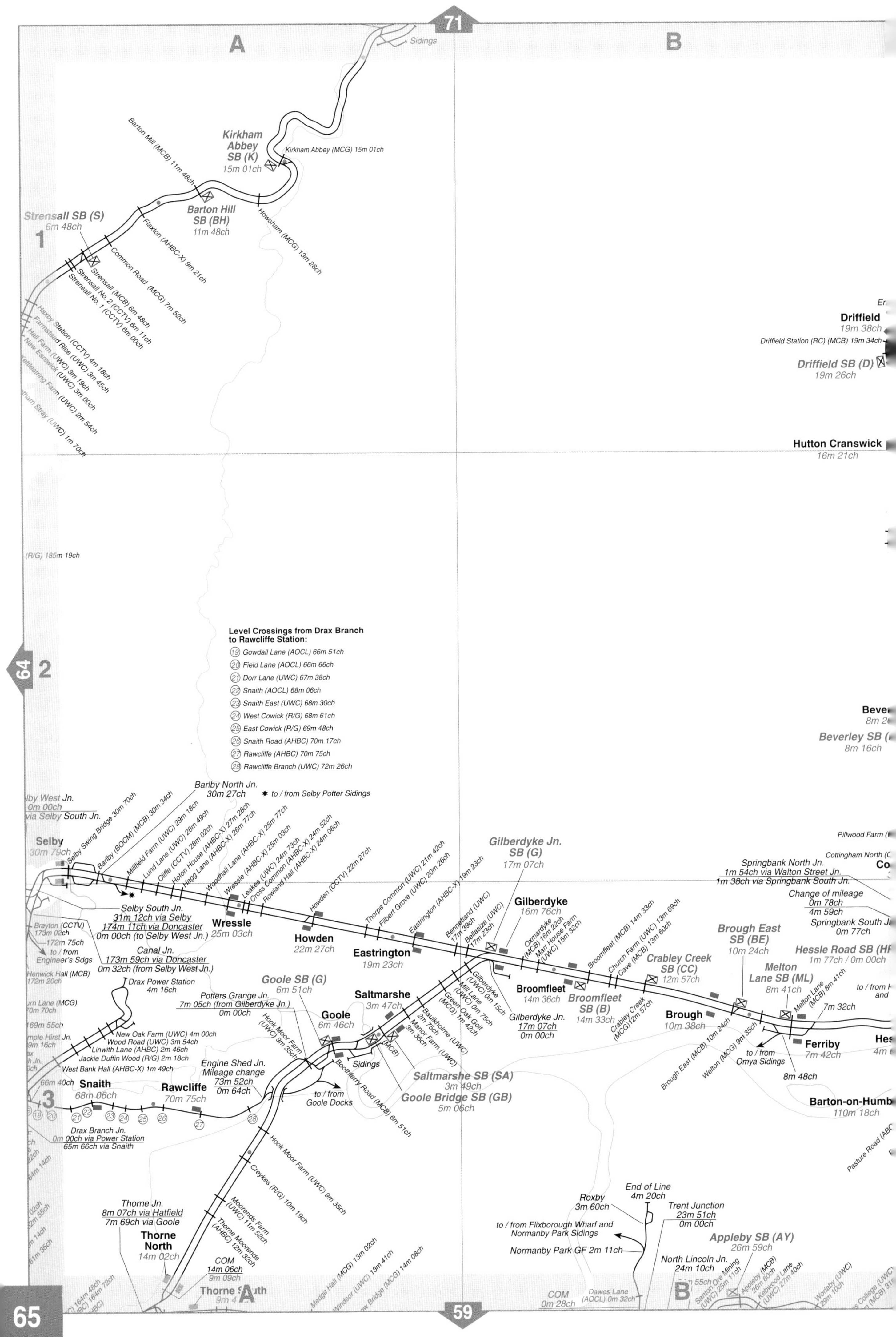
A
B
Sidings
Kirkham Abbey SB (K)
15m 01ch
Kirkham Abbey (MCG) 15m 01ch
Barton Mill (MCB) 11m 48ch
Barton Hill SB (BH)
11m 48ch
Howsham (MCG) 13m 28ch
Strensall SB (S)
6m 48ch
1
Flaxton (AHBC-X) 9m 21ch
Common Road (MCG) 7m 52ch
Strensall (MCB) 6m 48ch
Strensall No. 2 (CCTV) 6m 11ch
Strensall No. 1 (CCTV) 6m 00ch
Haxby Station (CCTV) 4m 18ch
Farmstead Rise (UWC) 3m 45ch
Hall Farm (UWC) 3m 19ch
New Earswick (UWC) 3m 00ch
Kettlestring Farm (UWC) 2m 54ch
Stray (UWC) 1m 70ch
Driffield
19m 38ch
Driffield Station (RC) (MCB) 19m 34ch
Driffield SB (D)
19m 26ch
Hutton Cranswick
16m 21ch
(R/G) 185m 19ch
Level Crossings from Drax Branch to Rawcliffe Station:
19 Gowdall Lane (AOCL) 66m 51ch
20 Field Lane (AOCL) 66m 66ch
21 Dorr Lane (UWC) 67m 38ch
22 Snaith (AOCL) 68m 06ch
23 Snaith East (UWC) 68m 30ch
24 West Cowick (R/G) 68m 61ch
25 East Cowick (R/G) 69m 48ch
26 Snaith Road (AHBC) 70m 17ch
27 Rawcliffe (AHBC) 70m 75ch
28 Rawcliffe Branch (UWC) 72m 26ch
2
Beverley SB
8m 16ch
Barlby North Jn.
30m 27ch
* to / from Selby Potter Sidings
0m 00ch
via Selby South Jn.
Selby
30m 79ch
Selby Swing Bridge 30m 70ch
Barlby (BOCM) (MCB) 30m 34ch
Millfield Farm (UWC) 29m 18ch
Lund Lane (UWC) 28m 49ch
Cliffe (CCTV) 28m 02ch
Hoton House (AHBC-X) 27m 28ch
Hagg Lane (AHBC-X) 26m 77ch
Woodhall Lane (AHBC-X) 25m 77ch
Wressle (AHBC-X) 25m 03ch
Leakes (UWC) 24m 73ch
Cross Common (AHBC-X) 24m 52ch
Rowland Hall (AHBC-X) 24m 06ch
Howden (CCTV) 22m 27ch
Thorpe Common (UWC) 21m 42ch
Filbert Grove (UWC) 20m 26ch
Eastrington (AHBC-X) 19m 23ch
Gilberdyke Jn. SB (G)
17m 07ch
Bennetland (UWC) 17m 39ch
Bellasize (UWC) 17m 23ch
Gilberdyke
16m 76ch
Oxmardyke (MCB) 16m 22ch
Marr House Farm (UWC) 15m 32ch
Broomfleet (MCB) 14m 33ch
Church Farm (UWC) 13m 69ch
Cave (MCB) 13m 60ch
Pillwood Farm
Cottingham North
Springbank North Jn.
1m 54ch via Walton Street Jn.
1m 38ch via Springbank South Jn.
Change of mileage
0m 78ch
4m 59ch
Springbank South Jn.
0m 77ch
Selby South Jn.
31m 12ch via Selby
174m 11ch via Doncaster
0m 00ch (to Selby West Jn.)
Wressle
25m 03ch
Howden
22m 27ch
Eastrington
19m 23ch
Brayton (CCTV) 173m 02ch
172m 75ch
to / from Engineer's Sdgs
Canal Jn.
173m 59ch via Doncaster
0m 32ch (from Selby West Jn.)
Henwick Hall (MCB) 172m 20ch
Drax Power Station
4m 16ch
Goole SB (G)
6m 51ch
Potters Grange Jn.
7m 05ch (from Gilberdyke Jn.)
0m 00ch
Saltmarshe
3m 47ch
Gilberdyke (UWC) 0m 15ch
Mill Lane (UWC) 0m 75ch
Green Oak Goit (MCG) 1m 42ch
Baulkholme (UWC) 2m 75ch
Manor Farm (UWC) 3m 36ch
Broomfleet
14m 36ch
Broomfleet SB (B)
14m 33ch
Gilberdyke Jn.
17m 07ch
0m 00ch
Crabley Creek SB (CC)
12m 57ch
Crabley Creek (MCG) 12m 57ch
Brough East SB (BE)
10m 24ch
Melton Lane SB (ML)
8m 41ch
Melton Lane (MCB) 8m 41ch
Hessle Road SB (HR)
1m 77ch / 0m 00ch
7m 32ch
Brough
10m 38ch
Ferriby
7m 42ch
Brough East (MCB) 10m 24ch
Welton (MCG) 9m 35ch
to / from Omya Sidings
8m 48ch
Barton-on-Humber
110m 18ch
Pasture Road
Lane (MCG) 70m 70ch
169m 55ch
Temple Hirst Jn.
New Oak Farm (UWC) 4m 00ch
Wood Road (UWC) 3m 54ch
Linwith Lane (AHBC) 2m 46ch
Jackie Duffin Wood (R/G) 2m 18ch
West Bank Hall (AHBC-X) 1m 49ch
Goole
6m 46ch
Hook Moor Farm (UWC) 9m 35ch
(MCB)
Sidings
Boothferry Road (MCB) 6m 51ch
Saltmarshe SB (SA)
3m 49ch
Goole Bridge SB (GB)
5m 06ch
Engine Shed Jn.
Mileage change
73m 52ch
0m 64ch
66m 40ch
Snaith
68m 06ch
Rawcliffe
70m 75ch
to / from Goole Docks
3
Drax Branch Jn.
0m 00ch via Power Station
65m 66ch via Snaith
Hook Moor Farm (UWC) 9m 35ch
Creykes (R/G) 10m 19ch
Moorends Farm (UWC) 11m 52ch
Thorne Moorends (AHBC) 12m 32ch
Thorne Jn.
8m 07ch via Hatfield
7m 69ch via Goole
Thorne North
14m 02ch
COM
14m 06ch
9m 09ch
Thorne South
Medge Hall (MCG) 13m 02ch
Windsor (UWC) 13m 41ch
Bridge (MCG) 14m 08ch
End of Line
4m 20ch
Roxby
3m 60ch
Trent Junction
23m 51ch
0m 00ch
to / from Flixborough Wharf and Normanby Park Sidings
Normanby Park GF 2m 11ch
Appleby SB (AY)
26m 59ch
North Lincoln Jn.
24m 10ch
COM
0m 28ch
Dawes Lane (AOCL) 0m 32ch
Santon Ore Mining (UWC) 25m 11ch
Appleby (MCB) 26m 60ch
Kelwood Lane (UWC) 27m 40ch
Worlaby (UWC) 29m 10ch

64

59

72

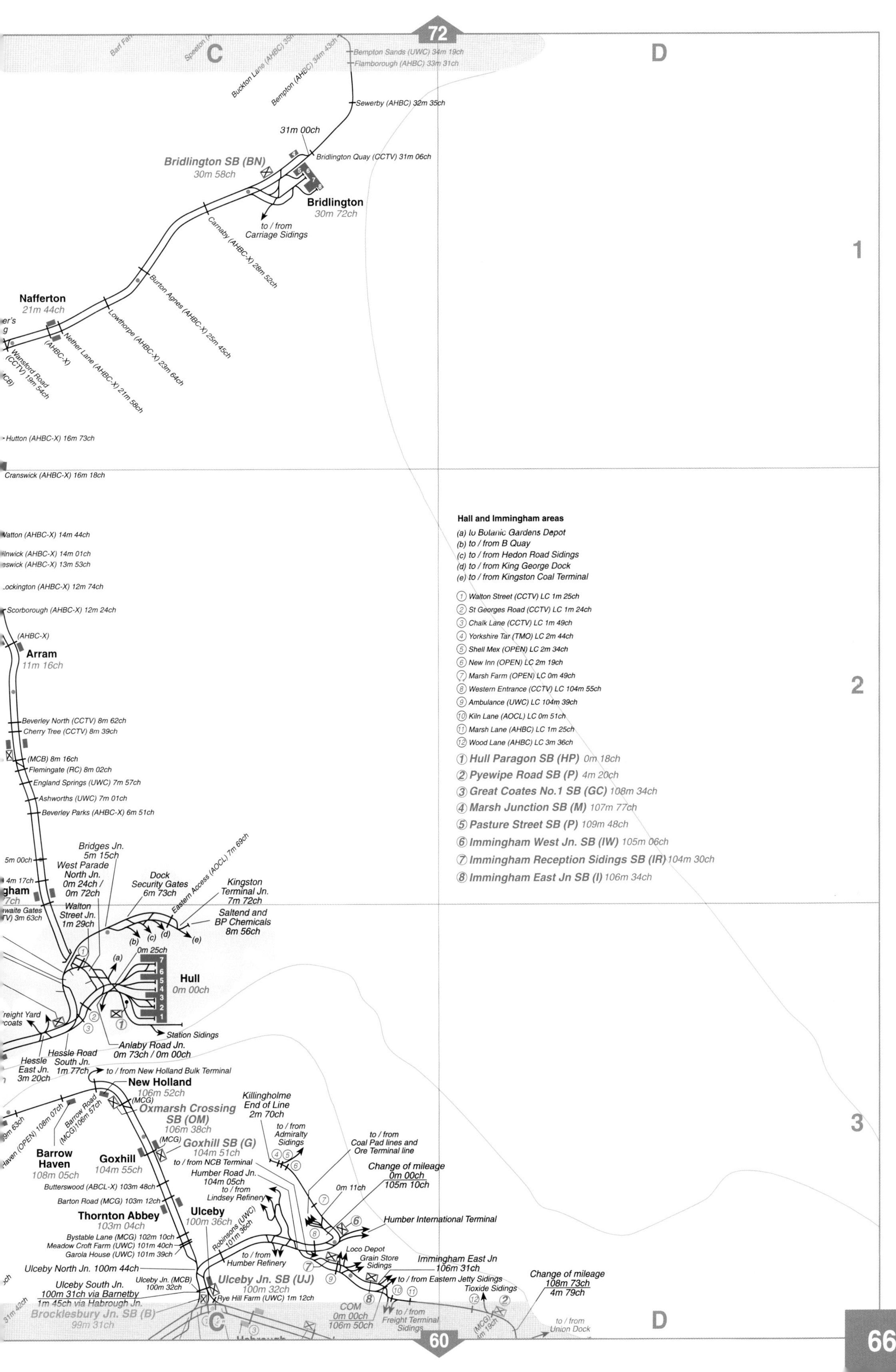
C
D
Bempton Sands (UWC) 34m 19ch
Flamborough (AHBC) 33m 31ch
Buckton Lane (AHBC) 35m
Bempton (AHBC) 34m 43ch
Sewerby (AHBC) 32m 35ch
31m 00ch
Bridlington Quay (CCTV) 31m 06ch
Bridlington SB (BN)
30m 58ch
Bridlington
30m 72ch
to / from
Carriage Sidings
Carnaby (AHBC-X) 28m 52ch
Burton Agnes (AHBC-X) 25m 45ch
Lowthorpe (AHBC-X) 23m 64ch
Nafferton
21m 44ch
Nether Lane (AHBC-X) 21m 58ch
Wansford Road (CCTV) 19m 54ch
Hutton (AHBC-X) 16m 73ch
Cranswick (AHBC-X) 16m 18ch
Scorborough (AHBC-X) 12m 24ch
(AHBC-X)
Arram
11m 16ch
Beverley North (CCTV) 8m 62ch
Cherry Tree (CCTV) 8m 39ch
(MCB) 8m 16ch
Flemingate (RC) 8m 02ch
England Springs (UWC) 7m 57ch
Ashworths (UWC) 7m 01ch
Beverley Parks (AHBC-X) 6m 51ch
1
2
3
Hall and Immingham areas
(a) to Botanic Gardens Depot
(b) to / from B Quay
(c) to / from Hedon Road Sidings
(d) to / from King George Dock
(e) to / from Kingston Coal Terminal
1 Walton Street (CCTV) LC 1m 25ch
2 St Georges Road (CCTV) LC 1m 24ch
3 Chalk Lane (CCTV) LC 1m 49ch
4 Yorkshire Tar (TMO) LC 2m 44ch
5 Shell Mex (OPEN) LC 2m 34ch
6 New Inn (OPEN) LC 2m 19ch
7 Marsh Farm (OPEN) LC 0m 49ch
8 Western Entrance (CCTV) LC 104m 55ch
9 Ambulance (UWC) LC 104m 39ch
10 Kiln Lane (AOCL) LC 0m 51ch
11 Marsh Lane (AHBC) LC 1m 25ch
12 Wood Lane (AHBC) LC 3m 36ch
1 Hull Paragon SB (HP) 0m 18ch
2 Pyewipe Road SB (P) 4m 20ch
3 Great Coates No.1 SB (GC) 108m 34ch
4 Marsh Junction SB (M) 107m 77ch
5 Pasture Street SB (P) 109m 48ch
6 Immingham West Jn. SB (IW) 105m 06ch
7 Immingham Reception Sidings SB (IR) 104m 30ch
8 Immingham East Jn SB (I) 106m 34ch
Bridges Jn.
5m 15ch
West Parade North Jn.
0m 24ch / 0m 72ch
Walton Street Jn.
1m 29ch
Dock Security Gates
6m 73ch
Eastern Access (AOCL) 7m 69ch
Kingston Terminal Jn.
7m 72ch
Saltend and BP Chemicals
8m 56ch
0m 25ch
Hull
0m 00ch
Station Sidings
Anlaby Road Jn.
0m 73ch / 0m 00ch
Hessle Road South Jn.
1m 77ch
Hessle East Jn.
3m 20ch
to / from New Holland Bulk Terminal
New Holland
106m 52ch
Oxmarsh Crossing SB (OM)
106m 38ch
Barrow Road (MCG) 106m 57ch
Goxhill SB (G)
104m 51ch
Barrow Haven
108m 05ch
Goxhill
104m 55ch
to / from NCB Terminal
Killingholme End of Line
2m 70ch
to / from Admiralty Sidings
to / from Coal Pad lines and Ore Terminal line
Change of mileage
0m 00ch
105m 10ch
Humber Road Jn.
104m 05ch
to / from Lindsey Refinery
0m 11ch
Humber International Terminal
Butterswood (ABCL-X) 103m 48ch
Barton Road (MCG) 103m 12ch
Thornton Abbey
103m 04ch
Ulceby
100m 36ch
Robinsons (UWC) 101m 36ch
Bystable Lane (MCG) 102m 10ch
Meadow Croft Farm (UWC) 101m 40ch
Garola House (UWC) 101m 39ch
to / from Humber Refinery
Loco Depot
Grain Store Sidings
Immingham East Jn
106m 31ch
to / from Eastern Jetty Sidings
Tioxide Sidings
Change of mileage
108m 73ch
4m 79ch
Ulceby North Jn. 100m 44ch
Ulceby Jn. (MCB) 100m 32ch
Ulceby Jn. SB (UJ)
100m 32ch
Ulceby South Jn.
100m 31ch via Barnetby
1m 45ch via Habrough Jn.
Rye Hill Farm (UWC) 1m 12ch
Brocklesbury Jn. SB (B)
99m 31ch
COM
0m 00ch
106m 50ch
to / from Freight Terminal Sidings
to / from Union Dock

60

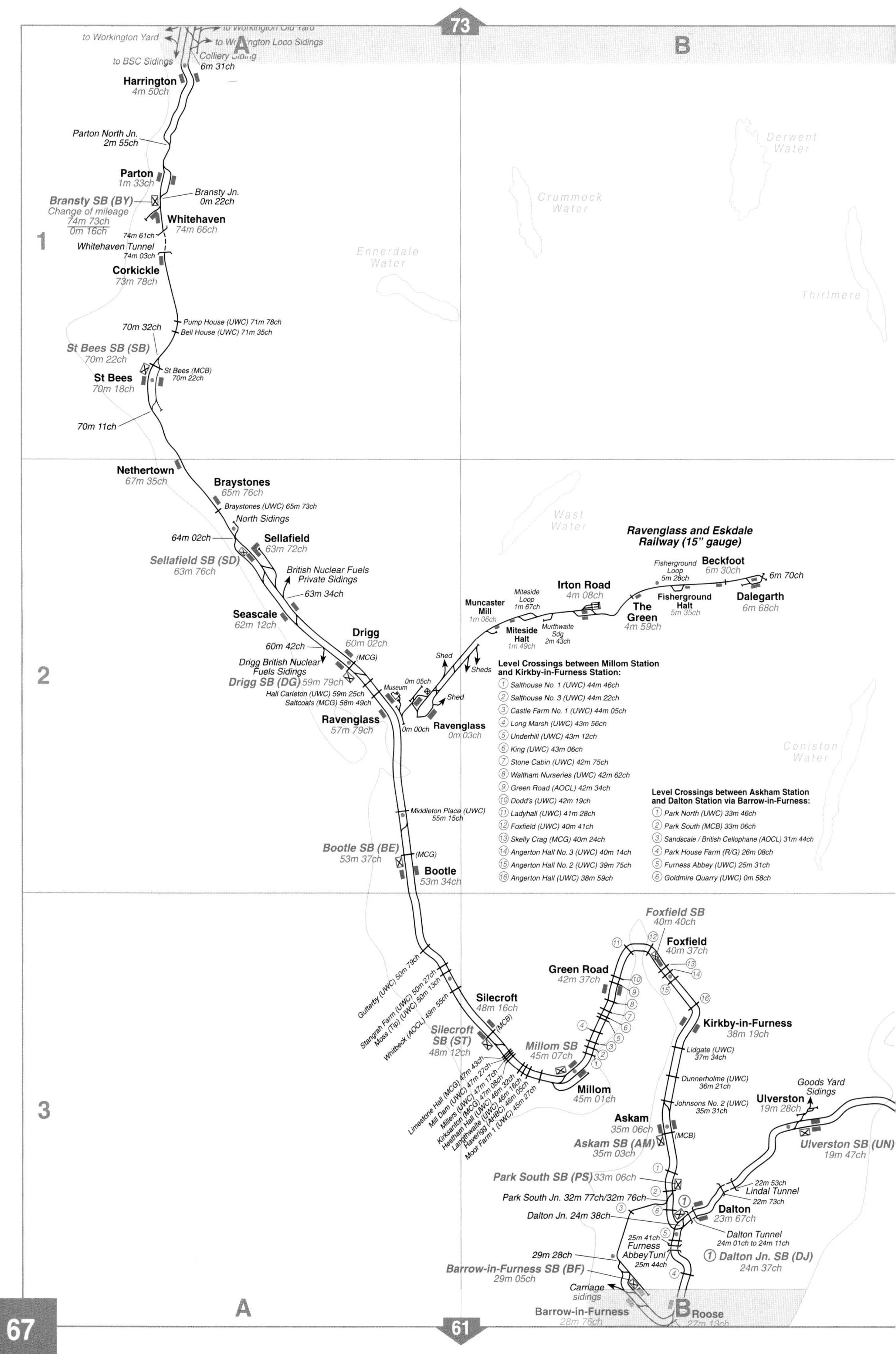

73
A
B
to Workington Old Yard
to Workington Yard
to Workington Loco Sidings
to BSC Sidings
Colliery Siding
6m 31ch
Harrington
4m 50ch
Parton North Jn.
2m 55ch
Parton
1m 33ch
Bransty SB (BY)
Change of mileage
74m 73ch
0m 16ch
Bransty Jn.
0m 22ch
Whitehaven
74m 66ch
74m 61ch
Whitehaven Tunnel
74m 03ch
Corkickle
73m 78ch
Pump House (UWC) 71m 78ch
Bell House (UWC) 71m 35ch
70m 32ch
St Bees SB (SB)
70m 22ch
St Bees (MCB)
70m 22ch
St Bees
70m 18ch
70m 11ch
Ennerdale Water
Crummock Water
Derwent Water
Thirlmere
1
Nethertown
67m 35ch
Braystones
65m 76ch
Braystones (UWC) 65m 73ch
North Sidings
64m 02ch
Sellafield
63m 72ch
Sellafield SB (SD)
63m 76ch
British Nuclear Fuels
Private Sidings
63m 34ch
Seascale
62m 12ch
Drigg
60m 02ch
60m 42ch
(MCG)
Drigg British Nuclear
Fuels Sidings
Drigg SB (DG) 59m 79ch
Hall Carleton (UWC) 59m 25ch
Saltcoats (MCG) 58m 49ch
Ravenglass
57m 79ch
Museum
0m 05ch
Shed
Sheds
Shed
0m 00ch
Ravenglass
0m 03ch
Wast Water
Ravenglass and Eskdale
Railway (15" gauge)
Muncaster
Mill
1m 06ch
Miteside
Loop
1m 67ch
Miteside
Halt
1m 49ch
Murthwaite
Sdg
2m 43ch
Irton Road
4m 08ch
The
Green
4m 59ch
Fisherground
Loop
5m 28ch
Fisherground
Halt
5m 35ch
Beckfoot
6m 30ch
6m 70ch
Dalegarth
6m 68ch
2
Level Crossings between Millom Station
and Kirkby-in-Furness Station:
1 Salthouse No. 1 (UWC) 44m 46ch
2 Salthouse No. 3 (UWC) 44m 22ch
3 Castle Farm No. 1 (UWC) 44m 05ch
4 Long Marsh (UWC) 43m 56ch
5 Underhill (UWC) 43m 12ch
6 King (UWC) 43m 06ch
7 Stone Cabin (UWC) 42m 75ch
8 Waltham Nurseries (UWC) 42m 62ch
9 Green Road (AOCL) 42m 34ch
10 Dodd's (UWC) 42m 19ch
11 Ladyhall (UWC) 41m 28ch
12 Foxfield (UWC) 40m 41ch
13 Skelly Crag (MCG) 40m 24ch
14 Angerton Hall No. 3 (UWC) 40m 14ch
15 Angerton Hall No. 2 (UWC) 39m 75ch
16 Angerton Hall (UWC) 38m 59ch
Coniston Water
Level Crossings between Askham Station
and Dalton Station via Barrow-in-Furness:
1 Park North (UWC) 33m 46ch
2 Park South (MCB) 33m 06ch
3 Sandscale / British Cellophane (AOCL) 31m 44ch
4 Park House Farm (R/G) 26m 08ch
5 Furness Abbey (UWC) 25m 31ch
6 Goldmire Quarry (UWC) 0m 58ch
Middleton Place (UWC)
55m 15ch
Bootle SB (BE)
53m 37ch
(MCG)
Bootle
53m 34ch
Foxfield SB
40m 40ch
Foxfield
40m 37ch
Green Road
42m 37ch
Gutterby (UWC) 50m 79ch
Stangrah Farm (UWC) 50m 27ch
Moss (Tip) (UWC) 50m 13ch
Whitbeck (AOCL) 49m 55ch
Silecroft
48m 16ch
(MCB)
Silecroft
SB (ST)
48m 12ch
Millom SB
45m 07ch
Millom
45m 01ch
Kirkby-in-Furness
38m 19ch
Limestone Hall (MCG) 47m 43ch
Mill Dam (UWC) 47m 27ch
Millers (UWC) 47m 17ch
Kirksanton (MCG) 47m 08ch
Hestham Hall (UWC) 46m 32ch
Langthwaite (UWC) 46m 16ch
Haverigg (AHBC) 46m 05ch
Moor Farm 1 (UWC) 45m 27ch
Lidgate (UWC)
37m 34ch
Dunnerholme (UWC)
36m 21ch
Johnsons No. 2 (UWC)
35m 31ch
3
Askam
35m 06ch
(MCB)
Askam SB (AM)
35m 03ch
Goods Yard
Sidings
Ulverston
19m 28ch
Ulverston SB (UN)
19m 47ch
Park South SB (PS) 33m 06ch
Park South Jn. 32m 77ch/32m 76ch
22m 53ch
Lindal Tunnel
22m 73ch
Dalton Jn. 24m 38ch
Dalton
23m 67ch
Dalton Tunnel
24m 01ch to 24m 11ch
25m 41ch
Furness
Abbey Tunl
25m 44ch
1 Dalton Jn. SB (DJ)
24m 37ch
29m 28ch
Barrow-in-Furness SB (BF)
29m 05ch
Carriage
sidings
A
B
Barrow-in-Furness
28m 76ch
Roose
27m 13ch
61

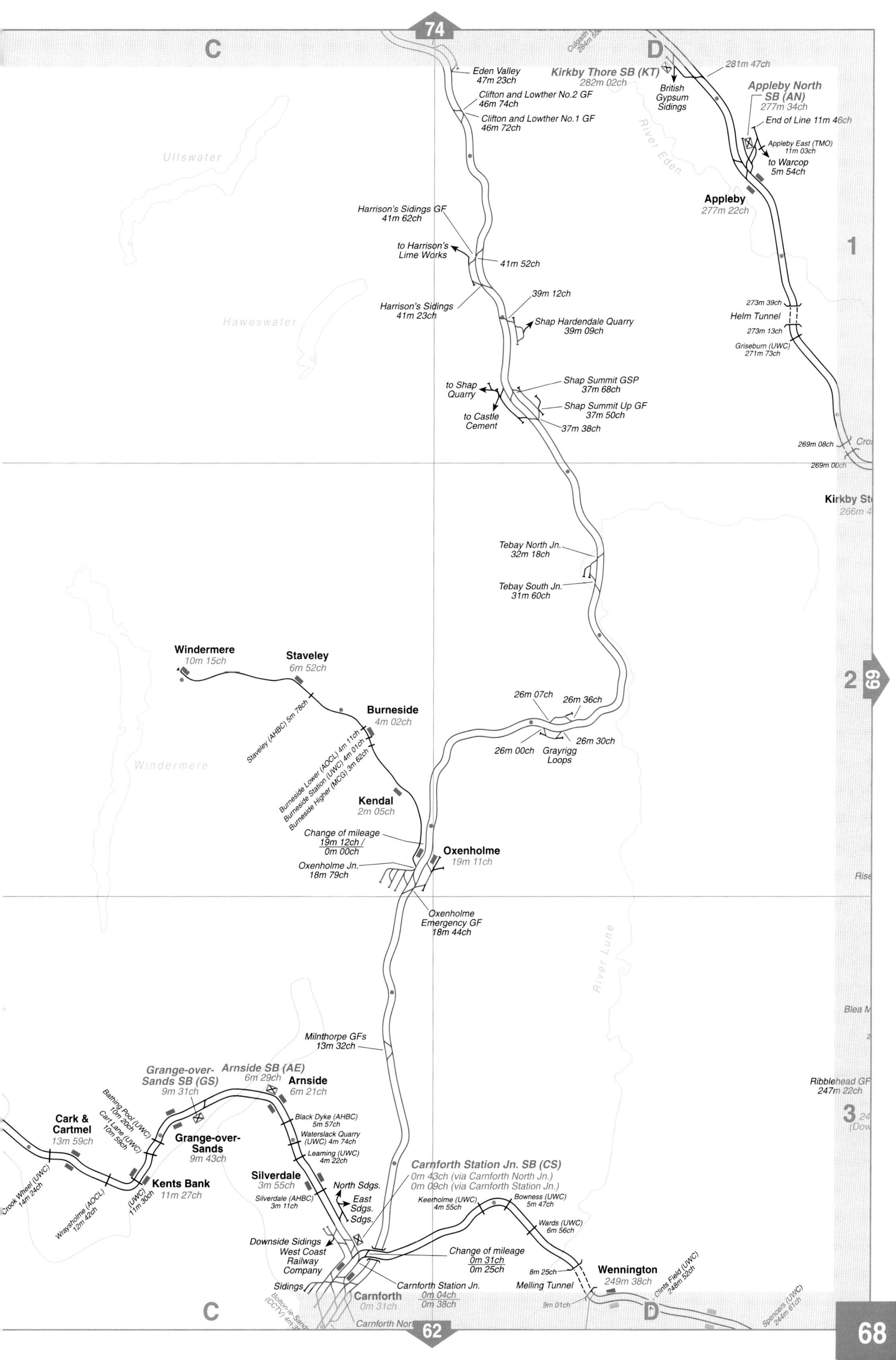
74
C
D
Eden Valley
47m 23ch
Kirkby Thore SB (KT)
282m 02ch
British
Gypsum
Sidings
281m 47ch
Appleby North
SB (AN)
277m 34ch
End of Line 11m 46ch
Appleby East (TMO)
11m 03ch
to Warcop
5m 54ch
Appleby
277m 22ch
Clifton and Lowther No.2 GF
46m 74ch
Clifton and Lowther No.1 GF
46m 72ch
Ullswater
River Eden
Harrison's Sidings GF
41m 62ch
to Harrison's
Lime Works
41m 52ch
Harrison's Sidings
41m 23ch
39m 12ch
Shap Hardendale Quarry
39m 09ch
Haweswater
273m 39ch
Helm Tunnel
273m 13ch
Griseburn (UWC)
271m 73ch
1
to Shap
Quarry
to Castle
Cement
Shap Summit GSP
37m 68ch
Shap Summit Up GF
37m 50ch
37m 38ch
269m 08ch
269m 00ch
Tebay North Jn.
32m 18ch
Tebay South Jn.
31m 60ch
Windermere
10m 15ch
Staveley
6m 52ch
Burneside
4m 02ch
Staveley (AHBC) 5m 78ch
Burneside Lower (AOCL) 4m 11ch
Burneside Station (UWC) 4m 01ch
Burneside Higher (MCG) 3m 62ch
Windermere
26m 07ch
26m 36ch
26m 00ch
Grayrigg
Loops
26m 30ch
2
69
Kendal
2m 05ch
Change of mileage
19m 12ch /
0m 00ch
Oxenholme Jn.
18m 79ch
Oxenholme
19m 11ch
Oxenholme
Emergency GF
18m 44ch
River Lune
Milnthorpe GFs
13m 32ch
Ribblehead GF
247m 22ch
3
Grange-over-
Sands SB (GS)
9m 31ch
Arnside SB (AE)
6m 29ch
Arnside
6m 21ch
Bathing Pool (UWC)
10m 20ch
Cart Lane (UWC)
10m 59ch
Cark &
Cartmel
13m 59ch
Grange-over-
Sands
9m 43ch
Black Dyke (AHBC)
5m 57ch
Waterslack Quarry
(UWC) 4m 74ch
Leaming (UWC)
4m 22ch
Carnforth Station Jn. SB (CS)
0m 43ch (via Carnforth North Jn.)
0m 09ch (via Carnforth Station Jn.)
Crook Wheel (UWC)
14m 24ch
Wraysholme (AOCL)
12m 42ch
(UWC)
11m 30ch
Kents Bank
11m 27ch
Silverdale
3m 55ch
Silverdale (AHBC)
3m 11ch
North Sdgs.
East
Sdgs.
Sdgs.
Keerholme (UWC)
4m 55ch
Bowness (UWC)
5m 47ch
Wards (UWC)
6m 56ch
Downside Sidings
West Coast
Railway
Company
Sidings
Change of mileage
0m 31ch
0m 25ch
Carnforth Station Jn.
0m 04ch
0m 38ch
8m 25ch
Melling Tunnel
9m 01ch
Wennington
249m 38ch
Clints Field (UWC)
248m 52ch
Carnforth
0m 31ch
Carnforth North
C
D
62

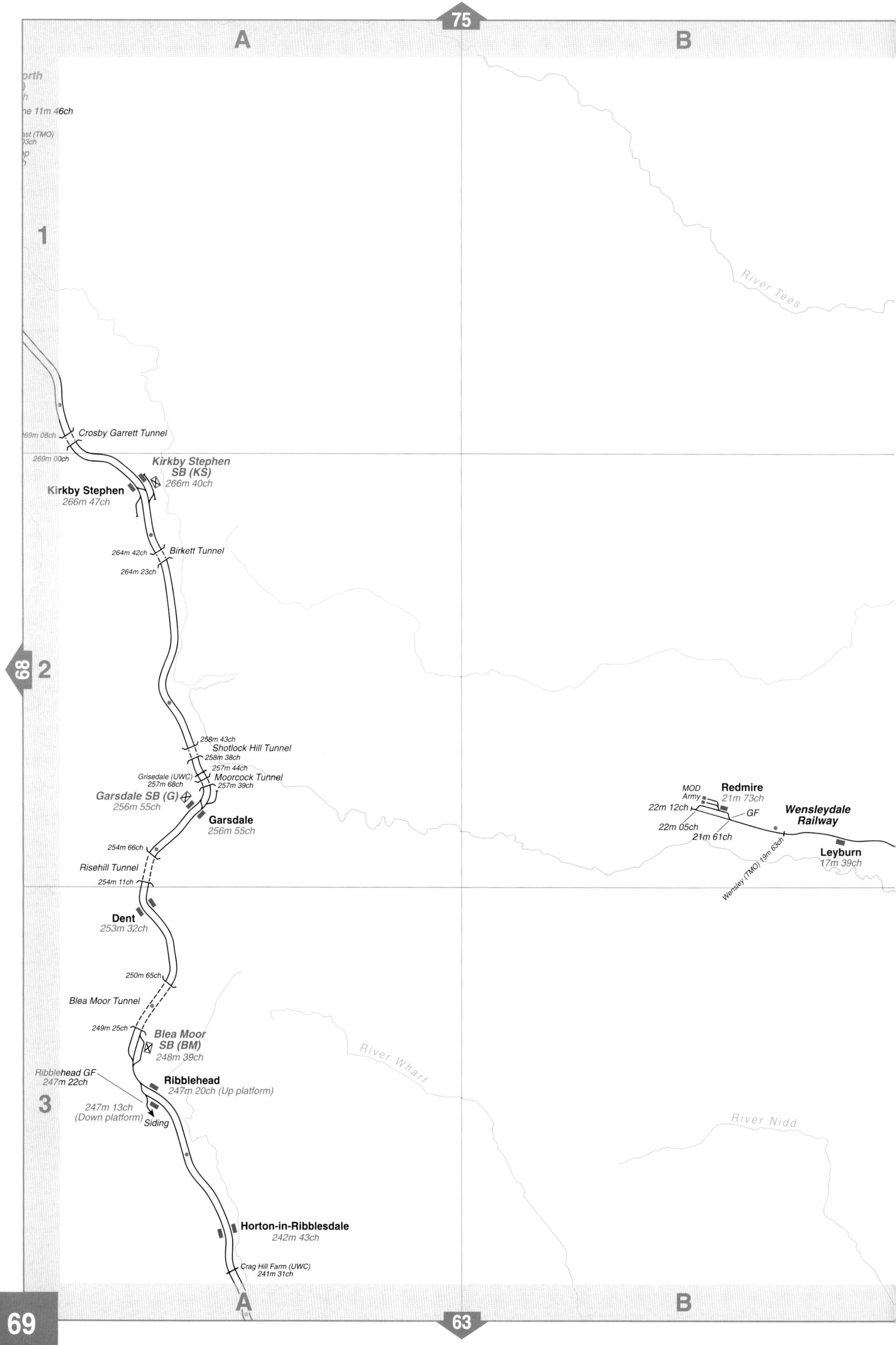
A
B
1
2
3
68
Crosby Garrett Tunnel
269m 08ch
269m 00ch
Kirkby Stephen SB (KS)
266m 40ch
Kirkby Stephen
266m 47ch
264m 42ch
Birkett Tunnel
264m 23ch
River Tees
258m 43ch
Shotlock Hill Tunnel
258m 38ch
257m 44ch
Grisedale (UWC)
257m 68ch
Moorcock Tunnel
257m 39ch
Garsdale SB (G)
256m 55ch
Garsdale
256m 55ch
254m 66ch
Risehill Tunnel
254m 11ch
Dent
253m 32ch
250m 65ch
Blea Moor Tunnel
249m 25ch
Blea Moor SB (BM)
248m 39ch
Ribblehead GF
247m 22ch
Ribblehead
247m 20ch (Up platform)
247m 13ch
(Down platform)
Siding
Horton-in-Ribblesdale
242m 43ch
Crag Hill Farm (UWC)
241m 31ch
River Wharf
River Nidd
MOD
Army
Redmire
21m 73ch
22m 12ch
GF
22m 05ch
21m 61ch
Wensleydale Railway
Wensley (TMO) 19m 63ch
Leyburn
17m 39ch
A
B

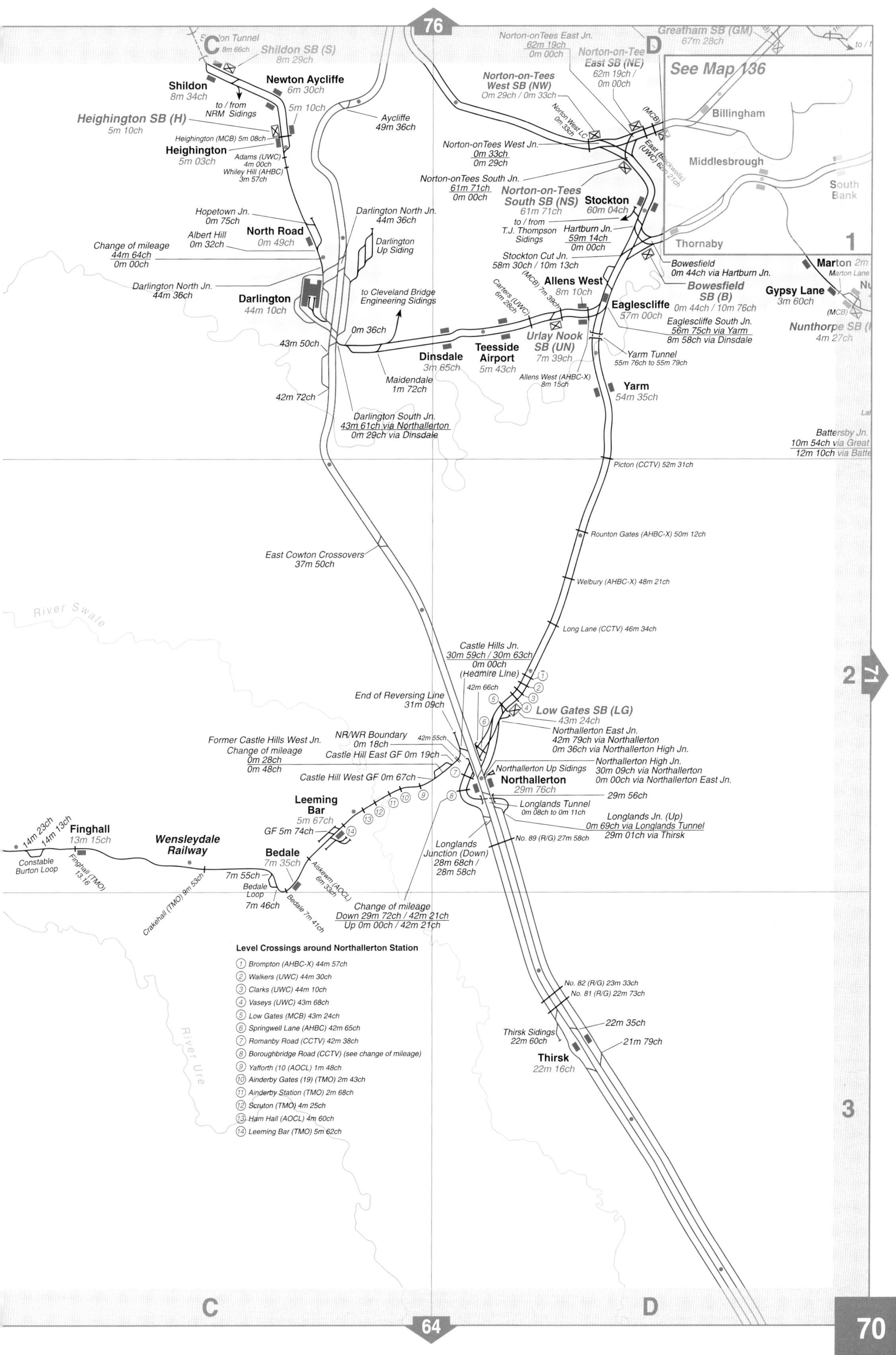
Shildon SB (S)
8m 29ch
Shildon
8m 34ch
Newton Aycliffe
6m 30ch
to / from NRM Sidings
5m 10ch
Heighington SB (H)
5m 10ch
Heighington (MCB) 5m 08ch
Heighington
5m 03ch
Adams (UWC) 4m 00ch
Whiley Hill (AHBC) 3m 57ch
Aycliffe
49m 36ch
Norton-onTees East Jn.
62m 19ch
0m 00ch
Norton-on-Tees East SB (NE)
62m 19ch / 0m 00ch
Norton-on-Tees West SB (NW)
0m 29ch / 0m 33ch
Norton West LC 0m 33ch
See Map 136
Billingham
Middlesbrough
South Bank
Thornaby
Norton-onTees West Jn.
0m 33ch
0m 29ch
Norton-onTees South Jn.
61m 71ch
0m 00ch
Norton-on-Tees South SB (NS)
61m 71ch
Stockton
60m 04ch
to / from T.J. Thompson Sidings
Hartburn Jn.
59m 14ch
0m 00ch
Stockton Cut Jn.
58m 30ch / 10m 13ch
Bowesfield
0m 44ch via Hartburn Jn.
Bowesfield SB (B)
0m 44ch / 10m 76ch
Marton
Gypsy Lane
3m 60ch
Nunthorpe SB
4m 27ch
Hopetown Jn.
0m 75ch
Albert Hill
0m 32ch
North Road
0m 49ch
Darlington North Jn.
44m 36ch
Darlington Up Siding
Change of mileage
44m 64ch
0m 00ch
Darlington North Jn.
44m 36ch
Darlington
44m 10ch
to Cleveland Bridge Engineering Sidings
0m 36ch
Carters (UWC) 6m 28ch
(MCB) 7m 39ch
Allens West
8m 10ch
Eaglescliffe
57m 00ch
Eaglescliffe South Jn.
56m 75ch via Yarm
8m 58ch via Dinsdale
Urlay Nook SB (UN)
7m 39ch
Teesside Airport
5m 43ch
Dinsdale
3m 65ch
Yarm Tunnel
55m 76ch to 55m 79ch
43m 50ch
Maidendale
1m 72ch
Allens West (AHBC-X)
8m 15ch
Yarm
54m 35ch
42m 72ch
Darlington South Jn.
43m 61ch via Northallerton
0m 29ch via Dinsdale
Battersby Jn.
10m 54ch via Great
12m 10ch via Batte
1
Picton (CCTV) 52m 31ch
Rounton Gates (AHBC-X) 50m 12ch
East Cowton Crossovers
37m 50ch
Welbury (AHBC-X) 48m 21ch
River Swale
Long Lane (CCTV) 46m 34ch
Castle Hills Jn.
30m 59ch / 30m 63ch
0m 00ch
(Hedmire Line)
42m 66ch
End of Reversing Line
31m 09ch
Low Gates SB (LG)
43m 24ch
2
71
Northallerton East Jn.
42m 79ch via Northallerton
0m 36ch via Northallerton High Jn.
Former Castle Hills West Jn.
Change of mileage
0m 28ch
0m 48ch
NR/WR Boundary
0m 18ch
42m 55ch
Castle Hill East GF 0m 19ch
Castle Hill West GF 0m 67ch
Northallerton Up Sidings
Northallerton High Jn.
30m 09ch via Northallerton
0m 00ch via Northallerton East Jn.
Northallerton
29m 76ch
29m 56ch
Longlands Tunnel
0m 08ch to 0m 11ch
Longlands Jn. (Up)
0m 69ch via Longlands Tunnel
29m 01ch via Thirsk
Leeming Bar
5m 67ch
GF 5m 74ch
Finghall
13m 15ch
14m 23ch
14m 13ch
Constable Burton Loop
Finghall (TMO) 13.16
Wensleydale Railway
Bedale
7m 35ch
7m 55ch
Bedale Loop
7m 46ch
Crakehall (TMO) 9m 53ch
Bedale 7m 41ch
Aiskew (AOCL) 6m 33ch
Longlands Junction (Down)
28m 68ch / 28m 58ch
No. 89 (R/G) 27m 58ch
Change of mileage
Down 29m 72ch / 42m 21ch
Up 0m 00ch / 42m 21ch
Level Crossings around Northallerton Station
1 Brompton (AHBC-X) 44m 57ch
2 Walkers (UWC) 44m 30ch
3 Clarks (UWC) 44m 10ch
4 Vaseys (UWC) 43m 68ch
5 Low Gates (MCB) 43m 24ch
6 Springwell Lane (AHBC) 42m 65ch
7 Romanby Road (CCTV) 42m 38ch
8 Boroughbridge Road (CCTV) (see change of mileage)
9 Yafforth (10 (AOCL) 1m 48ch
10 Ainderby Gates (19) (TMO) 2m 43ch
11 Ainderby Station (TMO) 2m 68ch
12 Scruton (TMO) 4m 25ch
13 Ham Hall (AOCL) 4m 60ch
14 Leeming Bar (TMO) 5m 62ch
River Ure
No. 82 (R/G) 23m 33ch
No. 81 (R/G) 22m 73ch
22m 35ch
Thirsk Sidings
22m 60ch
21m 79ch
Thirsk
22m 16ch
3
C
D

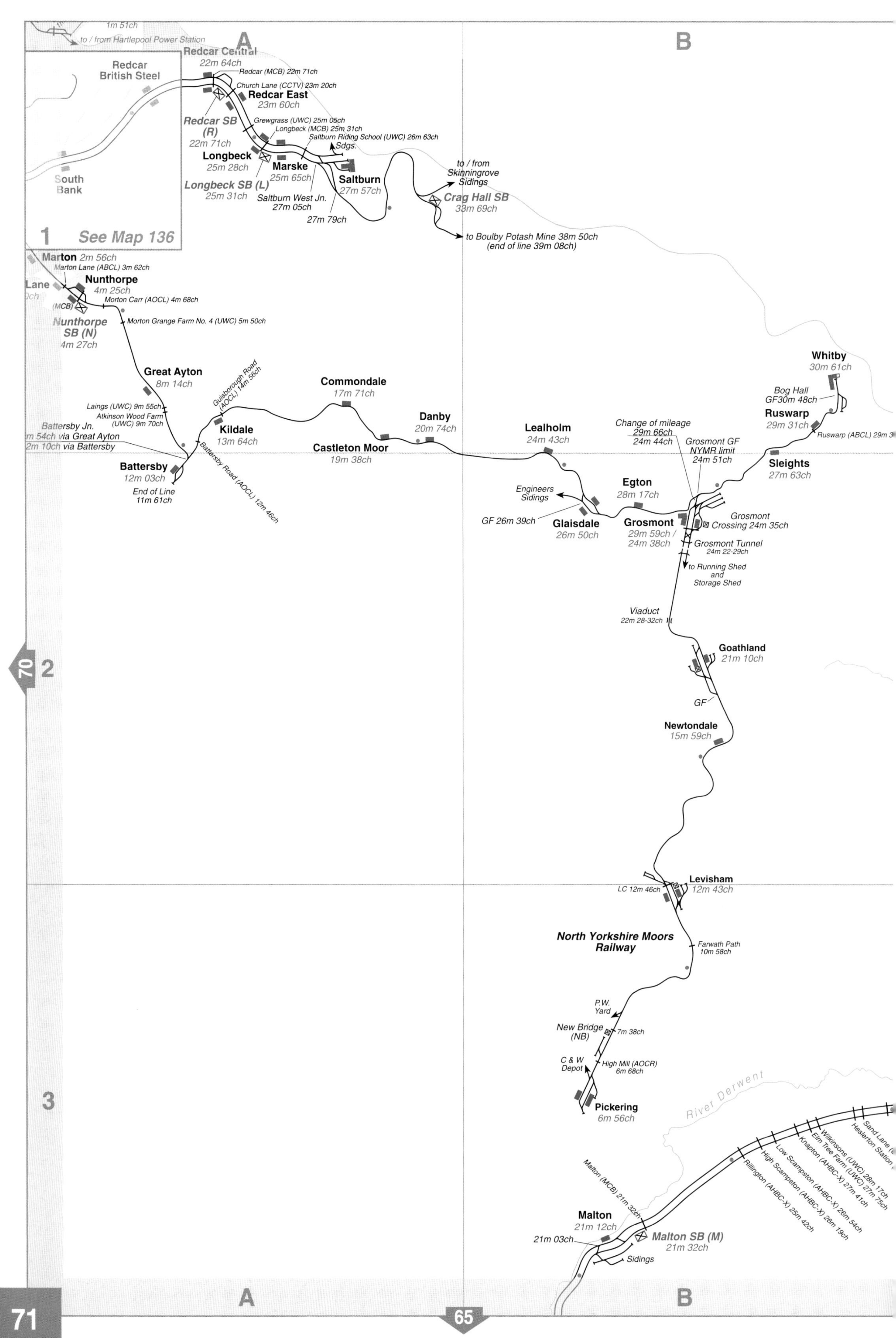
1m 51ch
to / from Hartlepool Power Station
A
B
Redcar British Steel
South Bank
1
See Map 136
Redcar Central
22m 64ch
Redcar (MCB) 22m 71ch
Church Lane (CCTV) 23m 20ch
Redcar East
23m 60ch
Redcar SB (R)
22m 71ch
Grewgrass (UWC) 25m 05ch
Longbeck (MCB) 25m 31ch
Saltburn Riding School (UWC) 26m 63ch
Sdgs.
Longbeck
25m 28ch
Marske
25m 65ch
Saltburn
27m 57ch
Longbeck SB (L)
25m 31ch
Saltburn West Jn.
27m 05ch
27m 79ch
to / from Skinningrove Sidings
Crag Hall SB
33m 69ch
to Boulby Potash Mine 38m 50ch
(end of line 39m 08ch)
Marton 2m 56ch
Marton Lane (ABCL) 3m 62ch
Lane
Nunthorpe
4m 25ch
Morton Carr (AOCL) 4m 68ch
(MCB)
Nunthorpe SB (N)
4m 27ch
Morton Grange Farm No. 4 (UWC) 5m 50ch
Great Ayton
8m 14ch
Guisborough Road (AOCL) 14m 56ch
Commondale
17m 71ch
Laings (UWC) 9m 55ch
Atkinson Wood Farm (UWC) 9m 70ch
Battersby Jn.
m 54ch via Great Ayton
2m 10ch via Battersby
Kildale
13m 64ch
Danby
20m 74ch
Castleton Moor
19m 38ch
Lealholm
24m 43ch
Change of mileage
29m 66ch
24m 44ch
Grosmont GF NYMR limit
24m 51ch
Whitby
30m 61ch
Bog Hall GF30m 48ch
Ruswarp
29m 31ch
Ruswarp (ABCL) 29m 3
Sleights
27m 63ch
Battersby
12m 03ch
End of Line
11m 61ch
Battersby Road (AOCL) 12m 46ch
Engineers Sidings
GF 26m 39ch
Glaisdale
26m 50ch
Egton
28m 17ch
Grosmont
29m 59ch /
24m 38ch
Grosmont Crossing 24m 35ch
Grosmont Tunnel
24m 22-29ch
to Running Shed and Storage Shed
Viaduct
22m 28-32ch
Goathland
21m 10ch
70
2
GF
Newtondale
15m 59ch
Levisham
12m 43ch
LC 12m 46ch
North Yorkshire Moors Railway
Farwath Path
10m 58ch
P.W. Yard
New Bridge (NB)
7m 38ch
C & W Depot
High Mill (AOCR)
6m 68ch
3
Pickering
6m 56ch
River Derwent
Sand Lane
Heslerton Station
Wilkinsons (UWC) 28m 17ch
Elm Tree Farm (UWC) 27m 75ch
Knapton (AHBC-X) 27m 41ch
Low Scampston (AHBC-X) 26m 54ch
High Scampston (AHBC-X) 26m 19ch
Rillington (AHBC-X) 25m 42ch
Malton (MCB) 21m 32ch
Malton
21m 12ch
21m 03ch
Malton SB (M)
21m 32ch
Sidings
A
B
65

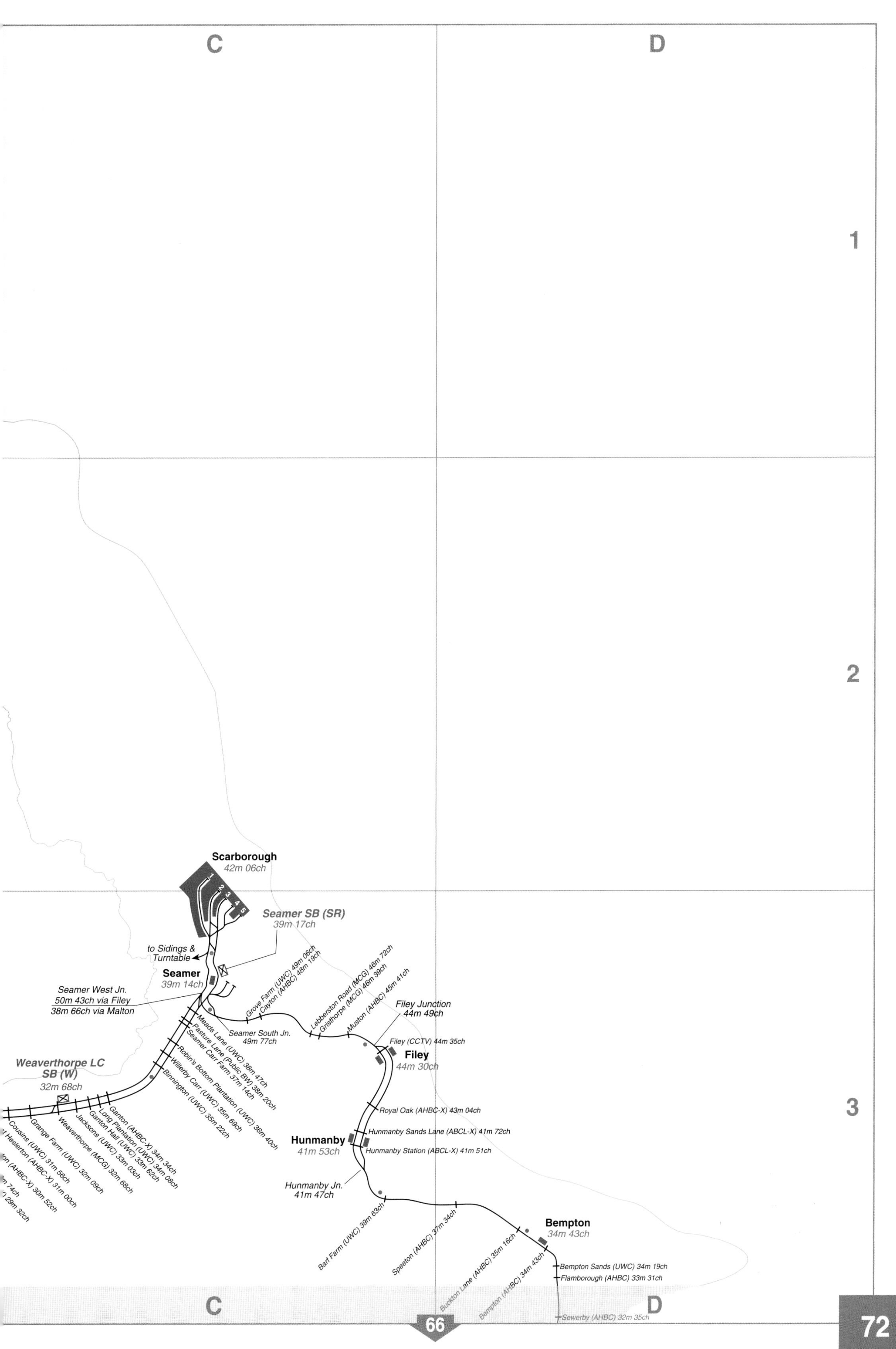
C
D
1
2
3
Scarborough
42m 06ch
1
2
3
4
5
Seamer SB (SR)
39m 17ch
to Sidings & Turntable
Seamer
39m 14ch
Seamer West Jn.
50m 43ch via Filey
38m 66ch via Malton
Grove Farm (UWC) 49m 06ch
Cayton (AHBC) 48m 19ch
Seamer South Jn.
49m 77ch
Lebberston Road (MCG) 46m 72ch
Gristhorpe (MCG) 46m 39ch
Muston (AHBC) 45m 41ch
Filey Junction
44m 49ch
Filey (CCTV) 44m 35ch
Filey
44m 30ch
Meads Lane (UWC) 38m 47ch
Pasture Lane (Public BW) 38m 20ch
Seamer Carr Farm 37m 14ch
Robin's Bottom Plantation (UWC) 36m 40ch
Willerby Carr (UWC) 35m 69ch
Binnington (UWC) 35m 22ch
Weaverthorpe LC
SB (W)
32m 68ch
Ganton (AHBC-X) 34m 34ch
Long Plantation (UWC) 34m 08ch
Ganton Hall (UWC) 33m 62ch
Jacksons (UWC) 33m 03ch
Weaverthorpe (MCG) 32m 68ch
Grange Farm (UWC) 32m 09ch
Cousins (UWC) 31m 56ch
Royal Oak (AHBC-X) 43m 04ch
Hunmanby Sands Lane (ABCL-X) 41m 72ch
Hunmanby
41m 53ch
Hunmanby Station (ABCL-X) 41m 51ch
Hunmanby Jn.
41m 47ch
Barf Farm (UWC) 39m 63ch
Speeton (AHBC) 37m 34ch
Bempton
34m 43ch
Buckton Lane (AHBC) 35m 16ch
Bempton (AHBC) 34m 43ch
Bempton Sands (UWC) 34m 19ch
Flamborough (AHBC) 33m 31ch
Sewerby (AHBC) 32m 35ch
C
D
66

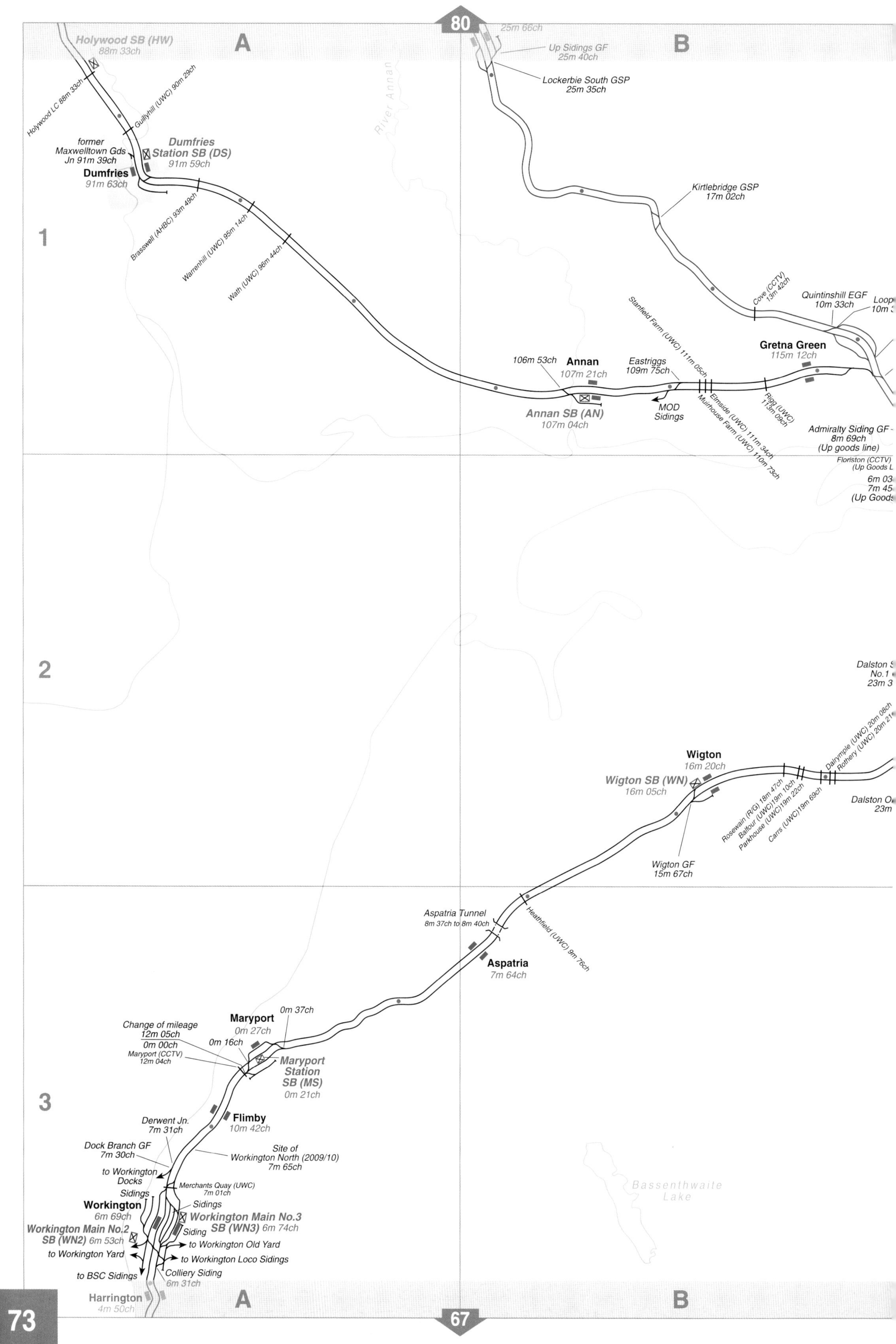

80
A
B
Holywood SB (HW)
88m 33ch
Holywood LC 88m 33ch
Gullyhill (UWC) 90m 29ch
former
Maxwelltown Gds
Jn 91m 39ch
Dumfries
Station SB (DS)
91m 59ch
Dumfries
91m 63ch
Brasswell (AHBC) 93m 49ch
Warrenhill (UWC) 95m 14ch
Wath (UWC) 96m 44ch
River Annan
25m 66ch
Up Sidings GF
25m 40ch
Lockerbie South GSP
25m 35ch
Kirtlebridge GSP
17m 02ch
Cove (CCTV)
13m 42ch
Quintinshill EGF
10m 33ch
Gretna Green
115m 12ch
Stanfield Farm (UWC) 111m 05ch
106m 53ch
Annan
107m 21ch
Eastriggs
109m 75ch
Annan SB (AN)
107m 04ch
MOD
Sidings
Elmside (UWC) 111m 34ch
Muirhouse Farm (UWC) 110m 73ch
Rigg (UWC)
113m 09ch
Admiralty Siding GF
8m 69ch
(Up goods line)
Floriston (CCTV)
1
2
3
Wigton
16m 20ch
Wigton SB (WN)
16m 05ch
Rosewain (R/G) 18m 47ch
Balfour (UWC)19m 10ch
Parkhouse (UWC)19m 22ch
Carrs (UWC)19m 69ch
Dalrymple (UWC) 20m 08ch
Wigton GF
15m 67ch
Aspatria Tunnel
8m 37ch to 8m 40ch
Heathfield (UWC) 9m 76ch
Aspatria
7m 64ch
Maryport
0m 27ch
0m 37ch
Change of mileage
12m 05ch
0m 00ch
0m 16ch
Maryport (CCTV)
12m 04ch
Maryport
Station
SB (MS)
0m 21ch
Flimby
10m 42ch
Derwent Jn.
7m 31ch
Dock Branch GF
7m 30ch
Site of
Workington North (2009/10)
7m 65ch
to Workington
Docks
Merchants Quay (UWC)
7m 01ch
Sidings
Workington
6m 69ch
Sidings
Workington Main No.3
SB (WN3) 6m 74ch
Workington Main No.2
SB (WN2) 6m 53ch
Siding
to Workington Old Yard
to Workington Yard
to Workington Loco Sidings
to BSC Sidings
Colliery Siding
6m 31ch
Bassenthwaite
Lake
Harrington
4m 50ch
67

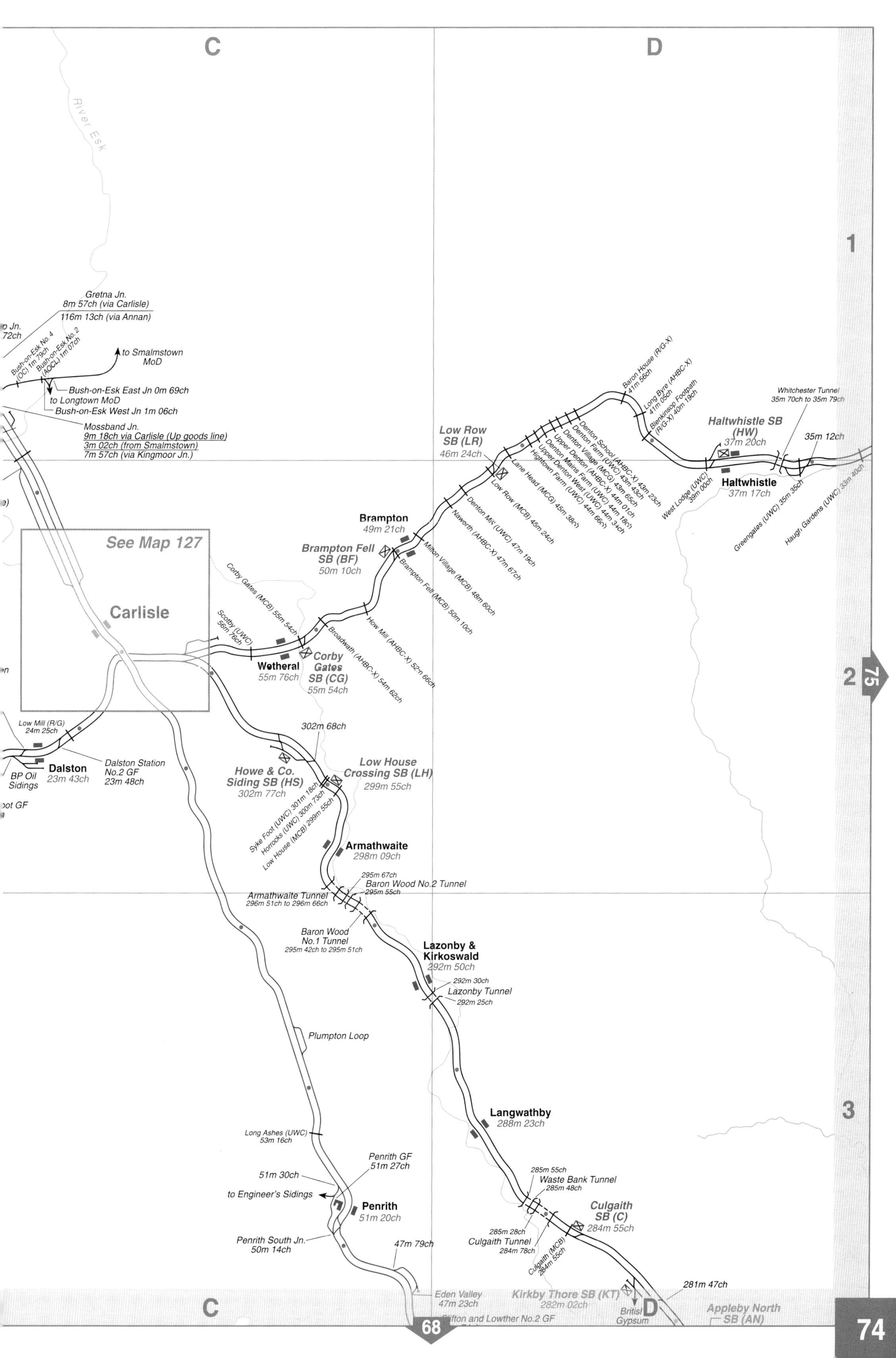

C
D
River Esk
1
2
3
Gretna Jn.
8m 57ch (via Carlisle)
116m 13ch (via Annan)
Bush-on-Esk No. 4 (OC) 1m 79ch
Bush-on-Esk No. 2 (AOCL) 1m 07ch
to Smalmstown MoD
Bush-on-Esk East Jn 0m 69ch
to Longtown MoD
Bush-on-Esk West Jn 1m 06ch
Mossband Jn.
9m 18ch via Carlisle (Up goods line)
3m 02ch (from Smalmstown)
7m 57ch (via Kingmoor Jn.)
See Map 127
Carlisle
Low Mill (R/G) 24m 25ch
Dalston
23m 43ch
Dalston Station No.2 GF
23m 48ch
BP Oil Sidings
Scotby (UWC) 56m 76ch
Corby Gates (MCB) 55m 54ch
Wetheral
55m 76ch
Corby Gates SB (CG)
55m 54ch
Broadwath (AHBC-X) 54m 62ch
How Mill (AHBC-X) 52m 66ch
Brampton Fell SB (BF)
50m 10ch
Brampton Fell (MCB) 50m 10ch
Brampton
49m 21ch
Milton Village (MCB) 48m 60ch
Naworth (AHBC-X) 47m 67ch
Denton Mill (UWC) 47m 19ch
Low Row SB (LR)
46m 24ch
Low Row (MCB) 45m 24ch
Lane Head (MCG) 45m 38ch
Hightown Farm (UWC) 44m 66ch
Upper Denton Mains Farm (UWC) 44m 34ch
Upper Denton West (UWC) 44m 18ch
Denton Village (AHBC-X) 44m 01ch
Denton Farm (UWC) 43m 65ch
Denton School (AHBC-X) 43m 23ch
Baron House (R/G-X) 41m 56ch
Long Byre (AHBC-X) 41m 05ch
Blenkinsop Footpath (R/G-X) 40m 19ch
West Lodge (UWC) 39m 00ch
Haltwhistle SB (HW)
37m 20ch
Haltwhistle
37m 17ch
Whitchester Tunnel
35m 70ch to 35m 79ch
35m 12ch
Greengates (UWC) 35m 35ch
Haugh Gardens (UWC) 33m 40ch
302m 68ch
Howe & Co. Siding SB (HS)
302m 77ch
Low House Crossing SB (LH)
299m 55ch
Syke Foot (UWC) 301m 18ch
Horrocks (UWC) 300m 73ch
Low House (MCB) 299m 55ch
Armathwaite
298m 09ch
295m 67ch
Baron Wood No.2 Tunnel
295m 55ch
Armathwaite Tunnel
296m 51ch to 296m 66ch
Baron Wood No.1 Tunnel
295m 42ch to 295m 51ch
Lazonby & Kirkoswald
292m 50ch
292m 30ch
Lazonby Tunnel
292m 25ch
Plumpton Loop
Langwathby
288m 23ch
Long Ashes (UWC) 53m 16ch
Penrith GF
51m 27ch
51m 30ch
to Engineer's Sidings
Penrith
51m 20ch
Penrith South Jn.
50m 14ch
47m 79ch
285m 55ch
Waste Bank Tunnel
285m 48ch
Culgaith SB (C)
284m 55ch
285m 28ch
Culgaith Tunnel
284m 78ch
Culgaith (MCB) 284m 55ch
Eden Valley
47m 23ch
Kirkby Thore SB (KT)
282m 02ch
British Gypsum
281m 47ch
Appleby North SB (AN)
Clifton and Lowther No.2 GF

75
68

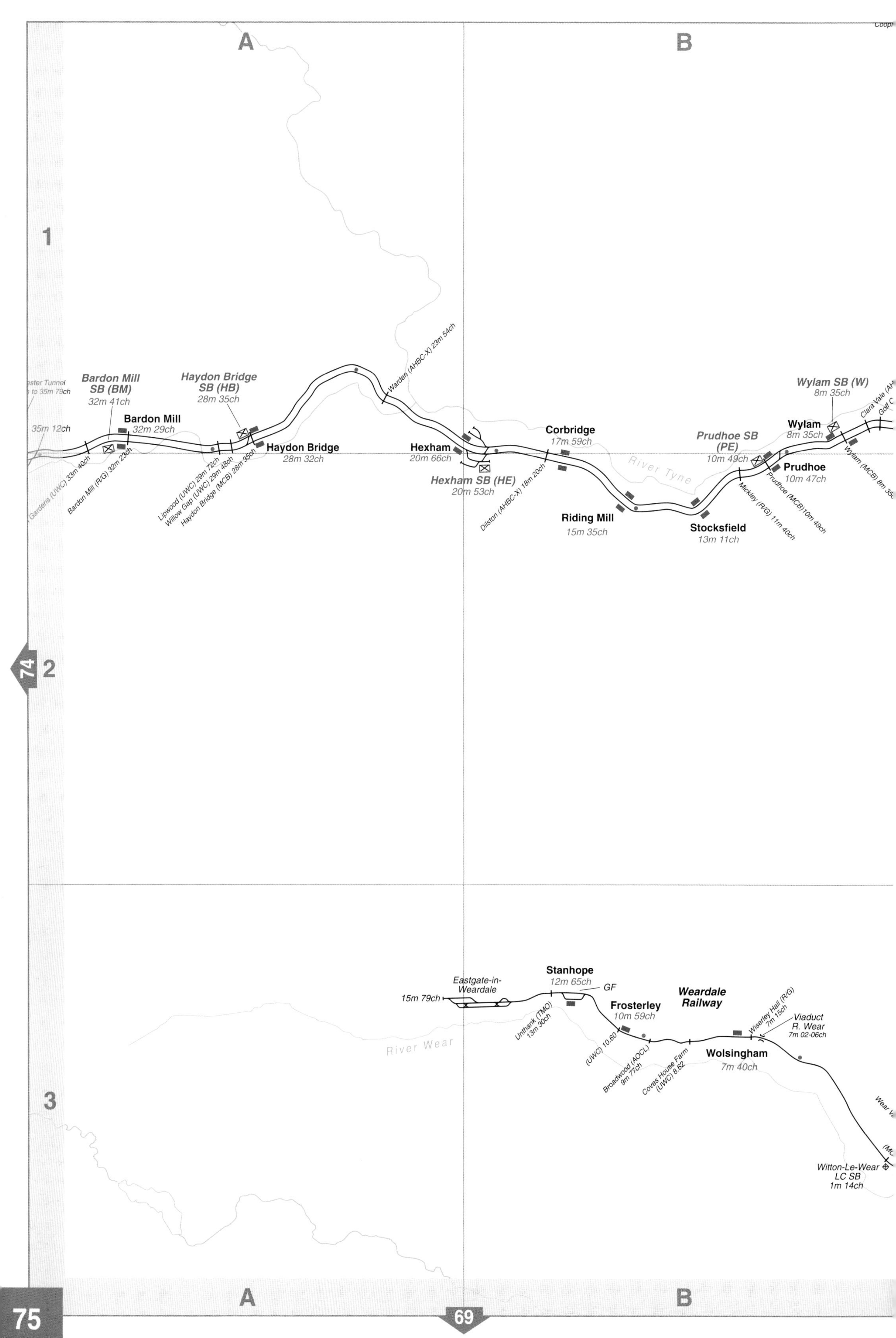
A
B
1
2
3
Bardon Mill SB (BM)
32m 41ch
Bardon Mill
32m 29ch
35m 12ch
Bardon Mill (R/G) 32m 23ch
Haydon Bridge SB (HB)
28m 35ch
Haydon Bridge
28m 32ch
Lipwood (UWC) 29m 72ch
Willow Gap (UWC) 29m 48ch
Haydon Bridge (MCB) 28m 35ch
Warden (AHBC-X) 23m 54ch
Hexham
20m 66ch
Hexham SB (HE)
20m 53ch
Corbridge
17m 59ch
Dilston (AHBC-X) 18m 20ch
Riding Mill
15m 35ch
River Tyne
Stocksfield
13m 11ch
Prudhoe SB (PE)
10m 49ch
Mickley (R/G) 11m 40ch
Prudhoe (MCB)10m 49ch
Prudhoe
10m 47ch
Wylam SB (W)
8m 35ch
Wylam
8m 35ch
Eastgate-in-Weardale
15m 79ch
Stanhope
12m 65ch
GF
Unthank (TMO) 13m 30ch
Frosterley
10m 59ch
Weardale Railway
(UWC) 10.60
Broadwood (AOCL) 9m 77ch
Coves House Farm (UWC) 8.62
Wolsingham
7m 40ch
Wiserley Hall (R/G) 7m 15ch
Viaduct R. Wear
7m 02-06ch
River Wear
Witton-Le-Wear LC SB
1m 14ch
74

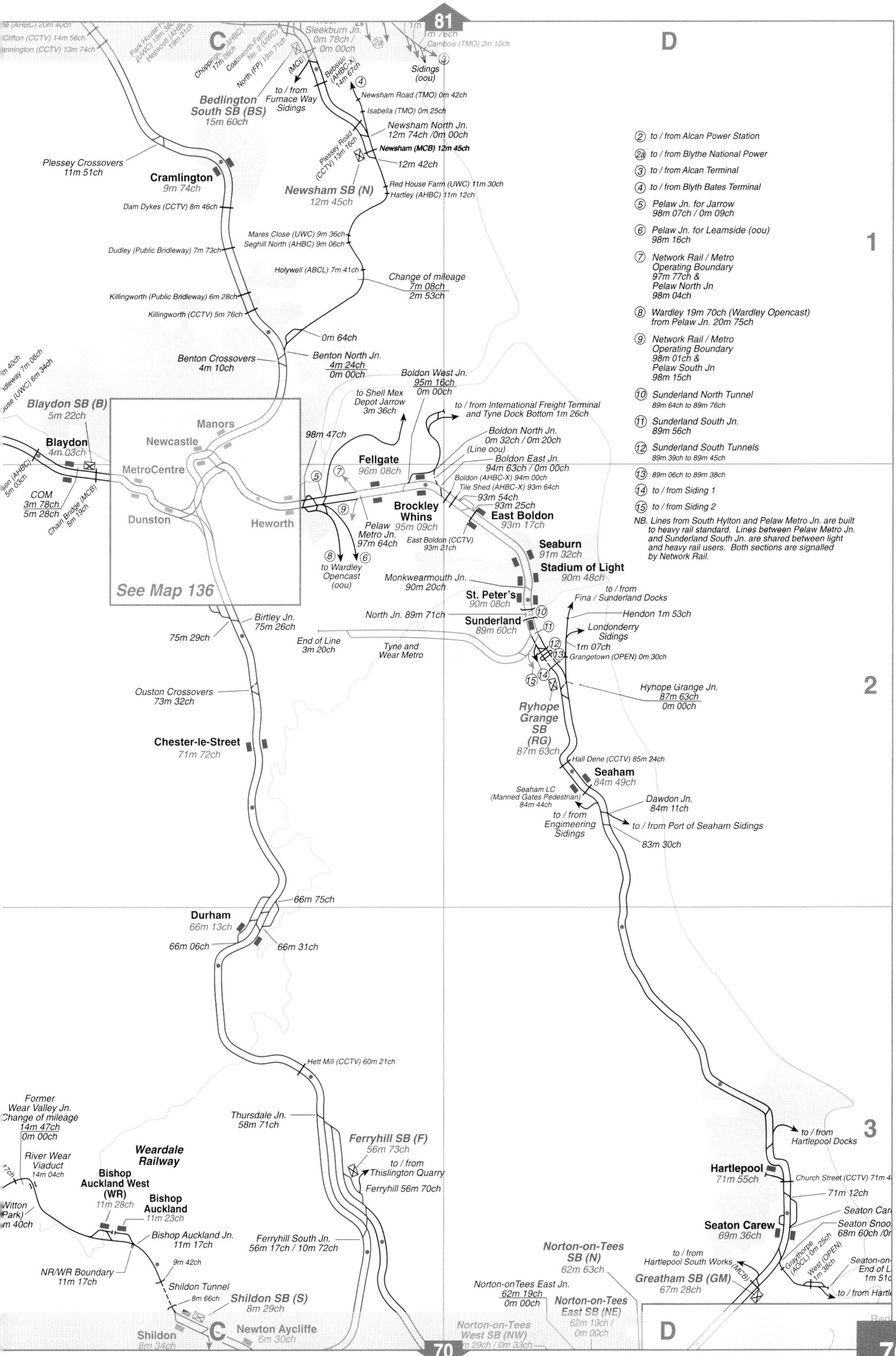
81
C
D
1
2
3
Sleekburn Jn.
0m 78ch / 0m 00ch
Cambois (TMO) 2m 10ch
Sidings (oou)
Clifton (CCTV) 14m 56ch
Cannington (CCTV) 13m 74ch
Choppington (AHBC) 17m 06ch
Coatsworth Farm No.2 (UWC)
North (FP) 15m 71ch
Park House Farm (UWC) 19m 38ch
Hepscott (AHBC) 19m 21ch
(MCB)
to / from Furnace Way Sidings
Bebside (AHBC-X) 14m 67ch
Bedlington South SB (BS)
15m 60ch
Newsham Road (TMO) 0m 42ch
Isabella (TMO) 0m 25ch
Newsham North Jn.
12m 74ch /0m 00ch
Newsham (MCB) 12m 45ch
Plessey Road (CCTV) 13m 16ch
12m 42ch
Newsham SB (N)
12m 45ch
Red House Farm (UWC) 11m 30ch
Hartley (AHBC) 11m 12ch
Plessey Crossovers
11m 51ch
Cramlington
9m 74ch
Dam Dykes (CCTV) 8m 46ch
Mares Close (UWC) 9m 36ch
Seghill North (AHBC) 9m 06ch
Dudley (Public Bridleway) 7m 73ch
Holywell (ABCL) 7m 41ch
Change of mileage
7m 08ch
2m 53ch
Killingworth (Public Bridleway) 6m 28ch
Killingworth (CCTV) 5m 76ch
0m 64ch
Benton Crossovers
4m 10ch
Benton North Jn.
4m 24ch
0m 00ch
Boldon West Jn.
95m 16ch
0m 00ch
to Shell Mex Depot Jarrow
3m 36ch
to / from International Freight Terminal and Tyne Dock Bottom 1m 26ch
Blaydon SB (B)
5m 22ch
Blaydon
4m 03ch
COM
3m 78ch
5m 28ch
Chain Bridge (MCB) 5m 19ch
Manors
Newcastle
MetroCentre
Dunston
Heworth
See Map 136
98m 47ch
Fellgate
96m 08ch
Boldon North Jn.
0m 32ch / 0m 20ch
(Line oou)
Boldon East Jn.
94m 63ch / 0m 00ch
Boldon (AHBC-X) 94m 00ch
Tile Shed (AHBC-X) 93m 64ch
93m 54ch
93m 25ch
Brockley Whins
95m 09ch
East Boldon
93m 17ch
East Boldon (CCTV) 93m 21ch
Pelaw Metro Jn.
97m 64ch
to Wardley Opencast (oou)
Seaburn
91m 32ch
Stadium of Light
90m 48ch
Monkwearmouth Jn.
90m 20ch
St. Peter's
90m 08ch
to / from Fina / Sunderland Docks
Hendon 1m 53ch
North Jn. 89m 71ch
Sunderland
89m 60ch
Londonderry Sidings
1m 07ch
Grangetown (OPEN) 0m 30ch
End of Line
3m 20ch
Tyne and Wear Metro
Birtley Jn.
75m 26ch
75m 29ch
Hyhope Grange Jn.
87m 63ch
0m 00ch
Ryhope Grange SB (RG)
87m 63ch
Ouston Crossovers
73m 32ch
Chester-le-Street
71m 72ch
Hall Dene (CCTV) 85m 24ch
Seaham
84m 49ch
Seaham LC (Manned Gates Pedestrian) 84m 44ch
Dawdon Jn.
84m 11ch
to / from Engimeering Sidings
to / from Port of Seaham Sidings
83m 30ch
66m 75ch
Durham
66m 13ch
66m 06ch
66m 31ch
Hett Mill (CCTV) 60m 21ch
Former Wear Valley Jn.
Change of mileage
14m 47ch
0m 00ch
Thursdale Jn.
58m 71ch
Ferryhill SB (F)
56m 73ch
to / from Thislington Quarry
Ferryhill 56m 70ch
River Wear Viaduct
14m 04ch
Weardale Railway
Bishop Auckland West (WR)
11m 28ch
Bishop Auckland
11m 23ch
Witton Park)
Bishop Auckland Jn.
11m 17ch
NR/WR Boundary
11m 17ch
Ferryhill South Jn.
56m 17ch / 10m 72ch
9m 42ch
Shildon Tunnel
8m 66ch
Shildon SB (S)
8m 29ch
Shildon
8m 34ch
Newton Aycliffe
6m 30ch
70
Norton-on-Tees West SB (NW)
Norton-onTees East Jn.
62m 19ch
0m 00ch
Norton-on-Tees East SB (NE)
62m 19ch / 0m 00ch
Norton-on-Tees SB (N)
62m 63ch
to / from Hartlepool South Works
Greatham SB (GM)
67m 28ch
to / from Hartlepool Docks
Hartlepool
71m 55ch
71m 12ch
Seaton Carew
69m 36ch
Graythorpe (AOCL) 0m 25ch
West (OPEN) 1m 38ch
2 to / from Alcan Power Station
2a to / from Blythe National Power
3 to / from Alcan Terminal
4 to / from Blyth Bates Terminal
5 Pelaw Jn. for Jarrow 98m 07ch / 0m 09ch
6 Pelaw Jn. for Leamside (oou) 98m 16ch
7 Network Rail / Metro Operating Boundary 97m 77ch & Pelaw North Jn 98m 04ch
8 Wardley 19m 70ch (Wardley Opencast) from Pelaw Jn. 20m 75ch
9 Network Rail / Metro Operating Boundary 98m 01ch & Pelaw South Jn 98m 15ch
10 Sunderland North Tunnel 89m 64ch to 89m 76ch
11 Sunderland South Jn. 89m 56ch
12 Sunderland South Tunnels 89m 39ch to 89m 45ch
13 89m 06ch to 89m 38ch
14 to / from Siding 1
15 to / from Siding 2
NB. Lines from South Hylton and Pelaw Metro Jn. are built to heavy rail standard. Lines between Pelaw Metro Jn. and Sunderland South Jn. are shared between light and heavy rail users. Both sections are signalled by Network Rail.

78

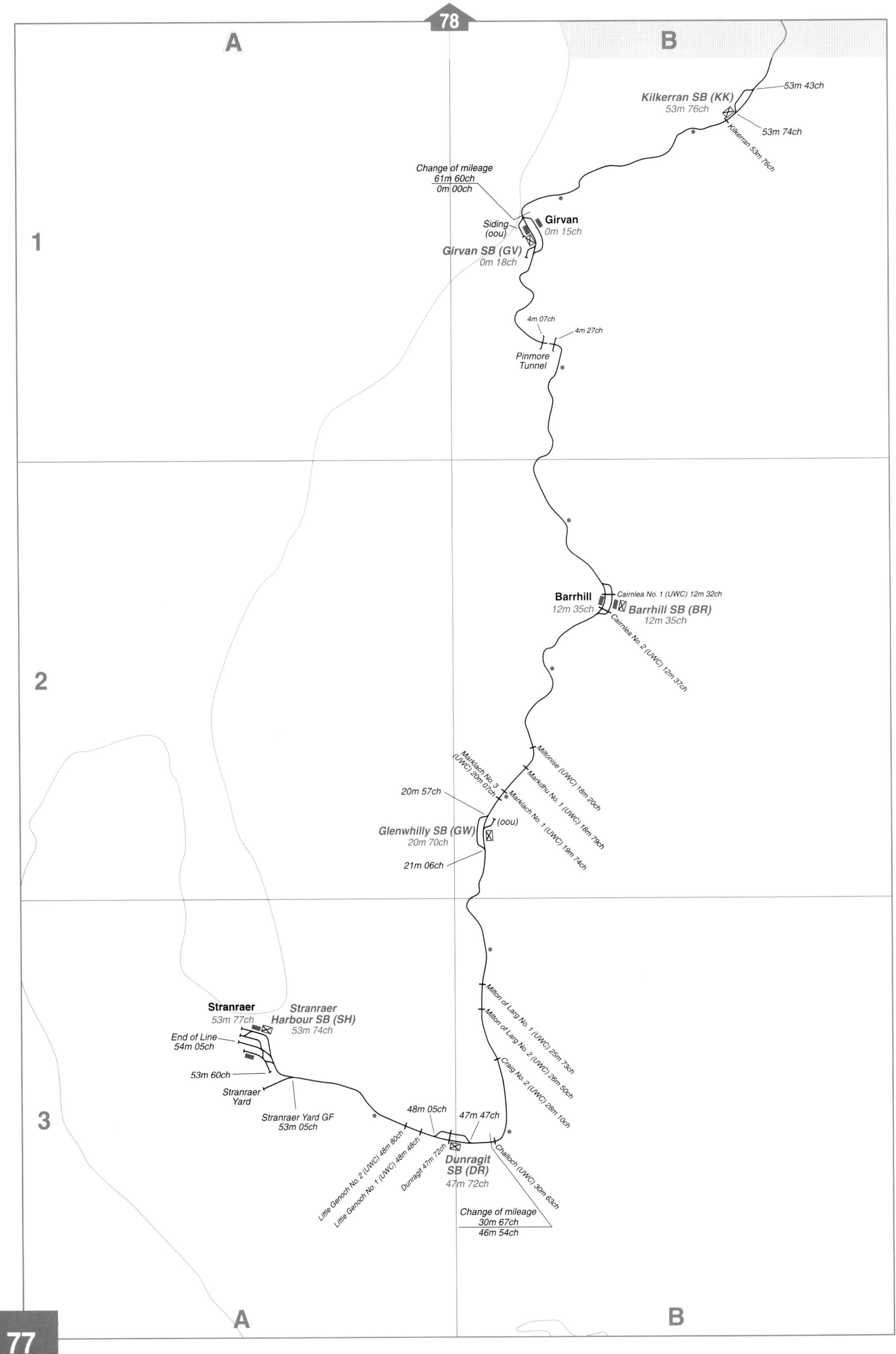

A
B
1
2
3
Kilkerran SB (KK)
53m 76ch
53m 43ch
53m 74ch
Kilkerran 53m 76ch
Change of mileage
61m 60ch
0m 00ch
Girvan
0m 15ch
Siding
(oou)
Girvan SB (GV)
0m 18ch
4m 07ch
4m 27ch
Pinmore
Tunnel
Barrhill
12m 35ch
Cairnlea No. 1 (UWC) 12m 32ch
Barrhill SB (BR)
12m 35ch
Cairnlea No. 2 (UWC) 12m 37ch
Miltonise (UWC) 18m 20ch
Markdhu No. 1 (UWC) 18m 79ch
Marklach No. 3
(UWC) 20m 07ch
Marklach No. 1 (UWC) 19m 74ch
20m 57ch
(oou)
Glenwhilly SB (GW)
20m 70ch
21m 06ch
Milton of Larg No. 1 (UWC) 25m 73ch
Milton of Larg No. 2 (UWC) 26m 50ch
Craig No. 2 (UWC) 28m 10ch
Stranraer
53m 77ch
Stranraer
Harbour SB (SH)
53m 74ch
End of Line
54m 05ch
53m 60ch
Stranraer
Yard
Stranraer Yard GF
53m 05ch
48m 05ch
47m 47ch
Little Genoch No. 2 (UWC) 48m 80ch
Little Genoch No. 1 (UWC) 48m 48ch
Dunragit 47m 72ch
Dunragit
SB (DR)
47m 72ch
Challoch (UWC) 30m 63ch
Change of mileage
30m 67ch
46m 54ch
A
B

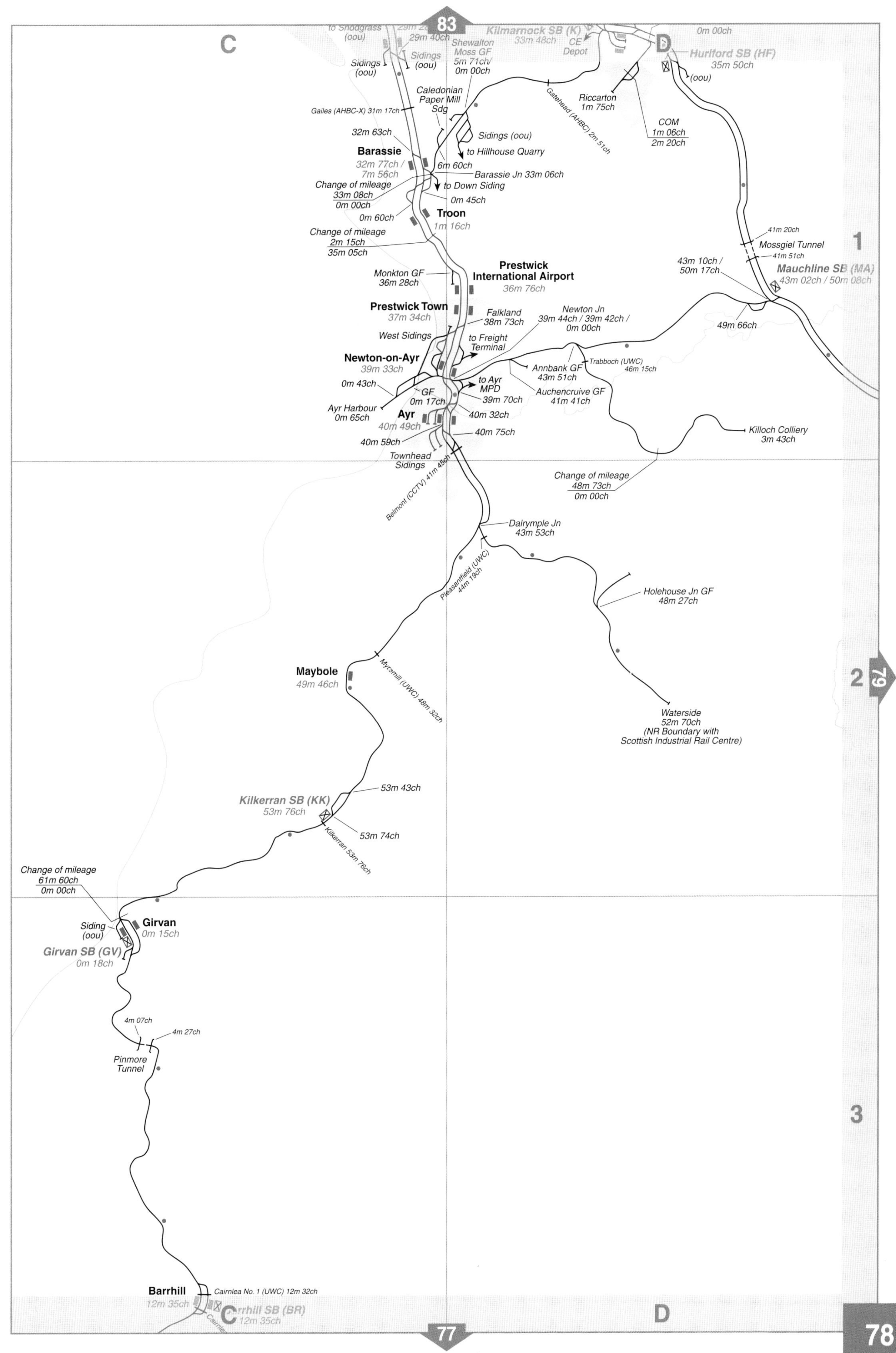

C
to Snodgrass (oou)
29m 40ch
Kilmarnock SB (K)
33m 48ch
CE Depot
D
0m 00ch
Hurlford SB (HF)
35m 50ch
(oou)
Sidings (oou)
Shewalton Moss GF
5m 71ch/
0m 00ch
Sidings (oou)
Gailes (AHBC-X) 31m 17ch
Caledonian Paper Mill Sdg
Gatehead (AHBC) 2m 51ch
Riccarton
1m 75ch
COM
1m 06ch
2m 20ch
32m 63ch
Sidings (oou)
to Hillhouse Quarry
Barassie
32m 77ch /
7m 56ch
6m 60ch
Barassie Jn 33m 06ch
Change of mileage
33m 08ch
0m 00ch
to Down Siding
0m 45ch
0m 60ch
Troon
1m 16ch
Change of mileage
2m 15ch
35m 05ch
41m 20ch
Mossgiel Tunnel
41m 51ch
43m 10ch /
50m 17ch
Mauchline SB (MA)
43m 02ch / 50m 08ch
1
Monkton GF
36m 28ch
Prestwick International Airport
36m 76ch
Prestwick Town
37m 34ch
Falkland
38m 73ch
Newton Jn
39m 44ch / 39m 42ch /
0m 00ch
49m 66ch
West Sidings
to Freight Terminal
Newton-on-Ayr
39m 33ch
Annbank GF
43m 51ch
Trabboch (UWC)
46m 15ch
0m 43ch
GF
0m 17ch
to Ayr MPD
39m 70ch
Auchencruive GF
41m 41ch
Ayr Harbour
0m 65ch
Ayr
40m 49ch
40m 32ch
40m 75ch
40m 59ch
Killoch Colliery
3m 43ch
Townhead Sidings
Belmont (CCTV) 41m 45ch
Change of mileage
48m 73ch
0m 00ch
Dalrymple Jn
43m 53ch
Pleasantfield (UWC)
44m 19ch
Holehouse Jn GF
48m 27ch
Maybole
49m 46ch
Myrsmill (UWC) 48m 32ch
2
79
Waterside
52m 70ch
(NR Boundary with
Scottish Industrial Rail Centre)
53m 43ch
Kilkerran SB (KK)
53m 76ch
53m 74ch
Kilkerran 53m 76ch
Change of mileage
61m 60ch
0m 00ch
Siding (oou)
Girvan
0m 15ch
Girvan SB (GV)
0m 18ch
4m 07ch
4m 27ch
Pinmore Tunnel
3
Barrhill
12m 35ch
Cairnlea No. 1 (UWC) 12m 32ch
Barrhill SB (BR)
12m 35ch
C
D

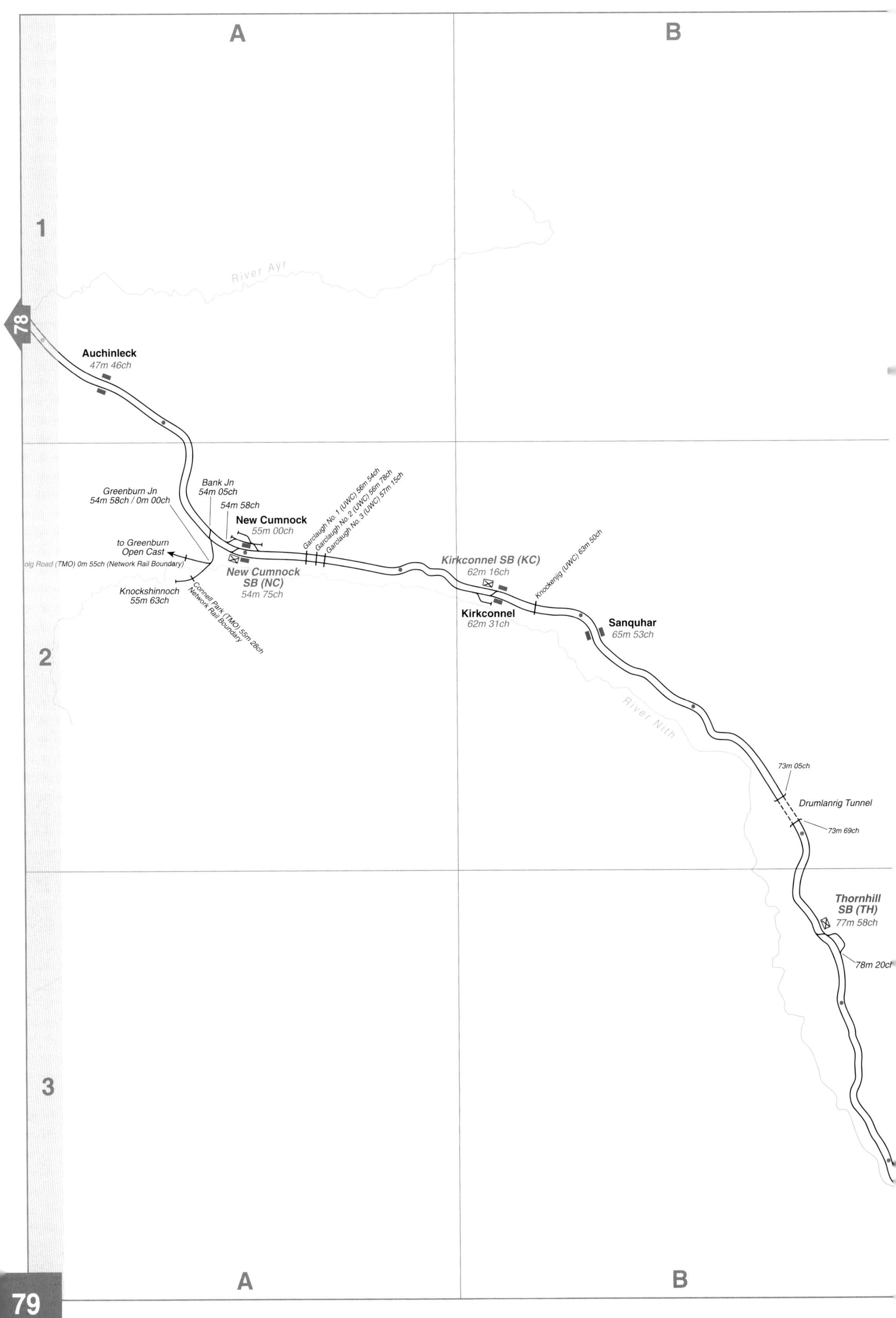
A
B
1
2
3
78
River Ayr
Auchinleck
47m 46ch
Greenburn Jn
54m 58ch / 0m 00ch
Bank Jn
54m 05ch
54m 58ch
New Cumnock
55m 00ch
to Greenburn
Open Cast
oig Road (TMO) 0m 55ch (Network Rail Boundary)
Knockshinnoch
55m 63ch
Connell Park (TMO) 55m 28ch
Network Rail Boundary
New Cumnock
SB (NC)
54m 75ch
Garclaugh No. 1 (UWC) 56m 54ch
Garclaugh No. 2 (UWC) 56m 78ch
Garclaugh No. 3 (UWC) 57m 15ch
Kirkconnel SB (KC)
62m 16ch
Kirkconnel
62m 31ch
Knockenjig (UWC) 63m 50ch
Sanquhar
65m 53ch
River Nith
73m 05ch
Drumlanrig Tunnel
73m 69ch
Thornhill
SB (TH)
77m 58ch
78m 20ch
A
B

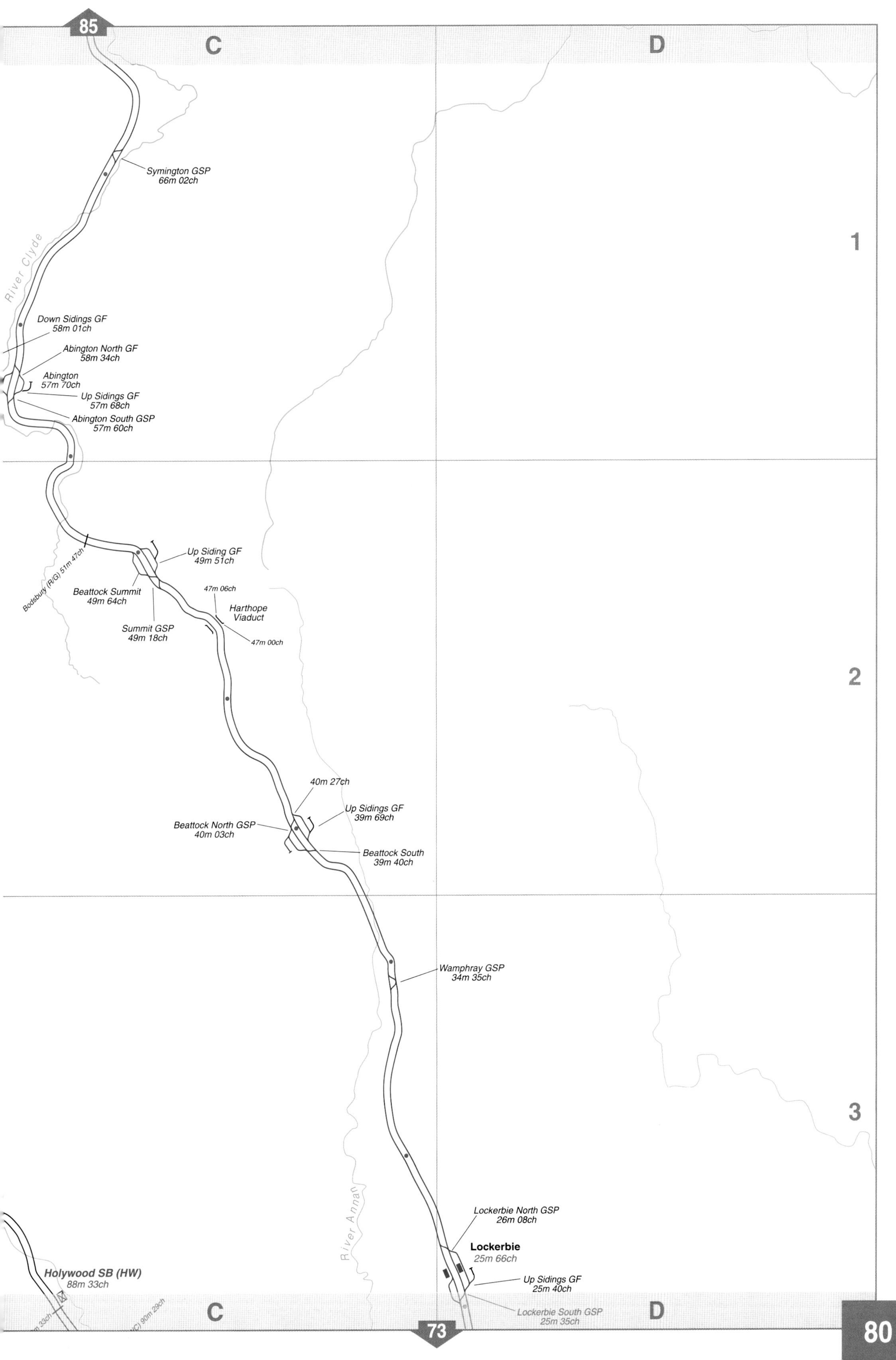
85
C
D
Symington GSP
66m 02ch
River Clyde
1
Down Sidings GF
58m 01ch
Abington North GF
58m 34ch
Abington
57m 70ch
Up Sidings GF
57m 68ch
Abington South GSP
57m 60ch
Up Siding GF
49m 51ch
Bodsbury (R/G) 51m 47ch
Beattock Summit
49m 64ch
47m 06ch
Harthope
Viaduct
Summit GSP
49m 18ch
47m 00ch
2
40m 27ch
Up Sidings GF
39m 69ch
Beattock North GSP
40m 03ch
Beattock South
39m 40ch
Wamphray GSP
34m 35ch
3
River Annan
Lockerbie North GSP
26m 08ch
Lockerbie
25m 66ch
Up Sidings GF
25m 40ch
Holywood SB (HW)
88m 33ch
Lockerbie South GSP
25m 35ch
C
D
73

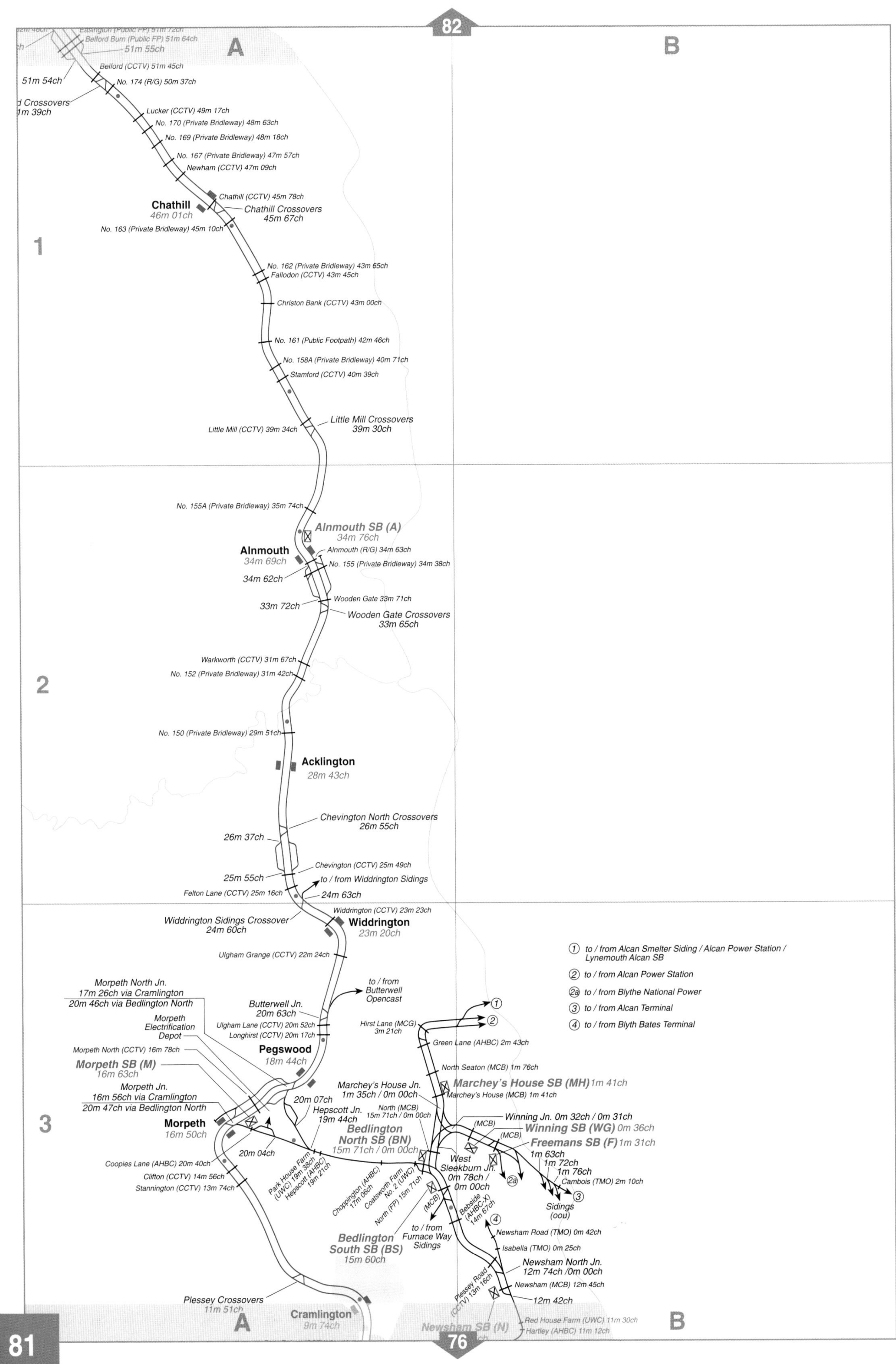

A
B
Easington (Public FP) 51m 72ch
Belford Burn (Public FP) 51m 64ch
51m 55ch
Belford (CCTV) 51m 45ch
51m 54ch
No. 174 (R/G) 50m 37ch
Lucker (CCTV) 49m 17ch
No. 170 (Private Bridleway) 48m 63ch
No. 169 (Private Bridleway) 48m 18ch
No. 167 (Private Bridleway) 47m 57ch
Newham (CCTV) 47m 09ch
Chathill (CCTV) 45m 78ch
Chathill
46m 01ch
Chathill Crossovers
45m 67ch
No. 163 (Private Bridleway) 45m 10ch
1
No. 162 (Private Bridleway) 43m 65ch
Fallodon (CCTV) 43m 45ch
Christon Bank (CCTV) 43m 00ch
No. 161 (Public Footpath) 42m 46ch
No. 158A (Private Bridleway) 40m 71ch
Stamford (CCTV) 40m 39ch
Little Mill Crossovers
39m 30ch
Little Mill (CCTV) 39m 34ch
No. 155A (Private Bridleway) 35m 74ch
Alnmouth SB (A)
34m 76ch
Alnmouth
34m 69ch
Alnmouth (R/G) 34m 63ch
No. 155 (Private Bridleway) 34m 38ch
34m 62ch
Wooden Gate 33m 71ch
33m 72ch
Wooden Gate Crossovers
33m 65ch
Warkworth (CCTV) 31m 67ch
No. 152 (Private Bridleway) 31m 42ch
2
No. 150 (Private Bridleway) 29m 51ch
Acklington
28m 43ch
Chevington North Crossovers
26m 55ch
26m 37ch
Chevington (CCTV) 25m 49ch
25m 55ch
to / from Widdrington Sidings
Felton Lane (CCTV) 25m 16ch
24m 63ch
Widdrington (CCTV) 23m 23ch
Widdrington Sidings Crossover
24m 60ch
Widdrington
23m 20ch
Ulgham Grange (CCTV) 22m 24ch
① to / from Alcan Smelter Siding / Alcan Power Station / Lynemouth Alcan SB
② to / from Alcan Power Station
②a to / from Blythe National Power
③ to / from Alcan Terminal
④ to / from Blyth Bates Terminal
Morpeth North Jn.
17m 26ch via Cramlington
20m 46ch via Bedlington North
to / from Butterwell Opencast
Butterwell Jn.
20m 63ch
Morpeth Electrification Depot
Ulgham Lane (CCTV) 20m 52ch
Longhirst (CCTV) 20m 17ch
Hirst Lane (MCG) 3m 21ch
Green Lane (AHBC) 2m 43ch
Morpeth North (CCTV) 16m 78ch
Pegswood
18m 44ch
Morpeth SB (M)
16m 63ch
North Seaton (MCB) 1m 76ch
Morpeth Jn.
16m 56ch via Cramlington
20m 47ch via Bedlington North
Marchey's House Jn.
1m 35ch / 0m 00ch
Marchey's House SB (MH) 1m 41ch
Marchey's House (MCB) 1m 41ch
20m 07ch
Hepscott Jn.
19m 44ch
North (MCB) 15m 71ch / 0m 00ch
Winning Jn. 0m 32ch / 0m 31ch
(MCB)
Winning SB (WG) 0m 36ch
(MCB)
Freemans SB (F) 1m 31ch
3
Morpeth
16m 50ch
Bedlington North SB (BN)
15m 71ch / 0m 00ch
20m 04ch
West Sleekburn Jn.
0m 78ch / 0m 00ch
1m 63ch
1m 72ch
1m 76ch
Cambois (TMO) 2m 10ch
Coopies Lane (AHBC) 20m 40ch
Clifton (CCTV) 14m 56ch
Stannington (CCTV) 13m 74ch
Park House Farm (UWC) 19m 38ch
Hepscott (AHBC) 19m 21ch
Choppington (AHBC) 17m 06ch
Coatsworth Farm No. 2 (UWC)
North (FP) 15m 71ch
(MCB)
Bebside (AHBC-X) 14m 67ch
Sidings (oou)
to / from Furnace Way Sidings
Bedlington South SB (BS)
15m 60ch
Newsham Road (TMO) 0m 42ch
Isabella (TMO) 0m 25ch
Newsham North Jn.
12m 74ch /0m 00ch
Newsham (MCB) 12m 45ch
Plessey Road (CCTV) 13m 16ch
12m 42ch
Plessey Crossovers
11m 51ch
Cramlington
9m 74ch
Newsham SB (N)
Red House Farm (UWC) 11m 30ch
Hartley (AHBC) 11m 12ch
A
B

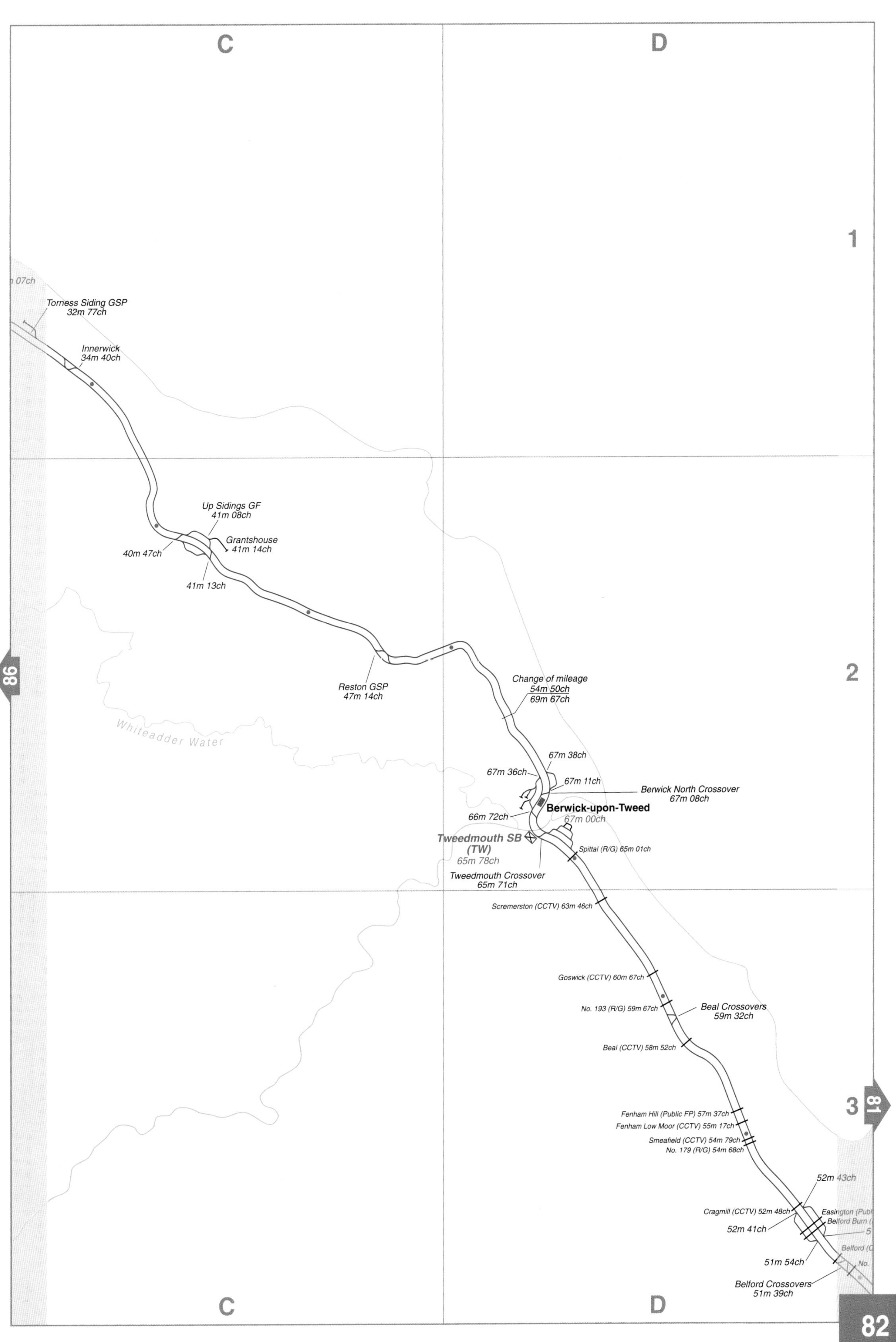
C
D
1
2
3
Torness Siding GSP
32m 77ch
Innerwick
34m 40ch
Up Sidings GF
41m 08ch
Grantshouse
41m 14ch
40m 47ch
41m 13ch
Reston GSP
47m 14ch
Whiteadder Water
Change of mileage
54m 50ch
69m 67ch
67m 38ch
67m 36ch
67m 11ch
Berwick North Crossover
67m 08ch
Berwick-upon-Tweed
67m 00ch
66m 72ch
Tweedmouth SB
(TW)
65m 78ch
Tweedmouth Crossover
65m 71ch
Spittal (R/G) 65m 01ch
Scremerston (CCTV) 63m 46ch
Goswick (CCTV) 60m 67ch
No. 193 (R/G) 59m 67ch
Beal Crossovers
59m 32ch
Beal (CCTV) 58m 52ch
Fenham Hill (Public FP) 57m 37ch
Fenham Low Moor (CCTV) 55m 17ch
Smeafield (CCTV) 54m 79ch
No. 179 (R/G) 54m 68ch
52m 43ch
Cragmill (CCTV) 52m 48ch
52m 41ch
51m 54ch
Belford Crossovers
51m 39ch
86
81

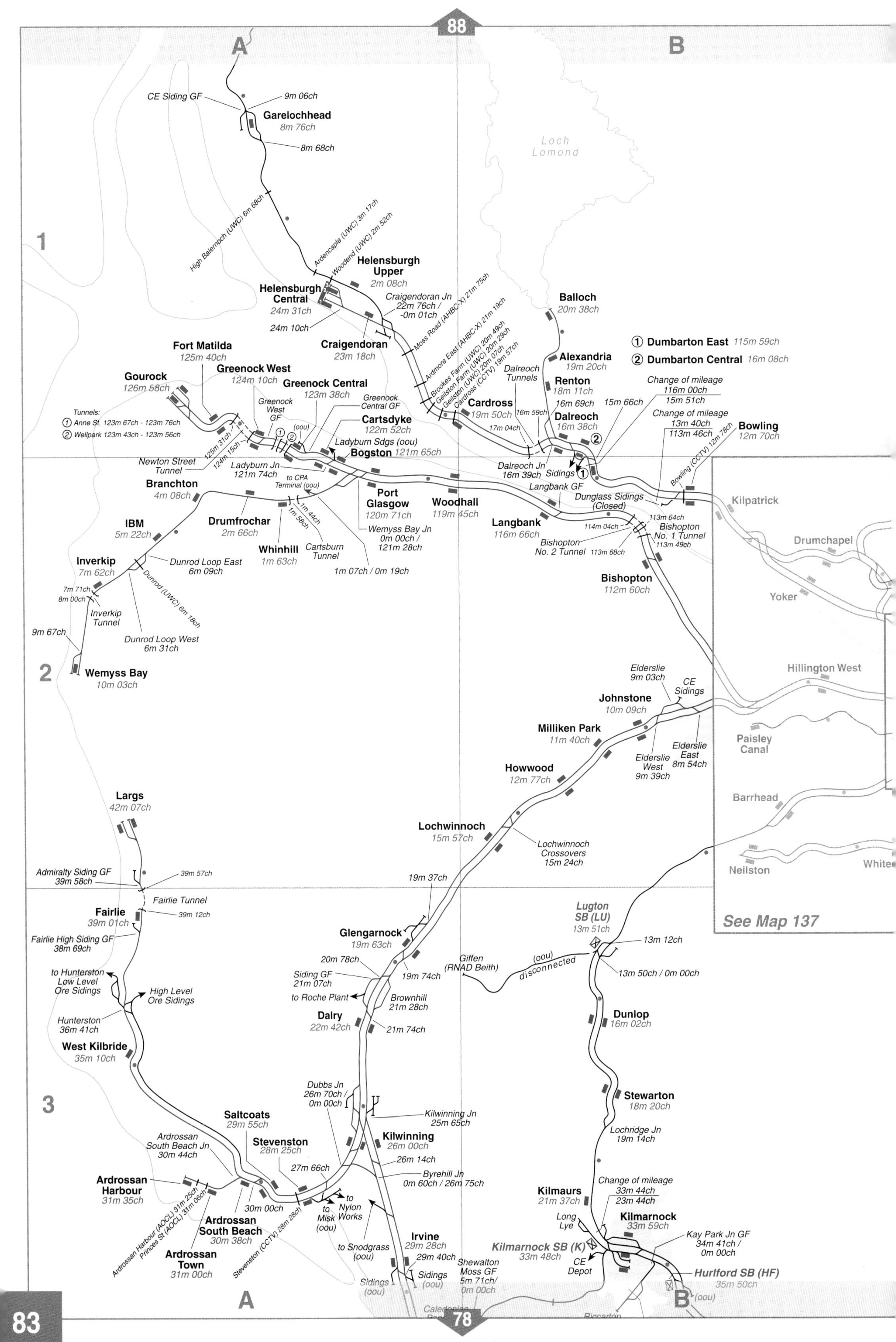
88
A
B
1
2
3
Loch Lomond
CE Siding GF
9m 06ch
Garelochhead
8m 76ch
8m 68ch
High Balernoch (UWC) 6m 68ch
Ardencaple (UWC) 3m 17ch
Woodend (UWC) 2m 52ch
Helensburgh Upper
2m 08ch
Helensburgh Central
24m 31ch
24m 10ch
Craigendoran Jn
22m 76ch / -0m 01ch
Craigendoran
23m 18ch
Moss Road (AHBC-X) 21m 75ch
Ardmore East (AHBC-X) 21m 19ch
Brookes Farm (UWC) 20m 49ch
Geilston Farm (UWC) 20m 29ch
Geilston (UWC) 20m 07ch
Cardross (CCTV) 19m 57ch
Cardross
19m 50ch
Balloch
20m 38ch
Alexandria
19m 20ch
Renton
18m 11ch
16m 69ch
Dalreoch
16m 38ch
Dalreoch Tunnels
16m 59ch
17m 04ch
Dalreoch Jn
16m 39ch
Sidings
15m 66ch
① Dumbarton East 115m 59ch
② Dumbarton Central 16m 08ch
Change of mileage
116m 00ch
15m 51ch
Change of mileage
13m 40ch
113m 46ch
Bowling
12m 70ch
Bowling (CCTV) 12m 78ch
Fort Matilda
125m 40ch
Gourock
126m 58ch
Greenock West
124m 10ch
Greenock Central
123m 38ch
Greenock West GF
Greenock Central GF
(oou)
Tunnels:
① Anne St. 123m 67ch - 123m 76ch
② Wellpark 123m 43ch - 123m 56ch
125m 31ch
124m 15ch
Newton Street Tunnel
Cartsdyke
122m 52ch
Ladyburn Sdgs (oou)
Bogston
121m 65ch
Ladyburn Jn
121m 74ch
to CPA Terminal (oou)
Port Glasgow
120m 71ch
Woodhall
119m 45ch
Langbank GF
Dunglass Sidings (Closed)
Langbank
116m 66ch
114m 04ch
113m 64ch
Bishopton No. 1 Tunnel
113m 49ch
Bishopton No. 2 Tunnel
113m 68ch
Bishopton
112m 60ch
Kilpatrick
Drumchapel
Yoker
Branchton
4m 08ch
Drumfrochar
2m 66ch
Whinhill
1m 63ch
1m 44ch
1m 58ch
Cartsburn Tunnel
Wemyss Bay Jn
0m 00ch /
121m 28ch
1m 07ch / 0m 19ch
IBM
5m 22ch
Inverkip
7m 62ch
Dunrod Loop East
6m 09ch
Dunrod (UWC) 6m 18ch
7m 71ch
8m 00ch
Inverkip Tunnel
Dunrod Loop West
6m 31ch
9m 67ch
Wemyss Bay
10m 03ch
Hillington West
Elderslie
9m 03ch
CE Sidings
Johnstone
10m 09ch
Milliken Park
11m 40ch
Elderslie West
9m 39ch
Elderslie East
8m 54ch
Paisley Canal
Howwood
12m 77ch
Barrhead
Lochwinnoch
15m 57ch
Lochwinnoch Crossovers
15m 24ch
Neilston
White
See Map 137
Largs
42m 07ch
Admiralty Siding GF
39m 58ch
39m 57ch
Fairlie Tunnel
39m 12ch
Fairlie
39m 01ch
Fairlie High Siding GF
38m 69ch
to Hunterston Low Level Ore Sidings
High Level Ore Sidings
Hunterston
36m 41ch
West Kilbride
35m 10ch
19m 37ch
Glengarnock
19m 63ch
20m 78ch
Siding GF
21m 07ch
to Roche Plant
19m 74ch
Brownhill
21m 28ch
Dalry
22m 42ch
21m 74ch
Giffen (RNAD Beith)
(oou)
disconnected
Lugton SB (LU)
13m 51ch
13m 12ch
13m 50ch / 0m 00ch
Dunlop
16m 02ch
Stewarton
18m 20ch
Lochridge Jn
19m 14ch
Dubbs Jn
26m 70ch /
0m 00ch
Kilwinning Jn
25m 65ch
Kilwinning
26m 00ch
26m 14ch
Byrehill Jn
0m 60ch / 26m 75ch
Saltcoats
29m 55ch
Stevenston
28m 25ch
Ardrossan South Beach Jn
30m 44ch
27m 66ch
Ardrossan Harbour
31m 35ch
30m 00ch
Ardrossan South Beach
30m 38ch
Ardrossan Town
31m 00ch
Ardrossan Harbour (AOCL) 31m 25ch
Princes St (AOCL) 31m 06ch
Stevenston (CCTV) 28m 28ch
to Misk (oou)
to Nylon Works
to Snodgrass (oou)
Irvine
29m 28ch
29m 40ch
Sidings (oou)
Sidings (oou)
Shewalton Moss GF
5m 71ch/
0m 00ch
Kilmaurs
21m 37ch
Change of mileage
33m 44ch
23m 44ch
Long Lye
Kilmarnock
33m 59ch
Kilmarnock SB (K)
33m 48ch
CE Depot
Kay Park Jn GF
34m 41ch /
0m 00ch
Hurlford SB (HF)
35m 50ch
(oou)
78

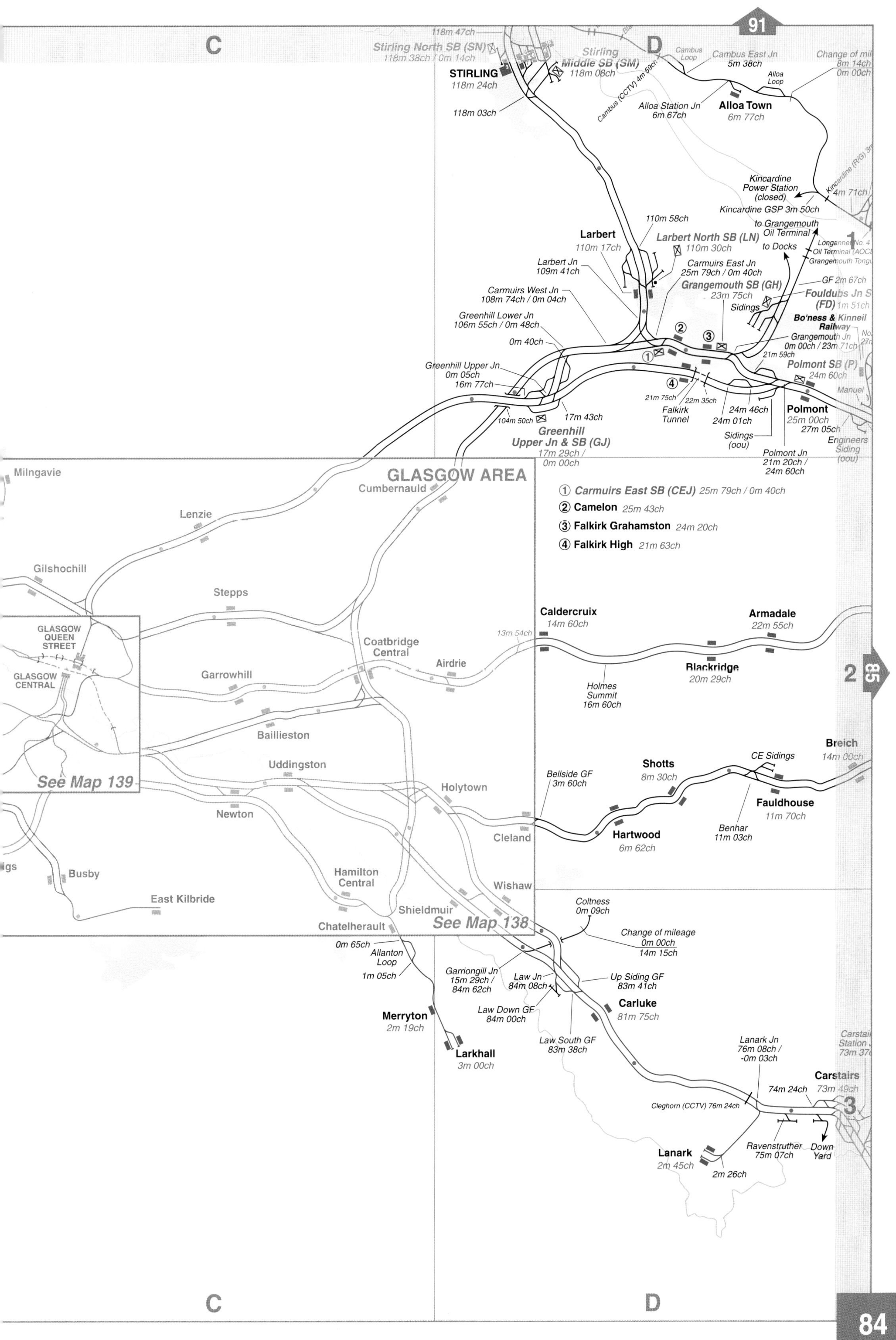
C
D
Stirling North SB (SN)
118m 38ch / 0m 14ch
118m 47ch
Stirling
Middle SB (SM)
118m 08ch
STIRLING
118m 24ch
118m 03ch
Cambus Loop
Cambus East Jn
5m 38ch
Change of mileage
8m 14ch
0m 00ch
Alloa Loop
Cambus (CCTV) 4m 59ch
Alloa Station Jn
6m 67ch
Alloa Town
6m 77ch
Kincardine Power Station (closed)
Kincardine GSP 3m 50ch
4m 71ch
to Grangemouth Oil Terminal
to Docks
Larbert
110m 17ch
110m 58ch
Larbert North SB (LN)
110m 30ch
Larbert Jn
109m 41ch
Carmuirs East Jn
25m 79ch / 0m 40ch
Carmuirs West Jn
108m 74ch / 0m 04ch
Grangemouth SB (GH)
23m 75ch
Sidings
GF 2m 67ch
Fouldubs Jn
(FD) 1m 51ch
Bo'ness & Kinneil Railway
Greenhill Lower Jn
106m 55ch / 0m 48ch
0m 40ch
Grangemouth Jn
0m 00ch / 23m 71ch
21m 59ch
Polmont SB (P)
24m 60ch
Greenhill Upper Jn
0m 05ch
16m 77ch
21m 75ch
22m 35ch
Falkirk Tunnel
24m 46ch
24m 01ch
Polmont
25m 00ch
27m 05ch
104m 50ch
17m 43ch
Greenhill Upper Jn & SB (GJ)
17m 29ch / 0m 00ch
Sidings (oou)
Polmont Jn
21m 20ch / 24m 60ch
Engineers Siding (oou)
Manuel
GLASGOW AREA
① Carmuirs East SB (CEJ) 25m 79ch / 0m 40ch
② Camelon 25m 43ch
③ Falkirk Grahamston 24m 20ch
④ Falkirk High 21m 63ch
Milngavie
Cumbernauld
Lenzie
Gilshochill
Stepps
GLASGOW QUEEN STREET
GLASGOW CENTRAL
Coatbridge Central
Airdrie
Garrowhill
Baillieston
Uddingston
See Map 139
Newton
Holytown
Cleland
Busby
East Kilbride
Hamilton Central
Wishaw
Shieldmuir
Chatelherault
See Map 138
Caldercruix
14m 60ch
13m 54ch
Armadale
22m 55ch
Blackridge
20m 29ch
Holmes Summit
16m 60ch
Bellside GF
3m 60ch
Shotts
8m 30ch
CE Sidings
Breich
14m 00ch
Fauldhouse
11m 70ch
Benhar
11m 03ch
Hartwood
6m 62ch
Coltness
0m 09ch
Change of mileage
0m 00ch
14m 15ch
0m 65ch
Allanton Loop
1m 05ch
Garriongill Jn
15m 29ch / 84m 62ch
Law Jn
84m 08ch
Up Siding GF
83m 41ch
Carluke
81m 75ch
Law Down GF
84m 00ch
Law South GF
83m 38ch
Merryton
2m 19ch
Larkhall
3m 00ch
Lanark Jn
76m 08ch / -0m 03ch
Carstairs
73m 49ch
74m 24ch
Cleghorn (CCTV) 76m 24ch
Ravenstruther
75m 07ch
Down Yard
Lanark
2m 45ch
2m 26ch
1
2
3
85

92

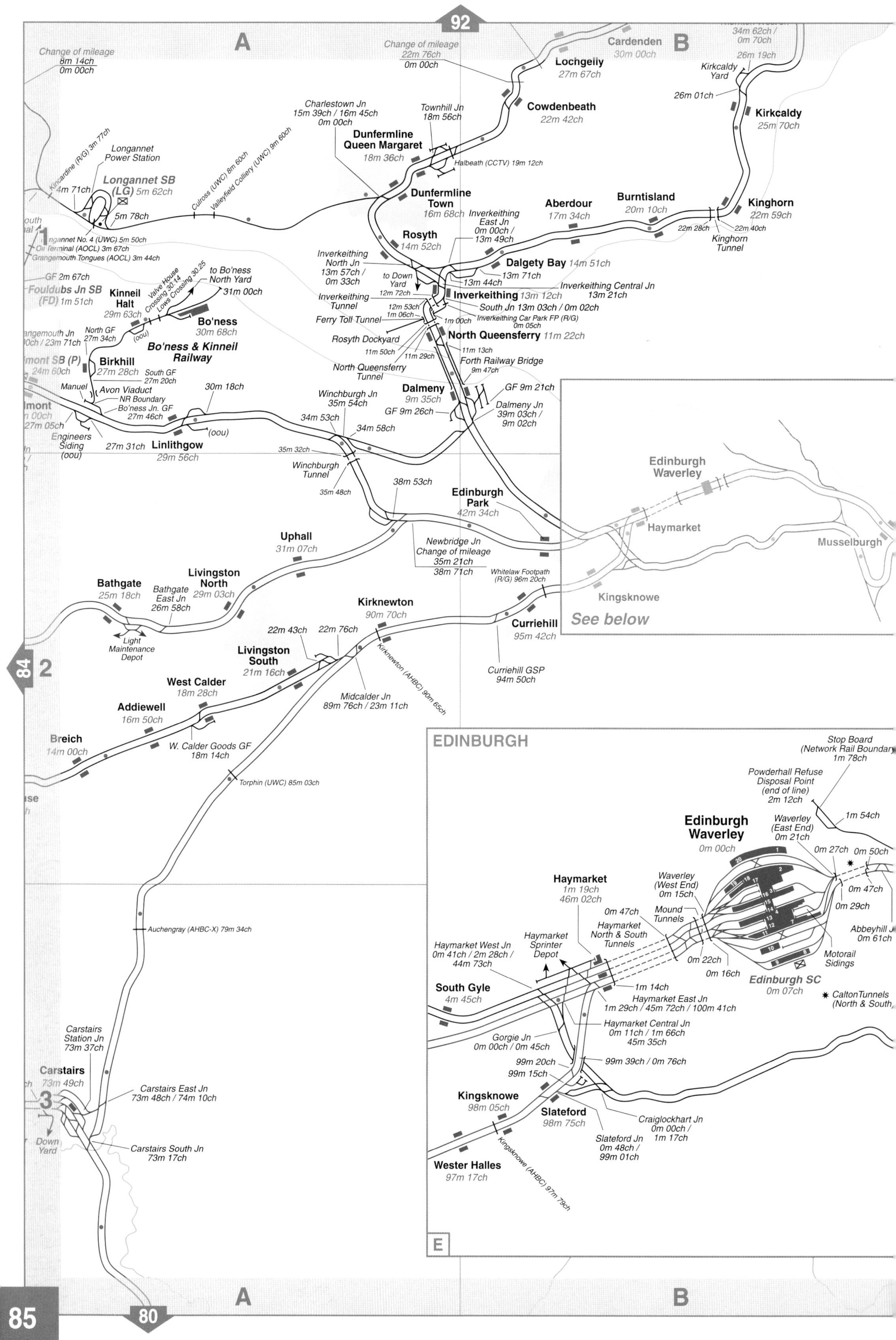
A
B
Change of mileage
8m 14ch
0m 00ch
Change of mileage
22m 76ch
0m 00ch
Cardenden
30m 00ch
Lochgelly
27m 67ch
Cowdenbeath
22m 42ch
34m 62ch /
0m 70ch
26m 19ch
Kirkcaldy Yard
26m 01ch
Kirkcaldy
25m 70ch
Charlestown Jn
15m 39ch / 16m 45ch
0m 00ch
Townhill Jn
18m 56ch
Dunfermline Queen Margaret
18m 36ch
Halbeath (CCTV) 19m 12ch
Longannet Power Station
Kincardine (R/G) 3m 77ch
Longannet SB (LG) 5m 62ch
4m 71ch
5m 78ch
Culross (UWC) 8m 60ch
Valleyfield Colliery (UWC) 9m 60ch
Longannet No. 4 (UWC) 5m 50ch
Oil Terminal (AOCL) 3m 67ch
Grangemouth Tongues (AOCL) 3m 44ch
GF 2m 67ch
Fouldubs Jn SB (FD) 1m 51ch
Dunfermline Town
16m 68ch
Rosyth
14m 52ch
Inverkeithing East Jn
0m 00ch /
13m 49ch
Aberdour
17m 34ch
Burntisland
20m 10ch
Kinghorn
22m 59ch
22m 28ch
22m 40ch
Kinghorn Tunnel
Inverkeithing North Jn
13m 57ch /
0m 33ch
Dalgety Bay 14m 51ch
13m 71ch
13m 44ch
to Down Yard
12m 72ch
Inverkeithing 13m 12ch
Inverkeithing Central Jn
13m 21ch
Inverkeithing Tunnel
12m 53ch
1m 06ch
South Jn 13m 03ch / 0m 02ch
Inverkeithing Car Park FP (R/G)
0m 05ch
Ferry Toll Tunnel
1m 00ch
Valve House Crossing 30.14
Lows Crossing 30.25
to Bo'ness North Yard
31m 00ch
Kinneil Halt
29m 63ch
Bo'ness
30m 68ch
Grangemouth Jn
23m 71ch
North GF
27m 34ch
(oou)
Bo'ness & Kinneil Railway
Rosyth Dockyard
North Queensferry 11m 22ch
11m 50ch
11m 29ch
11m 13ch
Forth Railway Bridge
9m 47ch
North Queensferry Tunnel
Birkhill
27m 28ch
South GF
27m 20ch
Polmont SB (P)
24m 60ch
Manuel
Avon Viaduct
NR Boundary
Bo'ness Jn. GF
27m 46ch
30m 18ch
Winchburgh Jn
35m 54ch
Dalmeny
9m 35ch
GF 9m 21ch
GF 9m 26ch
Dalmeny Jn
39m 03ch /
9m 02ch
27m 05ch
Engineers Siding (oou)
27m 31ch
Linlithgow
29m 56ch
(oou)
34m 53ch
34m 58ch
35m 32ch
Winchburgh Tunnel
35m 48ch
38m 53ch
Edinburgh Park
42m 34ch
Edinburgh Waverley
Haymarket
Musselburgh
Uphall
31m 07ch
Newbridge Jn
Change of mileage
35m 21ch
38m 71ch
Whitelaw Footpath (R/G) 96m 20ch
Kingsknowe
See below
Bathgate
25m 18ch
Bathgate East Jn
26m 58ch
Livingston North
29m 03ch
Light Maintenance Depot
Kirknewton
90m 70ch
Curriehill
95m 42ch
22m 43ch
22m 76ch
Livingston South
21m 16ch
Curriehill GSP
94m 50ch
84
2
West Calder
18m 28ch
Midcalder Jn
89m 76ch / 23m 11ch
Kirknewton (AHBC) 90m 65ch
Addiewell
16m 50ch
Breich
14m 00ch
W. Calder Goods GF
18m 14ch
Torphin (UWC) 85m 03ch
EDINBURGH
Stop Board
(Network Rail Boundary)
1m 78ch
Powderhall Refuse Disposal Point
(end of line)
2m 12ch
1m 54ch
Edinburgh Waverley
0m 00ch
Waverley (East End)
0m 21ch
0m 27ch
0m 50ch
0m 47ch
0m 29ch
Haymarket
1m 19ch
46m 02ch
Waverley (West End)
0m 15ch
0m 47ch
Mound Tunnels
Haymarket North & South Tunnels
Abbeyhill Jn
0m 61ch
Haymarket West Jn
0m 41ch / 2m 28ch /
44m 73ch
Haymarket Sprinter Depot
0m 22ch
0m 16ch
Motorail Sidings
Edinburgh SC
0m 07ch
South Gyle
4m 45ch
1m 14ch
Haymarket East Jn
1m 29ch / 45m 72ch / 100m 41ch
Calton Tunnels (North & South)
Auchengray (AHBC-X) 79m 34ch
Gorgie Jn
0m 00ch / 0m 45ch
Haymarket Central Jn
0m 11ch / 1m 66ch
45m 35ch
99m 20ch
99m 39ch / 0m 76ch
99m 15ch
Carstairs Station Jn
73m 37ch
Carstairs
73m 49ch
Carstairs East Jn
73m 48ch / 74m 10ch
3
Down Yard
Kingsknowe
98m 05ch
Slateford
98m 75ch
Craiglockhart Jn
0m 00ch /
1m 17ch
Carstairs South Jn
73m 17ch
Slateford Jn
0m 48ch /
99m 01ch
Wester Halles
97m 17ch
Kingsknowe (AHBC) 97m 79ch
E

80

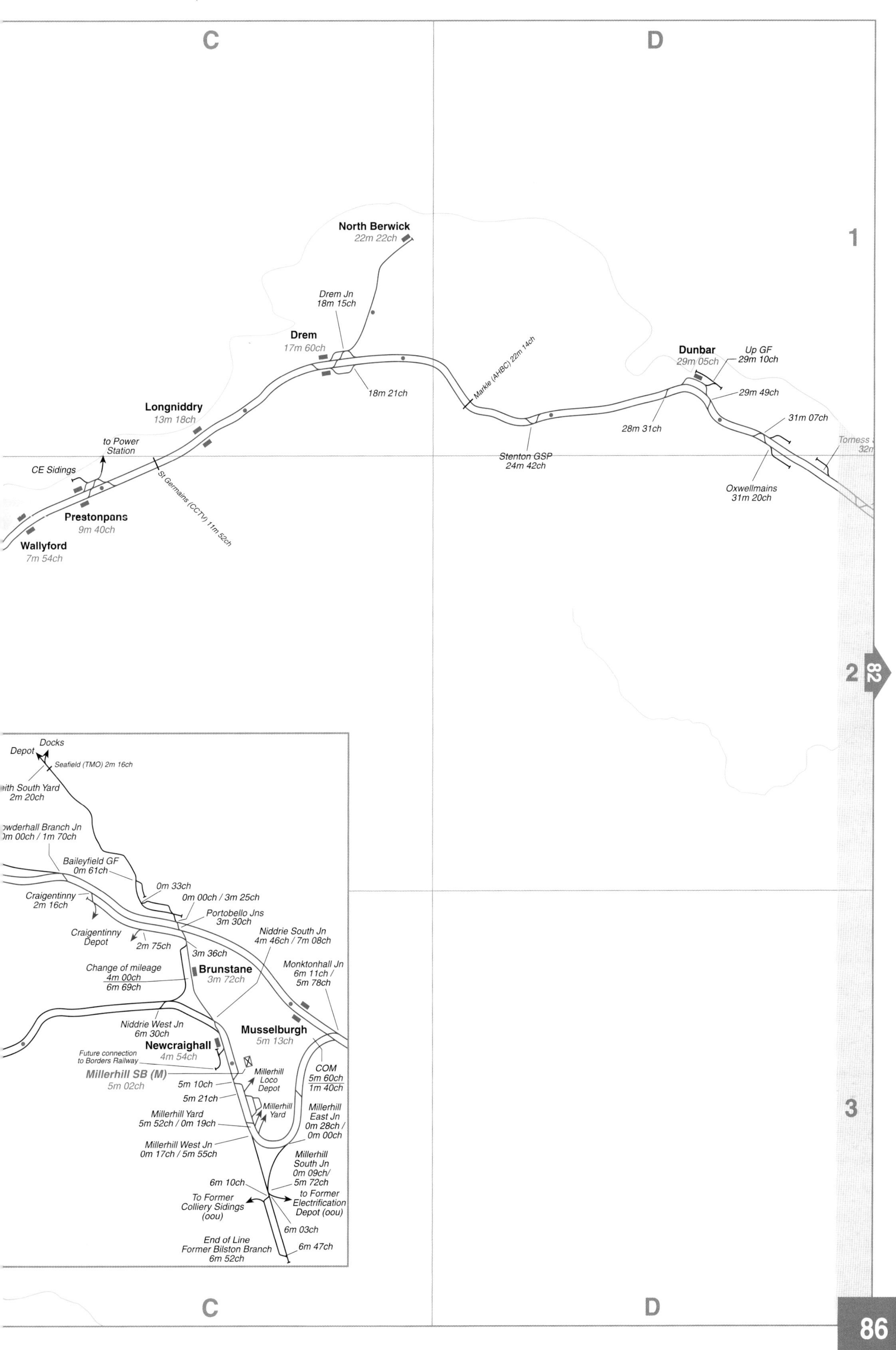
C
D
North Berwick
22m 22ch
Drem Jn
18m 15ch
Drem
17m 60ch
18m 21ch
Markle (AHBC) 22m 14ch
Dunbar
29m 05ch
Up GF
29m 10ch
29m 49ch
28m 31ch
31m 07ch
Stenton GSP
24m 42ch
Oxwellmains
31m 20ch
Longniddry
13m 18ch
to Power Station
CE Sidings
St Germains (CCTV) 11m 52ch
Prestonpans
9m 40ch
Wallyford
7m 54ch
1
2
82
3
Docks
Depot
Seafield (TMO) 2m 16ch
2m 20ch
Baileyfield GF
0m 61ch
0m 33ch
Craigentinny
2m 16ch
0m 00ch / 3m 25ch
Portobello Jns
3m 30ch
Craigentinny
Depot
2m 75ch
Niddrie South Jn
4m 46ch / 7m 08ch
3m 36ch
Change of mileage
4m 00ch
6m 69ch
Brunstane
3m 72ch
Monktonhall Jn
6m 11ch /
5m 78ch
Niddrie West Jn
6m 30ch
Musselburgh
5m 13ch
Future connection
to Borders Railway
Newcraighall
4m 54ch
Millerhill SB (M)
5m 02ch
Millerhill
Loco
Depot
COM
5m 60ch
1m 40ch
5m 10ch
5m 21ch
Millerhill
Yard
Millerhill Yard
5m 52ch / 0m 19ch
Millerhill
East Jn
0m 28ch /
0m 00ch
Millerhill West Jn
0m 17ch / 5m 55ch
Millerhill
South Jn
0m 09ch/
5m 72ch
6m 10ch
To Former
Colliery Sidings
(oou)
to Former
Electrification
Depot (oou)
6m 03ch
End of Line
Former Bilston Branch
6m 52ch
6m 47ch
C
D

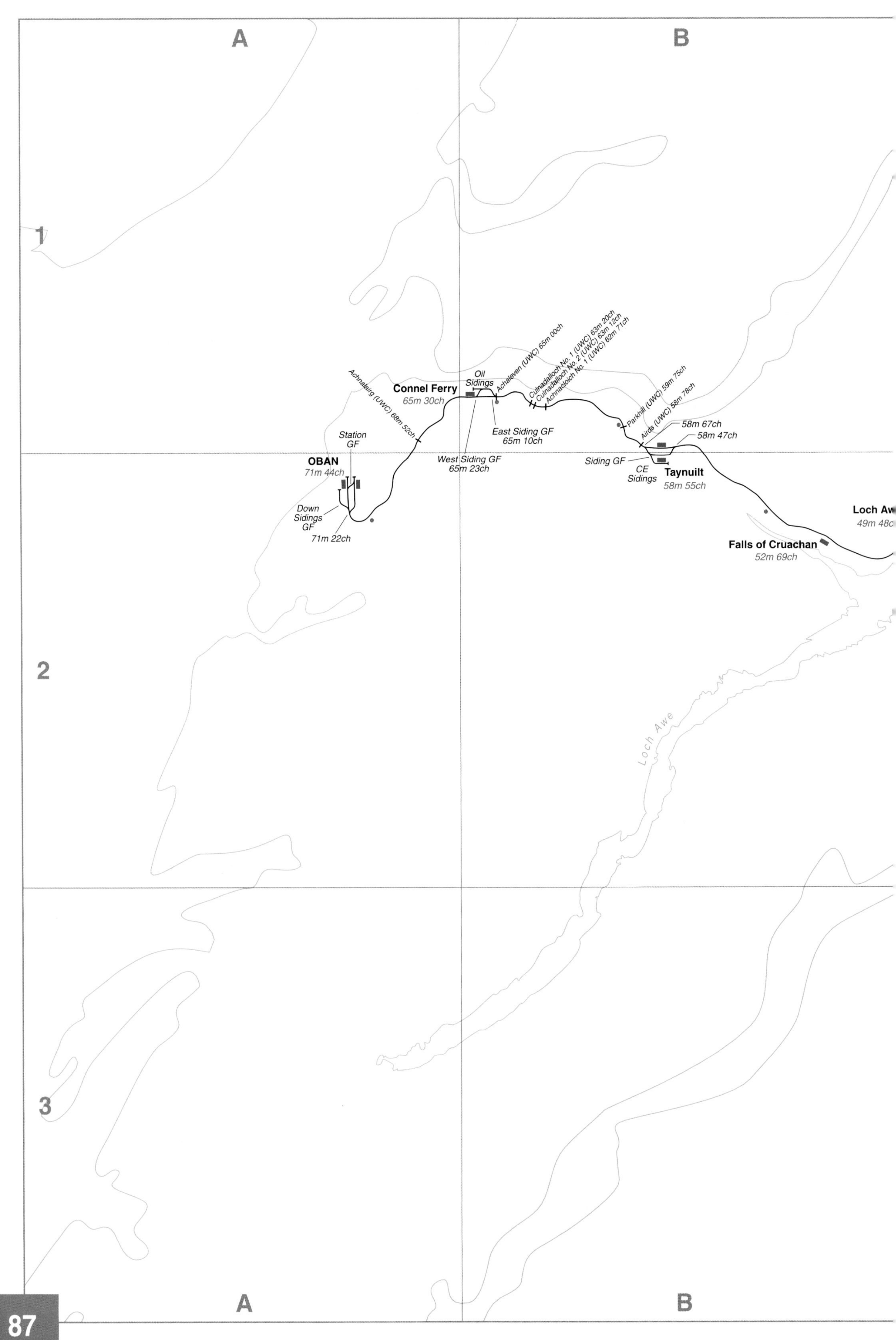
A
B
1
2
3
Achnalairg (UWC) 68m 52ch
Connel Ferry
65m 30ch
Oil
Sidings
Achaleven (UWC) 65m 00ch
Culnadalloch No. 1 (UWC) 63m 20ch
Culnadalloch No. 2 (UWC) 63m 12ch
Achnacloich No. 1 (UWC) 62m 71ch
Parkhill (UWC) 59m 75ch
Airds (UWC) 58m 78ch
58m 67ch
58m 47ch
Station
GF
OBAN
71m 44ch
East Siding GF
65m 10ch
West Siding GF
65m 23ch
Siding GF
CE
Sidings
Taynuilt
58m 55ch
Down
Sidings
GF
71m 22ch
Loch Aw
49m 48c
Falls of Cruachan
52m 69ch
Loch Awe
A
B

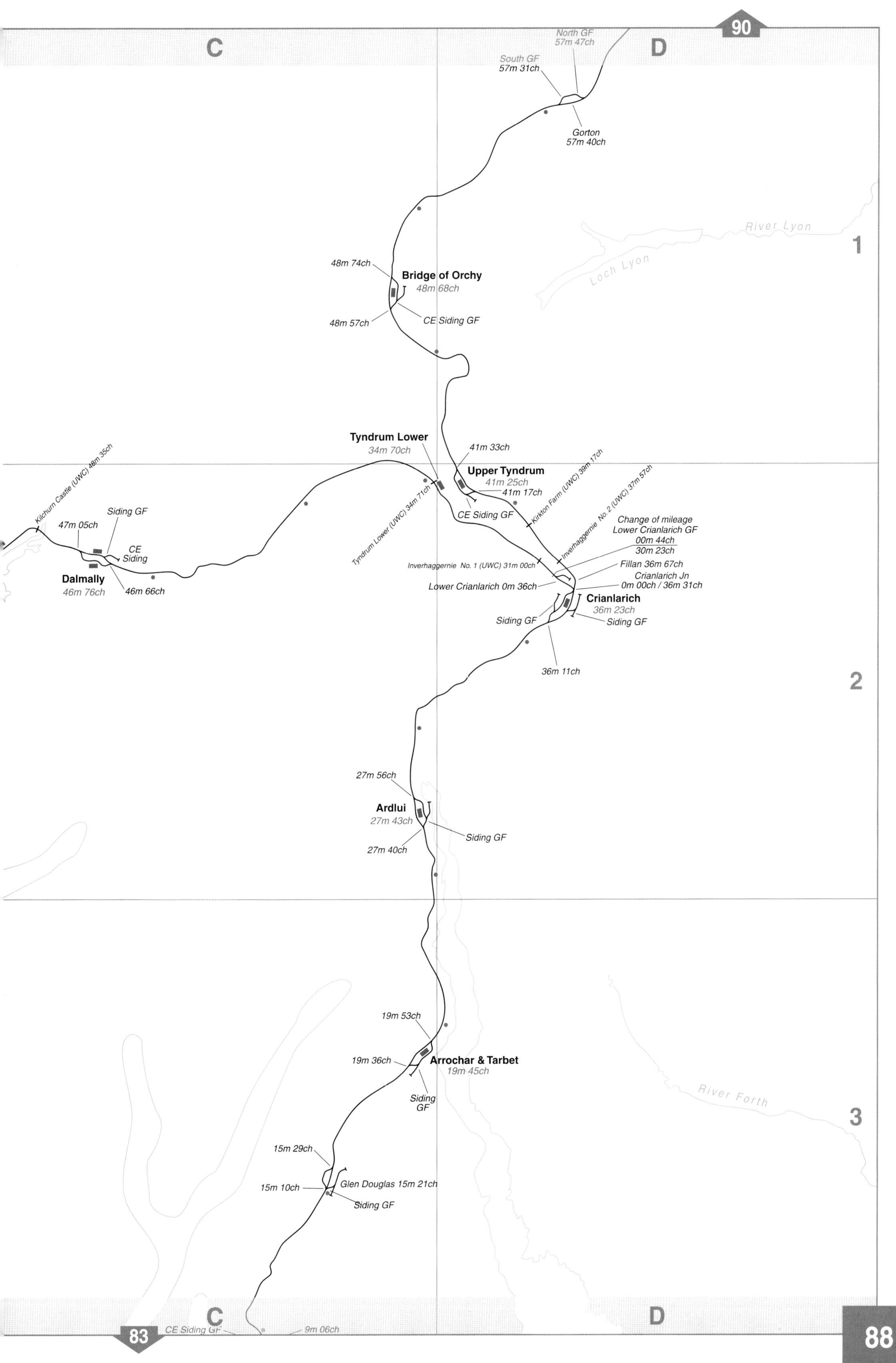
C
D
North GF
57m 47ch
South GF
57m 31ch
Gorton
57m 40ch
River Lyon
Loch Lyon
1
48m 74ch
Bridge of Orchy
48m 68ch
48m 57ch
CE Siding GF
Tyndrum Lower
34m 70ch
41m 33ch
Upper Tyndrum
41m 25ch
41m 17ch
CE Siding GF
Kirkton Farm (UWC) 39m 17ch
Inverhaggernie No. 2 (UWC) 37m 57ch
Tyndrum Lower (UWC) 34m 71ch
Kilchurn Castle (UWC) 48m 35ch
Siding GF
47m 05ch
CE
Siding
Dalmally
46m 76ch
46m 66ch
Change of mileage
Lower Crianlarich GF
00m 44ch
30m 23ch
Inverhaggernie No. 1 (UWC) 31m 00ch
Fillan 36m 67ch
Crianlarich Jn
Lower Crianlarich 0m 36ch
0m 00ch / 36m 31ch
Crianlarich
36m 23ch
Siding GF
Siding GF
36m 11ch
2
27m 56ch
Ardlui
27m 43ch
Siding GF
27m 40ch
19m 53ch
19m 36ch
Arrochar & Tarbet
19m 45ch
Siding
GF
River Forth
3
15m 29ch
15m 10ch
Glen Douglas 15m 21ch
Siding GF
C
D
CE Siding GF
9m 06ch

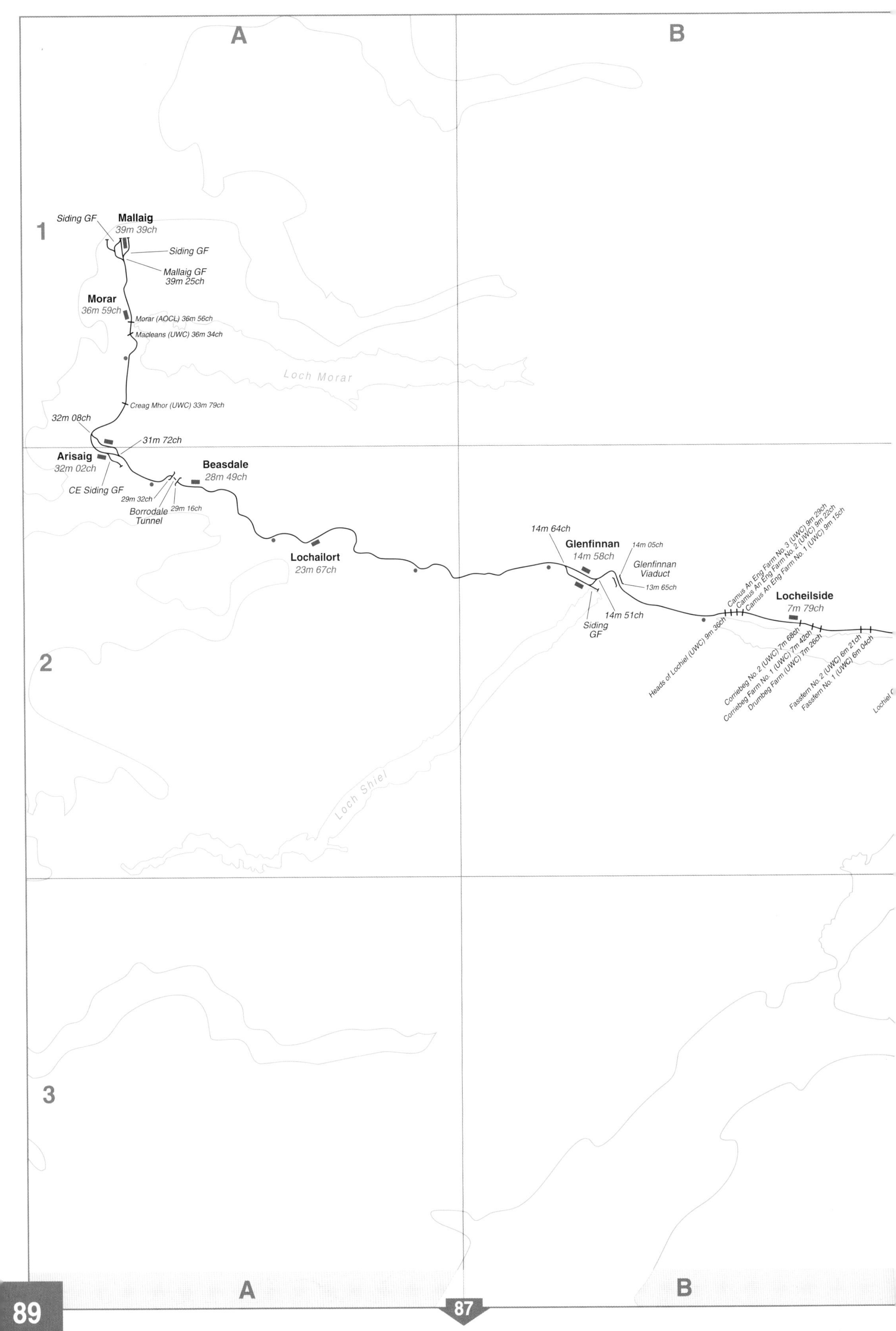
A
B
1
2
3
Siding GF
Mallaig
39m 39ch
Siding GF
Mallaig GF
39m 25ch
Morar
36m 59ch
Morar (AOCL) 36m 56ch
Macleans (UWC) 36m 34ch
Loch Morar
Creag Mhor (UWC) 33m 79ch
32m 08ch
31m 72ch
Arisaig
32m 02ch
CE Siding GF
Beasdale
28m 49ch
29m 32ch
Borrodale
Tunnel
29m 16ch
Lochailort
23m 67ch
14m 64ch
Glenfinnan
14m 58ch
14m 05ch
Glenfinnan
Viaduct
13m 65ch
14m 51ch
Siding
GF
Camus An Eng Farm No. 3 (UWC) 9m 29ch
Camus An Eng Farm No. 2 (UWC) 9m 22ch
Camus An Eng Farm No. 1 (UWC) 9m 15ch
Locheilside
7m 79ch
Heads of Lochiel (UWC) 9m 36ch
Corriebeg No. 2 (UWC) 7m 68ch
Corriebeg Farm No. 1 (UWC) 7m 42ch
Drumbeg Farm (UWC) 7m 26ch
Fassfern No. 2 (UWC) 6m 21ch
Fassfern No. 1 (UWC) 6m 04ch
Loch Shiel
A
B

87

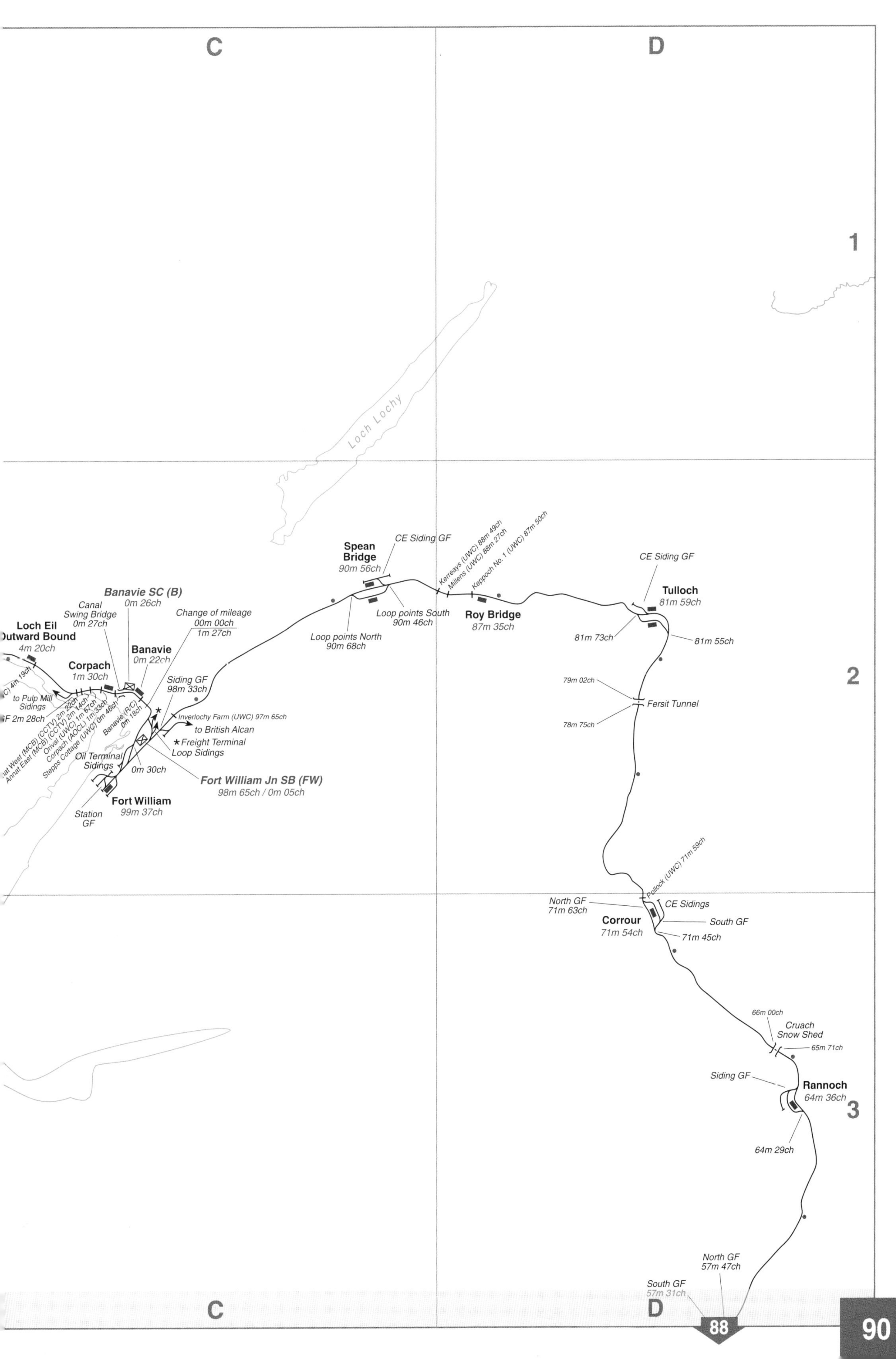
C
D
1
2
3
Loch Lochy
Spean Bridge
90m 56ch
CE Siding GF
Kerreays (UWC) 88m 49ch
Millens (UWC) 88m 27ch
Keppoch No. 1 (UWC) 87m 50ch
Loop points South
90m 46ch
Loop points North
90m 68ch
Roy Bridge
87m 35ch
CE Siding GF
Tulloch
81m 59ch
81m 73ch
81m 55ch
79m 02ch
Fersit Tunnel
78m 75ch
Banavie SC (B)
0m 26ch
Canal
Swing Bridge
0m 27ch
Change of mileage
00m 00ch
1m 27ch
Loch Eil
Outward Bound
4m 20ch
Banavie
0m 22ch
Corpach
1m 30ch
Siding GF
98m 33ch
to Pulp Mill
Sidings
Inverlochy Farm (UWC) 97m 65ch
to British Alcan
Freight Terminal
Loop Sidings
Oil Terminal
Sidings
0m 30ch
Fort William Jn SB (FW)
98m 65ch / 0m 05ch
Fort William
99m 37ch
Station
GF
Pollock (UWC) 71m 59ch
North GF
71m 63ch
CE Sidings
Corrour
71m 54ch
South GF
71m 45ch
66m 00ch
Cruach
Snow Shed
65m 71ch
Siding GF
Rannoch
64m 36ch
64m 29ch
North GF
57m 47ch
South GF
57m 31ch
88

A
B
1
2
3
Dunkeld & Birnam
15m 31ch
Inver Tunnel
15m 45ch
16m 72ch
16m 55ch
Dunkeld SB (DK)
15m 25ch
15m 16ch
13m 13ch
12m 78ch
Kingswood Tunnel
Stanley Jn SB (SJ)
7m 07ch
Stanley Jn
158m 35ch
to Perth Yard
PERTH
151m 25ch / 20m 64ch
Sidings (oou)
Perth SB (P)
151m 05ch
Change of mileage
151m 03ch
21m 01ch
✦ - Perth Central Jn
151m 05ch
Hilton Jn 45m 66ch / 149m 23ch
Kirkton of Mailer No. 2 (UWC) 148m 66ch
Hilton Jn SB (HJ) 149m 17ch
Forgandenny Ford (UWC) 147m 39ch
150m 04ch
149m 29ch
Moncrieffe Tunnel
45m 37ch
River Earn
Broombarns (UWC) 146m 31ch
Forteviot Farm (UWC) 145m 13ch
Forteviot (AHBC-X) 144m 44ch
Baldinnies No. 1 (UWC) 142m 70ch
Easter Balgour (UWC) 142m 36ch
Broadslap (UWC) 141m 02ch
Whitemoss (AHBC-X) 140m 24ch
Down Sidings
Up Sidings
Auchterarder SB (AR)
137m 41ch
Gleneagles
135m 50ch
Blackford SB (BK)
133m 28ch
Blackford (MCB) 133m 28ch
Boreland Farm (UWC) 132m 20ch
Carsebreck (UWC) 131m 07ch
Greenloaning SB (GL)
129m 17ch
129m 06ch
Quoiggs No. 1 (UWC) 128m 01ch
Drumallan (UWC) 126m 27ch
Dunblane
123m 19ch
Dunblane SB (DB)
123m 29ch
122m 66ch
Kippenross Tunnel
122m 38ch
Bridge of Allan
121m 10ch
Causewayhead Jn
1m 05ch
Waterside (CCTV) 1m 46ch
Manor Neuk (UWC) 2m 59ch
Manor Powis (UWC) 2m 71ch
Blackgrange (CCTV) 3m 43ch
Cornton (AHBC) 120m 10ch
Cornton No. 2 (R/G) 119m 60ch
118m 47ch
Stirling North SB (SN)
118m 38ch / 0m 14ch
STIRLING
118m 24ch
Stirling Middle SB (SM)
118m 08ch
Cambus Loop
Cambus East Jn
5m 38ch
Alloa Loop
Change of mileage
8m 14ch
0m 00ch

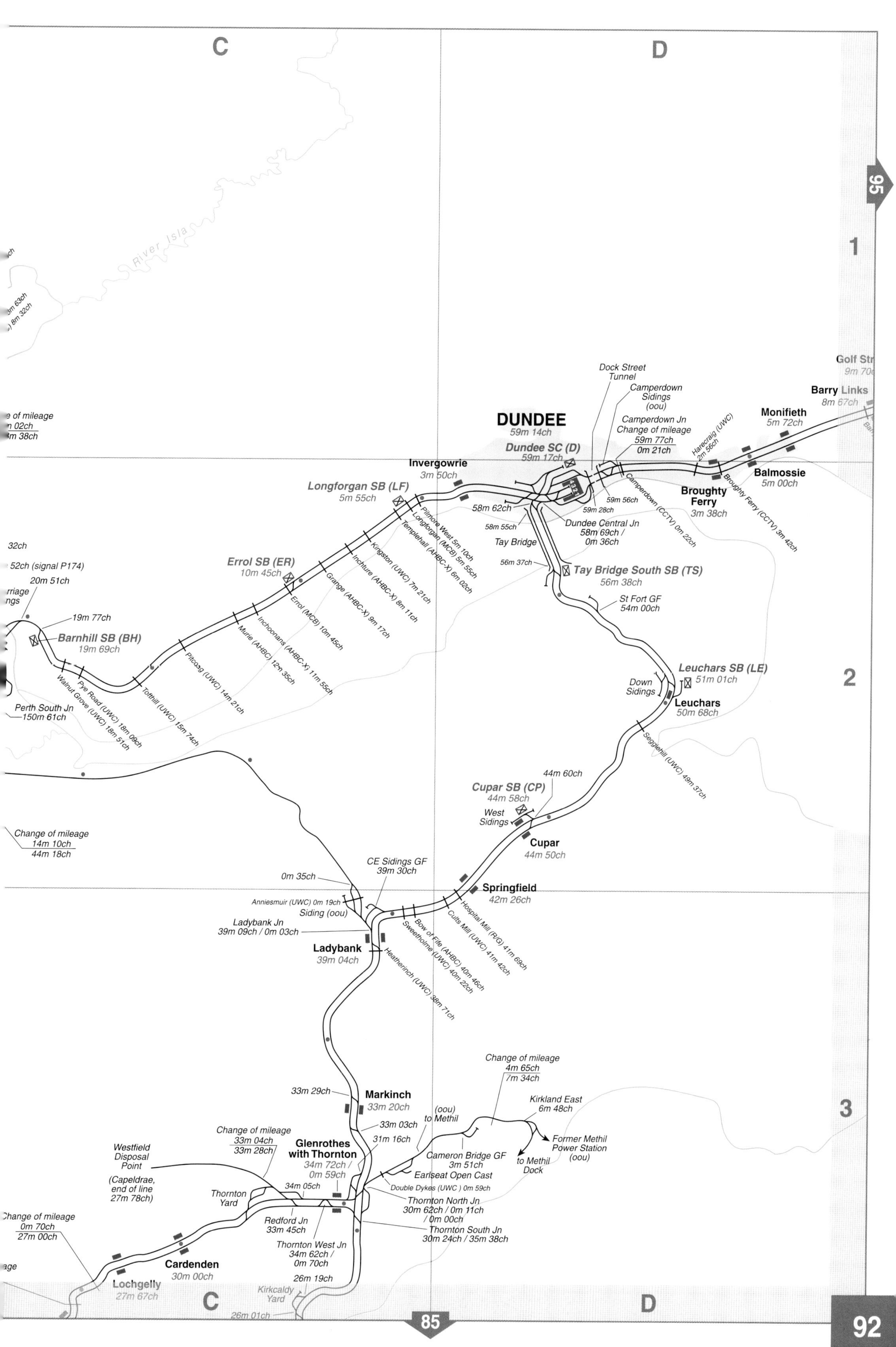

C
D
River Isla
DUNDEE
59m 14ch
Dundee SC (D)
59m 17ch
Dock Street Tunnel
Camperdown Sidings (oou)
Camperdown Jn
Change of mileage
59m 77ch
0m 21ch
Invergowrie
3m 50ch
Longforgan SB (LF)
5m 55ch
58m 62ch
58m 55ch
Tay Bridge
59m 28ch
59m 56ch
Dundee Central Jn
58m 69ch / 0m 36ch
Camperdown (CCTV) 0m 22ch
Harecraig (UWC) 2m 56ch
Broughty Ferry
3m 38ch
Broughty Ferry (CCTV) 3m 42ch
Monifieth
5m 72ch
Balmossie
5m 00ch
Barry Links
8m 67ch
Golf Str
9m 70
Errol SB (ER)
10m 45ch
Plimore West 5m 10ch
Longforgan (MCB) 5m 55ch
Templehall (AHBC-X) 6m 02ch
Kingston (UWC) 7m 21ch
Inchture (AHBC-X) 8m 11ch
Grange (AHBC-X) 9m 17ch
Errol (MCB) 10m 45ch
Inchcoonans (AHBC-X) 11m 55ch
Murie (AHBC) 12m 35ch
Pitcoag (UWC) 14m 21ch
Totthill (UWC) 15m 74ch
Pye Road (UWC) 18m 09ch
Walnut Grove (UWC) 18m 51ch
Barnhill SB (BH)
19m 69ch
19m 77ch
20m 51ch
32ch
52ch (signal P174)
Perth South Jn
150m 61ch
56m 37ch
Tay Bridge South SB (TS)
56m 38ch
St Fort GF
54m 00ch
Leuchars SB (LE)
51m 01ch
Down Sidings
Leuchars
50m 68ch
Seggiehill (UWC) 49m 37ch
44m 60ch
Cupar SB (CP)
44m 58ch
West Sidings
Cupar
44m 50ch
Springfield
42m 26ch
Change of mileage
14m 10ch
44m 18ch
CE Sidings GF
39m 30ch
0m 35ch
Anniesmuir (UWC) 0m 19ch
Siding (oou)
Ladybank Jn
39m 09ch / 0m 03ch
Ladybank
39m 04ch
Hospital Mill (R/G) 41m 69ch
Cults Mill (UWC) 41m 42ch
Bow of Fife (AHBC) 40m 46ch
Sweetholme (UWC) 40m 22ch
Heatherinch (UWC) 38m 71ch
Markinch
33m 20ch
33m 29ch
33m 03ch
31m 16ch
Change of mileage
4m 65ch
7m 34ch
Kirkland East
6m 48ch
(oou) to Methil
Former Methil Power Station (oou)
to Methil Dock
Cameron Bridge GF
3m 51ch
Earlseat Open Cast
Double Dykes (UWC) 0m 59ch
Change of mileage
33m 04ch
33m 28ch
Glenrothes with Thornton
34m 72ch / 0m 59ch
34m 05ch
Westfield Disposal Point
(Capeldrae, end of line 27m 78ch)
Thornton Yard
Thornton North Jn
30m 62ch / 0m 11ch / 0m 00ch
Thornton South Jn
30m 24ch / 35m 38ch
Redford Jn
33m 45ch
Thornton West Jn
34m 62ch / 0m 70ch
Change of mileage
0m 70ch
27m 00ch
Cardenden
30m 00ch
Lochgelly
27m 67ch
26m 19ch
Kirkcaldy Yard
26m 01ch
1
2
3
95
85

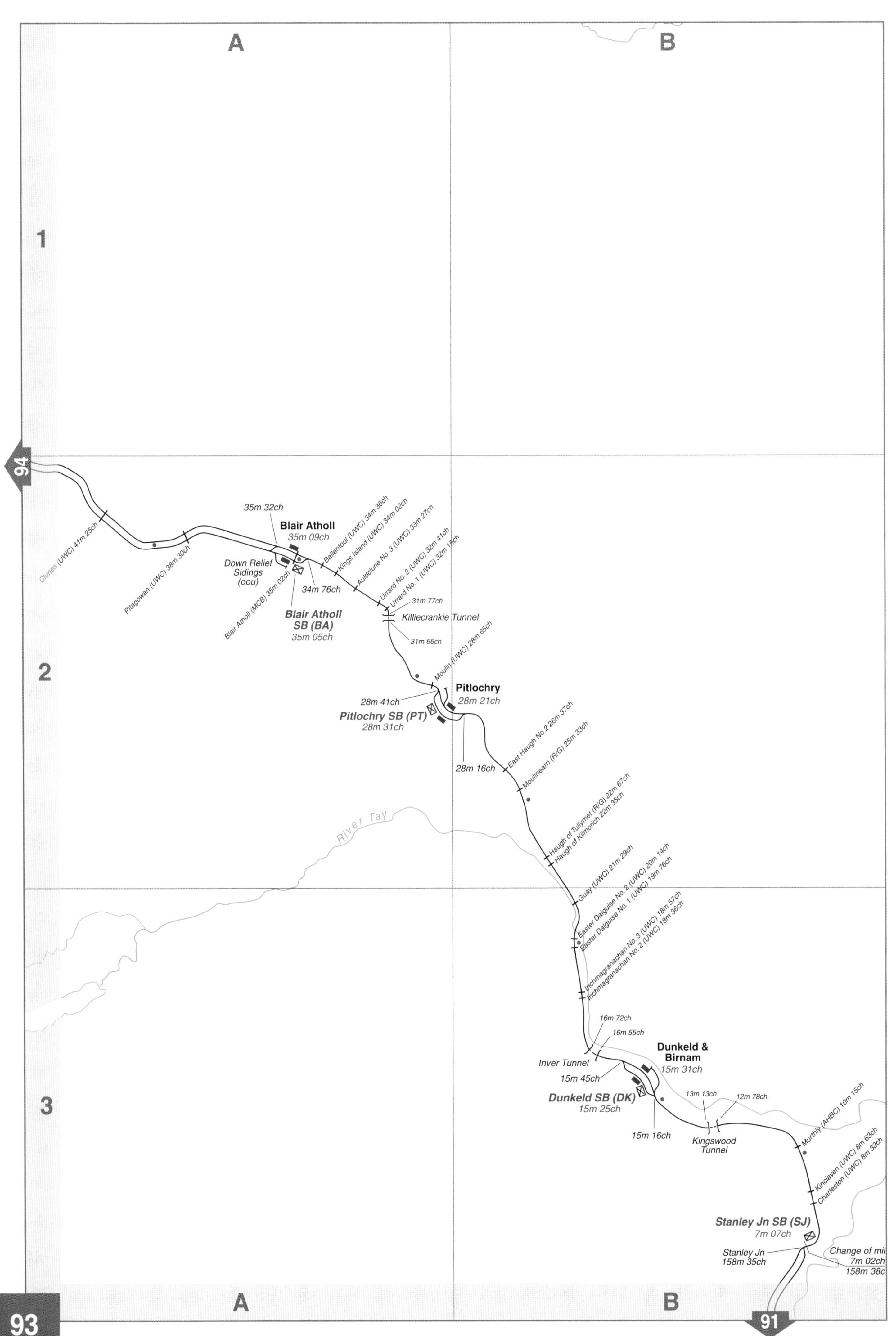
A
B
1
2
3
94
Clunes (UWC) 41m 25ch
Pitagowan (UWC) 38m 30ch
35m 32ch
Blair Atholl
35m 09ch
Down Relief
Sidings
(oou)
Blair Atholl (MCB) 35m 02ch
34m 76ch
Blair Atholl
SB (BA)
35m 05ch
Ballentoul (UWC) 34m 36ch
Kings Island (UWC) 34m 02ch
Auldclune No. 3 (UWC) 33m 27ch
Urrard No. 2 (UWC) 32m 41ch
Urrard No. 1 (UWC) 32m 15ch
31m 77ch
Killiecrankie Tunnel
31m 66ch
Moulin (UWC) 28m 65ch
Pitlochry
28m 21ch
28m 41ch
Pitlochry SB (PT)
28m 31ch
28m 16ch
East Haugh No.2 26m 37ch
Moulinearn (R/G) 25m 33ch
Haugh of Tullymet (R/G) 22m 67ch
Haugh of Kilmorich 22m 35ch
River Tay
Guay (UWC) 21m 29ch
Easter Dalguise No. 2 (UWC) 20m 14ch
Easter Dalguise No. 1 (UWC) 19m 76ch
Inchmagranachan No. 3 (UWC) 18m 57ch
Inchmagranachan No. 2 (UWC) 18m 36ch
16m 72ch
16m 55ch
Inver Tunnel
15m 45ch
Dunkeld &
Birnam
15m 31ch
Dunkeld SB (DK)
15m 25ch
15m 16ch
13m 13ch
12m 78ch
Kingswood
Tunnel
Murthly (AHBC) 10m 15ch
Kinclaven (UWC) 8m 63ch
Charleston (UWC) 8m 32ch
Stanley Jn SB (SJ)
7m 07ch
Stanley Jn
158m 35ch
Change of mil
7m 02ch
158m 38c
91

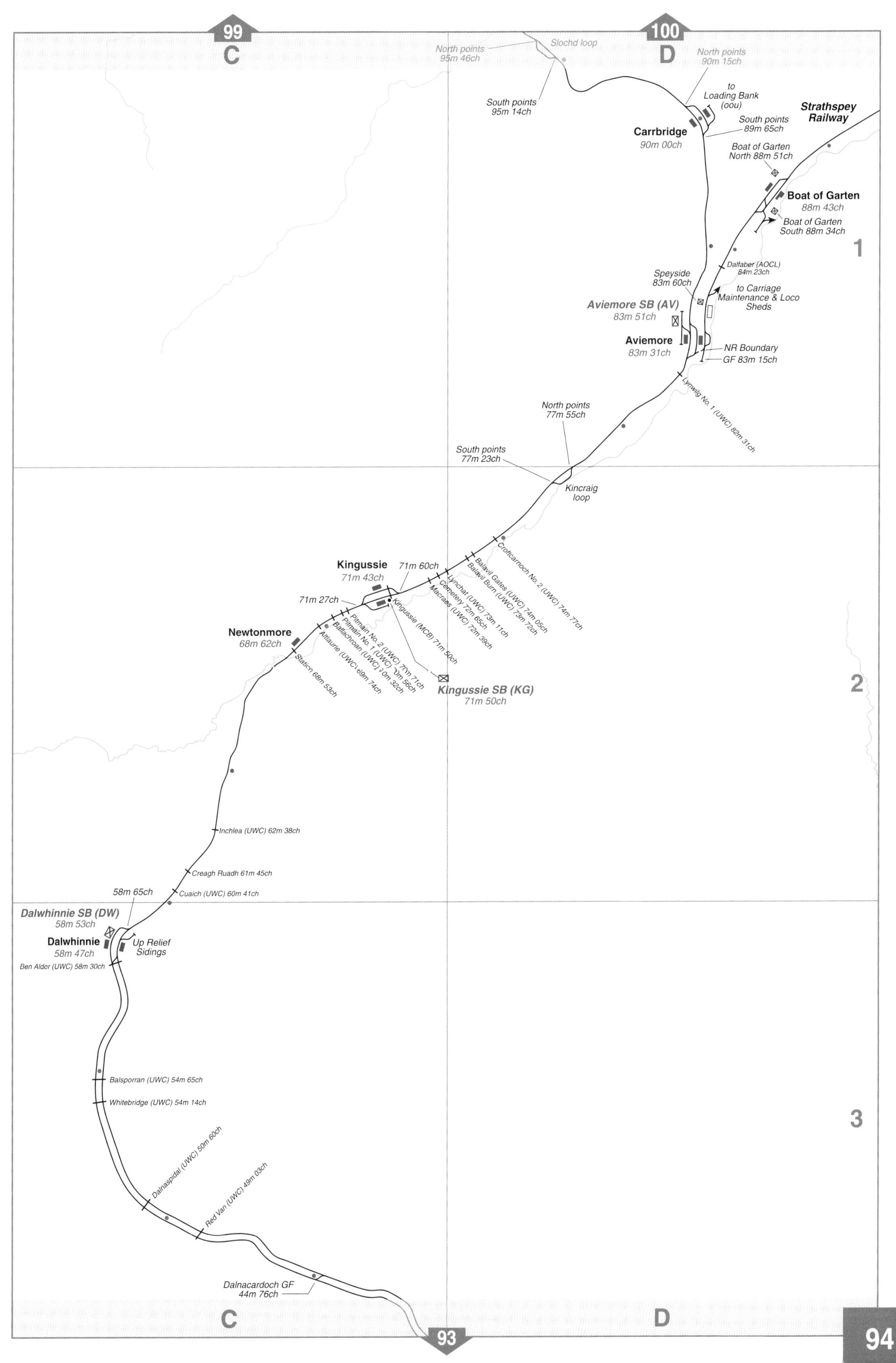
99
C
100
D
Slochd loop
North points
95m 46ch
South points
95m 14ch
North points
90m 15ch
to
Loading Bank
(oou)
Carrbridge
90m 00ch
South points
89m 65ch
Strathspey
Railway
Boat of Garten
North 88m 51ch
Boat of Garten
88m 43ch
Boat of Garten
South 88m 34ch
1
Dalfaber (AOCL)
84m 23ch
Speyside
83m 60ch
to Carriage
Maintenance & Loco
Sheds
Aviemore SB (AV)
83m 51ch
Aviemore
83m 31ch
NR Boundary
GF 83m 15ch
Lynwilg No. 1 (UWC) 82m 31ch
North points
77m 55ch
South points
77m 23ch
Kincraig
loop
Croftcarnoch No. 2 (UWC) 74m 77ch
Balavil Gates (UWC) 74m 05ch
Balavil Burn (UWC) 73m 72ch
Lynchat (UWC) 73m 11ch
Cemetery 72m 65ch
Macraes (UWC) 72m 39ch
Kingussie
71m 43ch
71m 60ch
71m 27ch
Kingussie (MCB) 71m 50ch
Kingussie SB (KG)
71m 50ch
Pitmain No. 2 (UWC) 70m 71ch
Pitmain No. 1 (UWC) 70m 56ch
Ballachroan (UWC) 70m 32ch
Altlaurie (UWC) 69m 74ch
Newtonmore
68m 62ch
Station 68m 53ch
2
Inchlea (UWC) 62m 38ch
Creagh Ruadh 61m 45ch
Cuaich (UWC) 60m 41ch
58m 65ch
Dalwhinnie SB (DW)
58m 53ch
Dalwhinnie
58m 47ch
Up Relief
Sidings
Ben Alder (UWC) 58m 30ch
Balsporran (UWC) 54m 65ch
Whitebridge (UWC) 54m 14ch
3
Dalnaspidal (UWC) 50m 60ch
Red Van (UWC) 49m 03ch
Dalnacardoch GF
44m 76ch
C
D
93

96

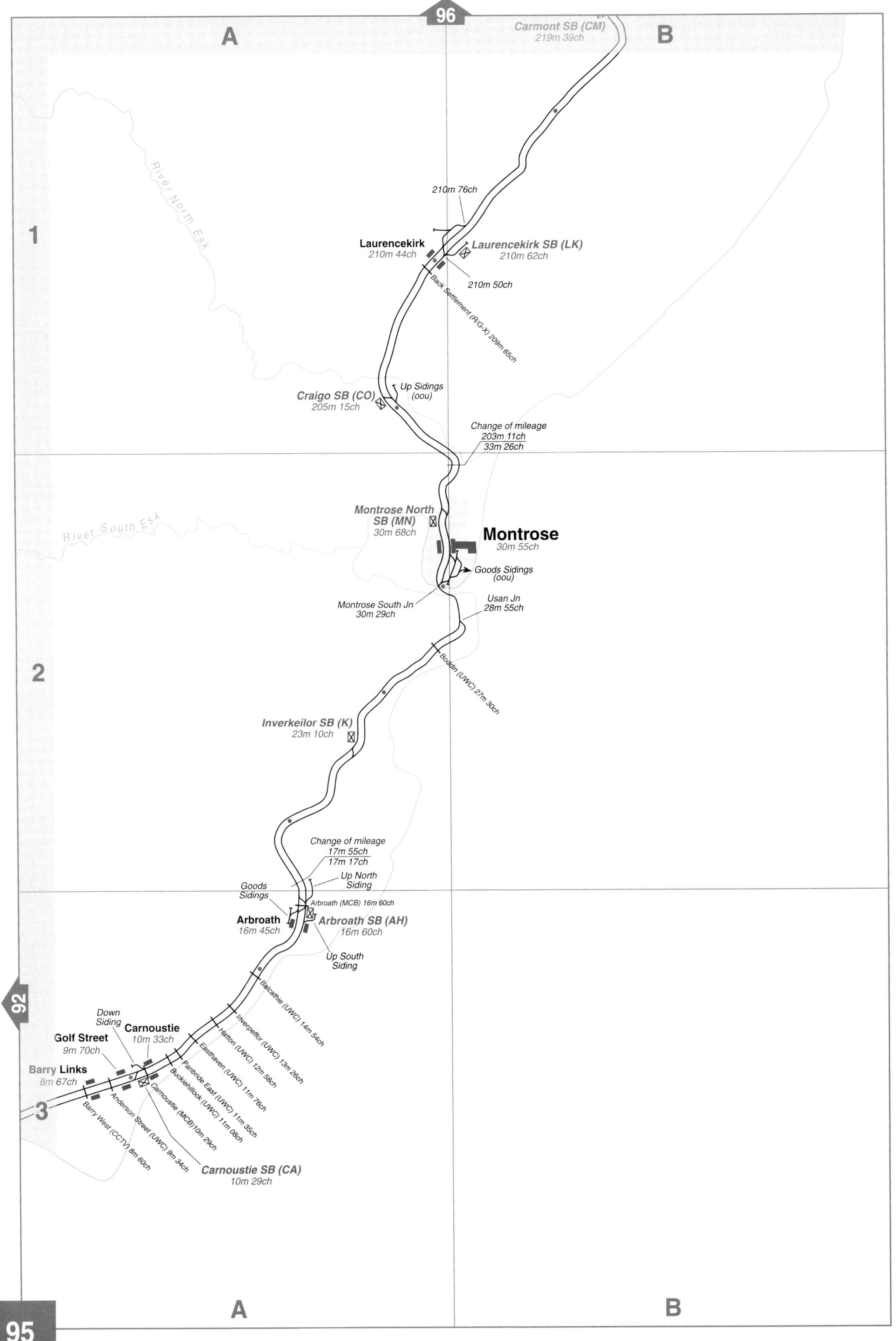

A
B
1
2
3
Carmont SB (CM)
219m 39ch
River North Esk
210m 76ch
Laurencekirk
210m 44ch
Laurencekirk SB (LK)
210m 62ch
210m 50ch
Back Settlement (R/G-X) 209m 65ch
Craigo SB (CO)
205m 15ch
Up Sidings
(oou)
Change of mileage
203m 11ch
33m 26ch
River South Esk
Montrose North
SB (MN)
30m 68ch
Montrose
30m 55ch
Goods Sidings
(oou)
Montrose South Jn
30m 29ch
Usan Jn
28m 55ch
Boddin (UWC) 27m 30ch
Inverkeilor SB (K)
23m 10ch
Change of mileage
17m 55ch
17m 17ch
Goods
Sidings
Up North
Siding
Arbroath (MCB) 16m 60ch
Arbroath
16m 45ch
Arbroath SB (AH)
16m 60ch
Up South
Siding
Balcathie (UWC) 14m 54ch
Inverpeffer (UWC) 13m 26ch
Hatton (UWC) 12m 58ch
Easthaven (UWC) 11m 76ch
Panbride East (UWC) 11m 35ch
Buckiehillock (UWC) 11m 08ch
Down
Siding
Carnoustie
10m 33ch
Golf Street
9m 70ch
Barry Links
8m 67ch
Carnoustie (MCB)10m 29ch
Anderson Street (UWC) 9m 34ch
Barry West (CCTV) 8m 60ch
Carnoustie SB (CA)
10m 29ch

92

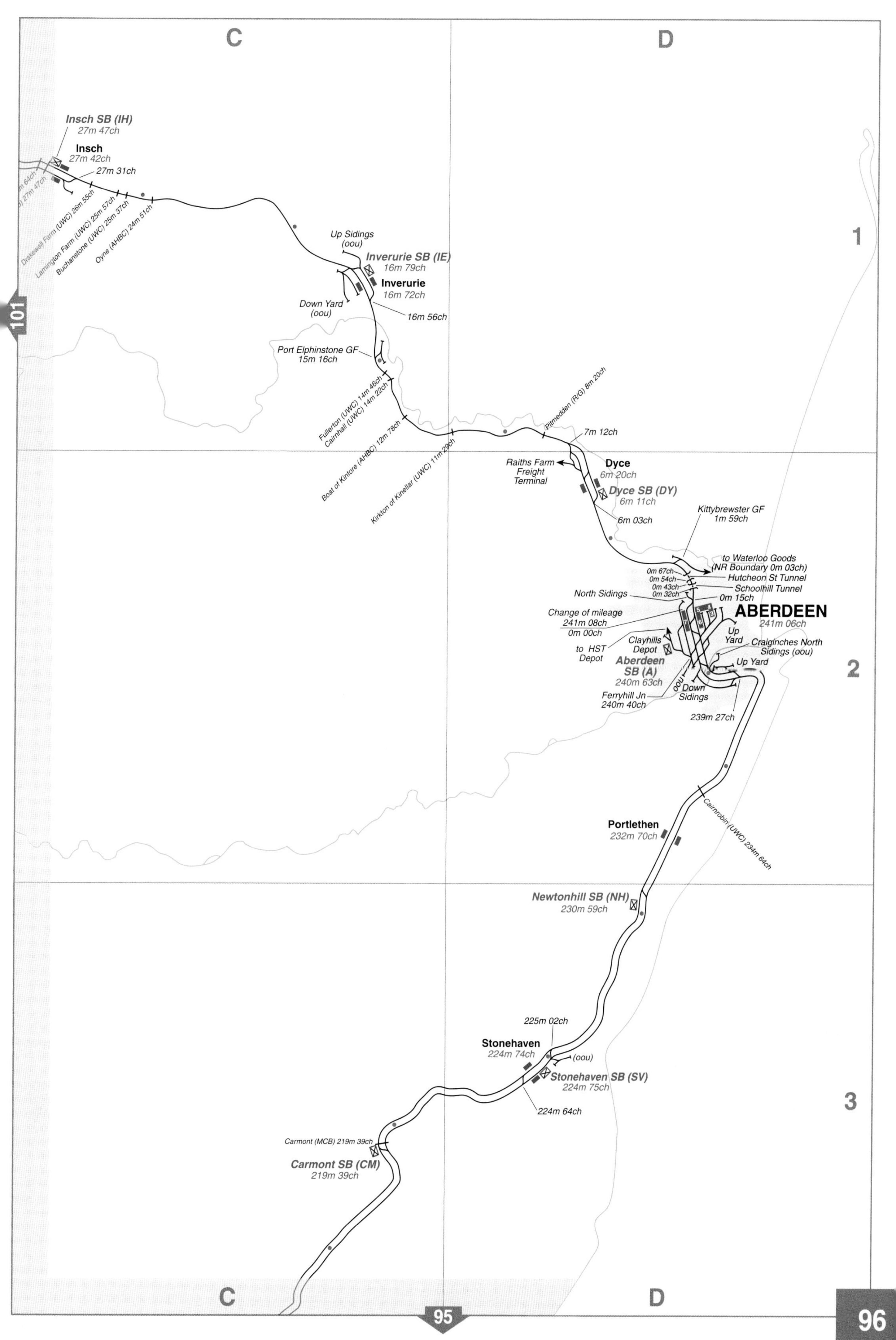
C
D
Insch SB (IH)
27m 47ch
Insch
27m 42ch
27m 31ch
Drakewell Farm (UWC) 26m 55ch
Lamington Farm (UWC) 25m 57ch
Buchanstone (UWC) 25m 37ch
Oyne (AHBC) 24m 51ch
Up Sidings
(oou)
Inverurie SB (IE)
16m 79ch
Inverurie
16m 72ch
Down Yard
(oou)
16m 56ch
Port Elphinstone GF
15m 16ch
Fullerton (UWC) 14m 46ch
Cairnhall (UWC) 14m 22ch
Boat of Kintore (AHBC) 12m 78ch
Kirkton of Kinellar (UWC) 11m 29ch
Pitmedden (R/G) 8m 20ch
7m 12ch
Raiths Farm
Freight
Terminal
Dyce
6m 20ch
Dyce SB (DY)
6m 11ch
6m 03ch
Kittybrewster GF
1m 59ch
to Waterloo Goods
(NR Boundary 0m 03ch)
0m 67ch
0m 54ch
0m 43ch
0m 32ch
Hutcheon St Tunnel
Schoolhill Tunnel
North Sidings
0m 15ch
ABERDEEN
241m 06ch
Change of mileage
241m 08ch
0m 00ch
to HST
Depot
Clayhills
Depot
Aberdeen
SB (A)
240m 63ch
Up
Yard
Craiginches North
Sidings (oou)
Up Yard
oou
Down
Sidings
Ferryhill Jn
240m 40ch
239m 27ch
1
2
3
Cairnrobin (UWC) 234m 64ch
Portlethen
232m 70ch
Newtonhill SB (NH)
230m 59ch
225m 02ch
Stonehaven
224m 74ch
(oou)
Stonehaven SB (SV)
224m 75ch
224m 64ch
Carmont (MCB) 219m 39ch
Carmont SB (CM)
219m 39ch
C
D

101

95

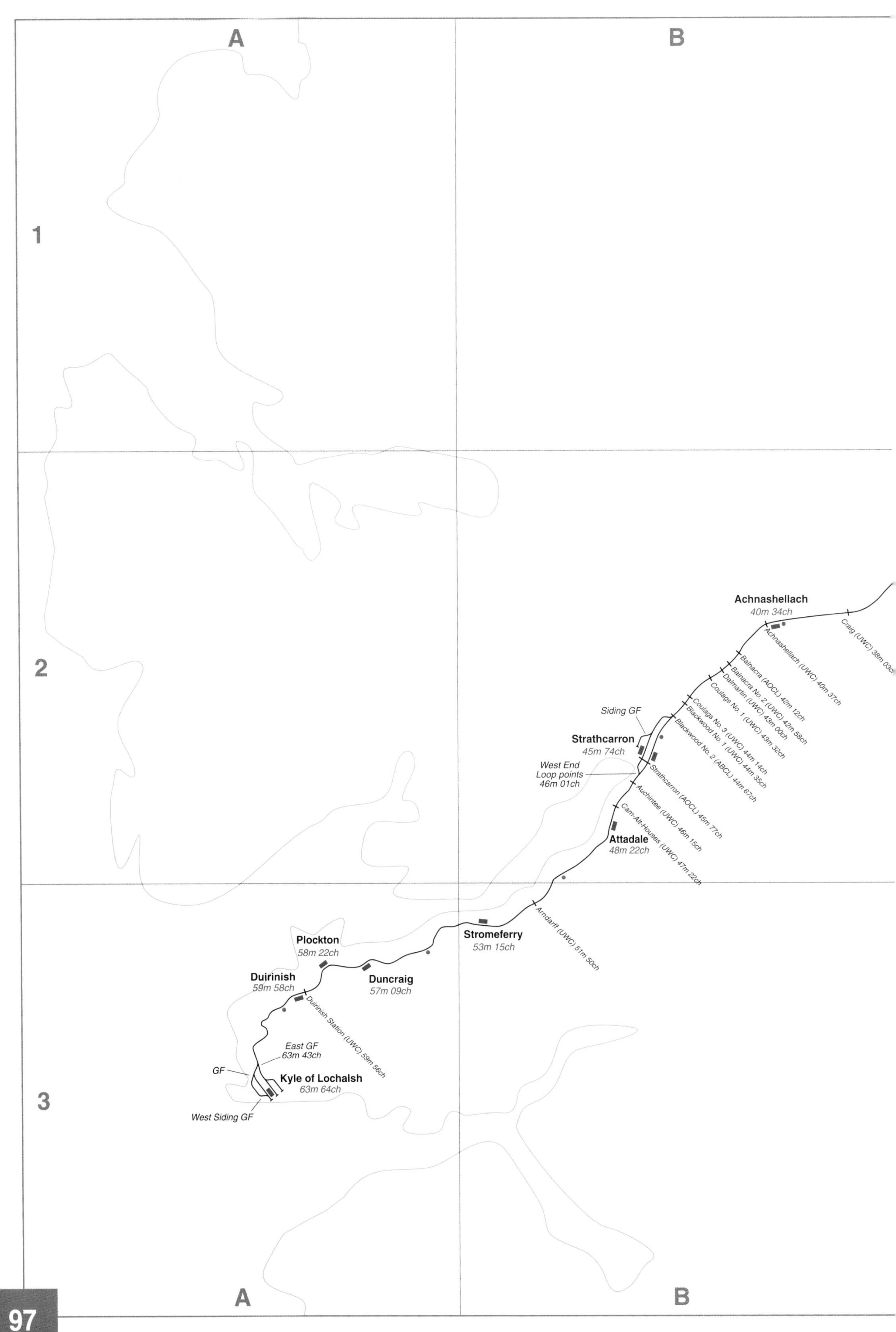
A
B
1
2
3
Achnashellach
40m 34ch
Craig (UWC) 38m 03ch
Achnashellach (UWC) 40m 37ch
Balnacra (AOCL) 42m 12ch
Balnacra No. 2 (UWC) 42m 58ch
Dalmartin (UWC) 43m 00ch
Coulags No. 1 (UWC) 43m 32ch
Coulags No. 3 (UWC) 44m 14ch
Blackwood No. 1 (UWC) 44m 35ch
Blackwood No. 2 (ABCL) 44m 67ch
Siding GF
Strathcarron
45m 74ch
West End
Loop points
46m 01ch
Strathcarron (AOCL) 45m 77ch
Auchintee (UWC) 46m 15ch
Cam-Alt-Houses (UWC) 47m 22ch
Attadale
48m 22ch
Arndarff (UWC) 51m 50ch
Stromeferry
53m 15ch
Plockton
58m 22ch
Duncraig
57m 09ch
Duirinish
59m 58ch
Duirinish Station (UWC) 59m 56ch
East GF
63m 43ch
GF
Kyle of Lochalsh
63m 64ch
West Siding GF
A
B

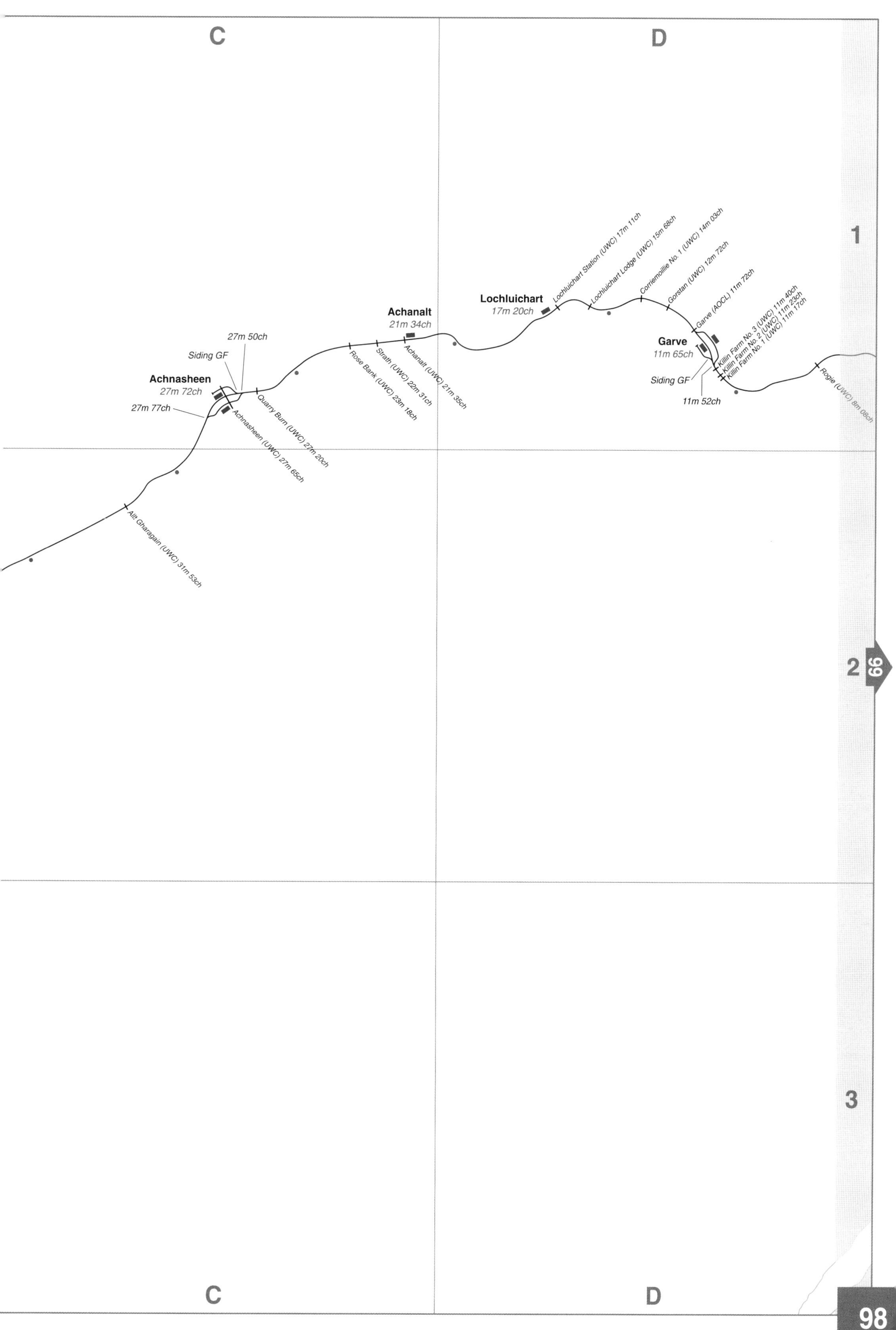
C
D
1
Lochluichart Station (UWC) 17m 11ch
Lochluichart Lodge (UWC) 15m 68ch
Corriemoillie No. 1 (UWC) 14m 03ch
Gorstan (UWC) 12m 72ch
Garve (AOCL) 11m 72ch
Killin Farm No. 3 (UWC) 11m 40ch
Killin Farm No. 2 (UWC) 11m 23ch
Killin Farm No. 1 (UWC) 11m 17ch
Lochluichart
17m 20ch
Achanalt
21m 34ch
Garve
11m 65ch
Siding GF
11m 52ch
Rogie (UWC) 8m 08ch
27m 50ch
Siding GF
Achnasheen
27m 72ch
27m 77ch
Rose Bank (UWC) 23m 18ch
Strath (UWC) 22m 31ch
Achanalt (UWC) 21m 35ch
Quarry Burn (UWC) 27m 20ch
Achnasheen (UWC) 27m 65ch
Allt Gharagain (UWC) 31m 53ch
2
99
3
C
D

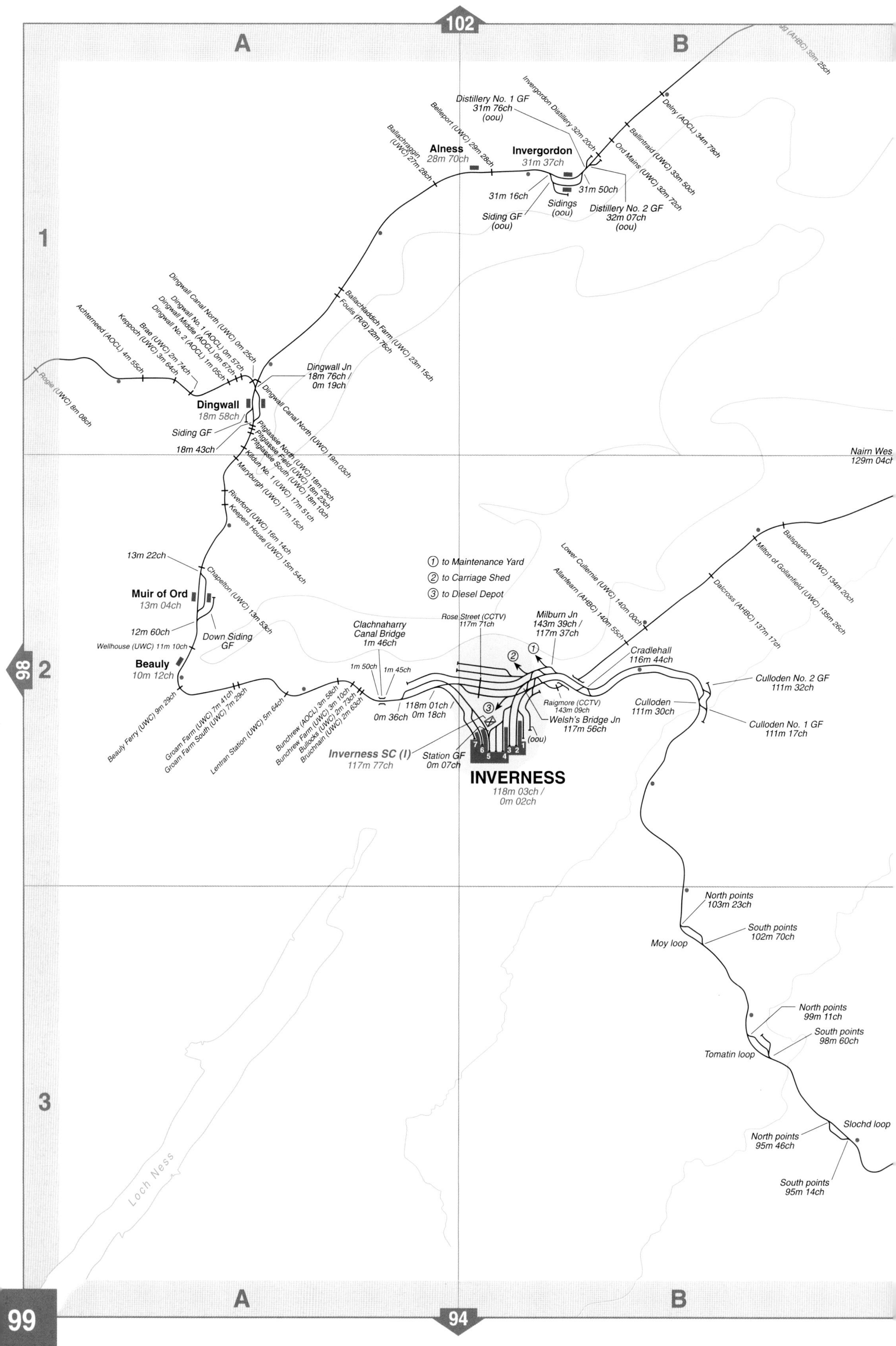

102
A
B
Distillery No. 1 GF
31m 76ch
(oou)
Invergordon Distillery 32m 20ch
Belleport (UWC) 29m 28ch
Ballachraggin (UWC) 27m 28ch
Alness
28m 70ch
Invergordon
31m 37ch
Delny (AOCL) 34m 79ch
Ballintraid (UWC) 33m 50ch
Ord Mains (UWC) 32m 72ch
31m 16ch
31m 50ch
Sidings
(oou)
Siding GF
(oou)
Distillery No. 2 GF
32m 07ch
(oou)
1
Ballachladdich Farm (UWC) 23m 15ch
Foulis (R/G) 22m 76ch
Dingwall Canal North (UWC) 0m 25ch
Dingwall No. 1 (AOCL) 0m 57ch
Dingwall Middle (AOCL) 0m 67ch
Dingwall No. 2 (AOCL) 1m 05ch
Brae (UWC) 2m 74ch
Keppoch (UWC) 3m 64ch
Achterneed (AOCL) 4m 55ch
Rogie (UWC) 8m 08ch
Dingwall Jn
18m 76ch /
0m 19ch
Dingwall
18m 58ch
Dingwall Canal North (UWC) 19m 03ch
Siding GF
18m 43ch
Pitglassie North (UWC) 18m 29ch
Pitglassie Field (UWC) 18m 23ch
Pitglassie South (UWC) 18m 10ch
Kildun No. 1 (UWC) 17m 51ch
Maryburgh (UWC) 17m 15ch
Riverford (UWC) 16m 14ch
Keepers House (UWC) 15m 54ch
Nairn Wes
129m 04ch
13m 22ch
Chapelton (UWC) 13m 53ch
Muir of Ord
13m 04ch
12m 60ch
Down Siding
GF
Wellhouse (UWC) 11m 10ch
Beauly
10m 12ch
2
98
1 to Maintenance Yard
2 to Carriage Shed
3 to Diesel Depot
Lower Cullernie (UWC) 140m 00ch
Allanfearn (AHBC) 140m 55ch
Balspardon (UWC) 134m 20ch
Milton of Gollanfield (UWC) 135m 26ch
Dalcross (AHBC) 137m 17ch
Clachnaharry
Canal Bridge
1m 46ch
Rose Street (CCTV)
117m 71ch
Milburn Jn
143m 39ch /
117m 37ch
Cradlehall
116m 44ch
1m 50ch
1m 45ch
0m 36ch
118m 01ch /
0m 18ch
Raigmore (CCTV)
143m 09ch
Welsh's Bridge Jn
117m 56ch
(oou)
Culloden No. 2 GF
111m 32ch
Culloden
111m 30ch
Culloden No. 1 GF
111m 17ch
Beauly Ferry (UWC) 9m 29ch
Groam Farm (UWC) 7m 41ch
Groam Farm South (UWC) 7m 29ch
Lentran Station (UWC) 5m 64ch
Bunchrew (AOCL) 3m 58ch
Bunchrew Farm (UWC) 3m 10ch
Bullocks (UWC) 2m 73ch
Bruichnain (UWC) 2m 63ch
Inverness SC (I)
117m 77ch
Station GF
0m 07ch
7 6 5 4 3 2 1
INVERNESS
118m 03ch /
0m 02ch
North points
103m 23ch
South points
102m 70ch
Moy loop
North points
99m 11ch
South points
98m 60ch
Tomatin loop
3
Slochd loop
North points
95m 46ch
South points
95m 14ch
Loch Ness
A
B
94

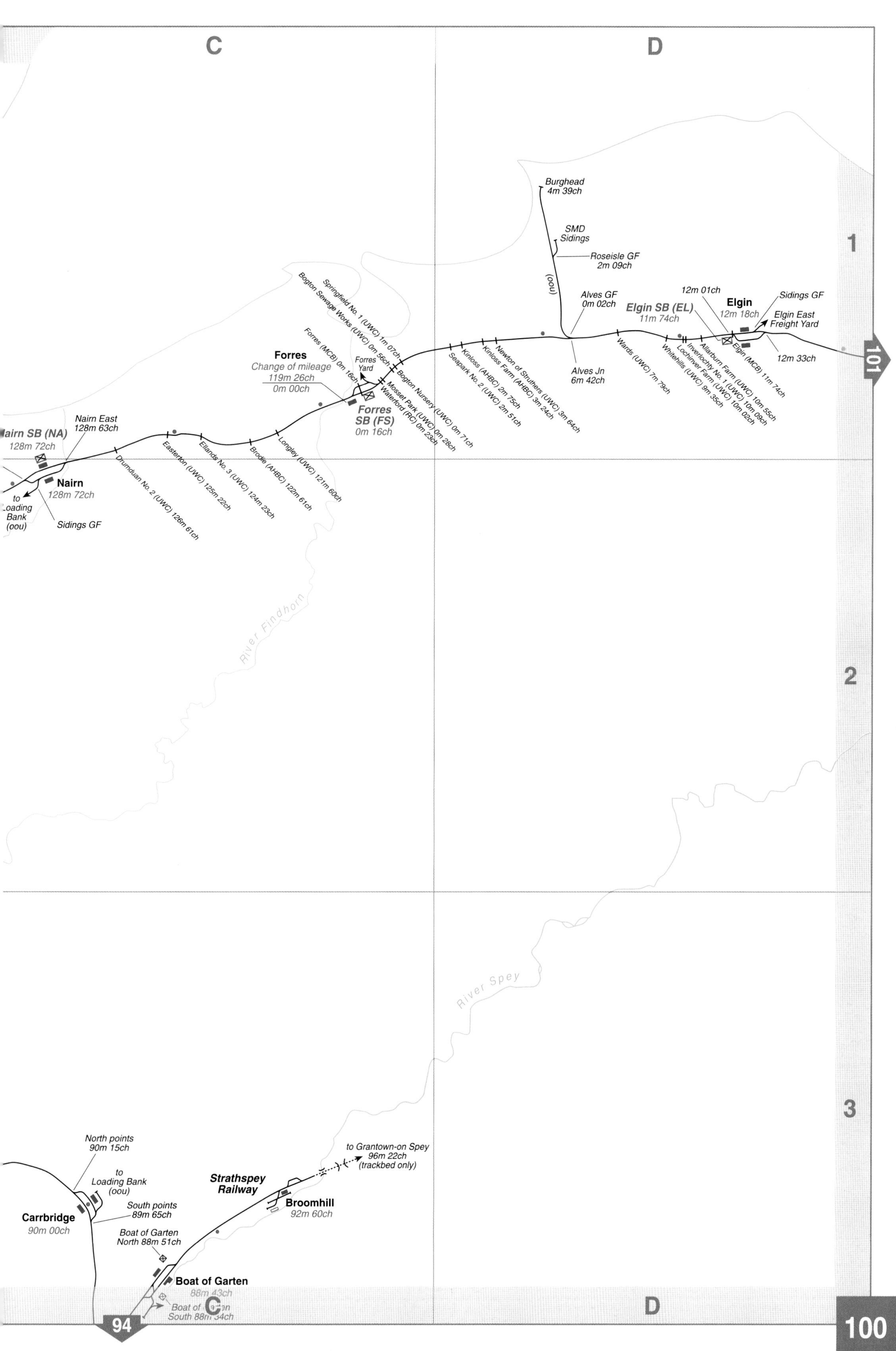
C
D
Burghead
4m 39ch
SMD
Sidings
Roseisle GF
2m 09ch
(oou)
Alves GF
0m 02ch
Alves Jn
6m 42ch
Elgin SB (EL)
11m 74ch
12m 01ch
Elgin
12m 18ch
Sidings GF
Elgin East
Freight Yard
12m 33ch
Forres
Change of mileage
119m 26ch
0m 00ch
Forres Yard
Forres
SB (FS)
0m 16ch
Nairn SB (NA)
128m 72ch
Nairn East
128m 63ch
Nairn
128m 72ch
to
Loading
Bank
(oou)
Sidings GF
Springfield No. 1 (UWC) 1m 07ch
Bogton Sewage Works (UWC) 0m 56ch
Forres (MCB) 0m 16ch
Bogton Nursery (UWC) 0m 71ch
Mosset Park (UWC) 0m 28ch
Waterford (RC) 0m 23ch
Seapark No. 2 (UWC) 2m 51ch
Kinloss (AHBC) 2m 75ch
Kinloss Farm (AHBC) 3m 24ch
Newton of Struthers (UWC) 3m 64ch
Wards (UWC) 7m 79ch
Whitehills (UWC) 9m 35ch
Lochinver Farm (UWC) 10m 02ch
Inverlochty No. 1 (UWC) 10m 09ch
Allarburn Farm (UWC) 10m 55ch
Elgin (MCB) 11m 74ch
Drumduan No. 2 (UWC) 126m 61ch
Easterton (UWC) 125m 22ch
Ellands No. 3 (UWC) 124m 23ch
Brodie (AHBC) 122m 61ch
Longley (UWC) 121m 60ch
River Findhorn
River Spey
1
2
3
North points
90m 15ch
to
Loading Bank
(oou)
South points
89m 65ch
Carrbridge
90m 00ch
Boat of Garten
North 88m 51ch
Boat of Garten
88m 43ch
Boat of Garten
South 88m 34ch
Strathspey
Railway
Broomhill
92m 60ch
to Grantown-on Spey
96m 22ch
(trackbed only)
101
94

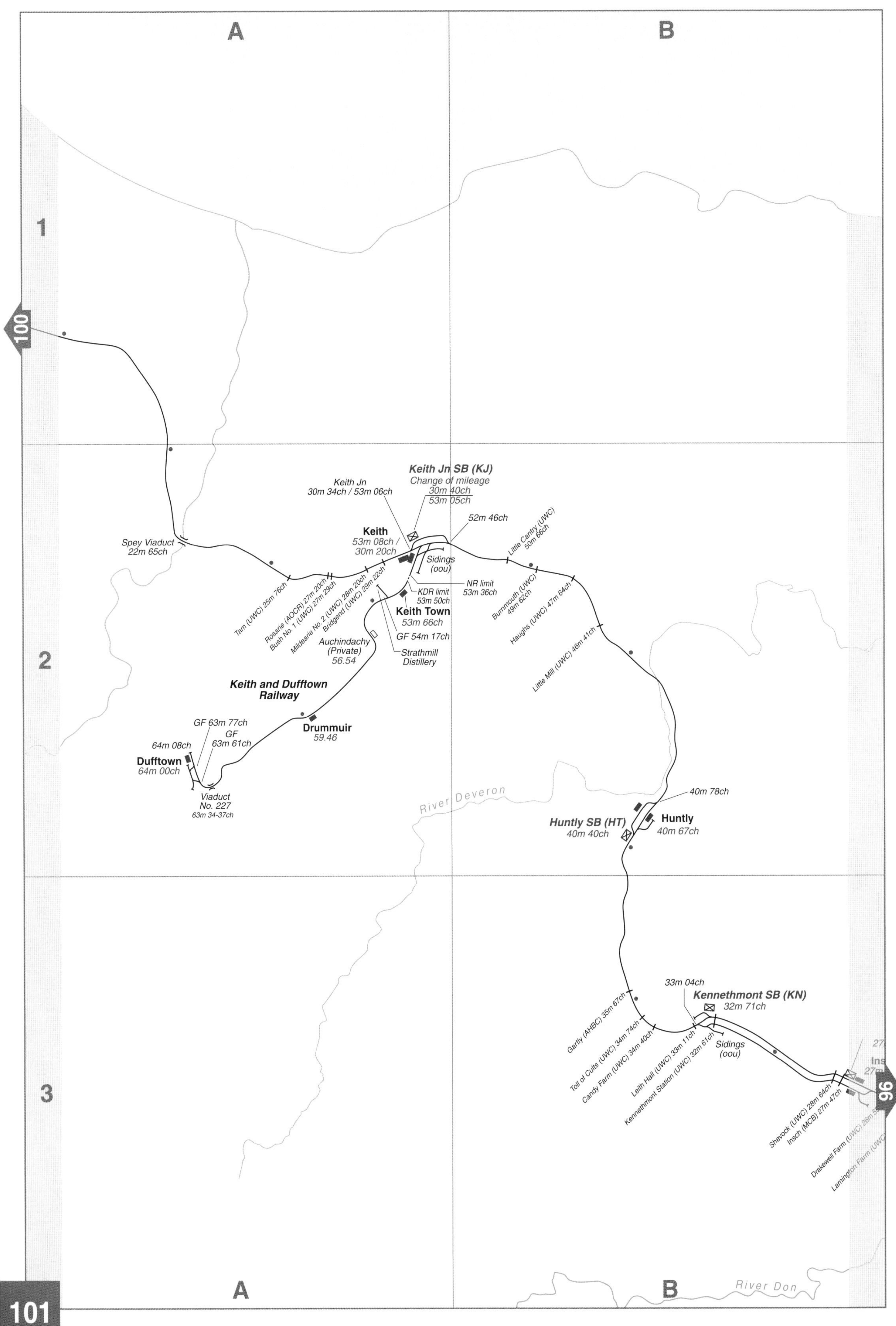
A
B
1
2
3
100
Keith Jn SB (KJ)
Change of mileage
30m 40ch
53m 05ch
Keith Jn
30m 34ch / 53m 06ch
52m 46ch
Keith
53m 08ch /
30m 20ch
Sidings
(oou)
Spey Viaduct
22m 65ch
Tam (UWC) 25m 76ch
Rosarie (AOCR) 27m 20ch
Bush No. 1 (UWC) 27m 29ch
Mildearie No. 2 (UWC) 28m 20ch
Bridgend (UWC) 29m 22ch
NR limit
53m 36ch
KDR limit
53m 50ch
Keith Town
53m 66ch
GF 54m 17ch
Strathmill
Distillery
Auchindachy
(Private)
56.54
Keith and Dufftown
Railway
Drummuir
59.46
GF 63m 77ch
GF
63m 61ch
64m 08ch
Dufftown
64m 00ch
Viaduct
No. 227
63m 34-37ch
Little Cantry (UWC)
50m 66ch
Burnmouth (UWC)
49m 62ch
Haughs (UWC) 47m 64ch
Little Mill (UWC) 46m 41ch
River Deveron
40m 78ch
Huntly SB (HT)
40m 40ch
Huntly
40m 67ch
33m 04ch
Kennethmont SB (KN)
32m 71ch
Gartly (AHBC) 35m 67ch
Toll of Cults (UWC) 34m 74ch
Candy Farm (UWC) 34m 40ch
Leith Hall (UWC) 33m 11ch
Kennethmont Station (UWC) 32m 61ch
Sidings
(oou)
Shevock (UWC) 28m 64ch
Insch (MCB) 27m 47ch
96
River Don

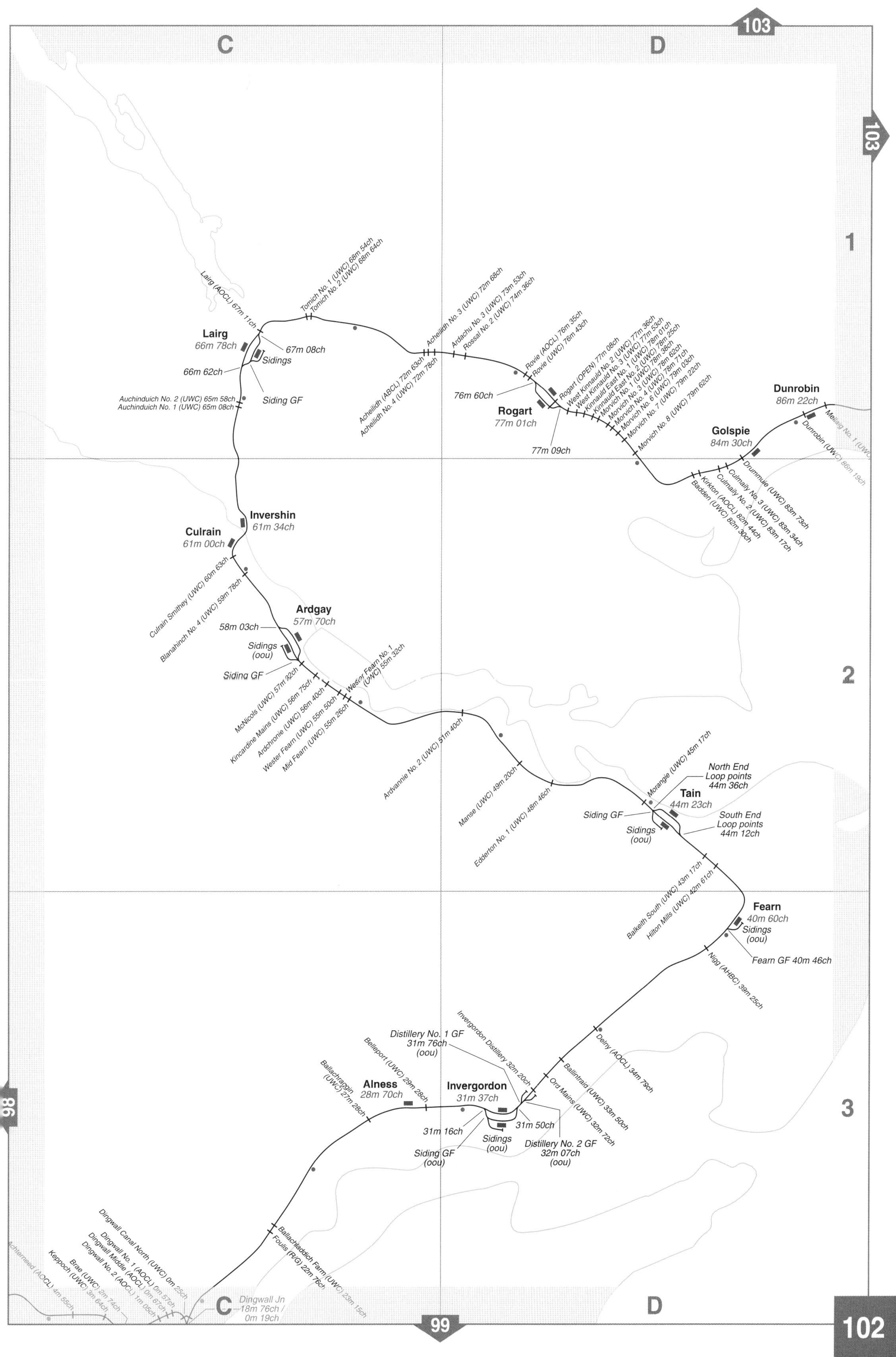
C
D
103
1
2
3
98
99
Lairg (AOCL) 67m 11ch
Tomich No. 1 (UWC) 68m 54ch
Tomich No. 2 (UWC) 68m 64ch
Lairg
66m 78ch
67m 08ch
Sidings
66m 62ch
Siding GF
Auchinduich No. 2 (UWC) 65m 58ch
Auchinduich No. 1 (UWC) 65m 08ch
Achellidh No. 3 (UWC) 72m 68ch
Ardachu No. 3 (UWC) 73m 53ch
Rossal No. 2 (UWC) 74m 36ch
Achellidh (ABCL) 72m 63ch
Achellidh No. 4 (UWC) 72m 78ch
Rovie (AOCL) 76m 35ch
Rovie (UWC) 76m 43ch
76m 60ch
Rogart
77m 01ch
Rogart (OPEN) 77m 08ch
West Kinnauld No. 2 (UWC) 77m 36ch
West Kinnauld No. 3 (UWC) 77m 53ch
Kinnauld East No. 1 (UWC) 78m 01ch
Kinnauld East No. 2 (UWC) 78m 25ch
Morvich No. 1 (UWC) 78m 38ch
Morvich No. 3 (UWC) 78m 62ch
Morvich No. 4 (UWC) 78m 71ch
Morvich No. 6 (UWC) 79m 03ch
Morvich No. 7 (UWC) 79m 22ch
Morvich No. 8 (UWC) 79m 62ch
77m 09ch
Dunrobin
86m 22ch
Golspie
84m 30ch
Dunrobin (UWC) 86m 19ch
Drummuie (UWC) 83m 73ch
Culmaily No. 3 (UWC) 83m 34ch
Culmaily No. 2 (UWC) 83m 17ch
Kirkton (AOCL) 82m 44ch
Badden (UWC) 82m 30ch
Invershin
61m 34ch
Culrain
61m 00ch
Culrain Smithey (UWC) 60m 63ch
Blanahinch No. 4 (UWC) 59m 78ch
Ardgay
57m 70ch
58m 03ch
Sidings
(oou)
Siding GF
McNicols (UWC) 57m 32ch
Kincardine Mains (UWC) 56m 75ch
Ardchronie (UWC) 56m 40ch
Wester Fearn (UWC) 55m 50ch
Mid Fearn (UWC) 55m 26ch
Wester Fearn No. 1 (UWC) 55m 32ch
Ardvannie No. 2 (UWC) 51m 40ch
Manse (UWC) 49m 20ch
Edderton No. 1 (UWC) 48m 46ch
Morangie (UWC) 45m 17ch
North End
Loop points
44m 36ch
Tain
44m 23ch
Siding GF
Sidings
(oou)
South End
Loop points
44m 12ch
Balkeith South (UWC) 43m 17ch
Hilton Mills (UWC) 42m 61ch
Fearn
40m 60ch
Sidings
(oou)
Fearn GF 40m 46ch
Nigg (AHBC) 39m 25ch
Delny (AOCL) 34m 79ch
Ballintraid (UWC) 33m 50ch
Ord Mains (UWC) 32m 72ch
Invergordon Distillery 32m 20ch
Distillery No. 1 GF
31m 76ch
(oou)
Distillery No. 2 GF
32m 07ch
(oou)
Invergordon
31m 37ch
31m 50ch
31m 16ch
Sidings
(oou)
Siding GF
(oou)
Belleport (UWC) 29m 28ch
Alness
28m 70ch
Ballachraggin (UWC) 27m 28ch
Ballachladdich Farm (UWC) 23m 15ch
Foulis (R/G) 22m 78ch
Dingwall Canal North (UWC) 0m 25ch
Dingwall No. 1 (AOCL) 0m 67ch
Dingwall Middle (AOCL) 0m 67ch
Dingwall No. 2 (AOCL) 1m 05ch
Brae (UWC) 2m 74ch
Keppoch (UWC) 3m 64ch
Achterneed (AOCL) 4m 55ch
Dingwall Jn
18m 76ch /
0m 19ch

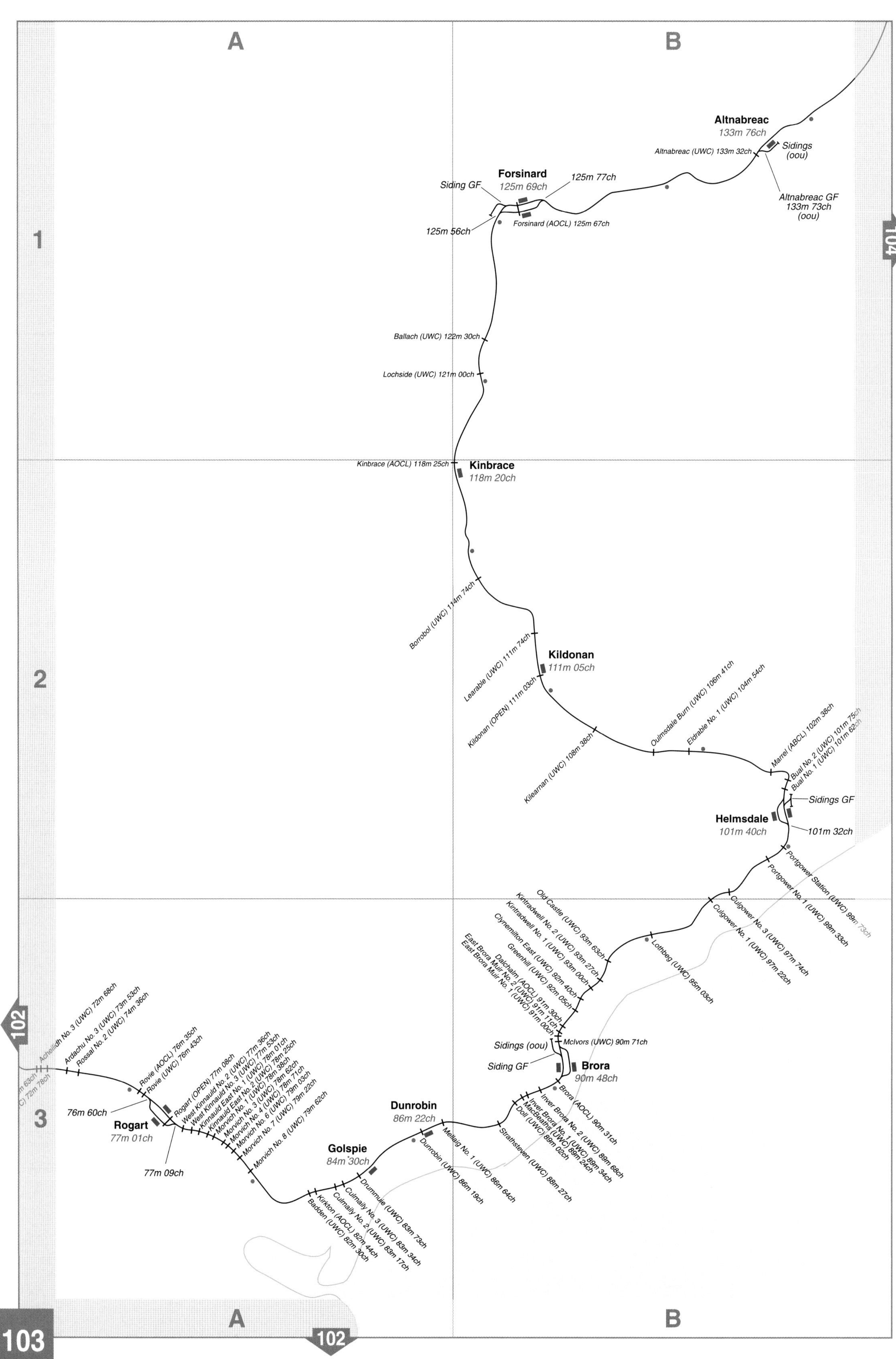

A
B
1
2
3
104
102
Altnabreac
133m 76ch
Altnabreac (UWC) 133m 32ch
Sidings
(oou)
Altnabreac GF
133m 73ch
(oou)
Forsinard
125m 69ch
125m 77ch
Siding GF
125m 56ch
Forsinard (AOCL) 125m 67ch
Ballach (UWC) 122m 30ch
Lochside (UWC) 121m 00ch
Kinbrace (AOCL) 118m 25ch
Kinbrace
118m 20ch
Borrobol (UWC) 114m 74ch
Kildonan
111m 05ch
Learable (UWC) 111m 74ch
Kildonan (OPEN) 111m 03ch
Kilearnan (UWC) 108m 38ch
Ouilmsdale Burn (UWC) 106m 41ch
Eldrable No. 1 (UWC) 104m 54ch
Marrel (ABCL) 102m 38ch
Bual No. 2 (UWC) 101m 75ch
Bual No. 1 (UWC) 101m 62ch
Sidings GF
Helmsdale
101m 40ch
101m 32ch
Portgower Station (UWC) 99m 73ch
Portgower No. 1 (UWC) 99m 33ch
Culgower No. 3 (UWC) 97m 74ch
Culgower No. 1 (UWC) 97m 22ch
Lothbeg (UWC) 95m 03ch
Old Castle (UWC) 93m 63ch
Kintradwell No. 2 (UWC) 93m 27ch
Kintradwell No. 1 (UWC) 93m 00ch
Clynemilton East (UWC) 92m 40ch
Greenhill (UWC) 92m 05ch
Dalchalm (AOCL) 91m 30ch
East Brora Muir No. 2 (UWC) 91m 11ch
East Brora Muir No. 1 (UWC) 91m 00ch
McIvors (UWC) 90m 71ch
Sidings (oou)
Siding GF
Brora
90m 48ch
Brora (AOCL) 90m 31ch
Inver Brora No. 2 (UWC) 89m 68ch
Inver Brora No. 1 (UWC) 89m 34ch
MacBeaths (UWC) 89m 24ch
Doll (UWC) 89m 02ch
Strathsteven (UWC) 88m 27ch
Meikle No. 1 (UWC) 86m 64ch
Dunrobin
86m 22ch
Dunrobin (UWC) 86m 19ch
Golspie
84m 30ch
Drummuie (UWC) 83m 73ch
Culmaily No. 3 (UWC) 83m 34ch
Culmaily No. 2 (UWC) 83m 17ch
Kirkton (AOCL) 82m 44ch
Badden (UWC) 82m 30ch
Achellidh No. 3 (UWC) 72m 68ch
Ardachu No. 3 (UWC) 73m 53ch
Rossal No. 2 (UWC) 74m 36ch
Rovie (AOCL) 76m 35ch
Rovie (UWC) 76m 43ch
76m 60ch
Rogart
77m 01ch
77m 09ch
Rogart (OPEN) 77m 08ch
West Kinnauld No. 2 (UWC) 77m 36ch
West Kinnauld No. 3 (UWC) 77m 53ch
Kinnauld East No. 1 (UWC) 78m 01ch
Kinnauld East No. 2 (UWC) 78m 25ch
Morvich No. 1 (UWC) 78m 38ch
Morvich No. 3 (UWC) 78m 62ch
Morvich No. 4 (UWC) 78m 71ch
Morvich No. 6 (UWC) 79m 03ch
Morvich No. 7 (UWC) 79m 22ch
Morvich No. 8 (UWC) 79m 62ch

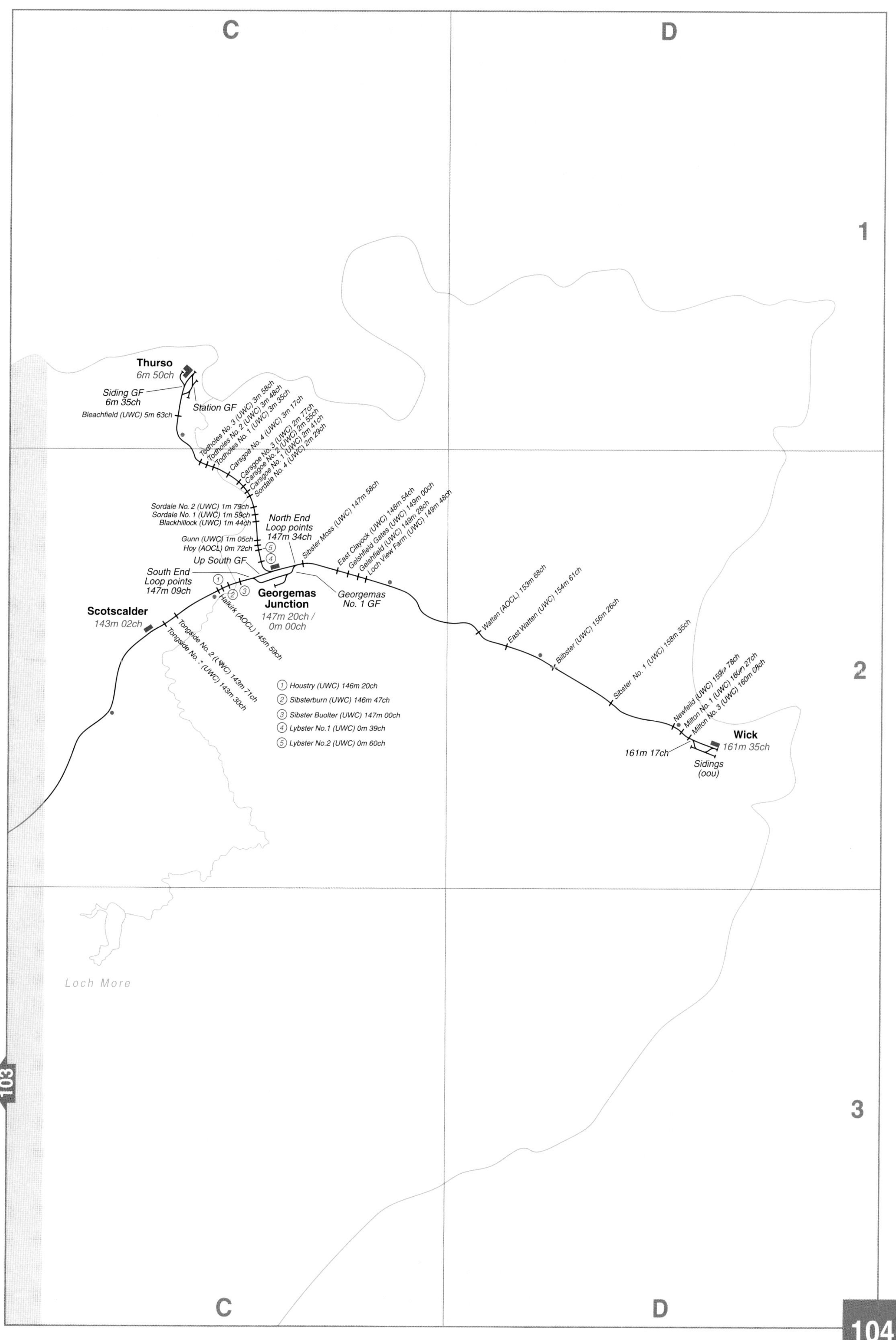
C
D
1
2
3
Thurso
6m 50ch
Siding GF
6m 35ch
Station GF
Bleachfield (UWC) 5m 63ch
Todholes No. 3 (UWC) 3m 58ch
Todholes No. 2 (UWC) 3m 48ch
Todholes No. 1 (UWC) 3m 35ch
Carsgoe No. 4 (UWC) 3m 17ch
Carsgoe No. 3 (UWC) 2m 77ch
Carsgoe No. 2 (UWC) 2m 55ch
Carsgoe No. 1 (UWC) 2m 41ch
Sordale No. 4 (UWC) 2m 29ch
Sordale No. 2 (UWC) 1m 79ch
Sordale No. 1 (UWC) 1m 59ch
Blackhillock (UWC) 1m 44ch
Gunn (UWC) 1m 05ch
Hoy (AOCL) 0m 72ch
North End
Loop points
147m 34ch
Up South GF
South End
Loop points
147m 09ch
Georgemas
Junction
147m 20ch /
0m 00ch
Georgemas
No. 1 GF
Scotscalder
143m 02ch
Halkirk (AOCL) 145m 59ch
Tongside No. 2 (UWC) 143m 71ch
Tongside No. 1 (UWC) 143m 30ch
Sibster Moss (UWC) 147m 58ch
East Clayock (UWC) 148m 54ch
Gelshfield Gates (UWC) 149m 00ch
Gelshfield (UWC) 149m 28ch
Loch View Farm (UWC) 149m 48ch
Watten (AOCL) 153m 68ch
East Watten (UWC) 154m 61ch
Bilbster (UWC) 156m 26ch
Sibster No. 1 (UWC) 158m 35ch
Newfeild (UWC) 159m 78ch
Milton No. 1 (UWC) 160m 27ch
Milton No. 3 (UWC) 160m 08ch
Wick
161m 35ch
161m 17ch
Sidings
(oou)
1 Houstry (UWC) 146m 20ch
2 Sibsterburn (UWC) 146m 47ch
3 Sibster Buolter (UWC) 147m 00ch
4 Lybster No.1 (UWC) 0m 39ch
5 Lybster No.2 (UWC) 0m 60ch
Loch More
C
D

Notes

Contents: Area and Inset Maps

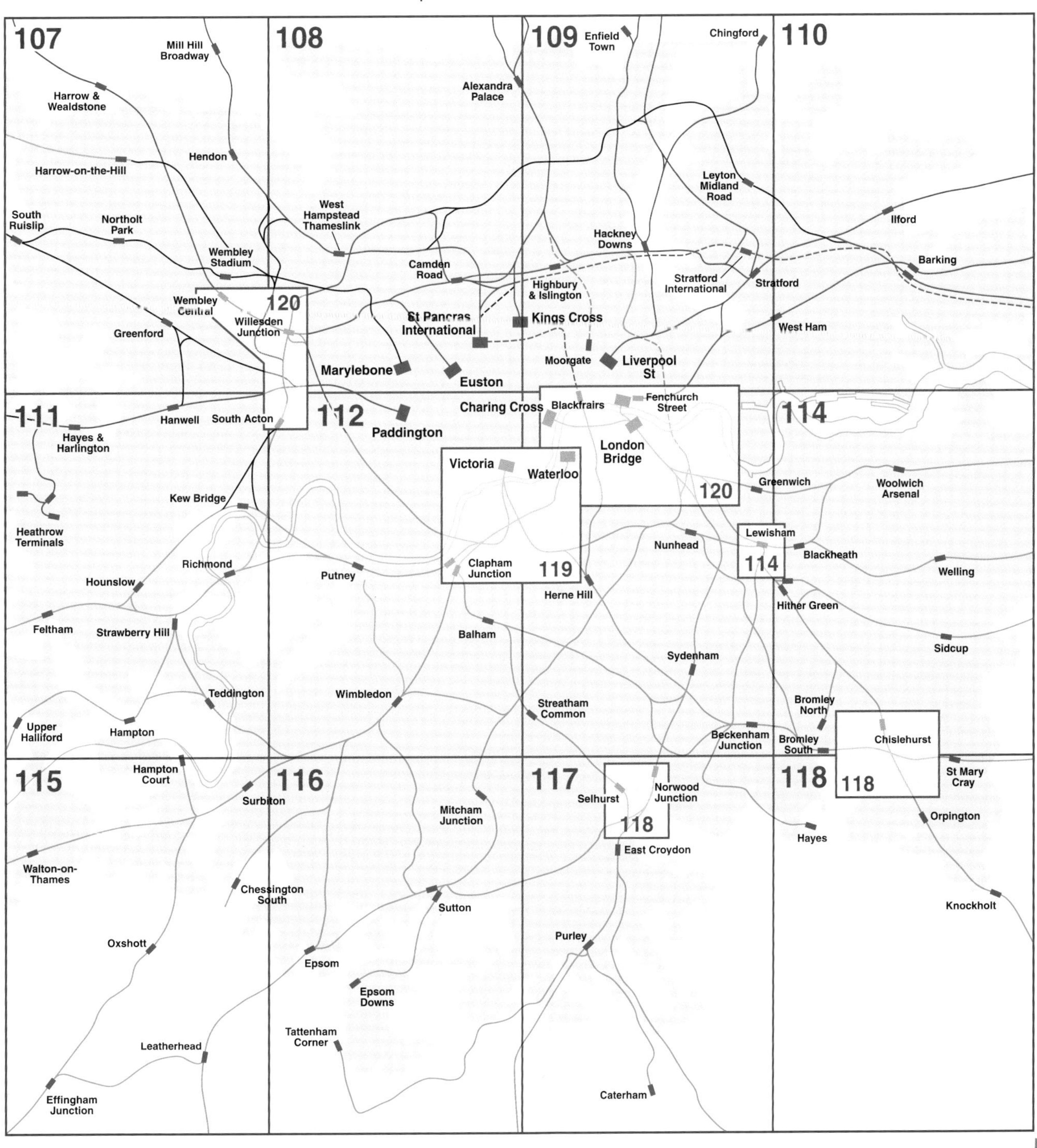

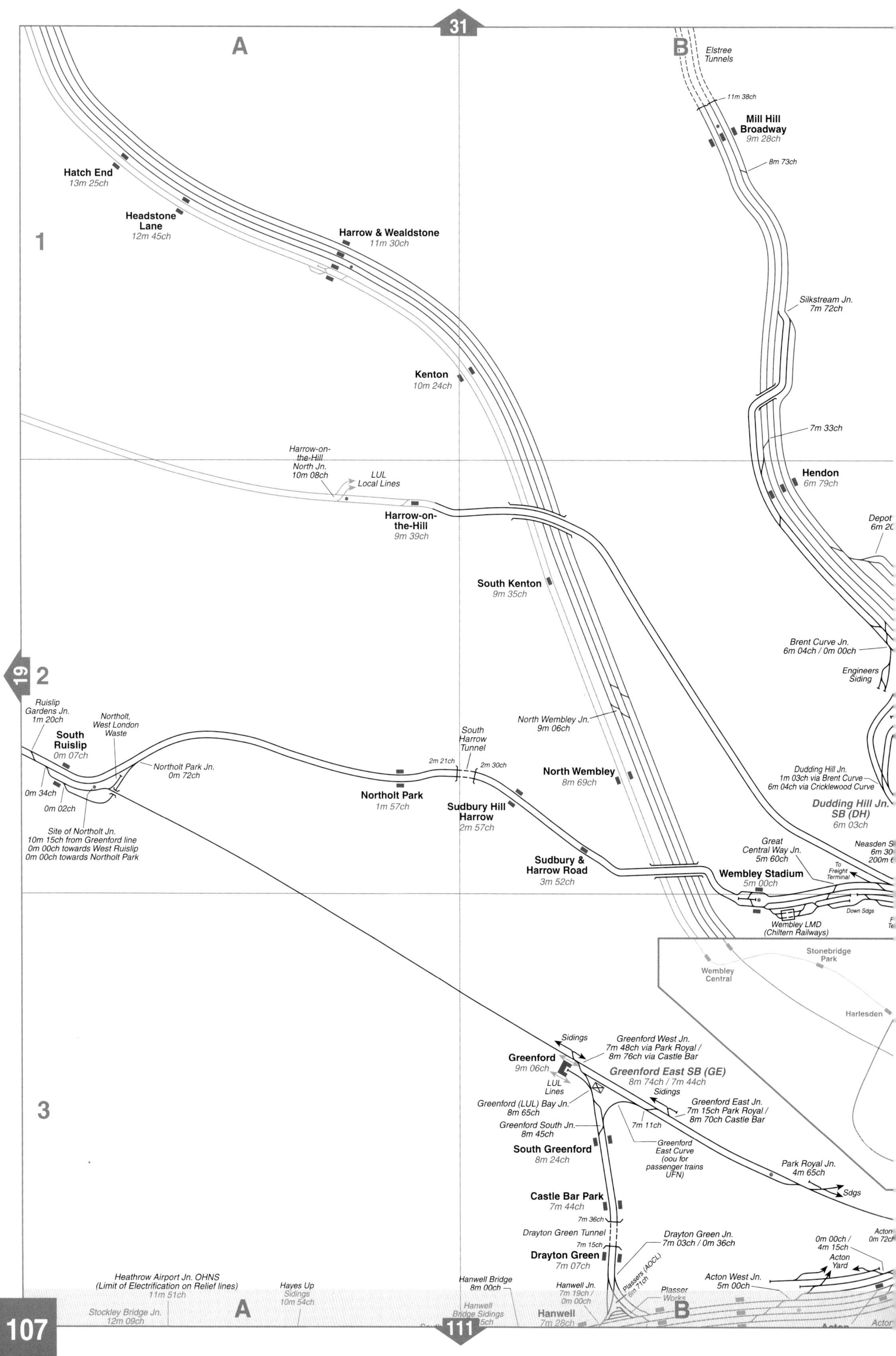
A
B
Elstree Tunnels
11m 38ch
Mill Hill Broadway
9m 28ch
8m 73ch
Hatch End
13m 25ch
Headstone Lane
12m 45ch
1
Harrow & Wealdstone
11m 30ch
Silkstream Jn.
7m 72ch
Kenton
10m 24ch
7m 33ch
Harrow-on-the-Hill North Jn.
10m 08ch
LUL Local Lines
Hendon
6m 79ch
Harrow-on-the-Hill
9m 39ch
Depot
South Kenton
9m 35ch
Brent Curve Jn.
6m 04ch / 0m 00ch
Engineers Siding
2
Ruislip Gardens Jn.
1m 20ch
South Ruislip
0m 07ch
Northolt, West London Waste
South Harrow Tunnel
North Wembley Jn.
9m 06ch
Northolt Park Jn.
0m 72ch
2m 21ch
2m 30ch
North Wembley
8m 69ch
Dudding Hill Jn.
1m 03ch via Brent Curve
6m 04ch via Cricklewood Curve
0m 34ch
0m 02ch
Northolt Park
1m 57ch
Sudbury Hill Harrow
2m 57ch
Dudding Hill Jn. SB (DH)
6m 03ch
Site of Northolt Jn.
10m 15ch from Greenford line
0m 00ch towards West Ruislip
0m 00ch towards Northolt Park
Sudbury & Harrow Road
3m 52ch
Great Central Way Jn.
5m 60ch
Wembley Stadium
5m 00ch
To Freight Terminal
Down Sdgs
Wembley LMD (Chiltern Railways)
Stonebridge Park
Wembley Central
Harlesden
Sidings
Greenford West Jn.
7m 48ch via Park Royal /
8m 76ch via Castle Bar
Greenford
9m 06ch
Greenford East SB (GE)
8m 74ch / 7m 44ch
LUL Lines
Sidings
Greenford (LUL) Bay Jn.
8m 65ch
Greenford East Jn.
7m 15ch Park Royal /
8m 70ch Castle Bar
3
Greenford South Jn.
8m 45ch
7m 11ch
Greenford East Curve (oou for passenger trains UFN)
South Greenford
8m 24ch
Park Royal Jn.
4m 65ch
Sdgs
Castle Bar Park
7m 44ch
7m 36ch
Drayton Green Tunnel
Drayton Green Jn.
7m 03ch / 0m 36ch
7m 15ch
Drayton Green
7m 07ch
0m 00ch /
4m 15ch
Acton Yard
Plassers (AOCL)
6m 71ch
Acton West Jn.
5m 00ch
Heathrow Airport Jn. OHNS
(Limit of Electrification on Relief lines)
11m 51ch
Hayes Up Sidings
10m 54ch
Hanwell Bridge
8m 00ch
Hanwell Jn.
7m 19ch /
0m 00ch
Plasser Works
Stockley Bridge Jn.
12m 09ch
Hanwell Bridge Sidings
Hanwell
7m 28ch

19
111

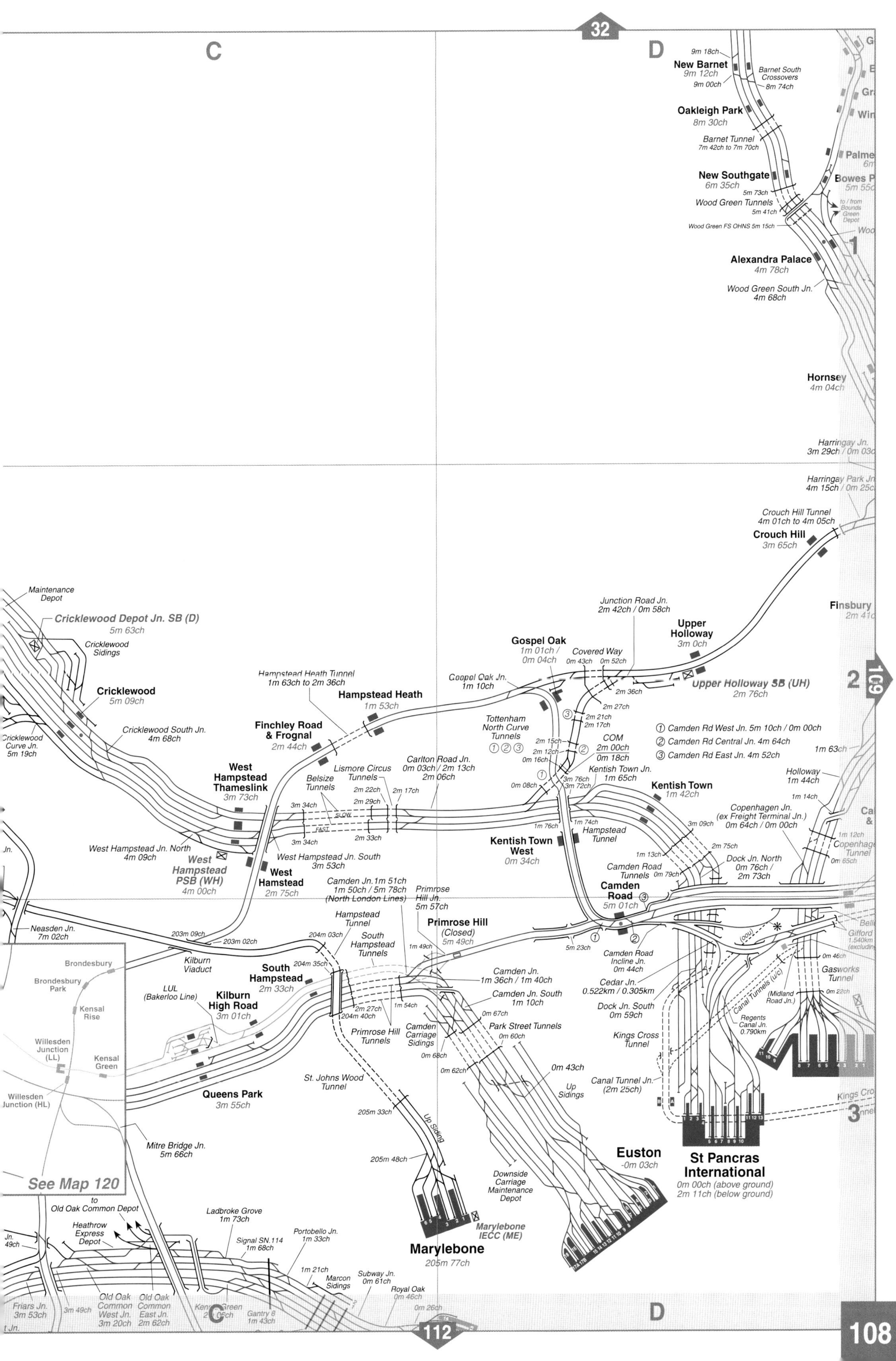
C
D
New Barnet
9m 12ch
9m 18ch
9m 00ch
Barnet South Crossovers
8m 74ch
Oakleigh Park
8m 30ch
Barnet Tunnel
7m 42ch to 7m 70ch
New Southgate
6m 35ch
5m 73ch
Wood Green Tunnels
5m 41ch
Wood Green FS OHNS 5m 15ch
Bowes Park
5m 55ch
to / from Bounds Green Depot
Alexandra Palace
4m 78ch
Wood Green South Jn.
4m 68ch
Hornsey
4m 04ch
Harringay Jn.
3m 29ch / 0m 03ch
Harringay Park Jn.
4m 15ch / 0m 25ch
Crouch Hill Tunnel
4m 01ch to 4m 05ch
Crouch Hill
3m 65ch
Maintenance Depot
Cricklewood Depot Jn. SB (D)
5m 63ch
Cricklewood Sidings
Cricklewood
5m 09ch
Cricklewood Curve Jn.
5m 19ch
Cricklewood South Jn.
4m 68ch
Junction Road Jn.
2m 42ch / 0m 58ch
Upper Holloway
3m 0ch
Gospel Oak
1m 01ch /
0m 04ch
Covered Way
0m 43ch
0m 52ch
Finsbury Park
2m 41ch
Hampstead Heath Tunnel
1m 63ch to 2m 36ch
Gospel Oak Jn.
1m 10ch
Upper Holloway SB (UH)
2m 76ch
2m 36ch
Hampstead Heath
1m 53ch
2m 27ch
2m 21ch
2m 17ch
Tottenham North Curve Tunnels
① ② ③
2m 15ch
2m 12ch
0m 16ch
① Camden Rd West Jn. 5m 10ch / 0m 00ch
② Camden Rd Central Jn. 4m 64ch
③ Camden Rd East Jn. 4m 52ch
Finchley Road & Frognal
2m 44ch
COM
2m 00ch
0m 18ch
Kentish Town Jn.
1m 65ch
1m 63ch
Holloway
1m 44ch
West Hampstead Thameslink
3m 73ch
Lismore Circus Tunnels
Carlton Road Jn.
0m 03ch / 2m 13ch
2m 06ch
Belsize Tunnels
2m 22ch
2m 17ch
2m 29ch
0m 08ch
3m 76ch
3m 72ch
Kentish Town
1m 42ch
1m 14ch
3m 34ch
SLOW
FAST
Copenhagen Jn.
(ex Freight Terminal Jn.)
0m 64ch / 0m 00ch
2m 33ch
1m 76ch
1m 74ch
Hampstead Tunnel
1m 12ch
Copenhagen Tunnel
0m 65ch
3m 34ch
Kentish Town West
0m 34ch
3m 09ch
West Hampstead Jn. North
4m 09ch
West Hampstead Jn. South
3m 53ch
2m 75ch
1m 13ch
West Hampstead PSB (WH)
4m 00ch
West Hamstead
2m 75ch
Camden Road Tunnels
0m 79ch
Dock Jn. North
0m 76ch /
2m 73ch
Camden Jn. 1m 51ch
1m 50ch / 5m 78ch
(North London Lines)
Primrose Hill Jn.
5m 57ch
Camden Road
5m 01ch
Hampstead Tunnel
Primrose Hill
(Closed)
5m 49ch
Neasden Jn.
7m 02ch
203m 09ch
203m 02ch
204m 03ch
South Hampstead Tunnels
1m 49ch
5m 23ch
Camden Road Incline Jn.
0m 44ch
Gifford
0m 46ch
Gasworks Tunnel
Brondesbury
Brondesbury Park
Kilburn Viaduct
South Hampstead
2m 33ch
204m 35ch
Camden Jn.
1m 36ch / 1m 40ch
Cedar Jn.
0.522km / 0.305km
(Midland Road Jn.)
0m 22ch
LUL
(Bakerloo Line)
Kilburn High Road
3m 01ch
Camden Jn. South
1m 10ch
Dock Jn. South
0m 59ch
Canal Tunnels (u/c)
Regents Canal Jn.
0.790km
Kensal Rise
2m 27ch
204m 40ch
1m 54ch
0m 67ch
Willesden Junction (LL)
Primrose Hill Tunnels
Camden Carriage Sidings
Park Street Tunnels
0m 60ch
Kings Cross Tunnel
Kensal Green
0m 68ch
0m 62ch
0m 43ch
Willesden Junction (HL)
Queens Park
3m 55ch
St. Johns Wood Tunnel
Up Sidings
Canal Tunnel Jn.
(2m 25ch)
Kings Cross
205m 33ch
Up Siding
Mitre Bridge Jn.
5m 66ch
205m 48ch
Euston
-0m 03ch
St Pancras International
0m 00ch (above ground)
2m 11ch (below ground)
See Map 120
Downside Carriage Maintenance Depot
to Old Oak Common Depot
Heathrow Express Depot
Ladbroke Grove
1m 73ch
Marylebone IECC (ME)
Marylebone
205m 77ch
Portobello Jn.
1m 33ch
Signal SN.114
1m 68ch
1m 21ch
Marcon Sidings
Subway Jn.
0m 61ch
Royal Oak
0m 46ch
0m 26ch
Friars Jn.
3m 53ch
3m 49ch
Old Oak Common West Jn.
3m 20ch
Old Oak Common East Jn.
2m 62ch
Kensal Green
2m 05ch
Gantry 8
1m 43ch
C
D
1
2
3

109

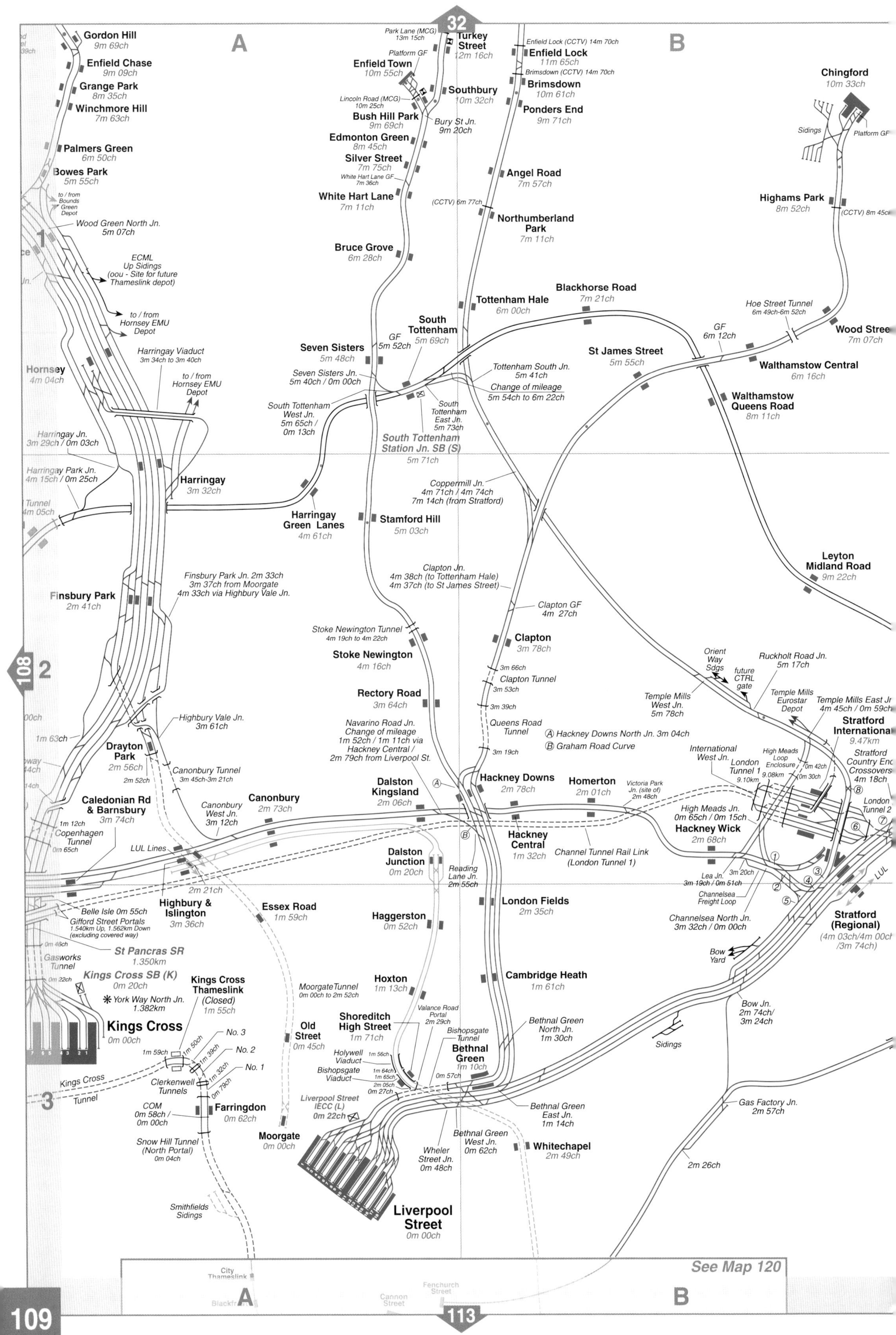

See Map 120
108

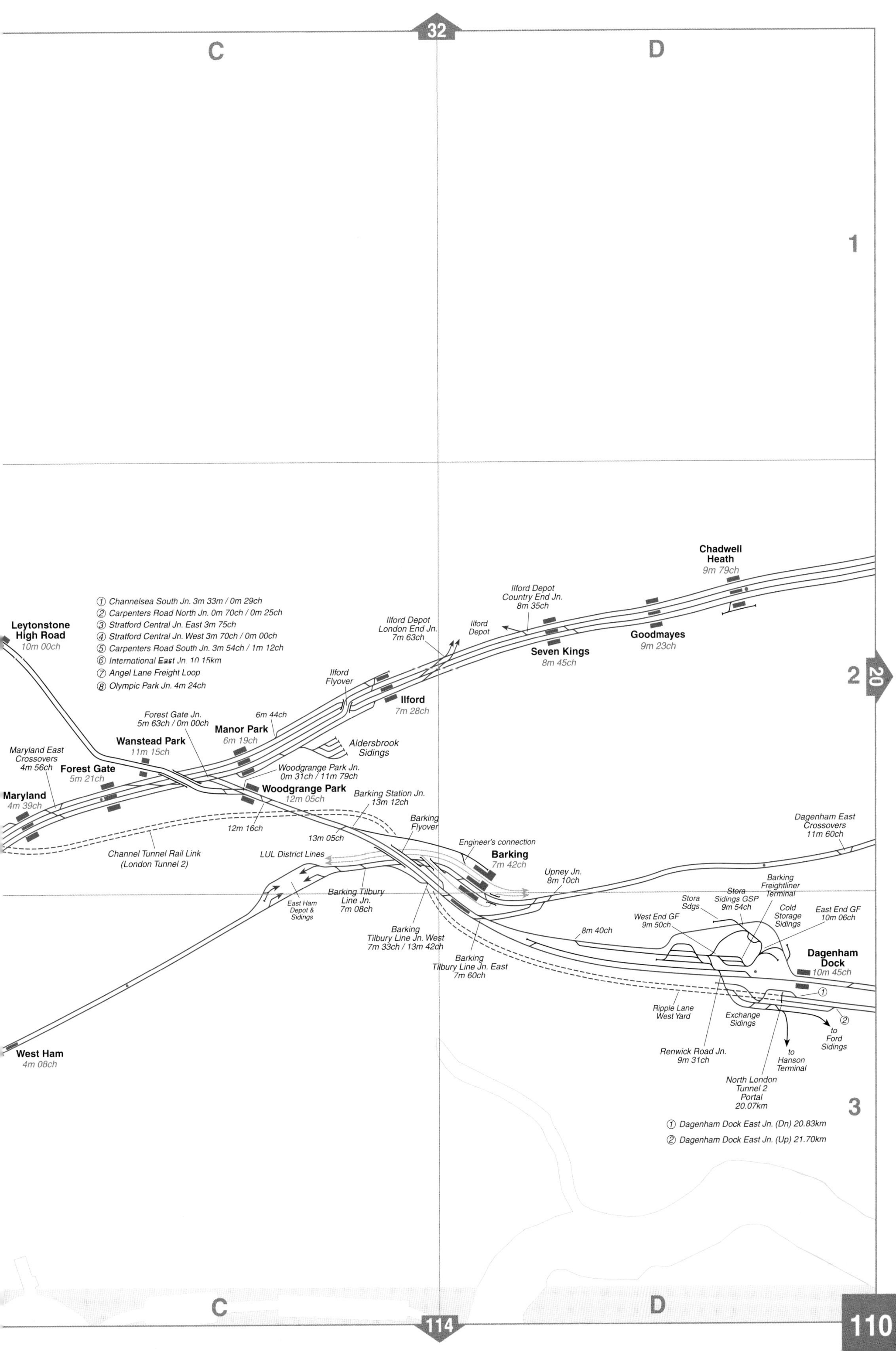
C
D
1
2
20
3
① Channelsea South Jn. 3m 33m / 0m 29ch
② Carpenters Road North Jn. 0m 70ch / 0m 25ch
③ Stratford Central Jn. East 3m 75ch
④ Stratford Central Jn. West 3m 70ch / 0m 00ch
⑤ Carpenters Road South Jn. 3m 54ch / 1m 12ch
⑥ International East Jn. 10.15km
⑦ Angel Lane Freight Loop
⑧ Olympic Park Jn. 4m 24ch
Leytonstone High Road
10m 00ch
Chadwell Heath
9m 79ch
Ilford Depot Country End Jn.
8m 35ch
Ilford Depot London End Jn.
7m 63ch
Ilford Depot
Seven Kings
8m 45ch
Goodmayes
9m 23ch
Ilford Flyover
Ilford
7m 28ch
Forest Gate Jn.
5m 63ch / 0m 00ch
6m 44ch
Manor Park
6m 19ch
Wanstead Park
11m 15ch
Aldersbrook Sidings
Maryland East Crossovers
4m 56ch
Forest Gate
5m 21ch
Woodgrange Park Jn.
0m 31ch / 11m 79ch
Woodgrange Park
12m 05ch
Maryland
4m 39ch
Barking Station Jn.
13m 12ch
Barking Flyover
12m 16ch
13m 05ch
Dagenham East Crossovers
11m 60ch
Engineer's connection
Barking
7m 42ch
Channel Tunnel Rail Link
(London Tunnel 2)
LUL District Lines
Upney Jn.
8m 10ch
Barking Freightliner Terminal
Stora Sdgs
Stora Sidings GSP
9m 54ch
Cold Storage Sidings
East End GF
10m 06ch
West End GF
9m 50ch
East Ham Depot & Sidings
Barking Tilbury Line Jn.
7m 08ch
Barking Tilbury Line Jn. West
7m 33ch / 13m 42ch
8m 40ch
Barking Tilbury Line Jn. East
7m 60ch
Dagenham Dock
10m 45ch
Ripple Lane West Yard
Exchange Sidings
to Ford Sidings
Renwick Road Jn.
9m 31ch
to Hanson Terminal
North London Tunnel 2 Portal
20.07km
West Ham
4m 08ch
① Dagenham Dock East Jn. (Dn) 20.83km
② Dagenham Dock East Jn. (Up) 21.70km
C
D

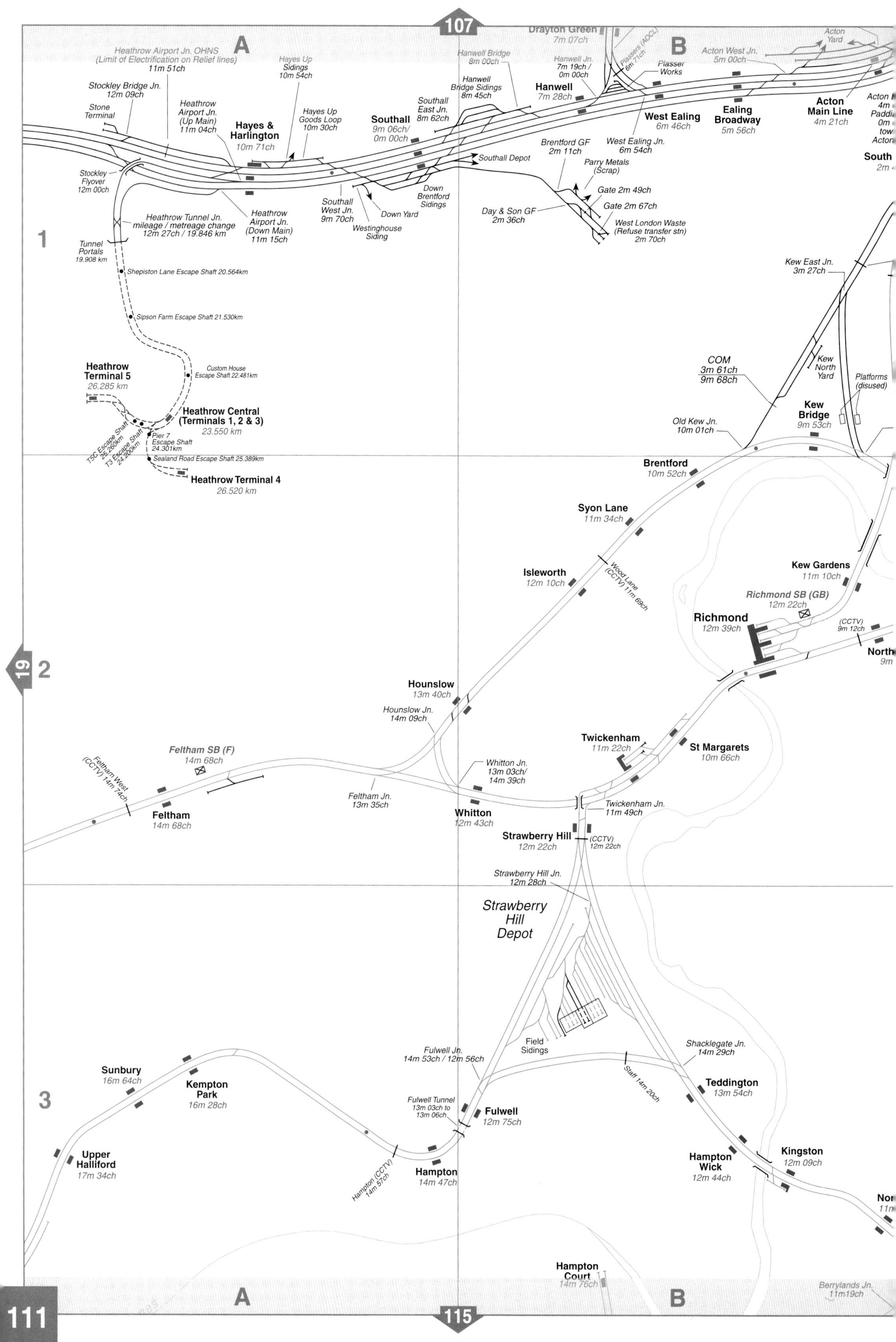
107
115
19
A
B
1
2
3
Drayton Green
7m 07ch
Heathrow Airport Jn. OHNS
(Limit of Electrification on Relief lines)
11m 51ch
Hayes Up Sidings
10m 54ch
Hanwell Bridge
8m 00ch
Hanwell Jn.
7m 19ch /
0m 00ch
Plassers (AOCL)
6m 71ch
Plasser Works
Acton West Jn.
5m 00ch
Acton Yard
Stockley Bridge Jn.
12m 09ch
Stone Terminal
Heathrow Airport Jn.
(Up Main)
11m 04ch
Hayes &
Harlington
10m 71ch
Hayes Up
Goods Loop
10m 30ch
Southall
9m 06ch/
0m 00ch
Southall
East Jn.
8m 62ch
Hanwell
Bridge Sidings
8m 45ch
Hanwell
7m 28ch
West Ealing
6m 46ch
Ealing
Broadway
5m 56ch
Acton
Main Line
4m 21ch
Southall Depot
Brentford GF
2m 11ch
West Ealing Jn.
6m 54ch
Parry Metals
(Scrap)
Gate 2m 49ch
Gate 2m 67ch
Day & Son GF
2m 36ch
West London Waste
(Refuse transfer stn)
2m 70ch
Stockley
Flyover
12m 00ch
Heathrow Airport Jn.
(Down Main)
11m 15ch
Southall
West Jn.
9m 70ch
Down Yard
Westinghouse
Siding
Down
Brentford
Sidings
Heathrow Tunnel Jn.
mileage / metreage change
12m 27ch / 19.846 km
Tunnel
Portals
19.908 km
Shepiston Lane Escape Shaft 20.564km
Sipson Farm Escape Shaft 21.530km
Custom House
Escape Shaft 22.481km
Heathrow
Terminal 5
26.285 km
Heathrow Central
(Terminals 1, 2 & 3)
23.550 km
T5C Escape Shaft
25.260km
T3 Escape Shaft
24.200km
Pier 7
Escape Shaft
24.301km
Sealand Road Escape Shaft 25.389km
Heathrow Terminal 4
26.520 km
Kew East Jn.
3m 27ch
COM
3m 61ch
9m 68ch
Kew
North
Yard
Platforms
(disused)
Kew
Bridge
9m 53ch
Old Kew Jn.
10m 01ch
Brentford
10m 52ch
Syon Lane
11m 34ch
Isleworth
12m 10ch
Wood Lane
(CCTV) 11m 69ch
Kew Gardens
11m 10ch
Richmond SB (GB)
12m 22ch
Richmond
12m 39ch
(CCTV)
9m 12ch
Hounslow
13m 40ch
Hounslow Jn.
14m 09ch
Feltham SB (F)
14m 68ch
Feltham West
(CCTV) 14m 74ch
Feltham
14m 68ch
Feltham Jn.
13m 35ch
Whitton Jn.
13m 03ch/
14m 39ch
Whitton
12m 43ch
Twickenham
11m 22ch
St Margarets
10m 66ch
Twickenham Jn.
11m 49ch
Strawberry Hill
12m 22ch
(CCTV)
12m 22ch
Strawberry Hill Jn.
12m 28ch
Strawberry
Hill
Depot
Field
Sidings
Fulwell Jn.
14m 53ch / 12m 56ch
Shacklegate Jn.
14m 29ch
Staff 14m 20ch
Sunbury
16m 64ch
Kempton
Park
16m 28ch
Fulwell Tunnel
13m 03ch to
13m 06ch
Fulwell
12m 75ch
Teddington
13m 54ch
Upper
Halliford
17m 34ch
Hampton (CCTV)
14m 57ch
Hampton
14m 47ch
Hampton
Wick
12m 44ch
Kingston
12m 09ch
Hampton
Court
14m 76ch
Berrylands Jn.
11m19ch

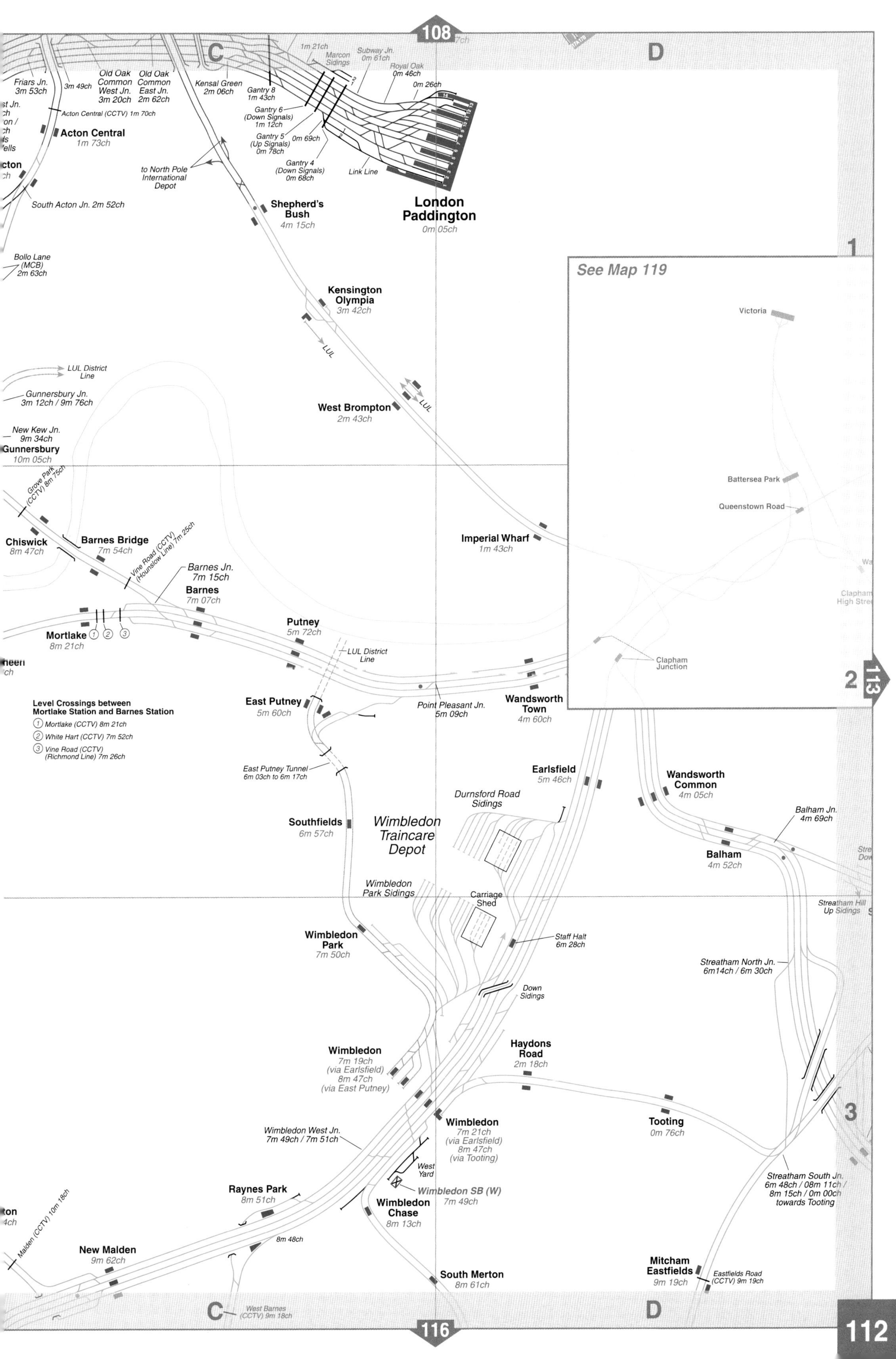
C
D
London Paddington
0m 05ch
Shepherd's Bush
4m 15ch
Acton Central
1m 73ch
Acton Central (CCTV) 1m 70ch
Friars Jn.
3m 53ch
3m 49ch
Old Oak Common West Jn.
3m 20ch
Old Oak Common East Jn.
2m 62ch
Kensal Green
2m 06ch
Gantry 8
1m 43ch
Gantry 6
(Down Signals)
1m 12ch
Gantry 5
(Up Signals)
0m 78ch
0m 69ch
Gantry 4
(Down Signals)
0m 68ch
Link Line
1m 21ch
Marcon Sidings
Subway Jn.
0m 61ch
Royal Oak
0m 46ch
0m 26ch
to North Pole International Depot
South Acton Jn. 2m 52ch
Bollo Lane
(MCB)
2m 63ch
Kensington Olympia
3m 42ch
LUL
LUL District Line
Gunnersbury Jn.
3m 12ch / 9m 76ch
New Kew Jn.
9m 34ch
Gunnersbury
10m 05ch
West Brompton
2m 43ch
See Map 119
Victoria
Battersea Park
Queenstown Road
Clapham High Street
Clapham Junction
Grove Park (CCTV) 8m 75ch
Chiswick
8m 47ch
Barnes Bridge
7m 54ch
Vine Road (CCTV) (Hounslow Line) 7m 25ch
Barnes Jn.
7m 15ch
Barnes
7m 07ch
Imperial Wharf
1m 43ch
Mortlake
8m 21ch
Putney
5m 72ch
LUL District Line
East Putney
5m 60ch
Point Pleasant Jn.
5m 09ch
Wandsworth Town
4m 60ch
Level Crossings between Mortlake Station and Barnes Station
(1) Mortlake (CCTV) 8m 21ch
(2) White Hart (CCTV) 7m 52ch
(3) Vine Road (CCTV) (Richmond Line) 7m 26ch
East Putney Tunnel
6m 03ch to 6m 17ch
Earlsfield
5m 46ch
Wandsworth Common
4m 05ch
Durnsford Road Sidings
Balham Jn.
4m 69ch
Southfields
6m 57ch
Wimbledon Traincare Depot
Balham
4m 52ch
Wimbledon Park Sidings
Carriage Shed
Streatham Hill Up Sidings
Wimbledon Park
7m 50ch
Staff Halt
6m 28ch
Streatham North Jn.
6m14ch / 6m 30ch
Down Sidings
Wimbledon
7m 19ch
(via Earlsfield)
8m 47ch
(via East Putney)
Haydons Road
2m 18ch
Tooting
0m 76ch
Wimbledon West Jn.
7m 49ch / 7m 51ch
Wimbledon
7m 21ch
(via Earlsfield)
8m 47ch
(via Tooting)
West Yard
Wimbledon SB (W)
7m 49ch
Streatham South Jn.
6m 48ch / 08m 11ch /
8m 15ch / 0m 00ch
towards Tooting
Raynes Park
8m 51ch
Wimbledon Chase
8m 13ch
Malden (CCTV) 10m 18ch
New Malden
9m 62ch
8m 48ch
South Merton
8m 61ch
Mitcham Eastfields
9m 19ch
Eastfields Road
(CCTV) 9m 19ch
West Barnes
(CCTV) 9m 18ch
1
2
3
113
116

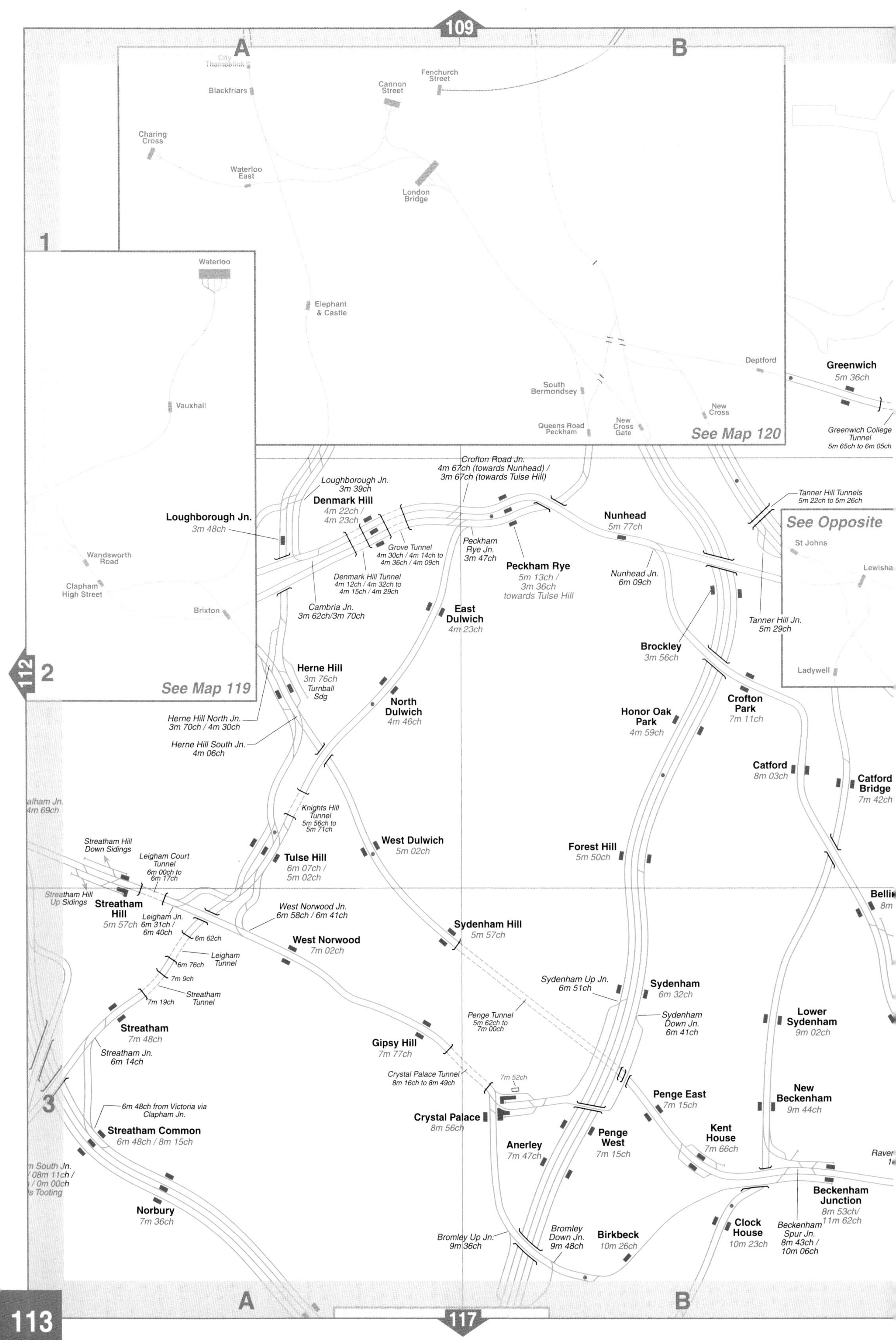

109
A
B
City Thameslink
Blackfriars
Cannon Street
Fenchurch Street
Charing Cross
Waterloo East
London Bridge
Waterloo
Elephant & Castle
Vauxhall
South Bermondsey
Deptford
Greenwich
5m 36ch
New Cross
Queens Road Peckham
New Cross Gate
See Map 120
Greenwich College Tunnel
5m 65ch to 6m 05ch
Crofton Road Jn.
4m 67ch (towards Nunhead) /
3m 67ch (towards Tulse Hill)
Loughborough Jn.
3m 39ch
Denmark Hill
4m 22ch /
4m 23ch
Loughborough Jn.
3m 48ch
Tanner Hill Tunnels
5m 22ch to 5m 26ch
Nunhead
5m 77ch
See Opposite
St Johns
Lewisha
Wandsworth Road
Clapham High Street
Brixton
Grove Tunnel
4m 30ch / 4m 14ch to
4m 36ch / 4m 09ch
Peckham Rye Jn.
3m 47ch
Peckham Rye
5m 13ch /
3m 36ch
towards Tulse Hill
Nunhead Jn.
6m 09ch
Denmark Hill Tunnel
4m 12ch / 4m 32ch to
4m 15ch / 4m 29ch
Cambria Jn.
3m 62ch/3m 70ch
East Dulwich
4m 23ch
Tanner Hill Jn.
5m 29ch
Brockley
3m 56ch
Ladywell
112
2
See Map 119
Herne Hill
3m 76ch
Turnball Sdg
North Dulwich
4m 46ch
Crofton Park
7m 11ch
Honor Oak Park
4m 59ch
Herne Hill North Jn.
3m 70ch / 4m 30ch
Herne Hill South Jn.
4m 06ch
Catford
8m 03ch
Catford Bridge
7m 42ch
alham Jn.
4m 69ch
Knights Hill Tunnel
5m 56ch to
5m 71ch
Streatham Hill Down Sidings
Leigham Court Tunnel
6m 00ch to
6m 17ch
Tulse Hill
6m 07ch /
5m 02ch
West Dulwich
5m 02ch
Forest Hill
5m 50ch
Streatham Hill Up Sidings
Streatham Hill
5m 57ch
Leigham Jn.
6m 31ch /
6m 40ch
West Norwood Jn.
6m 58ch / 6m 41ch
6m 62ch
Leigham Tunnel
6m 76ch
7m 9ch
Streatham Tunnel
7m 19ch
West Norwood
7m 02ch
Sydenham Hill
5m 57ch
Belli
8m
Sydenham Up Jn.
6m 51ch
Sydenham
6m 32ch
Sydenham Down Jn.
6m 41ch
Penge Tunnel
5m 62ch to
7m 00ch
Lower Sydenham
9m 02ch
Streatham
7m 48ch
Streatham Jn.
6m 14ch
Gipsy Hill
7m 77ch
Crystal Palace Tunnel
8m 16ch to 8m 49ch
7m 52ch
3
6m 48ch from Victoria via Clapham Jn.
Streatham Common
6m 48ch / 8m 15ch
Crystal Palace
8m 56ch
Penge East
7m 15ch
New Beckenham
9m 44ch
Penge West
7m 15ch
Anerley
7m 47ch
Kent House
7m 66ch
Raver
m South Jn.
08m 11ch /
/ 0m 00ch
s Tooting
Norbury
7m 36ch
Beckenham Junction
8m 53ch/
11m 62ch
Bromley Up Jn.
9m 36ch
Bromley Down Jn.
9m 48ch
Birkbeck
10m 26ch
Clock House
10m 23ch
Beckenham Spur Jn.
8m 43ch /
10m 06ch
A
B
117

110

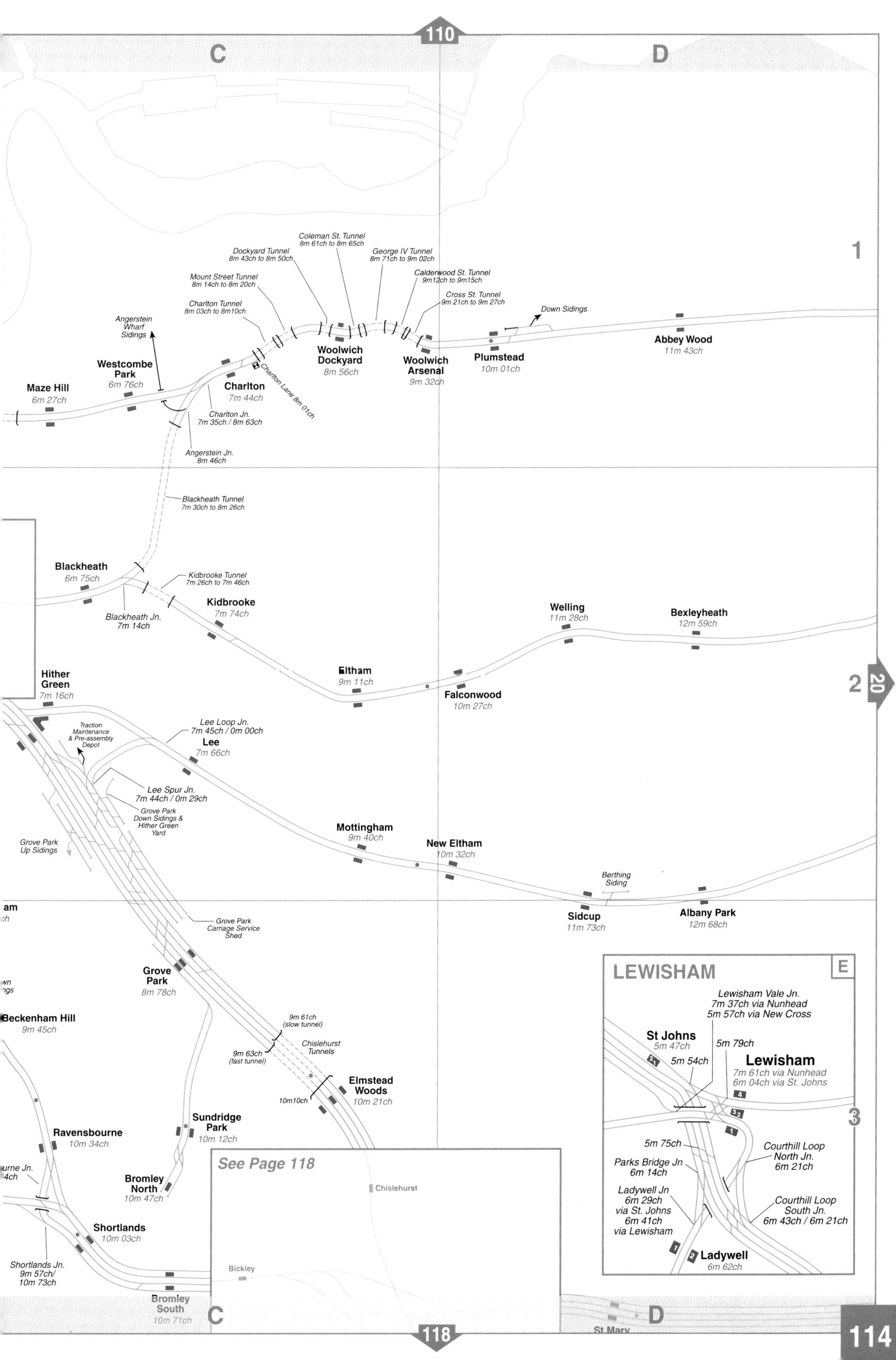
C
D
1
2
3
20
Coleman St. Tunnel
8m 61ch to 8m 65ch
Dockyard Tunnel
8m 43ch to 8m 50ch
George IV Tunnel
8m 71ch to 9m 02ch
Mount Street Tunnel
8m 14ch to 8m 20ch
Calderwood St. Tunnel
9m12ch to 9m15ch
Charlton Tunnel
8m 03ch to 8m10ch
Cross St. Tunnel
9m 21ch to 9m 27ch
Down Sidings
Angerstein Wharf Sidings
Westcombe Park
6m 76ch
Maze Hill
6m 27ch
Charlton
7m 44ch
Charlton Lane 8m 01ch
Woolwich Dockyard
8m 56ch
Woolwich Arsenal
9m 32ch
Plumstead
10m 01ch
Abbey Wood
11m 43ch
Charlton Jn.
7m 35ch / 8m 63ch
Angerstein Jn.
8m 46ch
Blackheath Tunnel
7m 30ch to 8m 26ch
Blackheath
6m 75ch
Kidbrooke Tunnel
7m 26ch to 7m 46ch
Kidbrooke
7m 74ch
Blackheath Jn.
7m 14ch
Welling
11m 28ch
Bexleyheath
12m 59ch
Eltham
9m 11ch
Falconwood
10m 27ch
Hither Green
7m 16ch
Traction Maintenance & Pre-assembly Depot
Lee Loop Jn.
7m 45ch / 0m 00ch
Lee
7m 66ch
Lee Spur Jn.
7m 44ch / 0m 29ch
Grove Park Down Sidings & Hither Green Yard
Grove Park Up Sidings
Mottingham
9m 40ch
New Eltham
10m 32ch
Berthing Siding
Sidcup
11m 73ch
Albany Park
12m 68ch
Grove Park Carriage Service Shed
Grove Park
8m 78ch
Beckenham Hill
9m 45ch
9m 61ch
(slow tunnel)
Chislehurst Tunnels
9m 63ch
(fast tunnel)
Elmstead Woods
10m 21ch
10m10ch
Ravensbourne
10m 34ch
Sundridge Park
10m 12ch
Bromley North
10m 47ch
Shortlands
10m 03ch
Shortlands Jn.
9m 57ch/
10m 73ch
Bromley South
10m 71ch
See Page 118
Chislehurst
Bickley
St Mary
LEWISHAM
E
Lewisham Vale Jn.
7m 37ch via Nunhead
5m 57ch via New Cross
St Johns
5m 47ch
5m 79ch
5m 54ch
Lewisham
7m 61ch via Nunhead
6m 04ch via St. Johns
5m 75ch
Parks Bridge Jn
6m 14ch
Courthill Loop North Jn.
6m 21ch
Ladywell Jn
6m 29ch
via St. Johns
6m 41ch
via Lewisham
Courthill Loop South Jn.
6m 43ch / 6m 21ch
Ladywell
6m 62ch

118

111

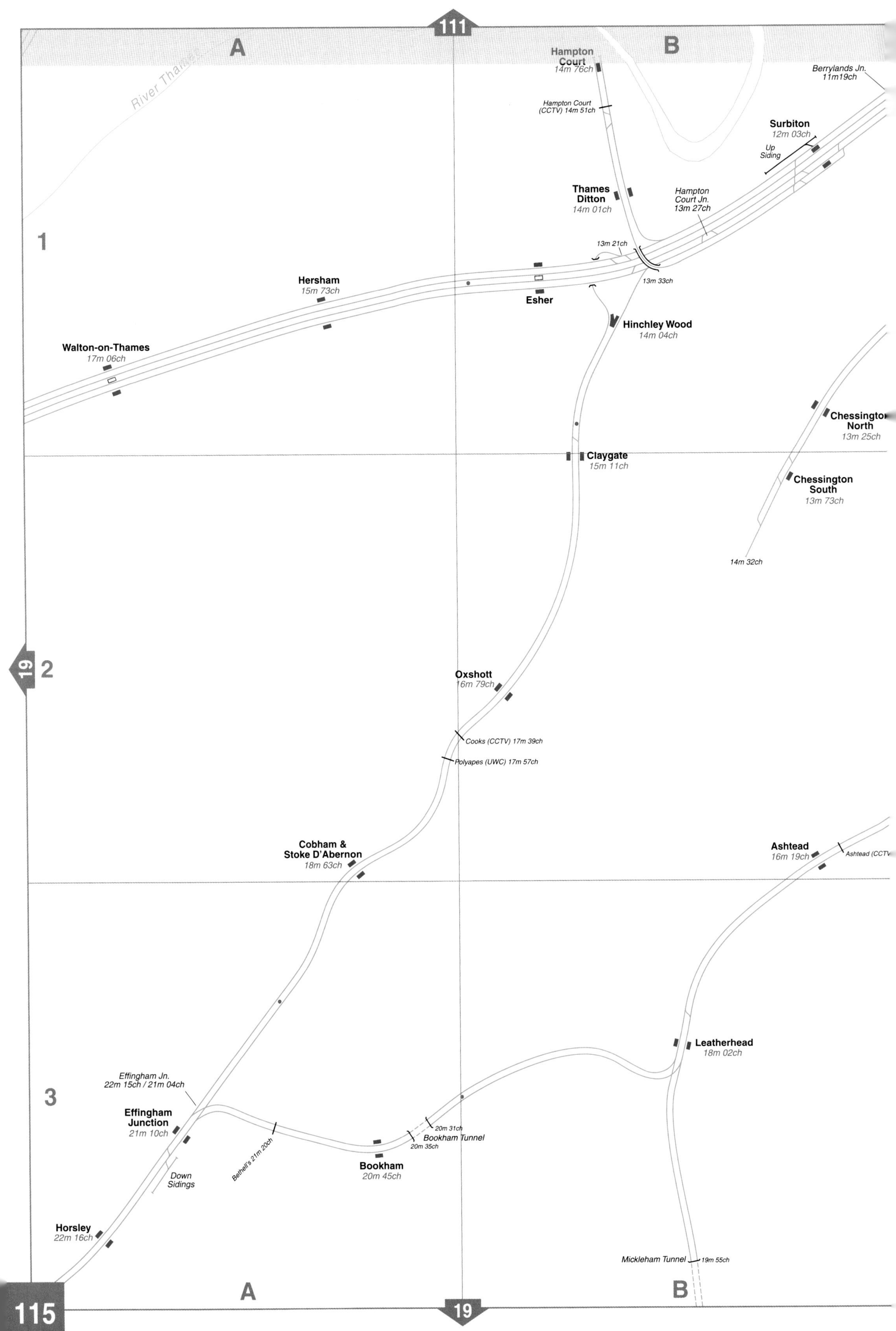
A
B
River Thames
Hampton Court
14m 76ch
Hampton Court (CCTV) 14m 51ch
Berrylands Jn.
11m19ch
Surbiton
12m 03ch
Up Siding
Thames Ditton
14m 01ch
Hampton Court Jn.
13m 27ch
1
13m 21ch
13m 33ch
Hersham
15m 73ch
Esher
Hinchley Wood
14m 04ch
Walton-on-Thames
17m 06ch
Chessington North
13m 25ch
Claygate
15m 11ch
Chessington South
13m 73ch
14m 32ch
2
Oxshott
16m 79ch
Cooks (CCTV) 17m 39ch
Polyapes (UWC) 17m 57ch
Cobham & Stoke D'Abernon
18m 63ch
Ashtead
16m 19ch
Ashtead (CCTV
Leatherhead
18m 02ch
Effingham Jn.
22m 15ch / 21m 04ch
3
Effingham Junction
21m 10ch
20m 31ch
Bookham Tunnel
20m 35ch
Bethell's 21m 20ch
Down Sidings
Bookham
20m 45ch
Horsley
22m 16ch
Mickleham Tunnel
19m 55ch
A
B

19

19

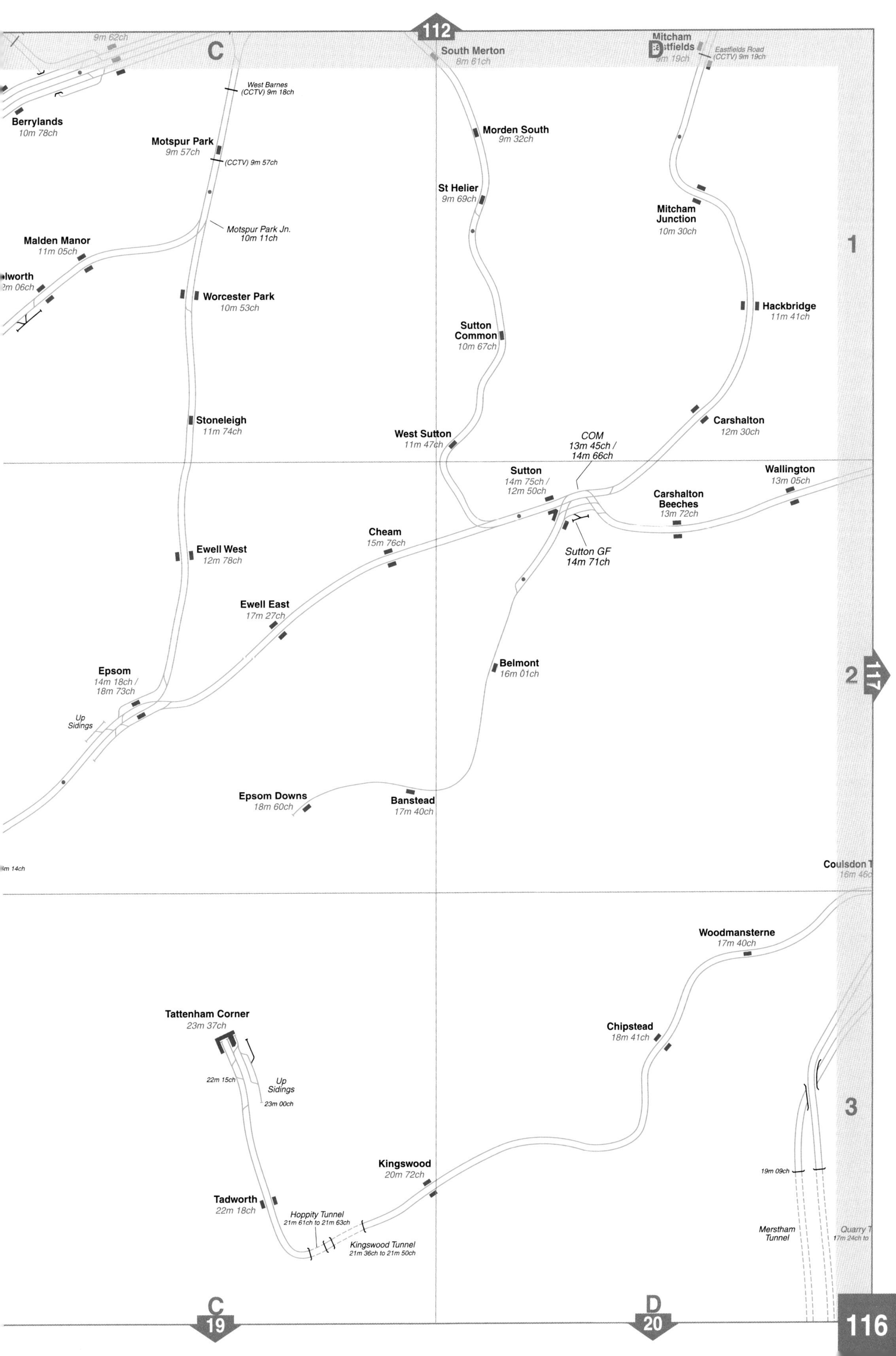
112
C
D
9m 62ch
South Merton
8m 61ch
Mitcham Eastfields
9m 19ch
Eastfields Road (CCTV) 9m 19ch
West Barnes (CCTV) 9m 18ch
Berrylands
10m 78ch
Motspur Park
9m 57ch
(CCTV) 9m 57ch
Morden South
9m 32ch
St Helier
9m 69ch
Mitcham Junction
10m 30ch
Motspur Park Jn.
10m 11ch
Malden Manor
11m 05ch
1
Worcester Park
10m 53ch
Hackbridge
11m 41ch
Sutton Common
10m 67ch
Stoneleigh
11m 74ch
West Sutton
11m 47ch
Carshalton
12m 30ch
COM
13m 45ch / 14m 66ch
Sutton
14m 75ch / 12m 50ch
Wallington
13m 05ch
Carshalton Beeches
13m 72ch
Cheam
15m 76ch
Sutton GF
14m 71ch
Ewell West
12m 78ch
Ewell East
17m 27ch
Belmont
16m 01ch
Epsom
14m 18ch / 18m 73ch
2
117
Up Sidings
Epsom Downs
18m 60ch
Banstead
17m 40ch
Coulsdon T
16m 46c
Woodmansterne
17m 40ch
Tattenham Corner
23m 37ch
Chipstead
18m 41ch
22m 15ch
Up Sidings
23m 00ch
3
19m 09ch
Kingswood
20m 72ch
Tadworth
22m 18ch
Hoppity Tunnel
21m 61ch to 21m 63ch
Kingswood Tunnel
21m 36ch to 21m 50ch
Merstham Tunnel
Quarry T
17m 24ch to
C
19
D
20

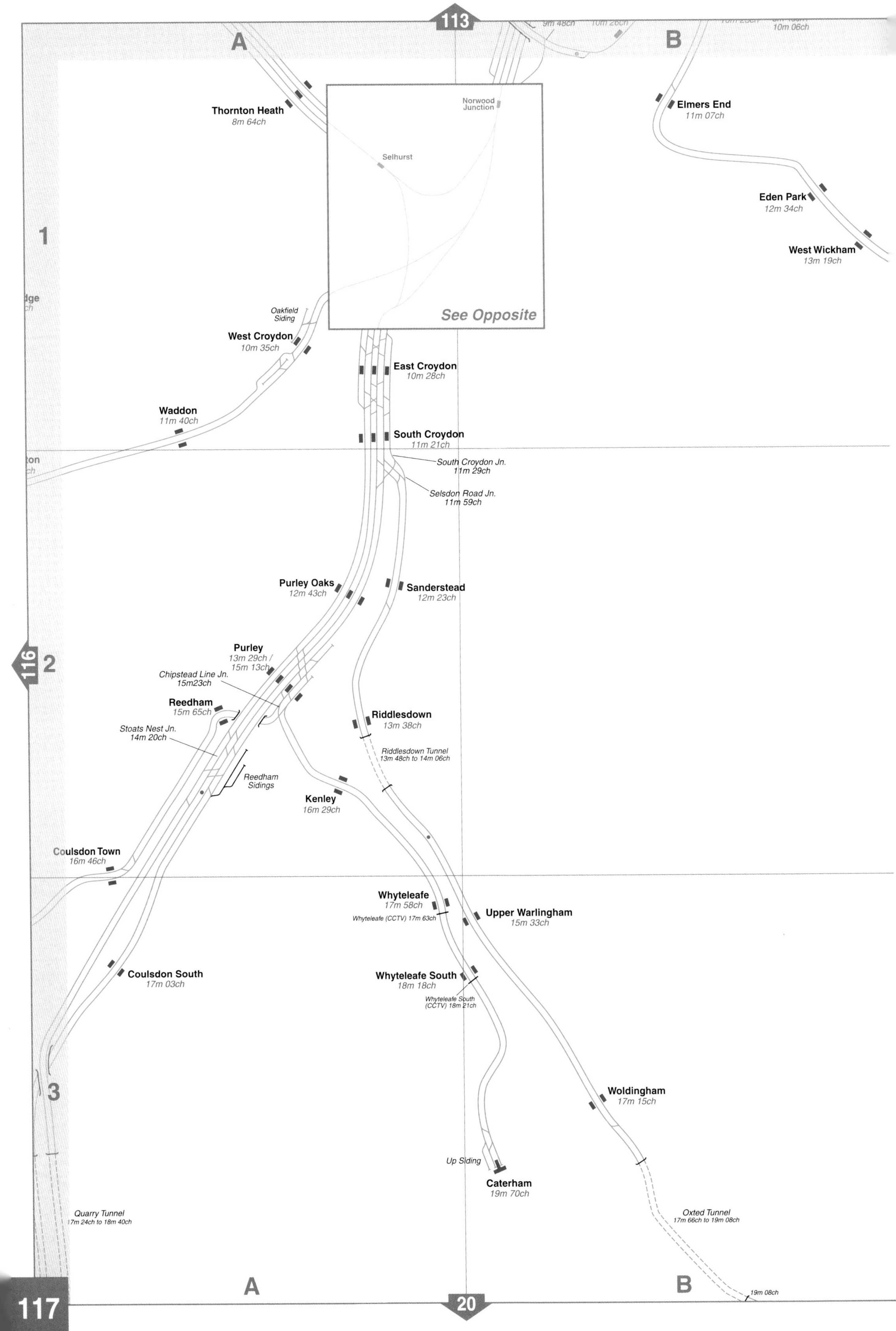
113
A
B
10m 06ch
Thornton Heath
8m 64ch
Norwood Junction
Selhurst
Elmers End
11m 07ch
Eden Park
12m 34ch
West Wickham
13m 19ch
1
See Opposite
Oakfield Siding
West Croydon
10m 35ch
East Croydon
10m 28ch
Waddon
11m 40ch
South Croydon
11m 21ch
South Croydon Jn.
11m 29ch
Selsdon Road Jn.
11m 59ch
Purley Oaks
12m 43ch
Sanderstead
12m 23ch
116
2
Purley
13m 29ch /
15m 13ch
Chipstead Line Jn.
15m23ch
Reedham
15m 65ch
Stoats Nest Jn.
14m 20ch
Riddlesdown
13m 38ch
Riddlesdown Tunnel
13m 48ch to 14m 06ch
Reedham Sidings
Kenley
16m 29ch
Coulsdon Town
16m 46ch
Whyteleafe
17m 58ch
Whyteleafe (CCTV) 17m 63ch
Upper Warlingham
15m 33ch
Coulsdon South
17m 03ch
Whyteleafe South
18m 18ch
Whyteleafe South (CCTV) 18m 21ch
3
Woldingham
17m 15ch
Up Siding
Caterham
19m 70ch
Quarry Tunnel
17m 24ch to 18m 40ch
Oxted Tunnel
17m 66ch to 19m 08ch
A
B
19m 08ch
20

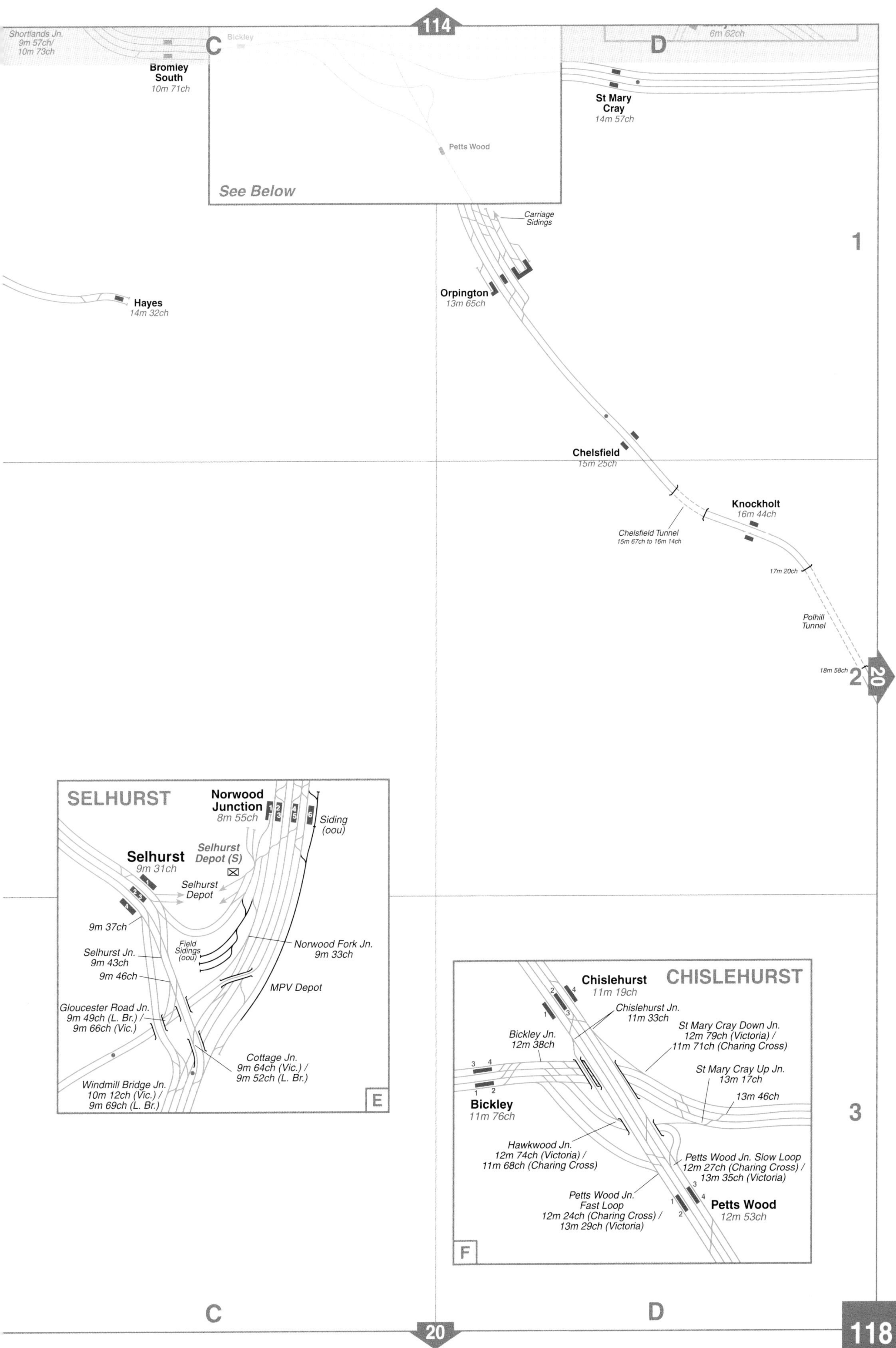
Shortlands Jn.
9m 57ch/
10m 73ch
Bromley
South
10m 71ch
C
Bickley
D
6m 62ch
St Mary
Cray
14m 57ch
Petts Wood
See Below
Carriage
Sidings
1
Hayes
14m 32ch
Orpington
13m 65ch
Chelsfield
15m 25ch
Knockholt
16m 44ch
Chelsfield Tunnel
15m 67ch to 16m 14ch
17m 20ch
Polhill
Tunnel
18m 58ch
2
20
SELHURST
Norwood
Junction
8m 55ch
1
2
3
4
5
6
Siding
(oou)
Selhurst
9m 31ch
Selhurst
Depot (S)
Selhurst
Depot
9m 37ch
Field
Sidings
(oou)
Norwood Fork Jn.
9m 33ch
Selhurst Jn.
9m 43ch
9m 46ch
MPV Depot
Gloucester Road Jn.
9m 49ch (L. Br.) /
9m 66ch (Vic.)
Cottage Jn.
9m 64ch (Vic.) /
9m 52ch (L. Br.)
Windmill Bridge Jn.
10m 12ch (Vic.) /
9m 69ch (L. Br.)
E
Chislehurst
11m 19ch
CHISLEHURST
Chislehurst Jn.
11m 33ch
Bickley Jn.
12m 38ch
St Mary Cray Down Jn.
12m 79ch (Victoria) /
11m 71ch (Charing Cross)
St Mary Cray Up Jn.
13m 17ch
13m 46ch
Bickley
11m 76ch
3
Hawkwood Jn.
12m 74ch (Victoria) /
11m 68ch (Charing Cross)
Petts Wood Jn. Slow Loop
12m 27ch (Charing Cross) /
13m 35ch (Victoria)
Petts Wood Jn.
Fast Loop
12m 24ch (Charing Cross) /
13m 29ch (Victoria)
Petts Wood
12m 53ch
F
C
D
20

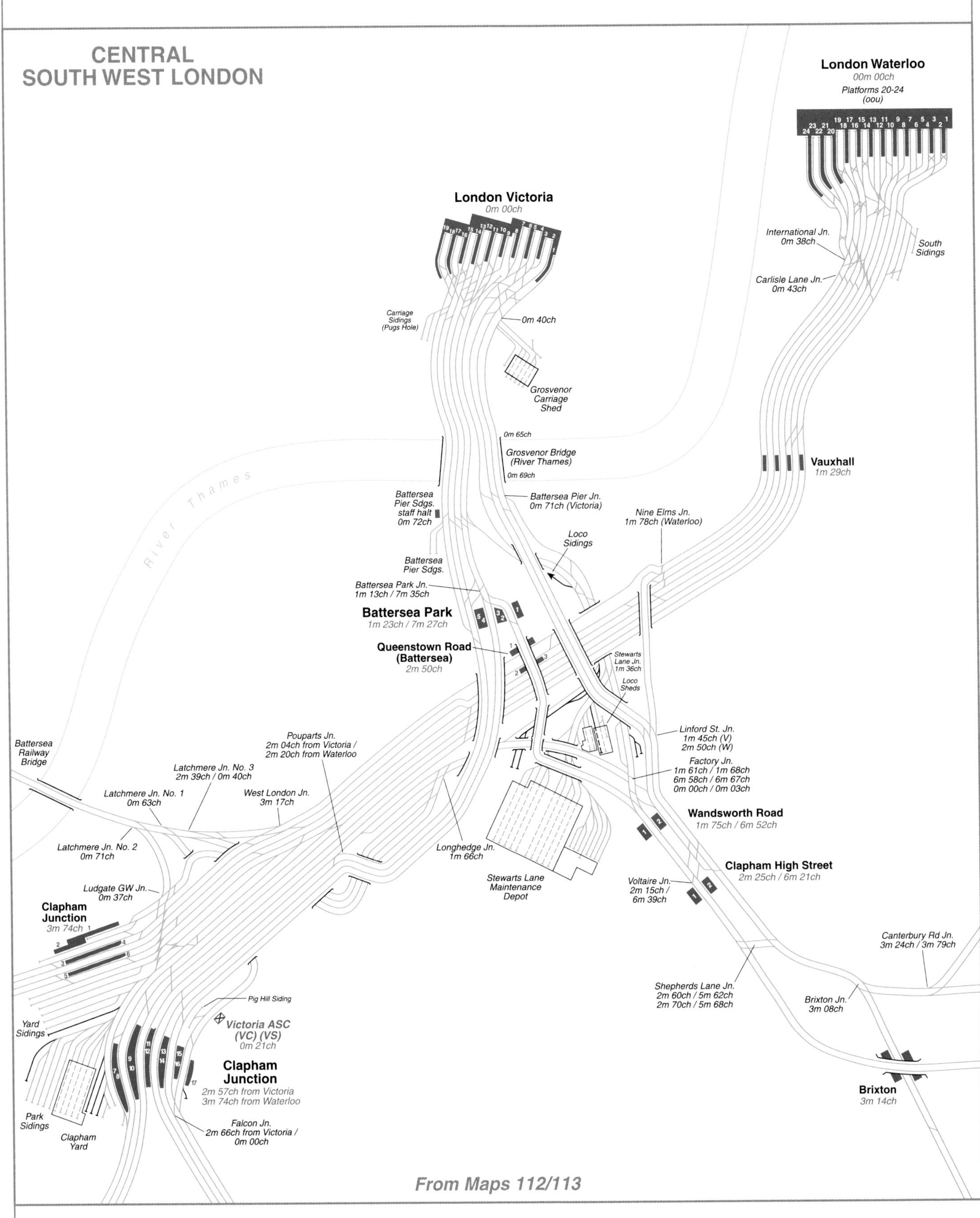

CENTRAL
SOUTH WEST LONDON
London Waterloo
00m 00ch
Platforms 20-24
(oou)
London Victoria
0m 00ch
International Jn.
0m 38ch
South Sidings
Carlisle Lane Jn.
0m 43ch
Carriage Sidings (Pugs Hole)
0m 40ch
Grosvenor Carriage Shed
0m 65ch
Grosvenor Bridge (River Thames)
0m 69ch
Vauxhall
1m 29ch
River Thames
Battersea Pier Jn.
0m 71ch (Victoria)
Battersea Pier Sdgs. staff halt
0m 72ch
Nine Elms Jn.
1m 78ch (Waterloo)
Loco Sidings
Battersea Pier Sdgs.
Battersea Park Jn.
1m 13ch / 7m 35ch
Battersea Park
1m 23ch / 7m 27ch
Queenstown Road (Battersea)
2m 50ch
Stewarts Lane Jn.
1m 36ch
Loco Sheds
Linford St. Jn.
1m 45ch (V)
2m 50ch (W)
Factory Jn.
1m 61ch / 1m 68ch
6m 58ch / 6m 67ch
0m 00ch / 0m 03ch
Battersea Railway Bridge
Pouparts Jn.
2m 04ch from Victoria /
2m 20ch from Waterloo
Latchmere Jn. No. 3
2m 39ch / 0m 40ch
Latchmere Jn. No. 1
0m 63ch
West London Jn.
3m 17ch
Wandsworth Road
1m 75ch / 6m 52ch
Latchmere Jn. No. 2
0m 71ch
Longhedge Jn.
1m 66ch
Clapham High Street
2m 25ch / 6m 21ch
Stewarts Lane Maintenance Depot
Voltaire Jn.
2m 15ch /
6m 39ch
Ludgate GW Jn.
0m 37ch
Clapham Junction
3m 74ch
Canterbury Rd Jn.
3m 24ch / 3m 79ch
Shepherds Lane Jn.
2m 60ch / 5m 62ch
2m 70ch / 5m 68ch
Brixton Jn.
3m 08ch
Pig Hill Siding
Yard Sidings
Victoria ASC
(VC) (VS)
0m 21ch
Clapham Junction
2m 57ch from Victoria
3m 74ch from Waterloo
Brixton
3m 14ch
Park Sidings
Clapham Yard
Falcon Jn.
2m 66ch from Victoria /
0m 00ch
From Maps 112/113

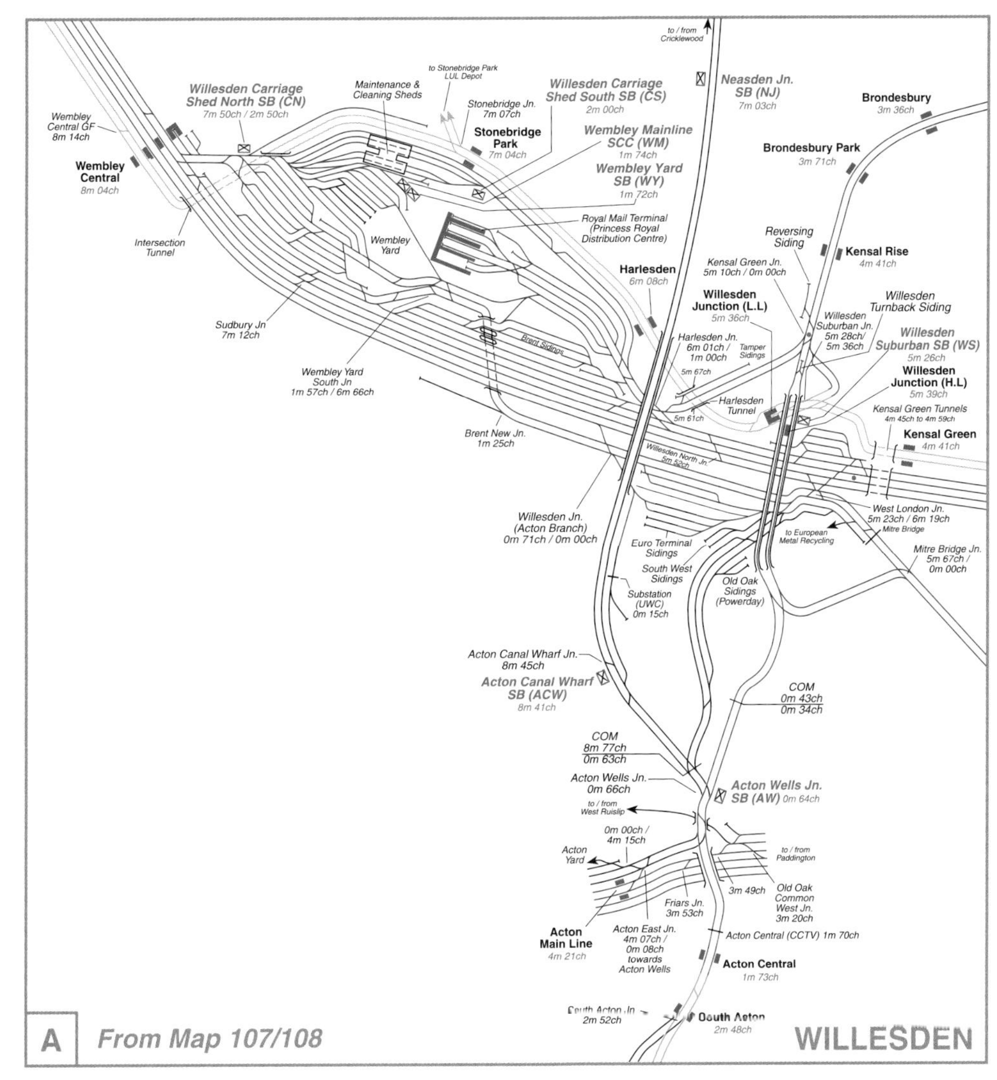

A
From Map 107/108
WILLESDEN
to / from Cricklewood
Neasden Jn. SB (NJ)
7m 03ch
Brondesbury
3m 36ch
Brondesbury Park
3m 71ch
Kensal Rise
4m 41ch
Willesden Carriage Shed North SB (CN)
7m 50ch / 2m 50ch
Maintenance & Cleaning Sheds
to Stonebridge Park LUL Depot
Stonebridge Jn.
7m 07ch
Stonebridge Park
7m 04ch
Willesden Carriage Shed South SB (CS)
2m 00ch
Wembley Mainline SCC (WM)
1m 74ch
Wembley Yard SB (WY)
1m 72ch
Wembley Central GF
8m 14ch
Wembley Central
8m 04ch
Intersection Tunnel
Royal Mail Terminal (Princess Royal Distribution Centre)
Wembley Yard
Harlesden
6m 08ch
Reversing Siding
Kensal Green Jn.
5m 10ch / 0m 00ch
Willesden Junction (L.L)
5m 36ch
Willesden Turnback Siding
Willesden Suburban Jn.
5m 28ch/ 5m 36ch
Willesden Suburban SB (WS)
5m 26ch
Willesden Junction (H.L)
5m 39ch
Sudbury Jn
7m 12ch
Brent Sidings
Wembley Yard South Jn
1m 57ch / 6m 66ch
Harlesden Jn.
6m 01ch / 1m 00ch
Tamper Sidings
5m 67ch
Harlesden Tunnel
5m 61ch
Kensal Green Tunnels
4m 45ch to 4m 59ch
Kensal Green
4m 41ch
Brent New Jn.
1m 25ch
Willesden North Jn.
5m 52ch
West London Jn.
5m 23ch / 6m 19ch
Mitre Bridge
Willesden Jn. (Acton Branch)
0m 71ch / 0m 00ch
Euro Terminal Sidings
to European Metal Recycling
Mitre Bridge Jn.
5m 67ch / 0m 00ch
South West Sidings
Old Oak Sidings (Powerday)
Substation (UWC)
0m 15ch
Acton Canal Wharf Jn.
8m 45ch
Acton Canal Wharf SB (ACW)
8m 41ch
COM
0m 43ch
0m 34ch
COM
8m 77ch
0m 63ch
Acton Wells Jn.
0m 66ch
Acton Wells Jn. SB (AW) 0m 64ch
to / from West Ruislip
0m 00ch / 4m 15ch
Acton Yard
to / from Paddington
3m 49ch
Old Oak Common West Jn.
3m 20ch
Friars Jn.
3m 53ch
Acton Main Line
4m 21ch
Acton East Jn.
4m 07ch / 0m 08ch towards Acton Wells
Acton Central (CCTV) 1m 70ch
Acton Central
1m 73ch
South Acton Jn.
2m 52ch
South Acton
2m 48ch

B
From Map 113
CENTRAL SOUTH EAST LONDON
City Thameslink
0m 28ch
Snow Hill Tunnel
Snow Hill Tunnel (South Portal)
0m 39ch
0m 21ch
Fenchurch Street
0m 00ch
East London Line
Christian Street Jn.
0m 61ch
Limehouse
1m 58ch
Shadwell
3m 18ch
River Thames
Cannon Street
0m 00ch / 1m 73ch from Charing Cross
Cannon Street Bridge
Blackfriars
0m 30ch
St. Pauls Bridge
Sidings
Wapping
3m 57ch
London Charing Cross
0m 00ch
Blackfriars Jn.
0m 63ch
Metropolitan Jn.
1m 31ch / 1m 14ch via Blackfriars
Borough Market Jn.
COM 0m 38ch / 1m 51ch
Rotherhithe
4m 03ch
Belvedere Rd Jn.
0m 45ch
London Bridge
1m 69ch from Charing Cross
Hungerford Bridge
Waterloo East
0m 61ch
2m 10ch
2m 14ch
Canada Water
4m 20ch
Union Street Jn.
0m 74ch
Southwark Bridge Jn.
1m 20ch
London Bridge
0m 00ch
London Bridge SB (L)
0m 13ch / 2m 02ch from Charing Cross
Surrey Quays Portal
4m 38ch
Surrey Quays
4m 46ch
Elephant & Castle
1m 47ch
Silwood Jn.
4m 66ch
Blue Anchor Jn. 3m 32ch (Deptford Line) / 1m 45ch (New Cross Gate Line)
Ⓐ East London Line SCC (EL)
5m 33ch
Ⓑ Canal Junction
5m 21ch
Ⓒ Rolt Street Junction
5m 45ch
Spa Road
2m 71ch (Deptford Line) / 1m 03ch (South Bermondsey Line) 1m 71ch (New Cross Line)
North Kent East Jn.
4m 25ch
Deptford
4m 76ch
South Bermondsey Jn.
1m 49ch
1m 55ch
1m 65ch
Bricklayers Arms Jn.
2m 12ch
South Bermondsey
1m 63ch
New Cross Gate Depot
4m 46ch
New Cross
4m 68ch
Old Kent Rd Jn.
2m 32ch
Queens Road Peckham
2m 58ch
New Cross Gate
2m 70ch
4m 60ch

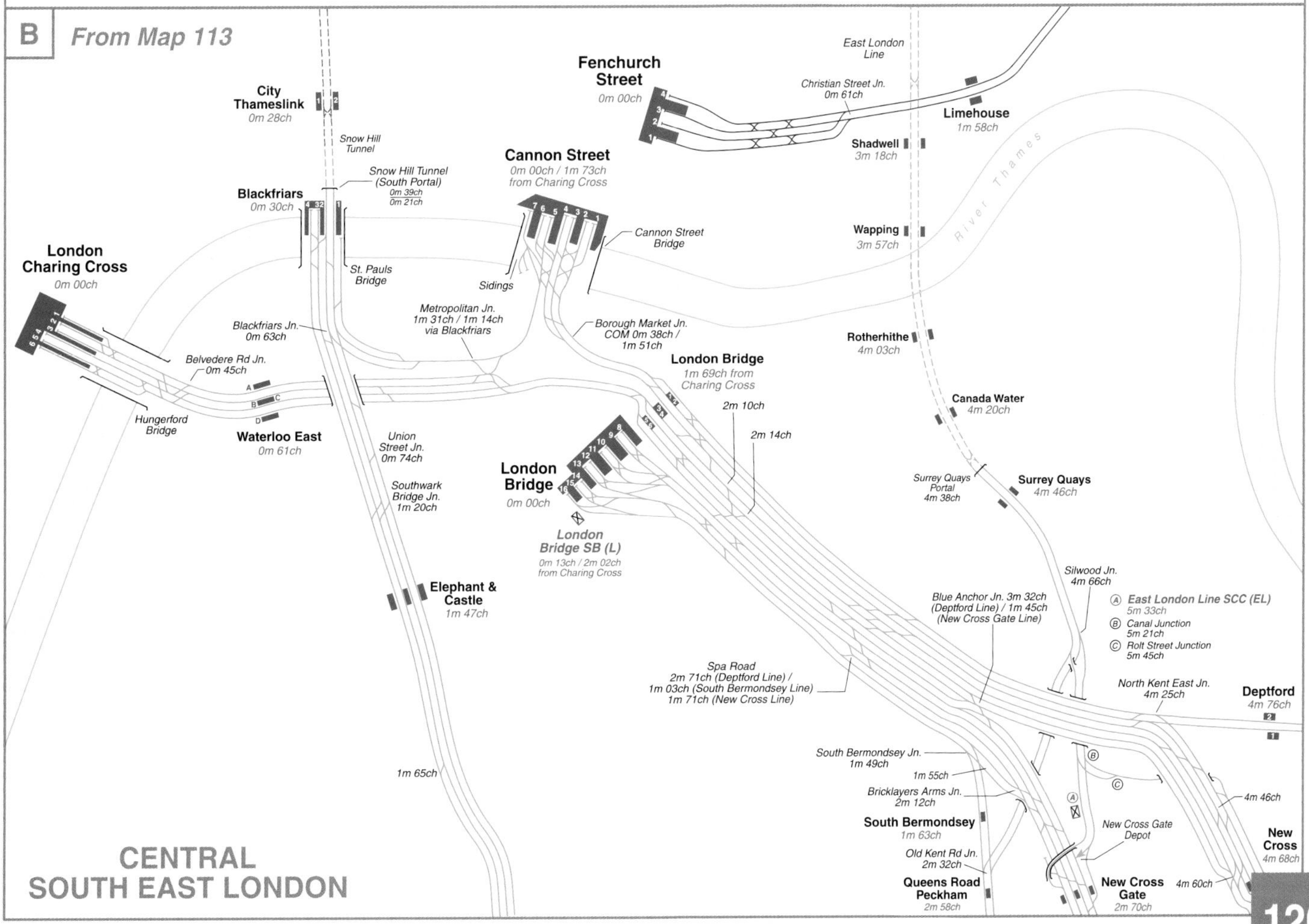

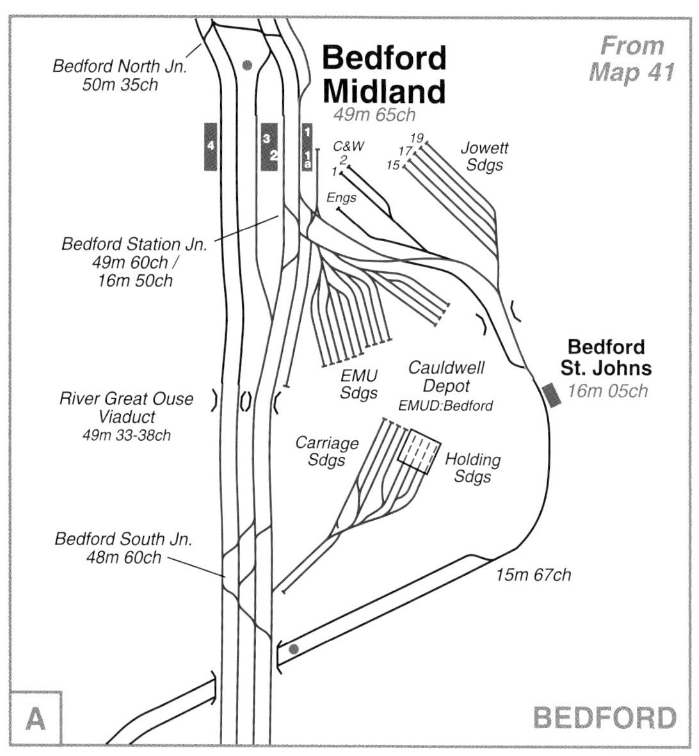
Bedford North Jn.
50m 35ch
Bedford
Midland
49m 65ch
From
Map 41
4
3
2
1
1a
C&W
2
1
Engs
19
17
15
Jowett
Sdgs
Bedford Station Jn.
49m 60ch /
16m 50ch
Bedford
St. Johns
16m 05ch
EMU
Sdgs
Cauldwell
Depot
EMUD:Bedford
River Great Ouse
Viaduct
49m 33-38ch
Carriage
Sdgs
Holding
Sdgs
Bedford South Jn.
48m 60ch
15m 67ch
A
BEDFORD

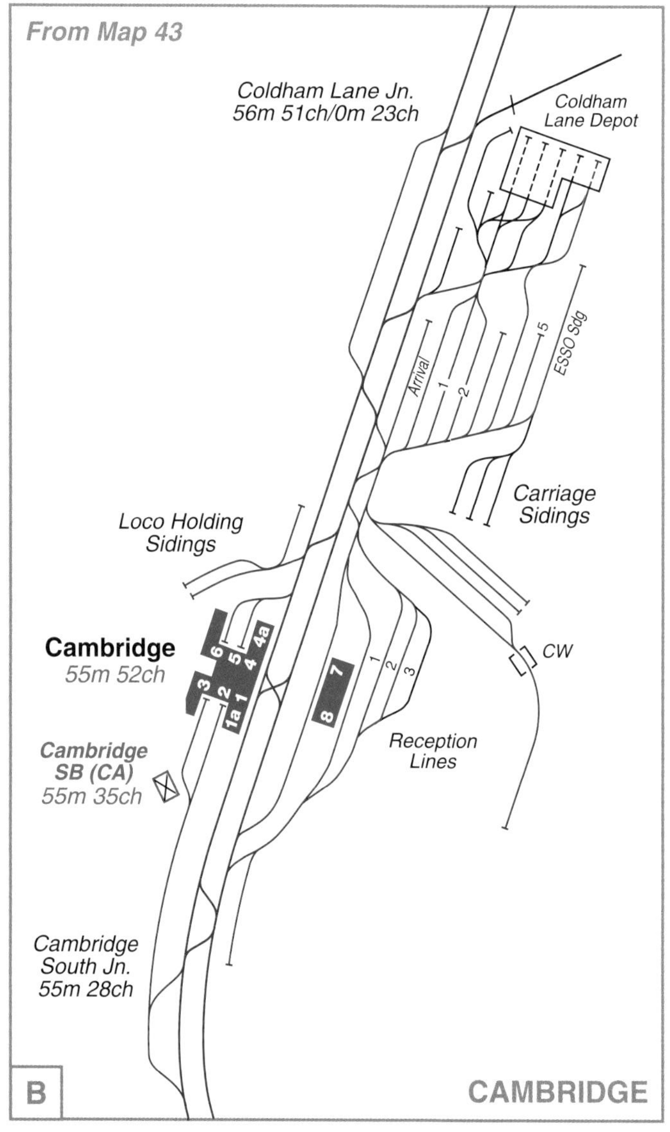
From Map 43
Coldham Lane Jn.
56m 51ch/0m 23ch
Coldham
Lane Depot
Arrival
1
2
5
ESSO Sdg
Carriage
Sidings
Loco Holding
Sidings
Cambridge
55m 52ch
6
5
4
4a
3
2
1a
1
7
8
1
2
3
CW
Reception
Lines
Cambridge
SB (CA)
55m 35ch
Cambridge
South Jn.
55m 28ch
B
CAMBRIDGE

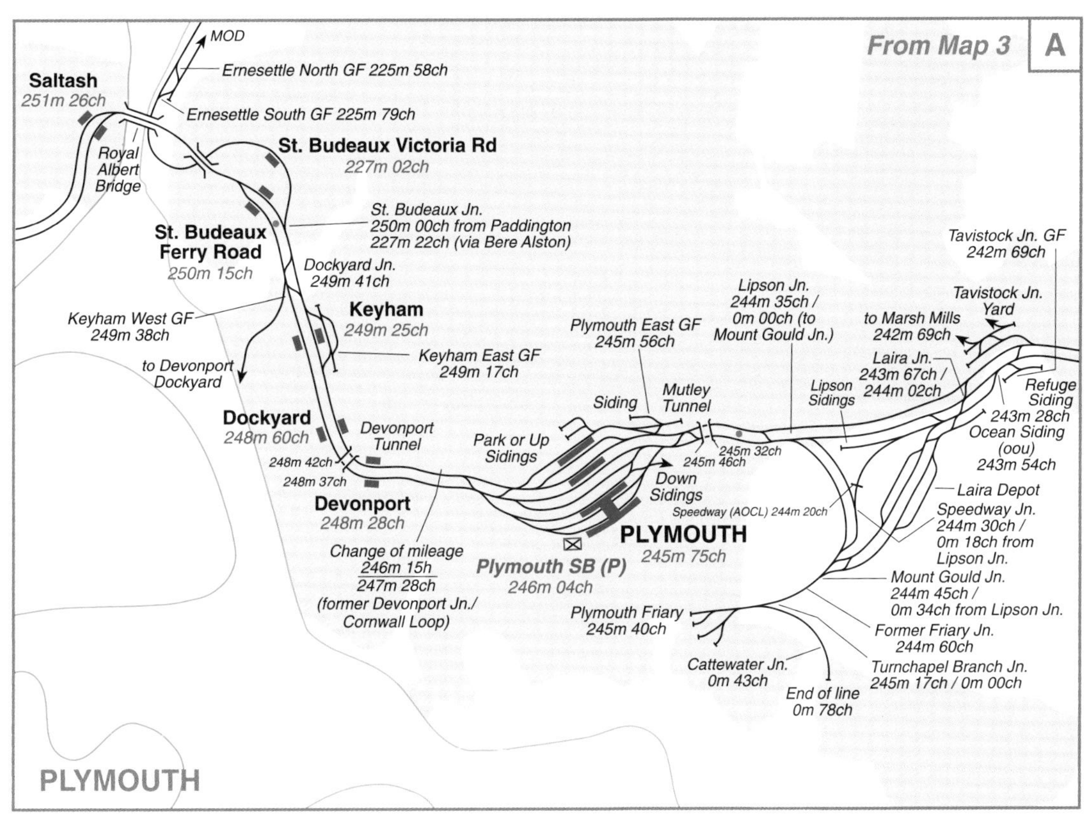

From Map 3
A
PLYMOUTH
MOD
Ernesettle North GF 225m 58ch
Ernesettle South GF 225m 79ch
Saltash
251m 26ch
Royal Albert Bridge
St. Budeaux Victoria Rd
227m 02ch
St. Budeaux Jn.
250m 00ch from Paddington
227m 22ch (via Bere Alston)
St. Budeaux Ferry Road
250m 15ch
Dockyard Jn.
249m 41ch
Keyham West GF
249m 38ch
Keyham
249m 25ch
Keyham East GF
249m 17ch
to Devonport Dockyard
Dockyard
248m 60ch
Devonport Tunnel
248m 42ch
248m 37ch
Devonport
248m 28ch
Change of mileage
246m 15h
247m 28ch
(former Devonport Jn./ Cornwall Loop)
Park or Up Sidings
Siding
Plymouth East GF
245m 56ch
Mutley Tunnel
245m 32ch
245m 46ch
Down Sidings
Speedway (AOCL) 244m 20ch
PLYMOUTH
245m 75ch
Plymouth SB (P)
246m 04ch
Plymouth Friary
245m 40ch
Cattewater Jn.
0m 43ch
End of line
0m 78ch
Lipson Jn.
244m 35ch /
0m 00ch (to Mount Gould Jn.)
Lipson Sidings
to Marsh Mills
242m 69ch
Tavistock Jn. GF
242m 69ch
Tavistock Jn. Yard
Laira Jn.
243m 67ch /
244m 02ch
Refuge Siding
243m 28ch
Ocean Siding (oou)
243m 54ch
Laira Depot
Speedway Jn.
244m 30ch /
0m 18ch from Lipson Jn.
Mount Gould Jn.
244m 45ch /
0m 34ch from Lipson Jn.
Former Friary Jn.
244m 60ch
Turnchapel Branch Jn.
245m 17ch / 0m 00ch

B
From Map 25
SWANSEA AND PORT TALBOT AREA
Tir Isaf (UWC) 2m 74ch
Glyn-y-Mul (UWC) 2m 52ch
5m 13ch
Llangyfelach Tunnel
4m 04ch
1m 50ch
Llansamlet
212m 08ch
Lonlas Tunnel
Skewen
210m 26ch
Neath and Brecon Jn.
41m 17ch /
0m 01ch
Neath and Brecon Jn. SB (NB)
41m 21ch
Barrow crossing (WL) 208m 26ch
Neath (Castell-Nedd)
208m 20ch
Jersey Marine Jn. North
208m 33ch / 1m 24ch
1m 08ch
Llandarcy GF
0m 37ch
BP
Change of mileage
208m 49ch /
0m 00ch
Dynevor GF 19m 36ch
Dynevor Jn.
207m 67ch / 19m 16ch
Penrhiwtyn (UWC) 207m 20ch
Court Sart Jn.
206m 58ch
Briton Ferry
206m 40ch
Briton Ferry Sidings
Briton Ferry Up Flying Loop Jn.
206m 14ch
Briton Ferry East
205m 29ch
Baglan
204m 53ch
Landore Viaduct
214m 39ch to 214m 57ch
Swansea Loop West Jn.
215m 14ch /
0m 00ch
Landore Depot
Landore Jn.
214m 62ch
216m 64ch
216m 28ch
Cockett Tunnel
Single Line Jn.
0m 48ch
Swansea Loop East Jn.
215m 43ch / 0m 53ch
Carriage Sidings
Down Sidings GF
215m 62ch
Jersey Marine Jn. South
2m 26ch / 44m 17ch
20m 24ch / 44m 19ch
Swansea Docks (Associated British Ports)
46m 30ch
Burrows Sidings
SWANSEA (Abertawe)
216m 07ch
Port Talbot Parkway
202m 59ch
Port Talbot SB (PT)
202m 49ch
Port Talbot (MCB)
202m 50ch
Port Talbot East (Taibach)
202m 10ch
Margam Middle Jn
201m 20ch
200m 63ch
0m 79ch
Margam Yard Jn

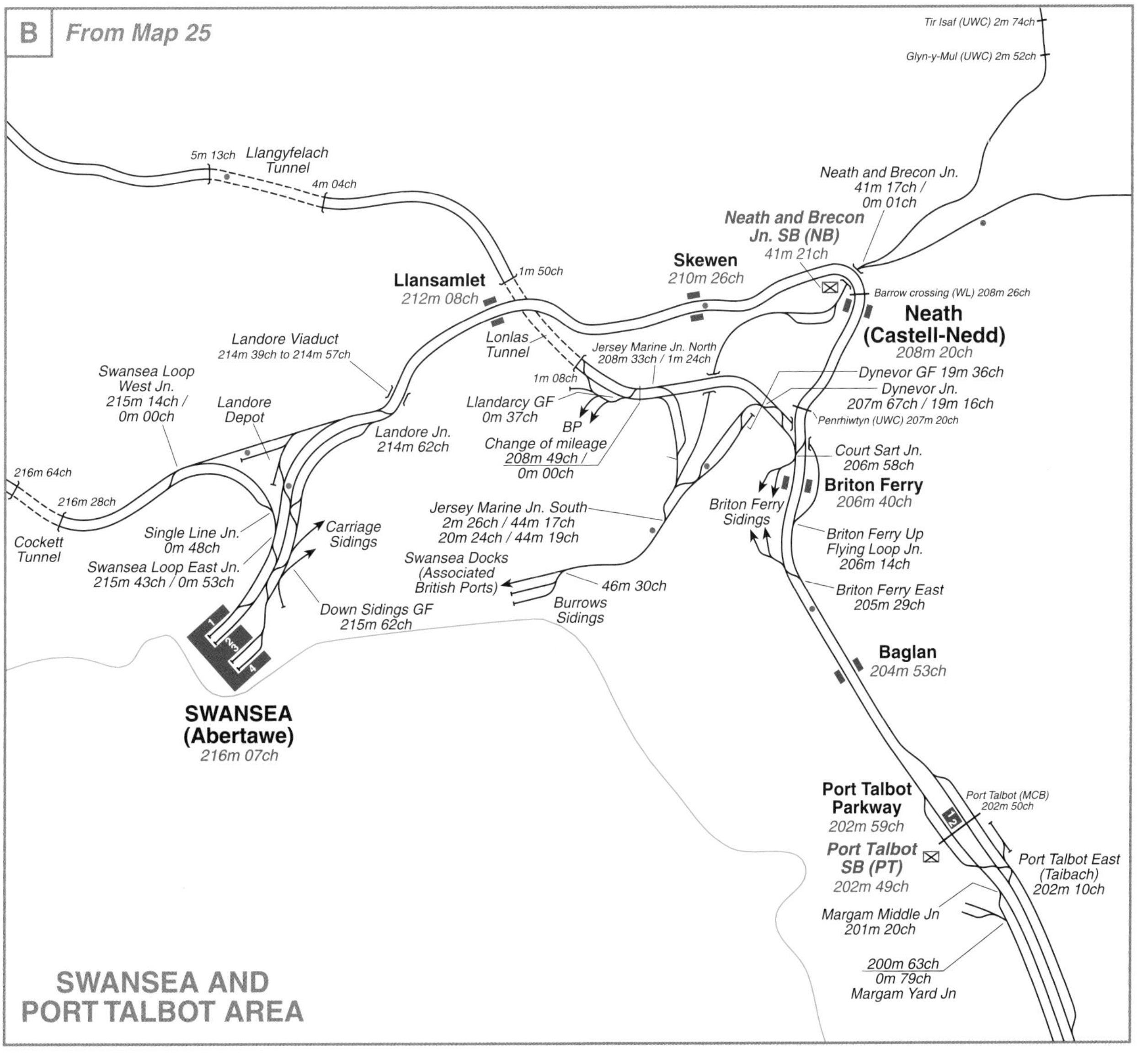

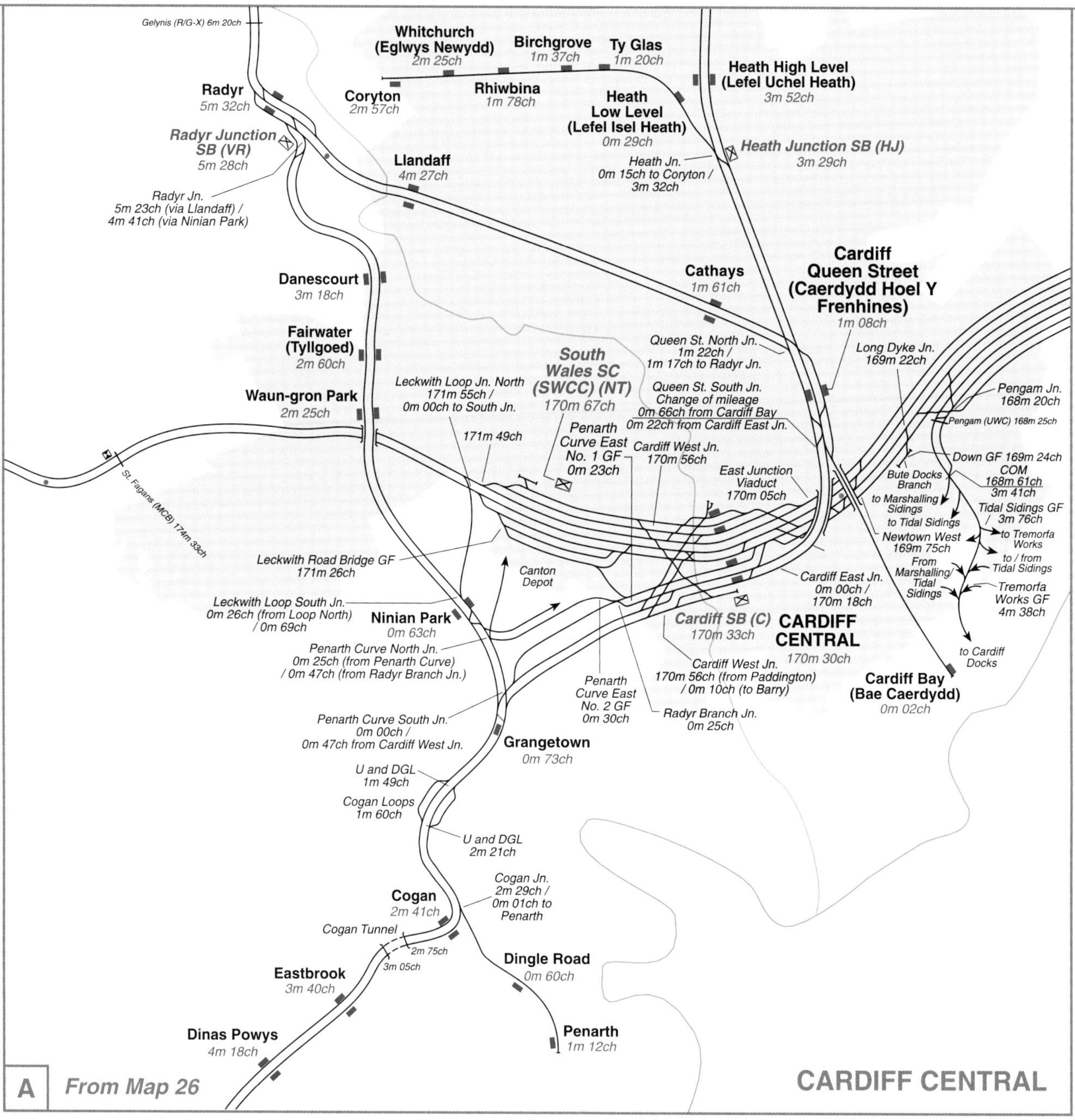

Gelynis (R/G-X) 6m 20ch
Whitchurch (Eglwys Newydd) 2m 25ch
Birchgrove 1m 37ch
Ty Glas 1m 20ch
Heath High Level (Lefel Uchel Heath) 3m 52ch
Radyr 5m 32ch
Coryton 2m 57ch
Rhiwbina 1m 78ch
Heath Low Level (Lefel Isel Heath) 0m 29ch
Radyr Junction SB (VR) 5m 28ch
Heath Junction SB (HJ) 3m 29ch
Llandaff 4m 27ch
Heath Jn. 0m 15ch to Coryton / 3m 32ch
Radyr Jn. 5m 23ch (via Llandaff) / 4m 41ch (via Ninian Park)
Cardiff Queen Street (Caerdydd Hoel Y Frenhines) 1m 08ch
Danescourt 3m 18ch
Cathays 1m 61ch
Fairwater (Tyllgoed) 2m 60ch
Queen St. North Jn. 1m 22ch / 1m 17ch to Radyr Jn.
Long Dyke Jn. 169m 22ch
South Wales SC (SWCC) (NT) 170m 67ch
Leckwith Loop Jn. North 171m 55ch / 0m 00ch to South Jn.
Queen St. South Jn. Change of mileage 0m 66ch from Cardiff Bay 0m 22ch from Cardiff East Jn.
Waun-gron Park 2m 25ch
Pengam Jn. 168m 20ch
Pengam (UWC) 168m 25ch
171m 49ch
Penarth Curve East No. 1 GF 0m 23ch
Cardiff West Jn. 170m 56ch
Down GF 169m 24ch
COM 168m 61ch
3m 41ch
St. Fagans (MCB) 174m 33ch
East Junction Viaduct 170m 05ch
Bute Docks Branch
to Marshalling Sidings
Tidal Sidings GF 3m 76ch
to Tidal Sidings
to Tremorfa Works
Newtown West 169m 75ch
to / from Tidal Sidings
From Marshalling/ Tidal Sidings
Tremorfa Works GF 4m 38ch
Leckwith Road Bridge GF 171m 26ch
Canton Depot
Cardiff East Jn. 0m 00ch / 170m 18ch
Leckwith Loop South Jn. 0m 26ch (from Loop North) / 0m 69ch
Ninian Park 0m 63ch
Cardiff SB (C) 170m 33ch
CARDIFF CENTRAL 170m 30ch
Penarth Curve North Jn. 0m 25ch (from Penarth Curve) / 0m 47ch (from Radyr Branch Jn.)
Cardiff West Jn. 170m 56ch (from Paddington) / 0m 10ch (to Barry)
to Cardiff Docks
Penarth Curve East No. 2 GF 0m 30ch
Cardiff Bay (Bae Caerdydd) 0m 02ch
Radyr Branch Jn. 0m 25ch
Penarth Curve South Jn. 0m 00ch / 0m 47ch from Cardiff West Jn.
Grangetown 0m 73ch
U and DGL 1m 49ch
Cogan Loops 1m 60ch
U and DGL 2m 21ch
Cogan Jn. 2m 29ch / 0m 01ch to Penarth
Cogan 2m 41ch
Cogan Tunnel
2m 75ch
3m 05ch
Eastbrook 3m 40ch
Dingle Road 0m 60ch
Dinas Powys 4m 18ch
Penarth 1m 12ch
A
From Map 26
CARDIFF CENTRAL

B
From Map 26
Maindee West Jn. 158m 16ch / 41m 66ch (Cwmbran Line)
End of Line 4m 69ch
Lime Kiln (CCTV) 6m 15ch
Risca South Jn. 4m 09ch
NEWPORT (Casnewydd) 158m 50ch
Maindee North Jn. 41m 03ch
Inlet GF 4m 45ch
Risca & Pontymister 5m 12ch
Rogerstone 3m 61ch
Gaer Jn. 159m 33ch
Tamper Sdg
Maindee East Jn. 157m 74ch / 41m 65ch (Cwmbran Line)
Machen Quarry 4m 19ch
156m 64ch
Outlet GF 3m 50ch
Machen Fach Farm (UWC) 2m 68ch
Keepers (UWC) 2m 49ch
Rhiwderin (AOCL) 1m 35ch
Park North Jn. 1m 45ch
Courtybella
157m 59ch
former Bassaleg Jn.COM 0m 00ch / 2m 05ch
0m 00ch
Low Level Sdgs
to Isis Link Orb Works
East Mendalgief 160m 24ch
Ord Works GF 1m 14ch
Park Jn. SB (N) 160m 06ch / 0m 74ch / 0m 54ch
Port Boundary 160m 27ch
Alpha Steel GF 2m 23ch
to Alpha Steel
Alexandra Dock Jn. 159m 60ch
to Fifoots Power Station / Uskmouth
Ebbw Jn. 160m 07ch /0m 00ch to Park Jn.
to Newport Docks
Down Sidings
Monsanto GF 2m 42ch
Rumney River Bridge Jn. 167m 61ch
Newport SB (N) 158m 36ch
① Newport Tunnel (New) 158m 71ch to 159m 25ch
② Newport Tunnel (Old) 158m 70ch to 159m 25ch
③ Gaer Tunnel 159m 47ch to 159m 65ch
④ Park Jn. 160m 24ch (from Paddington via Gaer Jn.) / 1m 20ch (towards Rogerstone)
⑤ Birdport Rail Terminal 2m 18ch to 2m 19ch
167m 40ch
Foot crossing (WL) 165m 19ch
Wentloog Freight Terminal East Jn. 165m 22ch
Wentloog Freight Terminal West Jn. 166m 01ch
No.1, No.2, No.3 Up / Down Reception 167m 74ch
NEWPORT

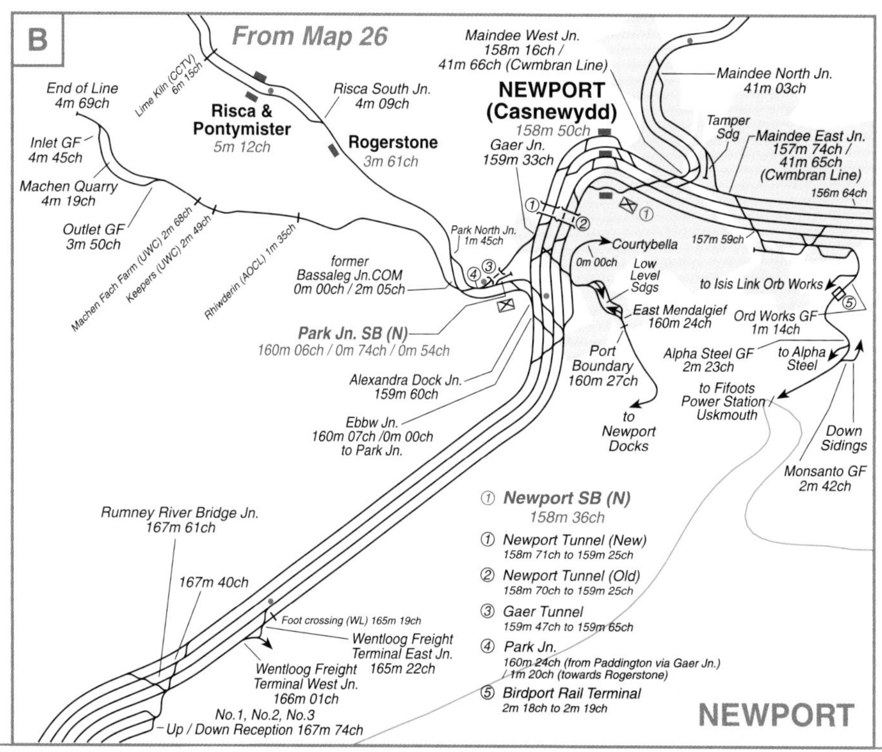

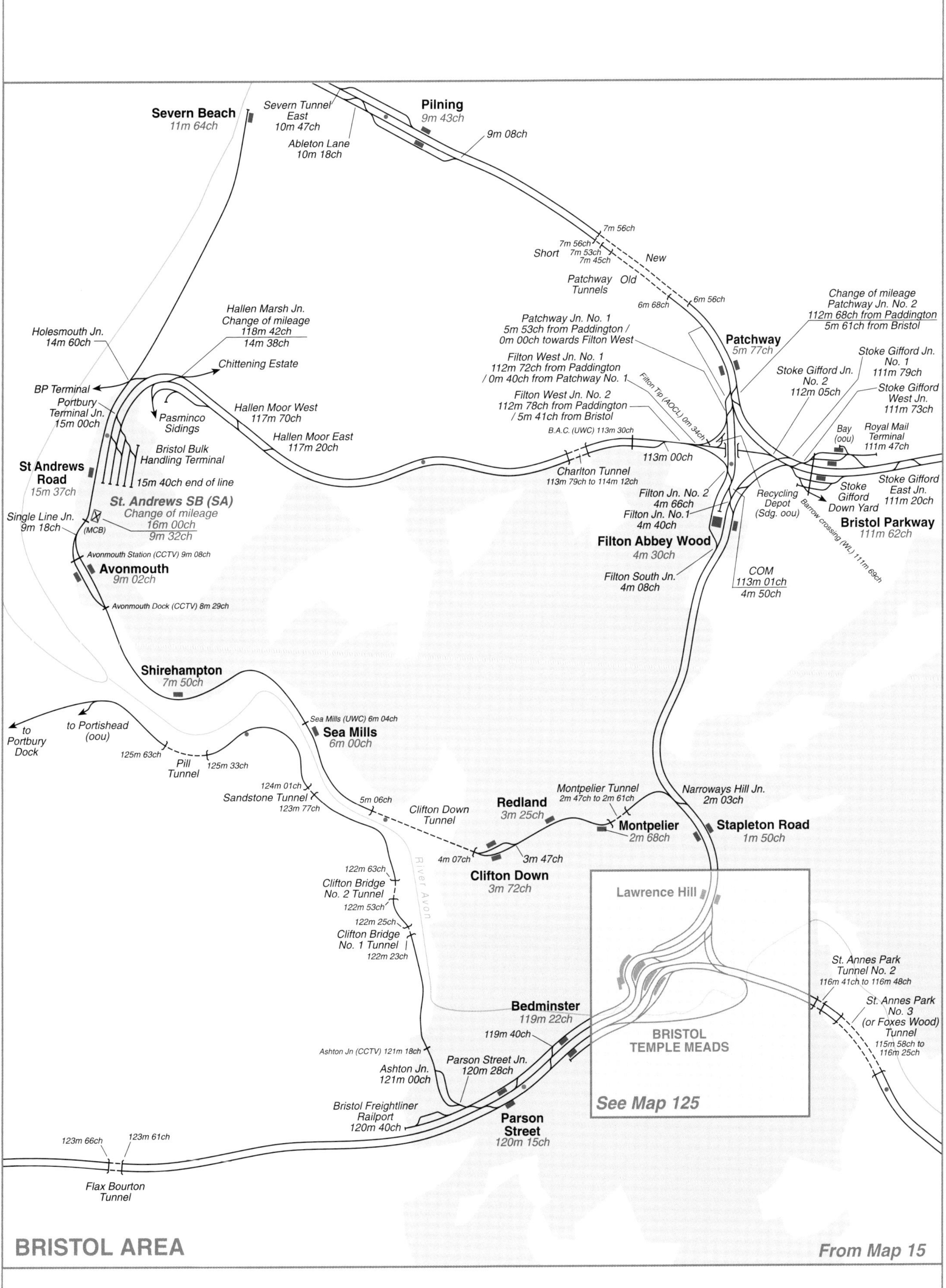

Severn Beach
11m 64ch
Severn Tunnel East
10m 47ch
Ableton Lane
10m 18ch
Pilning
9m 43ch
9m 08ch
7m 56ch
7m 56ch
7m 53ch
7m 45ch
Short
New
Patchway Old
Tunnels
6m 68ch
6m 56ch
Change of mileage
Patchway Jn. No. 2
112m 68ch from Paddington
5m 61ch from Bristol
Patchway Jn. No. 1
5m 53ch from Paddington /
0m 00ch towards Filton West
Patchway
5m 77ch
Filton West Jn. No. 1
112m 72ch from Paddington
/ 0m 40ch from Patchway No. 1
Filton West Jn. No. 2
112m 78ch from Paddington
/ 5m 41ch from Bristol
Filton Tip (AOCL) 0m 34ch
B.A.C. (UWC) 113m 30ch
Stoke Gifford Jn. No. 1
111m 79ch
Stoke Gifford Jn. No. 2
112m 05ch
Stoke Gifford West Jn.
111m 73ch
Bay (oou)
Royal Mail Terminal
111m 47ch
Stoke Gifford Down Yard
Stoke Gifford East Jn.
111m 20ch
Bristol Parkway
111m 62ch
Barrow crossing (WL) 111m 69ch
Recycling Depot (Sdg. oou)
Holesmouth Jn.
14m 60ch
Hallen Marsh Jn.
Change of mileage
118m 42ch
14m 38ch
Chittening Estate
BP Terminal
Portbury Terminal Jn.
15m 00ch
Pasminco Sidings
Hallen Moor West
117m 70ch
Hallen Moor East
117m 20ch
Bristol Bulk Handling Terminal
15m 40ch end of line
St Andrews Road
15m 37ch
St. Andrews SB (SA)
Change of mileage
16m 00ch
9m 32ch
Single Line Jn.
9m 18ch
(MCB)
Avonmouth Station (CCTV) 9m 08ch
Avonmouth
9m 02ch
Avonmouth Dock (CCTV) 8m 29ch
Charlton Tunnel
113m 79ch to 114m 12ch
113m 00ch
Filton Jn. No. 2
4m 66ch
Filton Jn. No. 1
4m 40ch
Filton Abbey Wood
4m 30ch
Filton South Jn.
4m 08ch
COM
113m 01ch
4m 50ch
Shirehampton
7m 50ch
to Portbury Dock
to Portishead (oou)
125m 63ch
Pill Tunnel
125m 33ch
Sea Mills (UWC) 6m 04ch
Sea Mills
6m 00ch
124m 01ch
Sandstone Tunnel
123m 77ch
5m 06ch
Clifton Down Tunnel
Redland
3m 25ch
Montpelier Tunnel
2m 47ch to 2m 61ch
Montpelier
2m 68ch
Narroways Hill Jn.
2m 03ch
Stapleton Road
1m 50ch
4m 07ch
3m 47ch
Clifton Down
3m 72ch
River Avon
122m 63ch
Clifton Bridge No. 2 Tunnel
122m 53ch
122m 25ch
Clifton Bridge No. 1 Tunnel
122m 23ch
Lawrence Hill
St. Annes Park Tunnel No. 2
116m 41ch to 116m 48ch
St. Annes Park No. 3 (or Foxes Wood) Tunnel
115m 58ch to 116m 25ch
Bedminster
119m 22ch
119m 40ch
BRISTOL TEMPLE MEADS
Ashton Jn (CCTV) 121m 18ch
Ashton Jn.
121m 00ch
Parson Street Jn.
120m 28ch
See Map 125
Bristol Freightliner Railport
120m 40ch
Parson Street
120m 15ch
123m 66ch
123m 61ch
Flax Bourton Tunnel
BRISTOL AREA
From Map 15

READING
River Thames
Final platform layout (2015)
IEP Depot
READING
35m 78ch
Reading East Jn.
14
15
12
13
10
11
8
9
3
7
2
6
5
1
4
Westbury Line Jn. 36m 17ch
Scours Lane
37m 60ch
Future Depot under construction
Reading SB (R)
36m 01ch
READING
35m 78ch
Reading West Jn.
36m 76ch /
0m 02ch
Reading New Jn.
35m 40ch /
68m 35ch
No. 1 and No.2 Engine Sidings
Depot
Engine Sidings
Reading Triangle Sidings
36m 40ch
Westbury Line Jn.
36m 17ch
Oxford Road Jn.
36m 67ch /
0m 43ch
Reading West
36m 75ch
A
From Map 18

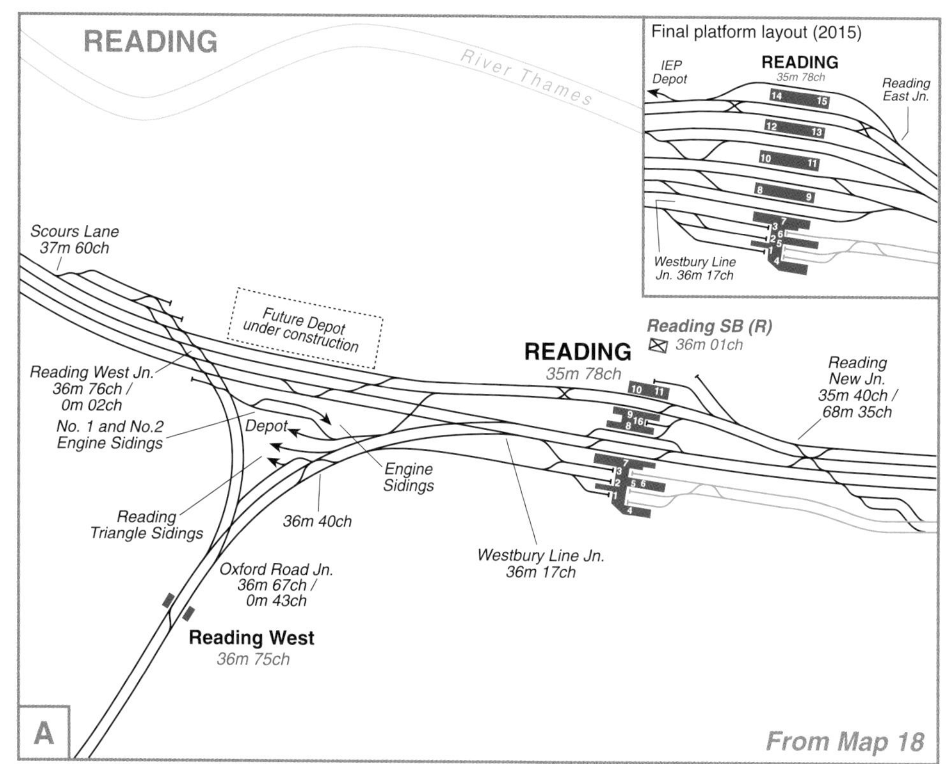

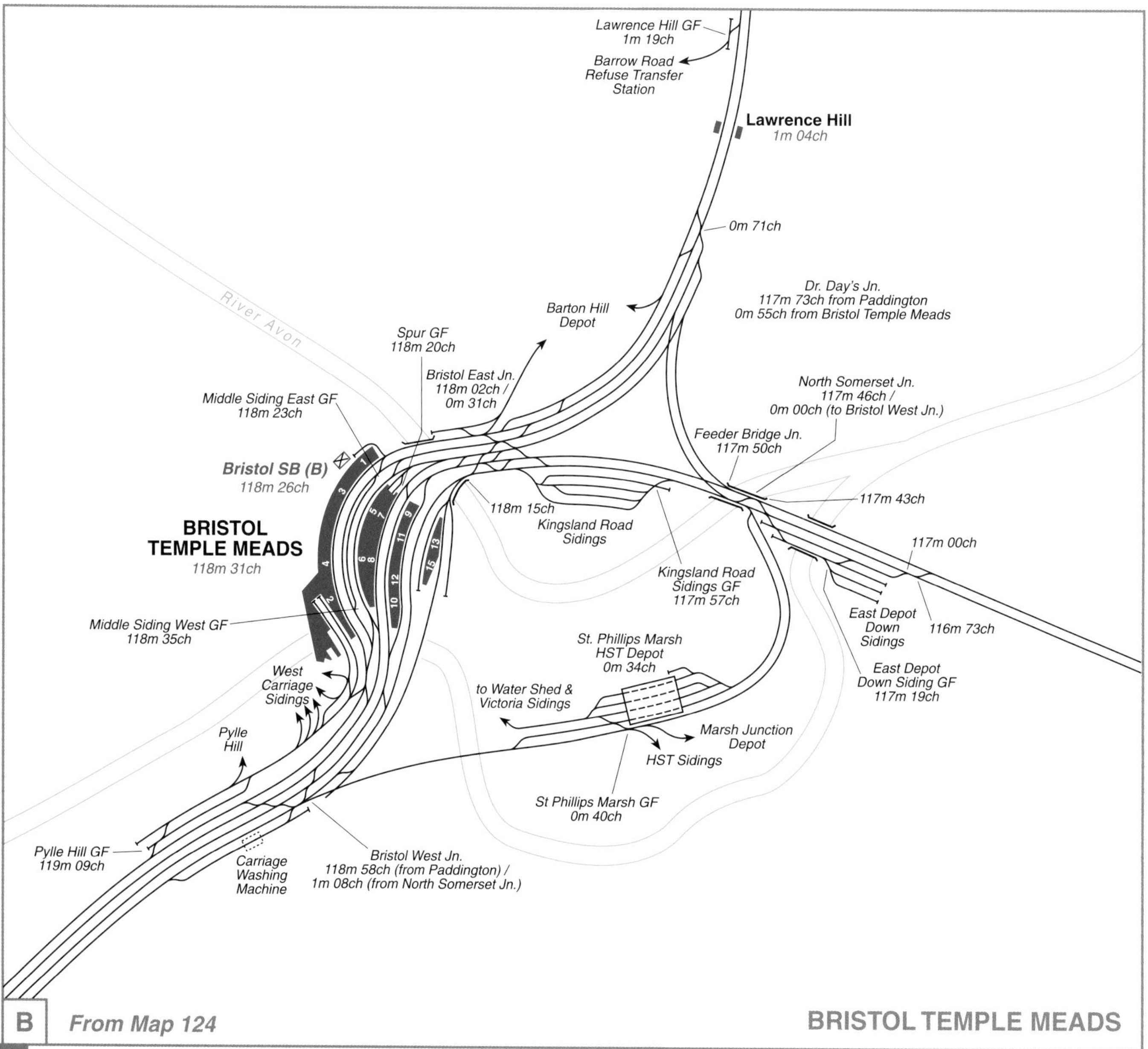

Lawrence Hill GF
1m 19ch
Barrow Road Refuse Transfer Station
Lawrence Hill
1m 04ch
0m 71ch
Dr. Day's Jn.
117m 73ch from Paddington
0m 55ch from Bristol Temple Meads
Barton Hill Depot
River Avon
Spur GF
118m 20ch
Bristol East Jn.
118m 02ch /
0m 31ch
Middle Siding East GF
118m 23ch
North Somerset Jn.
117m 46ch /
0m 00ch (to Bristol West Jn.)
Feeder Bridge Jn.
117m 50ch
Bristol SB (B)
118m 26ch
117m 43ch
118m 15ch
Kingsland Road Sidings
BRISTOL TEMPLE MEADS
118m 31ch
117m 00ch
Kingsland Road Sidings GF
117m 57ch
East Depot Down Sidings
116m 73ch
Middle Siding West GF
118m 35ch
St. Phillips Marsh HST Depot
0m 34ch
East Depot Down Siding GF
117m 19ch
West Carriage Sidings
to Water Shed & Victoria Sidings
Marsh Junction Depot
HST Sidings
Pylle Hill
St Phillips Marsh GF
0m 40ch
Pylle Hill GF
119m 09ch
Carriage Washing Machine
Bristol West Jn.
118m 58ch (from Paddington) /
1m 08ch (from North Somerset Jn.)
B
From Map 124
BRISTOL TEMPLE MEADS

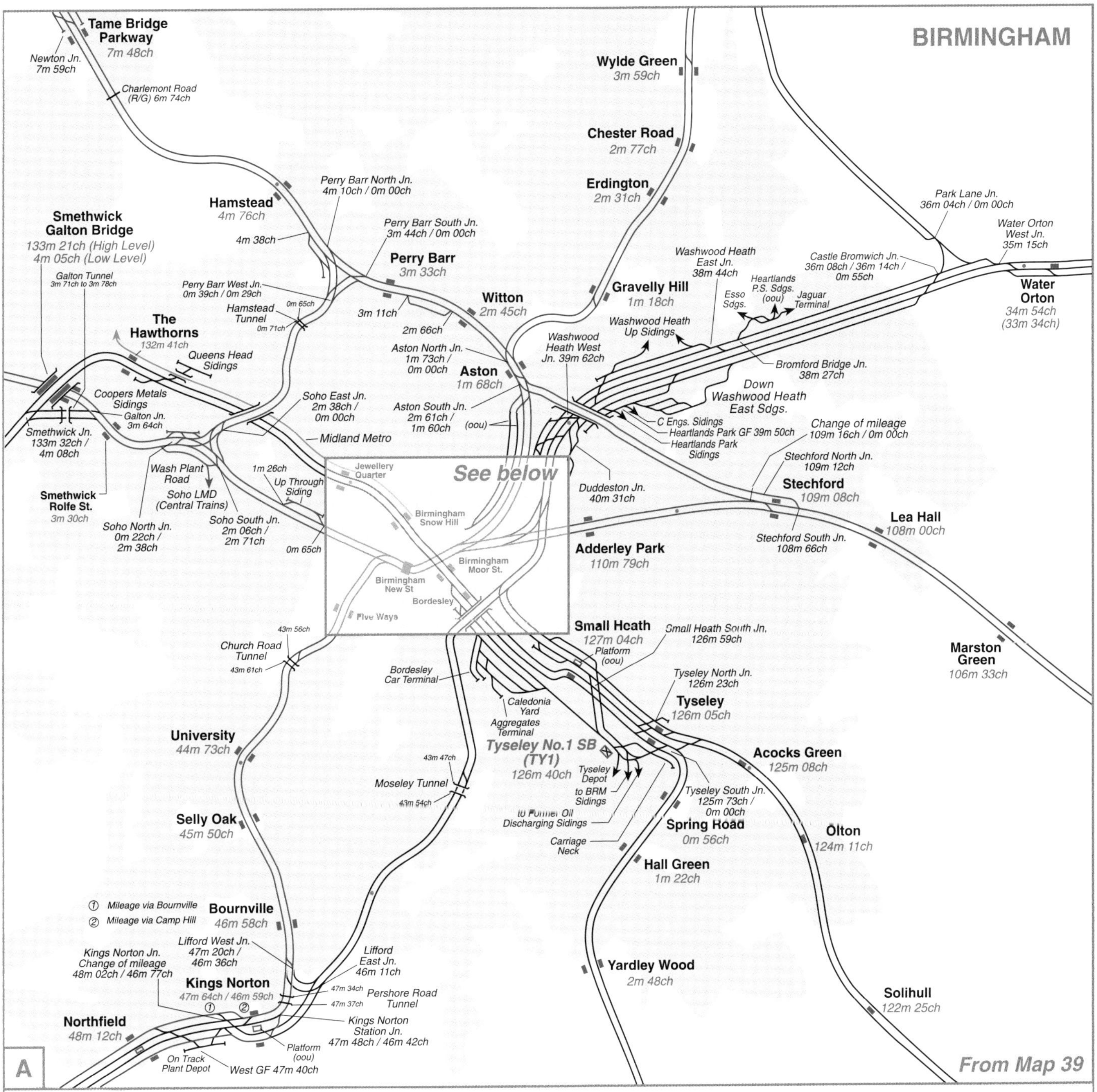

BIRMINGHAM
Tame Bridge Parkway
7m 48ch
Newton Jn.
7m 59ch
Charlemont Road
(R/G) 6m 74ch
Hamstead
4m 76ch
Perry Barr North Jn.
4m 10ch / 0m 00ch
Smethwick Galton Bridge
133m 21ch (High Level)
4m 05ch (Low Level)
Galton Tunnel
3m 71ch to 3m 78ch
4m 38ch
Perry Barr South Jn.
3m 44ch / 0m 00ch
Perry Barr
3m 33ch
Perry Barr West Jn.
0m 39ch / 0m 29ch
Hamstead Tunnel
0m 65ch
0m 71ch
3m 11ch
2m 66ch
Witton
2m 45ch
The Hawthorns
132m 41ch
Queens Head Sidings
Aston North Jn.
1m 73ch /
0m 00ch
Aston
1m 68ch
Coopers Metals Sidings
Galton Jn.
3m 64ch
Soho East Jn.
2m 38ch /
0m 00ch
Aston South Jn.
2m 61ch /
1m 60ch
(oou)
Smethwick Jn.
133m 32ch /
4m 08ch
Midland Metro
Wash Plant Road
1m 26ch
Up Through Siding
Soho LMD
(Central Trains)
Smethwick Rolfe St.
3m 30ch
Soho North Jn.
0m 22ch /
2m 38ch
Soho South Jn.
2m 06ch /
2m 71ch
0m 65ch
Wylde Green
3m 59ch
Chester Road
2m 77ch
Erdington
2m 31ch
Gravelly Hill
1m 18ch
Washwood Heath East Jn.
38m 44ch
Heartlands P.S. Sdgs.
(oou)
Esso Sdgs.
Jaguar Terminal
Washwood Heath Up Sidings
Washwood Heath West Jn. 39m 62ch
Bromford Bridge Jn.
38m 27ch
Down Washwood Heath East Sdgs.
C Engs. Sidings
Heartlands Park GF 39m 50ch
Heartlands Park Sidings
Park Lane Jn.
36m 04ch / 0m 00ch
Water Orton West Jn.
35m 15ch
Castle Bromwich Jn.
36m 08ch / 36m 14ch /
0m 55ch
Water Orton
34m 54ch
(33m 34ch)
Change of mileage
109m 16ch / 0m 00ch
Stechford North Jn.
109m 12ch
Stechford
109m 08ch
Lea Hall
108m 00ch
Stechford South Jn.
108m 66ch
Jewellery Quarter
See below
Birmingham Snow Hill
Birmingham Moor St.
Birmingham New St
Bordesley
Five Ways
Duddeston Jn.
40m 31ch
Adderley Park
110m 79ch
Small Heath
127m 04ch
Platform
(oou)
Small Heath South Jn.
126m 59ch
Marston Green
106m 33ch
43m 56ch
Church Road Tunnel
43m 61ch
Bordesley Car Terminal
Tyseley North Jn.
126m 23ch
Caledonia Yard
Aggregates Terminal
Tyseley
126m 05ch
University
44m 73ch
Tyseley No.1 SB (TY1)
126m 40ch
Acocks Green
125m 08ch
43m 47ch
Moseley Tunnel
43m 54ch
Tyseley Depot
to BRM Sidings
to Former Oil Discharging Sidings
Tyseley South Jn.
125m 73ch /
0m 00ch
Selly Oak
45m 50ch
Spring Road
0m 56ch
Olton
124m 11ch
Carriage Neck
Hall Green
1m 22ch
① Mileage via Bournville
② Mileage via Camp Hill
Bournville
46m 58ch
Lifford West Jn.
47m 20ch /
46m 36ch
Kings Norton Jn.
Change of mileage
48m 02ch / 46m 77ch
Lifford East Jn.
46m 11ch
Yardley Wood
2m 48ch
Kings Norton
47m 64ch / 46m 59ch
47m 34ch
47m 37ch
Pershore Road Tunnel
Solihull
122m 25ch
Northfield
48m 12ch
Kings Norton Station Jn.
47m 48ch / 46m 42ch
Platform
(oou)
On Track Plant Depot
West GF 47m 40ch
A
From Map 39

BIRMINGHAM CENTRE
Jewellery Quarter
130m 04ch
130m 02ch
129m 75ch
129m 72ch
Hockley No. 2 Tunnel
129m 66ch
Hockley No.1 Tunnel
Midland Metro
West Midlands SC (#)
40m 38ch
Duddeston
0m 53ch
Saltley PSB (SY)
40m 48ch
(oou)
Lawley Street Sidings
(Freightliner)
to EMR Sidings
to former Saltley Depot
Aston SB (AN)
0m 37ch
Vauxhall Jn.
0m 31ch
Landor St. Jn.
40m 60ch / 40m 62ch
Sidings
(oou)
Birmingham Snow Hill
129m 36ch
Curzon St. Jn.
112m 07ch /
0m 00ch
Grand Jn.
111m 72ch / 41m 26ch
0m 52ch
BIRMINGHAM NEW ST.
112m 73ch /
0m 05ch
Down Sidings
129m 21ch
Snow Hill Tunnel
New St. South Tunnel
Arena Tunnel
New St. North Tunnel
0m 61ch
0m 53ch
0m 19ch
112m 47ch
St. Andrews Jn.
41m 18ch /
0m 00ch
Holliday St. Tunnel
42m 40ch
Canal Tunnel
42m 57ch
42m 67ch
42m 68ch
42m 72ch
42m 54ch
Proof House Jn
112m 19ch / 41m 51ch
1
4
5
8
9
12
(oou)
128m 72ch
112m 58ch
42m 78ch
Granville St. Tunnel
43m 07ch
Bath Row Tunnel
Engine Sidings
Birmingham New St. PSB (NS)
0m 16ch / 42m 35ch
Birmingham Moor St.
128m 66ch
128m 35ch
Bordesley
128m 03ch
Bordesley Jn.
41m 44ch /
128m 11ch
Five Ways
43m 18ch
B

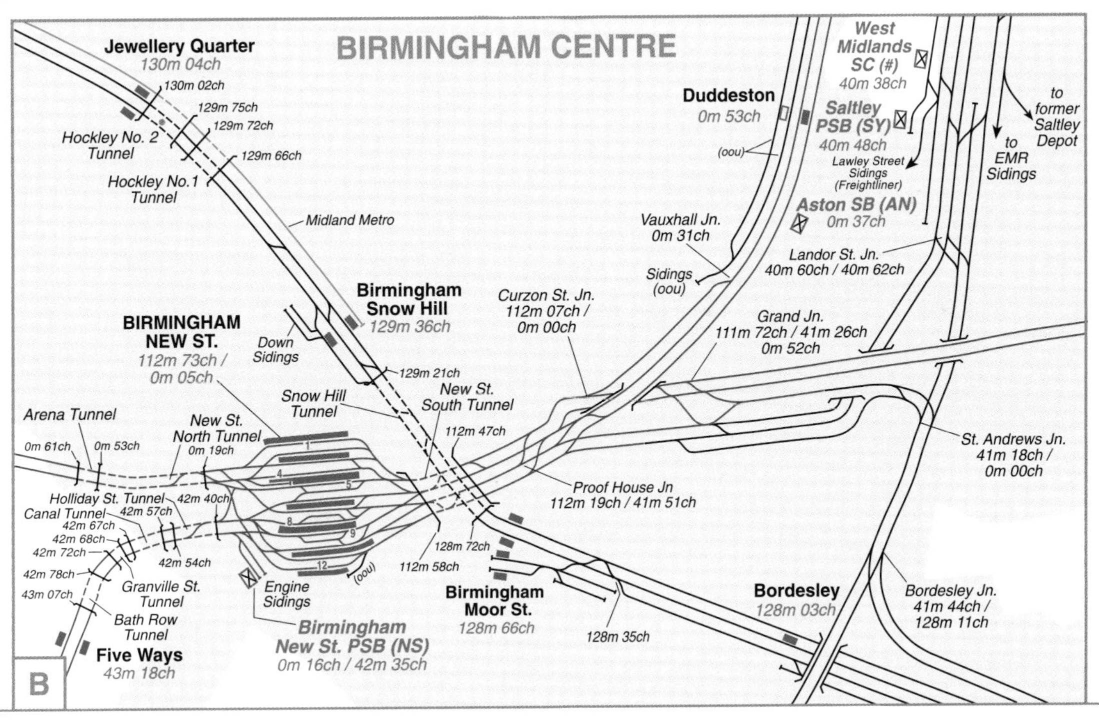

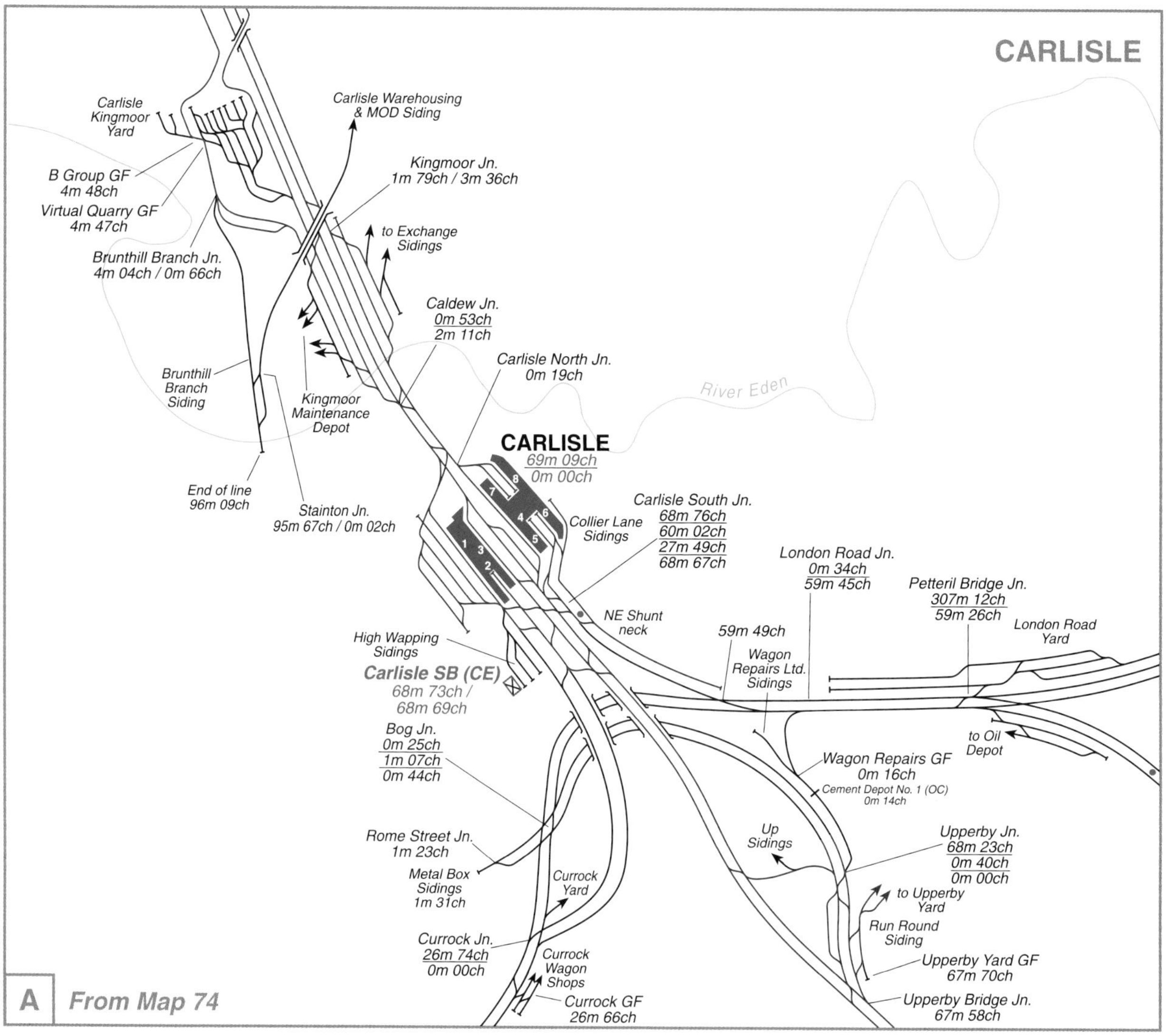

CARLISLE
Carlisle Kingmoor Yard
B Group GF
4m 48ch
Virtual Quarry GF
4m 47ch
Brunthill Branch Jn.
4m 04ch / 0m 66ch
Carlisle Warehousing & MOD Siding
Kingmoor Jn.
1m 79ch / 3m 36ch
to Exchange Sidings
Caldew Jn.
0m 53ch
2m 11ch
Carlisle North Jn.
0m 19ch
River Eden
Brunthill Branch Siding
Kingmoor Maintenance Depot
CARLISLE
69m 09ch
0m 00ch
End of line
96m 09ch
Stainton Jn.
95m 67ch / 0m 02ch
Collier Lane Sidings
Carlisle South Jn.
68m 76ch
60m 02ch
27m 49ch
68m 67ch
London Road Jn.
0m 34ch
59m 45ch
Petteril Bridge Jn.
307m 12ch
59m 26ch
London Road Yard
NE Shunt neck
59m 49ch
High Wapping Sidings
Carlisle SB (CE)
68m 73ch /
68m 69ch
Wagon Repairs Ltd. Sidings
to Oil Depot
Bog Jn.
0m 25ch
1m 07ch
0m 44ch
Wagon Repairs GF
0m 16ch
Cement Depot No. 1 (OC)
0m 14ch
Up Sidings
Rome Street Jn.
1m 23ch
Metal Box Sidings
1m 31ch
Currock Yard
Upperby Jn.
68m 23ch
0m 40ch
0m 00ch
to Upperby Yard
Run Round Siding
Currock Jn.
26m 74ch
0m 00ch
Currock Wagon Shops
Upperby Yard GF
67m 70ch
Currock GF
26m 66ch
Upperby Bridge Jn.
67m 58ch
A From Map 74

WOLVERHAMPTON & WALSALL
Bloxwich North
3m 01ch
Bloxwich
2m 32ch
Bloxwich (MCB)
2m 05ch
Bloxwich SB (BH)
2m 06ch
COM
0m 00ch
6m 79ch
Bushbury Jn.
15m 32ch /
14m 43ch
Limit of Electrification
143m 65ch
Oxley SB (OY)
143m 14ch
WALSALL
6m 29ch
6m 34ch
6m 40ch
Ryecroft Jn.
6m 76ch (via Walsall)
47m 53ch (via Aldridge)
Permanent Way Depot
Walsall Pleck Jn.
5m 42ch / 1m 16ch /
0m 65ch
Walsall North Jn.
6m 40ch
Bushbury (Oxley) Jn.
0m 00ch / 15m 23ch
Park St.Tunnel
Wednesfield Heath Tunnel
Noose Lane (CCTV)
12m 47ch
Walsall South Jn.
6m 18ch
Oxley Carriage Sidings
Darlaston Jn.
9m 65ch / 0m 15ch
Tasker Street Yard
Oxley (Stafford Road) Jn.
142m 79ch / 143m 02ch /
1m 02ch
13m 73ch
13m 65ch
Midland Yard
Walsall Pleck Jn.
COM 0m 65ch / 5m 42ch
Carriage Sidings
Walsall PSB (WL)
0m 63ch
Portobello Jn.
12m 64ch/0m 04ch
Bescot Jn. 8m 50ch / 0m 00ch
Wolverhampton North Jn.
13m 32ch / 143m 52ch
Bescot Stadium
8m 47ch
Bescot Middle Jn.
8m 25ch
Ballast Sidings
WOLVERHAMPTON
12m 75ch
Wolverhampton PSB (WN)
12m 69ch
Up Engineering Sidings
Crane St. Jn.
12m 60ch / 1m 59ch
to Diesel Depot Roads
Disused
Wolverhampton Steel Terminal
Bescot Down Tower SB (DT)
8m 20ch
Bescot Yard
B From Map 39

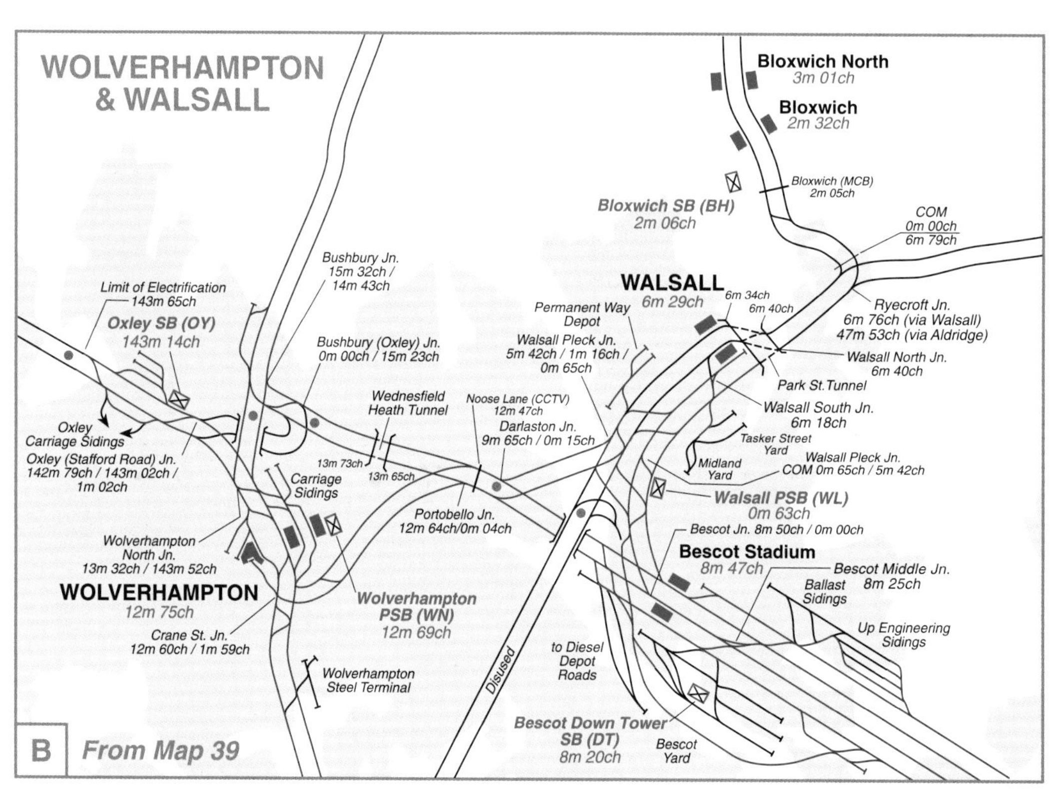

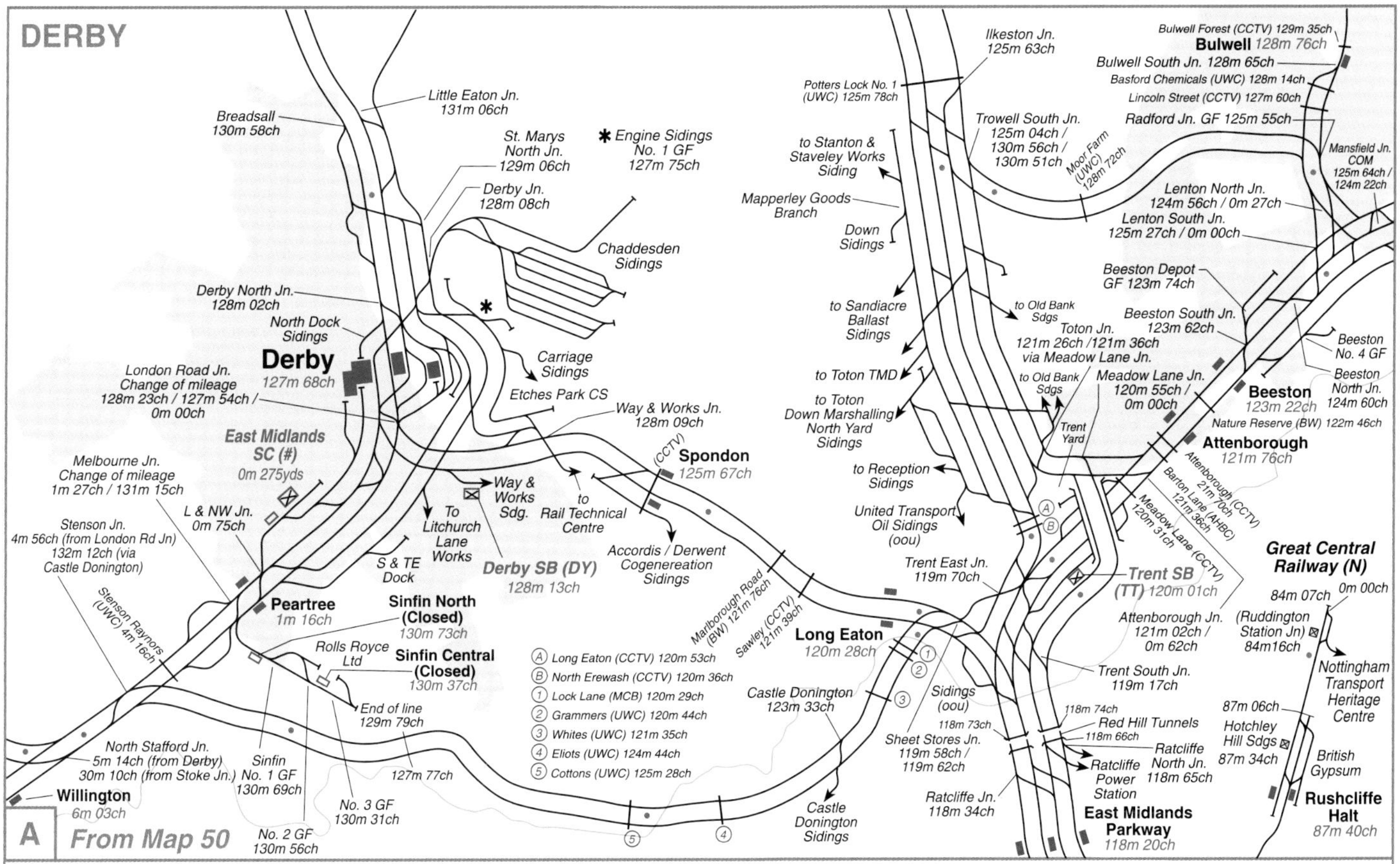

DERBY

Little Eaton Jn. 131m 06ch
Breadsall 130m 58ch
St. Marys North Jn. 129m 06ch
✱ Engine Sidings No. 1 GF 127m 75ch
Derby Jn. 128m 08ch
Chaddesden Sidings
Derby North Jn. 128m 02ch
North Dock Sidings
Derby 127m 68ch
London Road Jn. Change of mileage 128m 23ch / 127m 54ch / 0m 00ch
Carriage Sidings
Etches Park CS
Way & Works Jn. 128m 09ch
East Midlands SC (#) 0m 275yds
Melbourne Jn. Change of mileage 1m 27ch / 131m 15ch
L & NW Jn. 0m 75ch
Stenson Jn. 4m 56ch (from London Rd Jn) 132m 12ch (via Castle Donington)
To Litchurch Lane Works
Way & Works Sdg.
to Rail Technical Centre
Spondon 125m 67ch
(CCTV)
Accordis / Derwent Cogenereation Sidings
S & TE Dock
Derby SB (DY) 128m 13ch
Peartree 1m 16ch
Sinfin North (Closed) 130m 73ch
Sinfin Central (Closed) 130m 37ch
Rolls Royce Ltd
Stenson Raynors (UWC) 4m 16ch
End of line 129m 79ch
North Stafford Jn. 5m 14ch (from Derby) 30m 10ch (from Stoke Jn.)
Sinfin No. 1 GF 130m 69ch
127m 77ch
No. 3 GF 130m 31ch
No. 2 GF 130m 56ch
Willington 6m 03ch
Marlborough Road (BW) 121m 76ch
Sawley (CCTV) 121m 39ch
Long Eaton 120m 28ch
Castle Donington 123m 33ch
Castle Donington Sidings

Ⓐ Long Eaton (CCTV) 120m 53ch
Ⓑ North Erewash (CCTV) 120m 36ch
① Lock Lane (MCB) 120m 29ch
② Grammers (UWC) 120m 44ch
③ Whites (UWC) 121m 35ch
④ Eliots (UWC) 124m 44ch
⑤ Cottons (UWC) 125m 28ch

Potters Lock No. 1 (UWC) 125m 78ch
Ilkeston Jn. 125m 63ch
to Stanton & Staveley Works Siding
Trowell South Jn. 125m 04ch / 130m 56ch / 130m 51ch
Moor Farm (UWC) 128m 72ch
Mapperley Goods Branch
Down Sidings
to Sandiacre Ballast Sidings
to Old Bank Sdgs
Toton Jn. 121m 26ch / 121m 36ch via Meadow Lane Jn.
to Toton TMD
to Old Bank Sdgs
to Toton Down Marshalling North Yard Sidings
Trent Yard
to Reception Sidings
United Transport Oil Sidings (oou)
Trent East Jn. 119m 70ch
Sidings (oou)
118m 73ch
Sheet Stores Jn. 119m 58ch / 119m 62ch
Ratcliffe Jn. 118m 34ch
118m 74ch
Red Hill Tunnels
118m 66ch
Ratcliffe Power Station
Ratcliffe North Jn. 118m 65ch
East Midlands Parkway 118m 20ch
Trent SB (TT) 120m 01ch
Trent South Jn. 119m 17ch
Attenborough Jn. 121m 02ch / 0m 62ch
Meadow Lane (CCTV) 120m 31ch
Barton Lane (AHBC) 121m 36ch
Attenborough (CCTV) 21m 70ch

Bulwell Forest (CCTV) 129m 35ch
Bulwell 128m 76ch
Bulwell South Jn. 128m 65ch
Basford Chemicals (UWC) 128m 14ch
Lincoln Street (CCTV) 127m 60ch
Radford Jn. GF 125m 55ch
Mansfield Jn. COM 125m 64ch / 124m 22ch
Lenton North Jn. 124m 56ch / 0m 27ch
Lenton South Jn. 125m 27ch / 0m 00ch
Beeston Depot GF 123m 74ch
Beeston South Jn. 123m 62ch
Beeston No. 4 GF
Meadow Lane Jn. 120m 55ch / 0m 00ch
Beeston 123m 22ch
Beeston North Jn. 124m 60ch
Nature Reserve (BW) 122m 46ch
Attenborough 121m 76ch
Great Central Railway (N)
84m 07ch
0m 00ch
(Ruddington Station Jn) 84m16ch
Nottingham Transport Heritage Centre
87m 06ch
Hotchley Hill Sdgs 87m 34ch
British Gypsum
Rushcliffe Halt 87m 40ch

A From Map 50

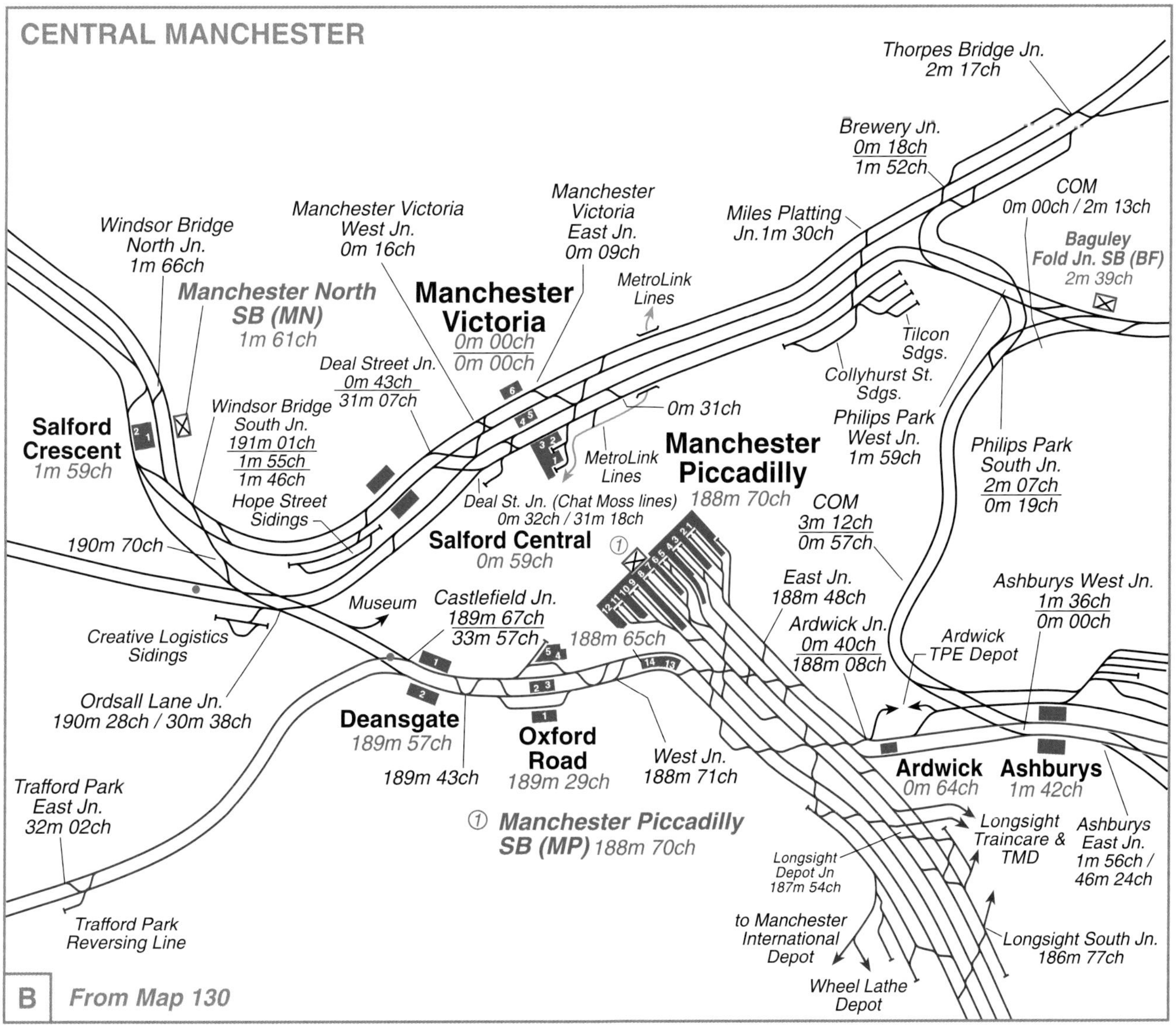

CENTRAL MANCHESTER

Thorpes Bridge Jn. 2m 17ch
Brewery Jn. 0m 18ch 1m 52ch
COM 0m 00ch / 2m 13ch
Baguley Fold Jn. SB (BF) 2m 39ch
Miles Platting Jn. 1m 30ch
Manchester Victoria East Jn. 0m 09ch
Manchester Victoria West Jn. 0m 16ch
Windsor Bridge North Jn. 1m 66ch
Manchester North SB (MN) 1m 61ch
Manchester Victoria 0m 00ch 0m 00ch
MetroLink Lines
Tilcon Sdgs.
Collyhurst St. Sdgs.
Deal Street Jn. 0m 43ch 31m 07ch
Windsor Bridge South Jn. 191m 01ch 1m 55ch 1m 46ch
Salford Crescent 1m 59ch
0m 31ch
Philips Park West Jn. 1m 59ch
Philips Park South Jn. 2m 07ch 0m 19ch
MetroLink Lines
Manchester Piccadilly 188m 70ch
Hope Street Sidings
Deal St. Jn. (Chat Moss lines) 0m 32ch / 31m 18ch
COM 3m 12ch 0m 57ch
190m 70ch
Salford Central 0m 59ch
East Jn. 188m 48ch
Ardwick Jn. 0m 40ch 188m 08ch
Ashburys West Jn. 1m 36ch 0m 00ch
Museum
Castlefield Jn. 189m 67ch 33m 57ch
188m 65ch
Creative Logistics Sidings
Ardwick TPE Depot
Ordsall Lane Jn. 190m 28ch / 30m 38ch
Deansgate 189m 57ch
Oxford Road 189m 29ch
West Jn. 188m 71ch
Ardwick 0m 64ch
Ashburys 1m 42ch
189m 43ch
Trafford Park East Jn. 32m 02ch
① Manchester Piccadilly SB (MP) 188m 70ch
Longsight Traincare & TMD
Ashburys East Jn. 1m 56ch / 46m 24ch
Longsight Depot Jn 187m 54ch
Trafford Park Reversing Line
to Manchester International Depot
Longsight South Jn. 186m 77ch
Wheel Lathe Depot

B From Map 130

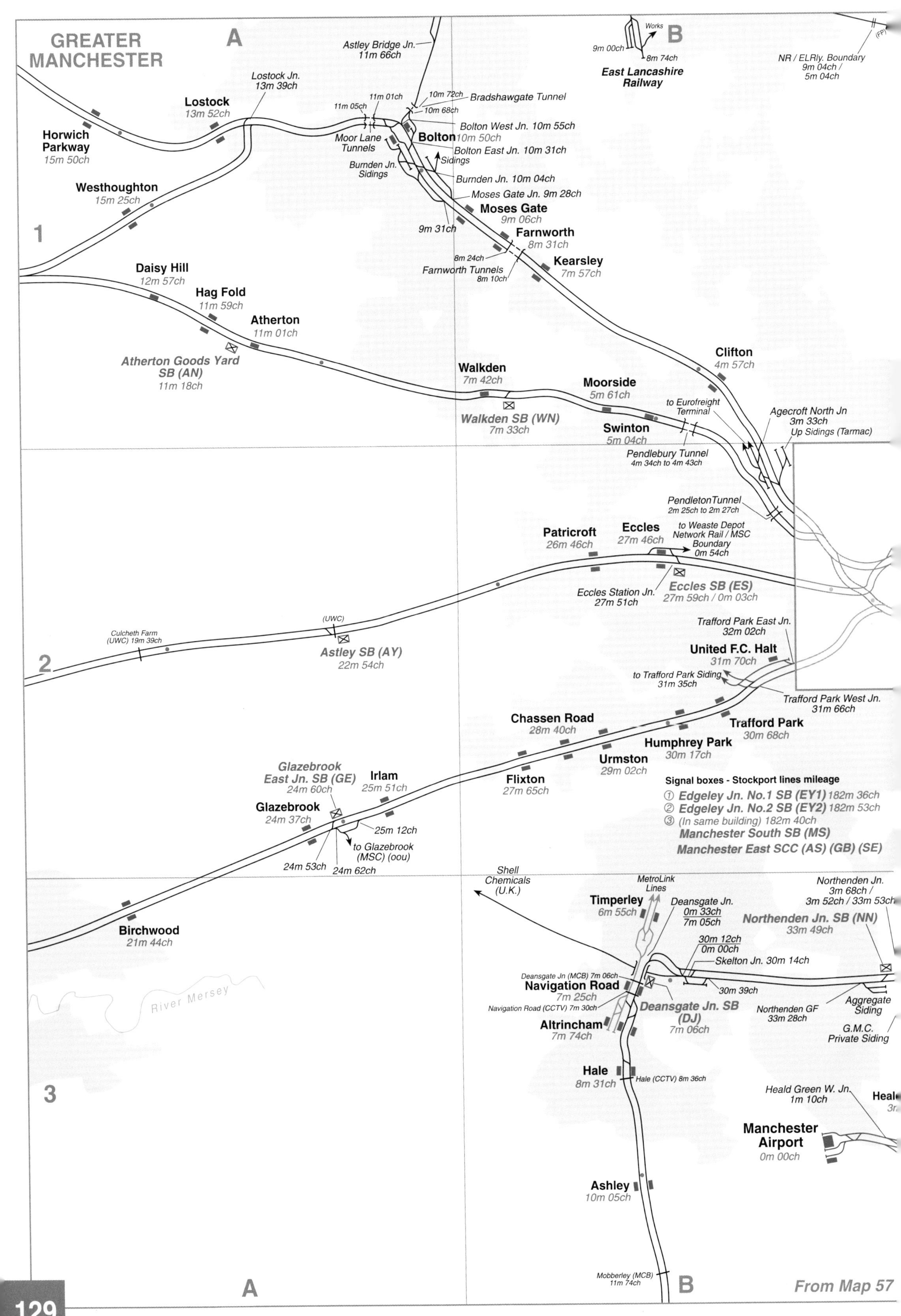
GREATER MANCHESTER
A
B
1
2
3
Horwich Parkway 15m 50ch
Lostock 13m 52ch
Lostock Jn. 13m 39ch
Westhoughton 15m 25ch
Astley Bridge Jn. 11m 66ch
11m 05ch
11m 01ch
10m 72ch
Bradshawgate Tunnel
10m 68ch
Bolton West Jn. 10m 55ch
Moor Lane Tunnels
Bolton 10m 50ch
Bolton East Jn. 10m 31ch
Sidings
Burnden Jn. Sidings
Burnden Jn. 10m 04ch
Moses Gate Jn. 9m 28ch
Moses Gate 9m 06ch
9m 31ch
Farnworth 8m 31ch
8m 24ch
Farnworth Tunnels 8m 10ch
Kearsley 7m 57ch
Works
9m 00ch
8m 74ch
East Lancashire Railway
NR / ELRly. Boundary 9m 04ch / 5m 04ch
(FP)
Daisy Hill 12m 57ch
Hag Fold 11m 59ch
Atherton 11m 01ch
Atherton Goods Yard SB (AN) 11m 18ch
Walkden 7m 42ch
Walkden SB (WN) 7m 33ch
Moorside 5m 61ch
Swinton 5m 04ch
to Eurofreight Terminal
Pendlebury Tunnel 4m 34ch to 4m 43ch
Clifton 4m 57ch
Agecroft North Jn 3m 33ch
Up Sidings (Tarmac)
PendletonTunnel 2m 25ch to 2m 27ch
Patricroft 26m 46ch
Eccles 27m 46ch
to Weaste Depot
Network Rail / MSC Boundary 0m 54ch
Eccles SB (ES) 27m 59ch / 0m 03ch
Eccles Station Jn. 27m 51ch
Culcheth Farm (UWC) 19m 39ch
(UWC)
Astley SB (AY) 22m 54ch
Trafford Park East Jn. 32m 02ch
United F.C. Halt 31m 70ch
to Trafford Park Siding 31m 35ch
Trafford Park West Jn. 31m 66ch
Chassen Road 28m 40ch
Trafford Park 30m 68ch
Humphrey Park 30m 17ch
Urmston 29m 02ch
Flixton 27m 65ch
Glazebrook East Jn. SB (GE) 24m 60ch
Irlam 25m 51ch
Glazebrook 24m 37ch
25m 12ch
to Glazebrook (MSC) (oou)
24m 53ch
24m 62ch
Signal boxes - Stockport lines mileage
① Edgeley Jn. No.1 SB (EY1) 182m 36ch
② Edgeley Jn. No.2 SB (EY2) 182m 53ch
③ (In same building) 182m 40ch
Manchester South SB (MS)
Manchester East SCC (AS) (GB) (SE)
Birchwood 21m 44ch
River Mersey
Shell Chemicals (U.K.)
MetroLink Lines
Timperley 6m 55ch
Deansgate Jn. 0m 33ch 7m 05ch
30m 12ch 0m 00ch
Skelton Jn. 30m 14ch
30m 39ch
Northenden Jn. 3m 68ch / 3m 52ch / 33m 53ch
Northenden Jn. SB (NN) 33m 49ch
Deansgate Jn (MCB) 7m 06ch
Navigation Road 7m 25ch
Navigation Road (CCTV) 7m 30ch
Deansgate Jn. SB (DJ) 7m 06ch
Altrincham 7m 74ch
Northenden GF 33m 28ch
Aggregate Siding
G.M.C. Private Siding
Hale 8m 31ch
Hale (CCTV) 8m 36ch
Heald Green W. Jn. 1m 10ch
Manchester Airport 0m 00ch
Ashley 10m 05ch
Mobberley (MCB) 11m 74ch
From Map 57

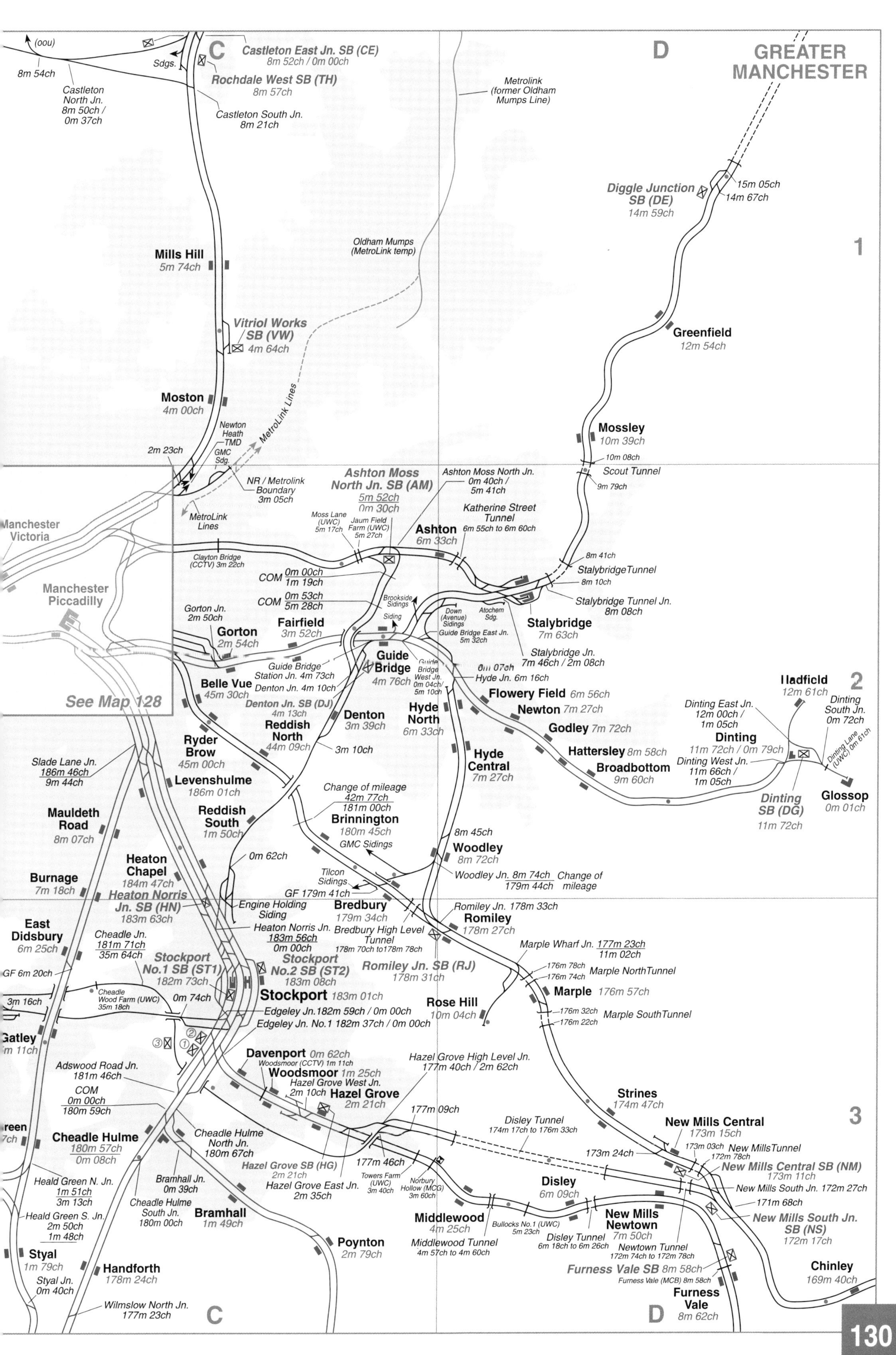
GREATER MANCHESTER
C
D
Castleton East Jn. SB (CE)
8m 52ch / 0m 00ch
Rochdale West SB (TH)
8m 57ch
Castleton South Jn.
8m 21ch
Castleton North Jn.
8m 50ch / 0m 37ch
8m 54ch
Sdgs.
Metrolink (former Oldham Mumps Line)
Diggle Junction SB (DE)
14m 59ch
15m 05ch
14m 67ch
Oldham Mumps (MetroLink temp)
Mills Hill
5m 74ch
Vitriol Works SB (VW)
4m 64ch
Greenfield
12m 54ch
Moston
4m 00ch
MetroLink Lines
Mossley
10m 39ch
Scout Tunnel
Manchester Victoria
Manchester Piccadilly
Ashton Moss North Jn. SB (AM)
Ashton
6m 33ch
Stalybridge
7m 63ch
Gorton
2m 54ch
Fairfield
3m 52ch
Guide Bridge
4m 76ch
Belle Vue
45m 30ch
See Map 128
Denton
3m 39ch
Hyde North
6m 33ch
Flowery Field 6m 56ch
Newton 7m 27ch
Godley 7m 72ch
Hattersley 8m 58ch
Broadbottom 9m 60ch
Hadfield 12m 61ch
Dinting 11m 72ch / 0m 79ch
Glossop 0m 01ch
Dinting SB (DG) 11m 72ch
Reddish North 44m 09ch
Ryder Brow 45m 00ch
Levenshulme 186m 01ch
Hyde Central 7m 27ch
Brinnington 180m 45ch
Reddish South 1m 50ch
Woodley 8m 72ch
Mauldeth Road 8m 07ch
Heaton Chapel 184m 47ch
Burnage 7m 18ch
Heaton Norris Jn. SB (HN) 183m 63ch
Bredbury 179m 34ch
Romiley 178m 27ch
East Didsbury 6m 25ch
Stockport No.1 SB (ST1) 182m 73ch
Stockport No.2 SB (ST2) 183m 08ch
Romiley Jn. SB (RJ) 178m 31ch
Stockport 183m 01ch
Rose Hill 10m 04ch
Marple 176m 57ch
Davenport 0m 62ch
Woodsmoor 1m 25ch
Hazel Grove 2m 21ch
Strines 174m 47ch
New Mills Central 173m 15ch
New Mills Central SB (NM) 173m 11ch
Cheadle Hulme 180m 57ch 0m 08ch
Hazel Grove SB (HG) 2m 21ch
Disley 6m 09ch
New Mills South Jn. SB (NS) 172m 17ch
Middlewood 4m 25ch
New Mills Newtown 7m 50ch
Bramhall 1m 49ch
Styal 1m 79ch
Handforth 178m 24ch
Poynton 2m 79ch
Furness Vale SB 8m 58ch
Furness Vale 8m 62ch
Chinley 169m 40ch
1
2
3

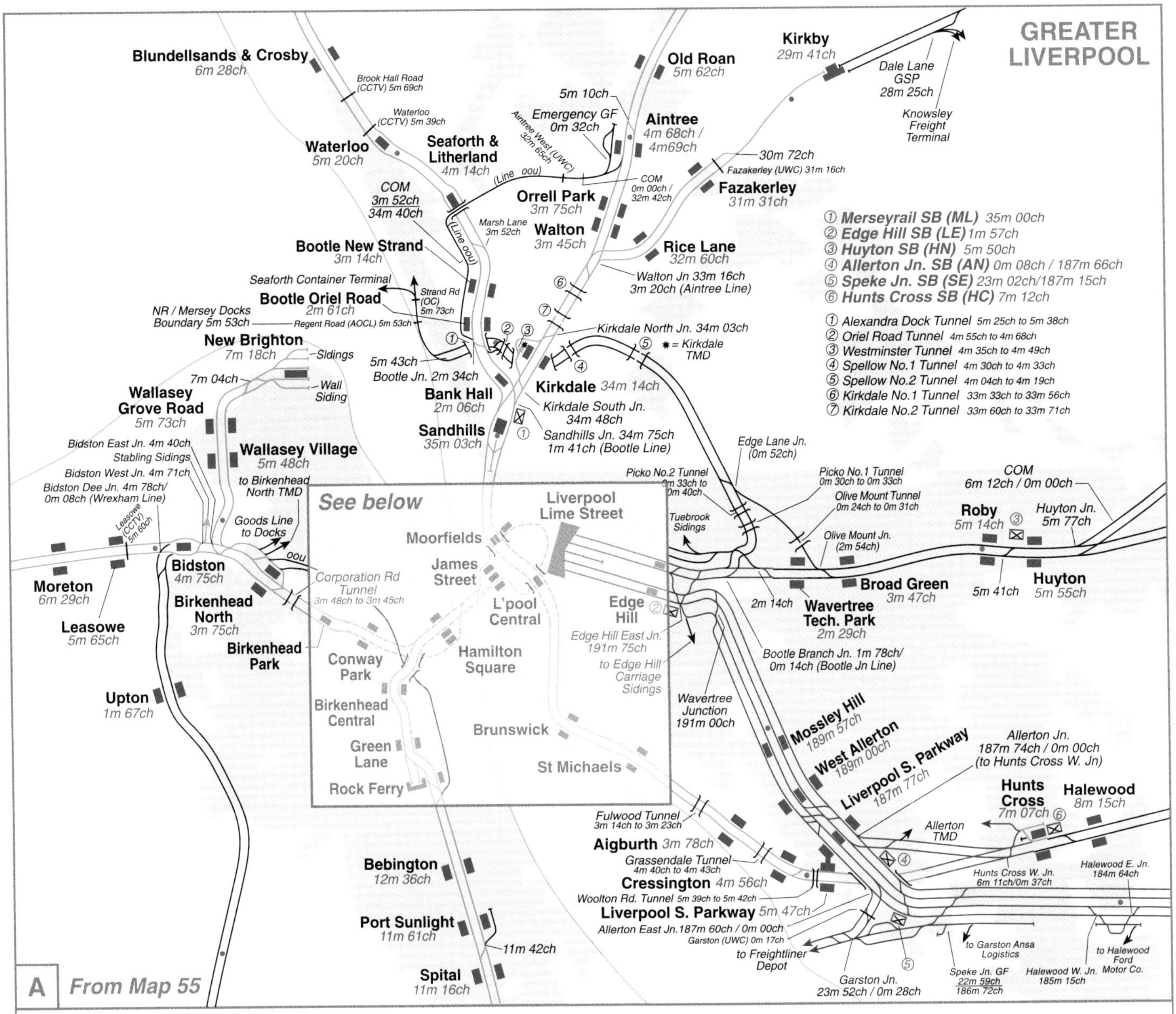

GREATER LIVERPOOL

Blundellsands & Crosby 6m 28ch
Brook Hall Road (CCTV) 5m 69ch
Waterloo (CCTV) 5m 39ch
Waterloo 5m 20ch
Seaforth & Litherland 4m 14ch
COM 3m 52ch 34m 40ch
(Line oou)
Marsh Lane 3m 52ch
Bootle New Strand 3m 14ch
Seaforth Container Terminal
Strand Rd (OC) 5m 73ch
Bootle Oriel Road 2m 61ch
NR / Mersey Docks Boundary 5m 53ch
Regent Road (AOCL) 5m 53ch
New Brighton 7m 18ch
Sidings
7m 04ch
Wall Siding
5m 43ch
Bootle Jn. 2m 34ch
Bank Hall 2m 06ch
Sandhills 35m 03ch
Wallasey Grove Road 5m 73ch
Bidston East Jn. 4m 40ch
Stabling Sidings
Bidston West Jn. 4m 71ch
Bidston Dee Jn. 4m 78ch/ 0m 08ch (Wrexham Line)
Wallasey Village 5m 48ch
to Birkenhead North TMD
Goods Line to Docks
Leasowe (CCTV) 5m 60ch
Moreton 6m 29ch
Leasowe 5m 65ch
Bidston 4m 75ch
Birkenhead North 3m 75ch
Birkenhead Park
Corporation Rd Tunnel 3m 48ch to 3m 45ch
Upton 1m 67ch
Old Roan 5m 62ch
5m 10ch
Emergency GF 0m 32ch
Aintree West (UWC) 32m 65ch
Aintree 4m 68ch / 4m69ch
COM 0m 00ch / 32m 42ch
Orrell Park 3m 75ch
Walton 3m 45ch
Kirkby 29m 41ch
Dale Lane GSP 28m 25ch
Knowsley Freight Terminal
30m 72ch
Fazakerley (UWC) 31m 16ch
Fazakerley 31m 31ch
Rice Lane 32m 60ch
Walton Jn 33m 16ch
3m 20ch (Aintree Line)
Kirkdale North Jn. 34m 03ch
✱ = Kirkdale TMD
Kirkdale 34m 14ch
Kirkdale South Jn. 34m 48ch
Sandhills Jn. 34m 75ch
1m 41ch (Bootle Line)

① Merseyrail SB (ML) 35m 00ch
② Edge Hill SB (LE) 1m 57ch
③ Huyton SB (HN) 5m 50ch
④ Allerton Jn. SB (AN) 0m 08ch / 187m 66ch
⑤ Speke Jn. SB (SE) 23m 02ch/187m 15ch
⑥ Hunts Cross SB (HC) 7m 12ch

① Alexandra Dock Tunnel 5m 25ch to 5m 38ch
② Oriel Road Tunnel 4m 55ch to 4m 68ch
③ Westminster Tunnel 4m 35ch to 4m 49ch
④ Spellow No.1 Tunnel 4m 30ch to 4m 33ch
⑤ Spellow No.2 Tunnel 4m 04ch to 4m 19ch
⑥ Kirkdale No.1 Tunnel 33m 33ch to 33m 56ch
⑦ Kirkdale No.2 Tunnel 33m 60ch to 33m 71ch

See below
Liverpool Lime Street
Moorfields
James Street
L'pool Central
Edge Hill
Hamilton Square
Conway Park
Birkenhead Central
Brunswick
Green Lane
St Michaels
Rock Ferry
Edge Hill East Jn. 191m 75ch
to Edge Hill Carriage Sidings
Edge Lane Jn. (0m 52ch)
Picko No.2 Tunnel 0m 33ch to 0m 40ch
Tuebrook Sidings
Picko No.1 Tunnel 0m 30ch to 0m 33ch
Olive Mount Tunnel 0m 24ch to 0m 31ch
Olive Mount Jn. (2m 54ch)
COM 6m 12ch / 0m 00ch
Roby 5m 14ch
Huyton Jn. 5m 77ch
Broad Green 3m 47ch
Huyton 5m 55ch
5m 41ch
2m 14ch
Wavertree Tech. Park 2m 29ch
Bootle Branch Jn. 1m 78ch/ 0m 14ch (Bootle Jn Line)
Wavertree Junction 191m 00ch
Mossley Hill 189m 57ch
West Allerton 189m 00ch
Liverpool S. Parkway 187m 77ch
Allerton Jn. 187m 74ch / 0m 00ch (to Hunts Cross W. Jn)
Hunts Cross 7m 07ch
Halewood 8m 15ch
Allerton TMD
Hunts Cross W. Jn. 6m 11ch/0m 37ch
Halewood E. Jn. 184m 64ch
Fulwood Tunnel 3m 14ch to 3m 23ch
Aigburth 3m 78ch
Grassendale Tunnel 4m 40ch to 4m 43ch
Cressington 4m 56ch
Woolton Rd. Tunnel 5m 39ch to 5m 42ch
Liverpool S. Parkway 5m 47ch
Allerton East Jn. 187m 60ch / 0m 00ch
Garston (UWC) 0m 17ch
to Freightliner Depot
Garston Jn. 23m 52ch / 0m 28ch
to Garston Ansa Logistics
Speke Jn. GF 22m 59ch 186m 72ch
Halewood W. Jn. 185m 15ch
to Halewood Ford Motor Co.
Bebington 12m 36ch
Port Sunlight 11m 61ch
11m 42ch
Spital 11m 16ch

A From Map 55

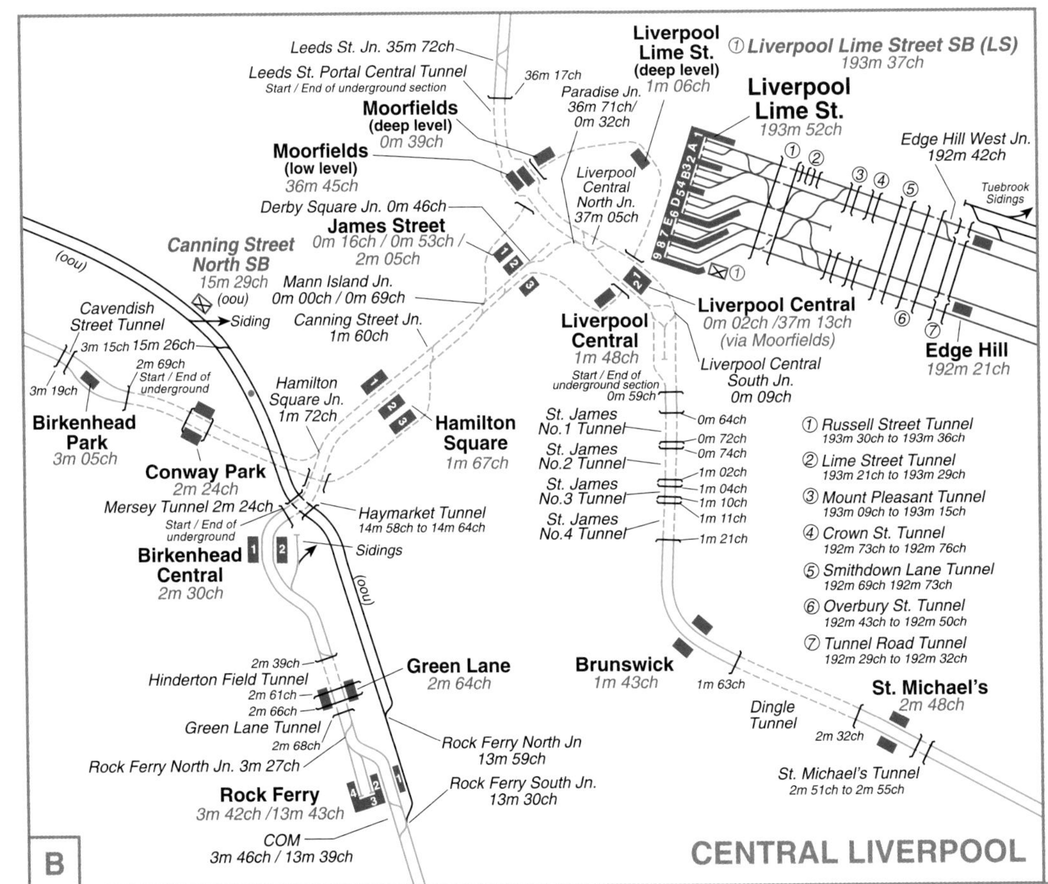

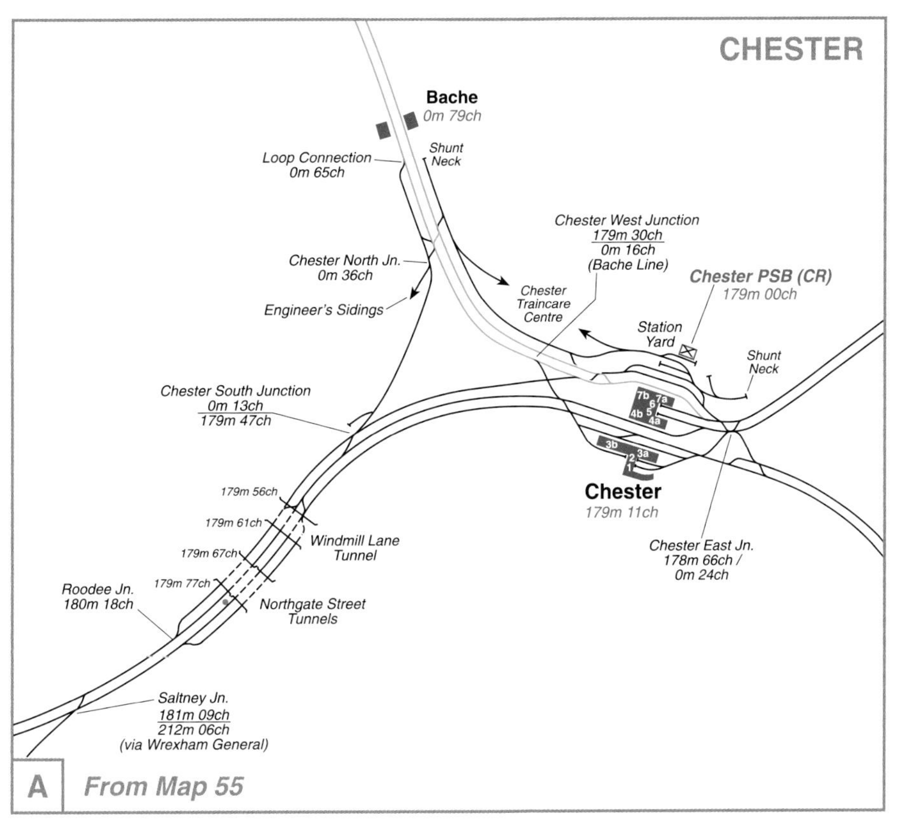
CHESTER
Bache
0m 79ch
Shunt Neck
Loop Connection
0m 65ch
Chester West Junction
179m 30ch
0m 16ch
(Bache Line)
Chester PSB (CR)
179m 00ch
Chester North Jn.
0m 36ch
Engineer's Sidings
Chester Traincare Centre
Station Yard
Shunt Neck
Chester South Junction
0m 13ch
179m 47ch
7b
7a
6
4b
5
4a
3b
3a
2
1
Chester
179m 11ch
179m 56ch
179m 61ch
179m 67ch
179m 77ch
Windmill Lane Tunnel
Chester East Jn.
178m 66ch /
0m 24ch
Roodee Jn.
180m 18ch
Northgate Street Tunnels
Saltney Jn.
181m 09ch
212m 06ch
(via Wrexham General)
A
From Map 55

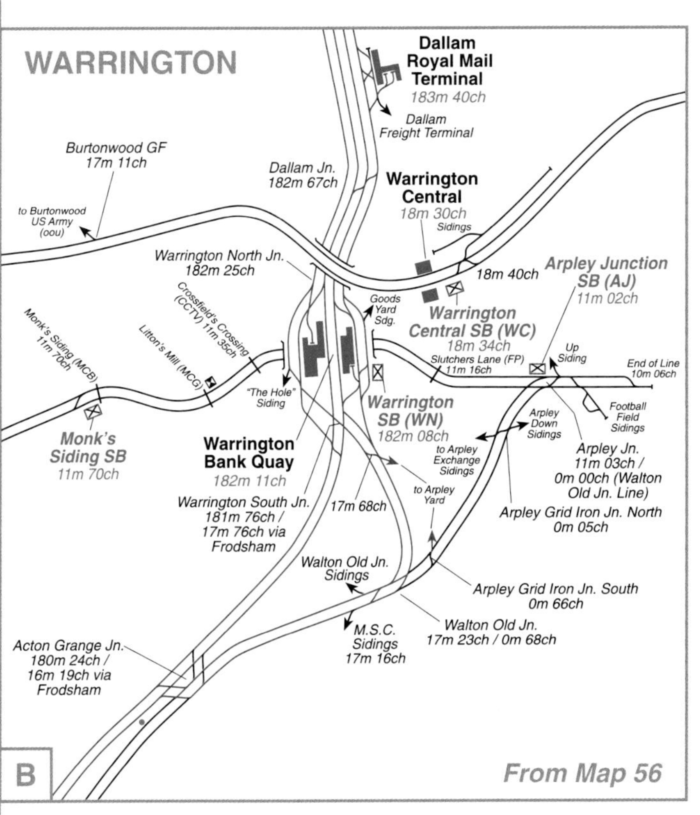
WARRINGTON
Dallam Royal Mail Terminal
183m 40ch
Dallam Freight Terminal
Burtonwood GF
17m 11ch
Dallam Jn.
182m 67ch
Warrington Central
18m 30ch
Sidings
to Burtonwood US Army (oou)
Warrington North Jn.
182m 25ch
18m 40ch
Arpley Junction SB (AJ)
11m 02ch
Goods Yard Sdg.
Warrington Central SB (WC)
18m 34ch
Crosfield's Crossing (CCTV) 11m 35ch
Litton's Mill (MCG)
Monk's Siding (MCB) 11m 70ch
Slutchers Lane (FP) 11m 16ch
Up Siding
End of Line 10m 06ch
"The Hole" Siding
Warrington SB (WN)
182m 08ch
Football Field Sidings
Arpley Down Sidings
Monk's Siding SB
11m 70ch
Warrington Bank Quay
182m 11ch
to Arpley Exchange Sidings
Arpley Jn.
11m 03ch /
0m 00ch (Walton Old Jn. Line)
Warrington South Jn.
181m 76ch /
17m 76ch via Frodsham
17m 68ch
to Arpley Yard
Arpley Grid Iron Jn. North
0m 05ch
Walton Old Jn. Sidings
Arpley Grid Iron Jn. South
0m 66ch
M.S.C. Sidings
17m 16ch
Walton Old Jn.
17m 23ch / 0m 68ch
Acton Grange Jn.
180m 24ch /
16m 19ch via Frodsham
B
From Map 56

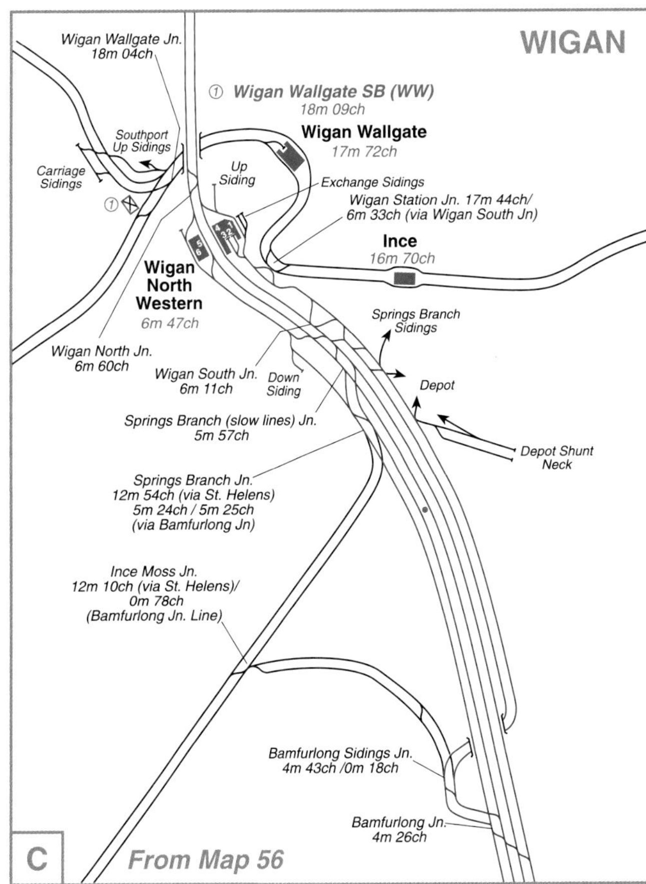
WIGAN
Wigan Wallgate Jn.
18m 04ch
① Wigan Wallgate SB (WW)
18m 09ch
Southport Up Sidings
Wigan Wallgate
17m 72ch
Carriage Sidings
Up Siding
Exchange Sidings
①
Wigan Station Jn. 17m 44ch/
6m 33ch (via Wigan South Jn)
Ince
16m 70ch
Wigan North Western
6m 47ch
Springs Branch Sidings
Wigan North Jn.
6m 60ch
Wigan South Jn.
6m 11ch
Down Siding
Depot
Springs Branch (slow lines) Jn.
5m 57ch
Depot Shunt Neck
Springs Branch Jn.
12m 54ch (via St. Helens)
5m 24ch / 5m 25ch
(via Bamfurlong Jn)
Ince Moss Jn.
12m 10ch (via St. Helens)/
0m 78ch
(Bamfurlong Jn. Line)
Bamfurlong Sidings Jn.
4m 43ch /0m 18ch
Bamfurlong Jn.
4m 26ch
C
From Map 56

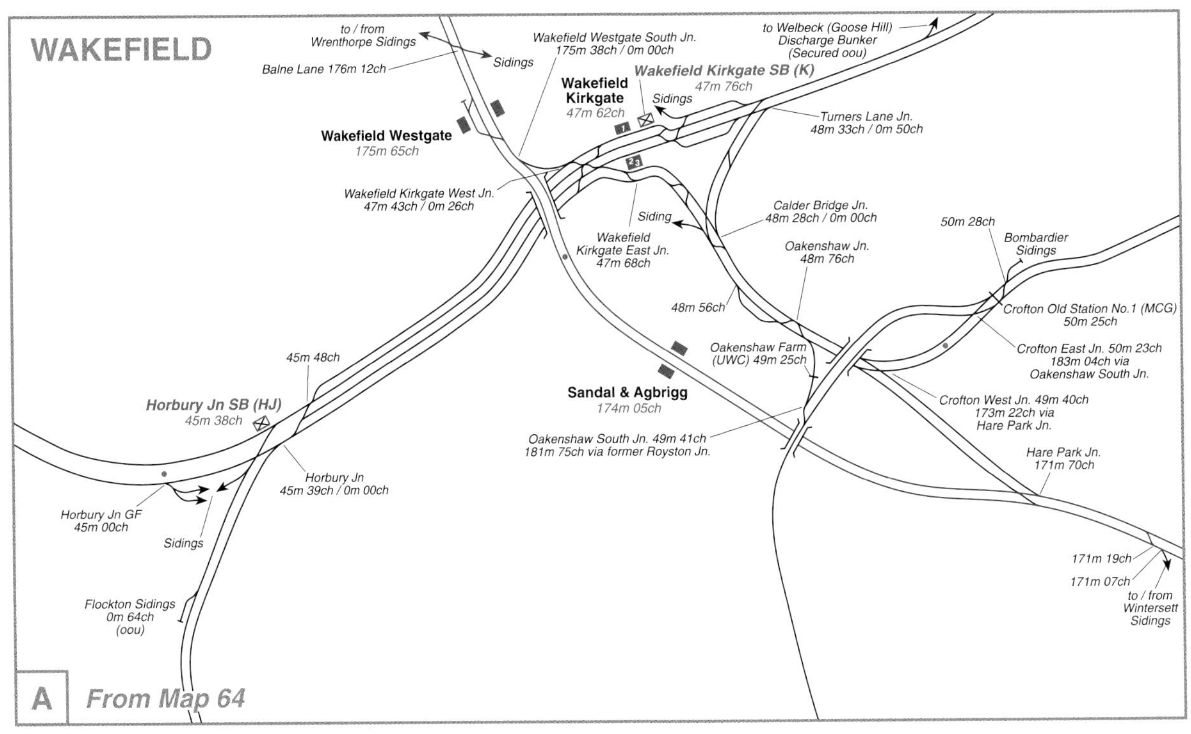
WAKEFIELD
to / from
Wrenthorpe Sidings
Sidings
Balne Lane 176m 12ch
Wakefield Westgate South Jn.
175m 38ch / 0m 00ch
to Welbeck (Goose Hill)
Discharge Bunker
(Secured oou)
Wakefield
Kirkgate
47m 62ch
Wakefield Kirkgate SB (K)
47m 76ch
Sidings
Wakefield Westgate
175m 65ch
Turners Lane Jn.
48m 33ch / 0m 50ch
Wakefield Kirkgate West Jn.
47m 43ch / 0m 26ch
Siding
Wakefield
Kirkgate East Jn.
47m 68ch
Calder Bridge Jn.
48m 28ch / 0m 00ch
50m 28ch
Bombardier
Sidings
Oakenshaw Jn.
48m 76ch
48m 56ch
Crofton Old Station No.1 (MCG)
50m 25ch
45m 48ch
Oakenshaw Farm
(UWC) 49m 25ch
Crofton East Jn. 50m 23ch
183m 04ch via
Oakenshaw South Jn.
Horbury Jn SB (HJ)
45m 38ch
Sandal & Agbrigg
174m 05ch
Crofton West Jn. 49m 40ch
173m 22ch via
Hare Park Jn.
Oakenshaw South Jn. 49m 41ch
181m 75ch via former Royston Jn.
Hare Park Jn.
171m 70ch
Horbury Jn
45m 39ch / 0m 00ch
Horbury Jn GF
45m 00ch
Sidings
171m 19ch
171m 07ch
to / from
Wintersett
Sidings
Flockton Sidings
0m 64ch
(oou)
A From Map 64

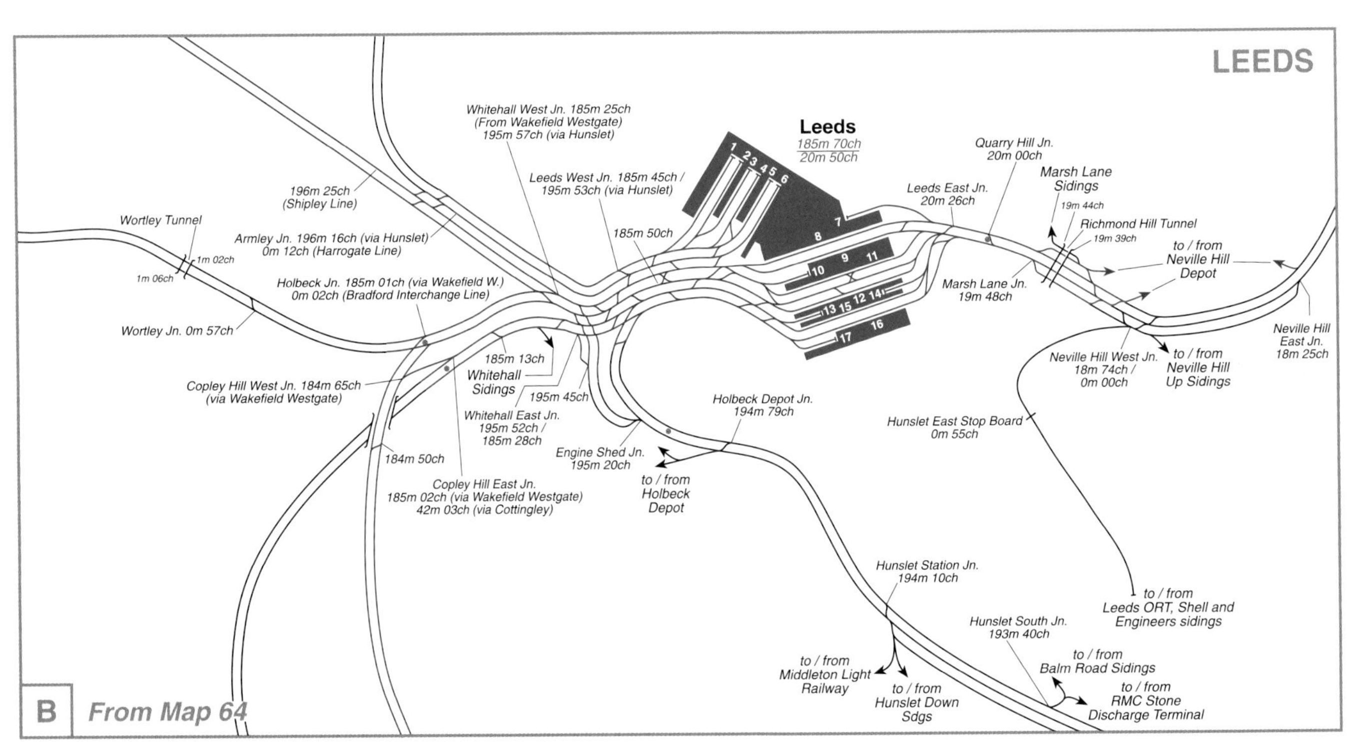
LEEDS
Whitehall West Jn. 185m 25ch
(From Wakefield Westgate)
195m 57ch (via Hunslet)
Leeds
185m 70ch
20m 50ch
Quarry Hill Jn.
20m 00ch
Leeds West Jn. 185m 45ch /
195m 53ch (via Hunslet)
Leeds East Jn.
20m 26ch
Marsh Lane
Sidings
196m 25ch
(Shipley Line)
19m 44ch
Richmond Hill Tunnel
19m 39ch
Wortley Tunnel
185m 50ch
to / from
Neville Hill
Depot
Armley Jn. 196m 16ch (via Hunslet)
0m 12ch (Harrogate Line)
1m 02ch
1m 06ch
Holbeck Jn. 185m 01ch (via Wakefield W.)
0m 02ch (Bradford Interchange Line)
Marsh Lane Jn.
19m 48ch
Neville Hill
East Jn.
18m 25ch
Wortley Jn. 0m 57ch
185m 13ch
Whitehall
Sidings
Neville Hill West Jn.
18m 74ch /
0m 00ch
to / from
Neville Hill
Up Sidings
Copley Hill West Jn. 184m 65ch
(via Wakefield Westgate)
195m 45ch
Whitehall East Jn.
195m 52ch /
185m 28ch
Holbeck Depot Jn.
194m 79ch
Hunslet East Stop Board
0m 55ch
184m 50ch
Engine Shed Jn.
195m 20ch
Copley Hill East Jn.
185m 02ch (via Wakefield Westgate)
42m 03ch (via Cottingley)
to / from
Holbeck
Depot
Hunslet Station Jn.
194m 10ch
to / from
Leeds ORT, Shell and
Engineers sidings
Hunslet South Jn.
193m 40ch
to / from
Middleton Light
Railway
to / from
Hunslet Down
Sdgs
to / from
Balm Road Sidings
to / from
RMC Stone
Discharge Terminal
B From Map 64

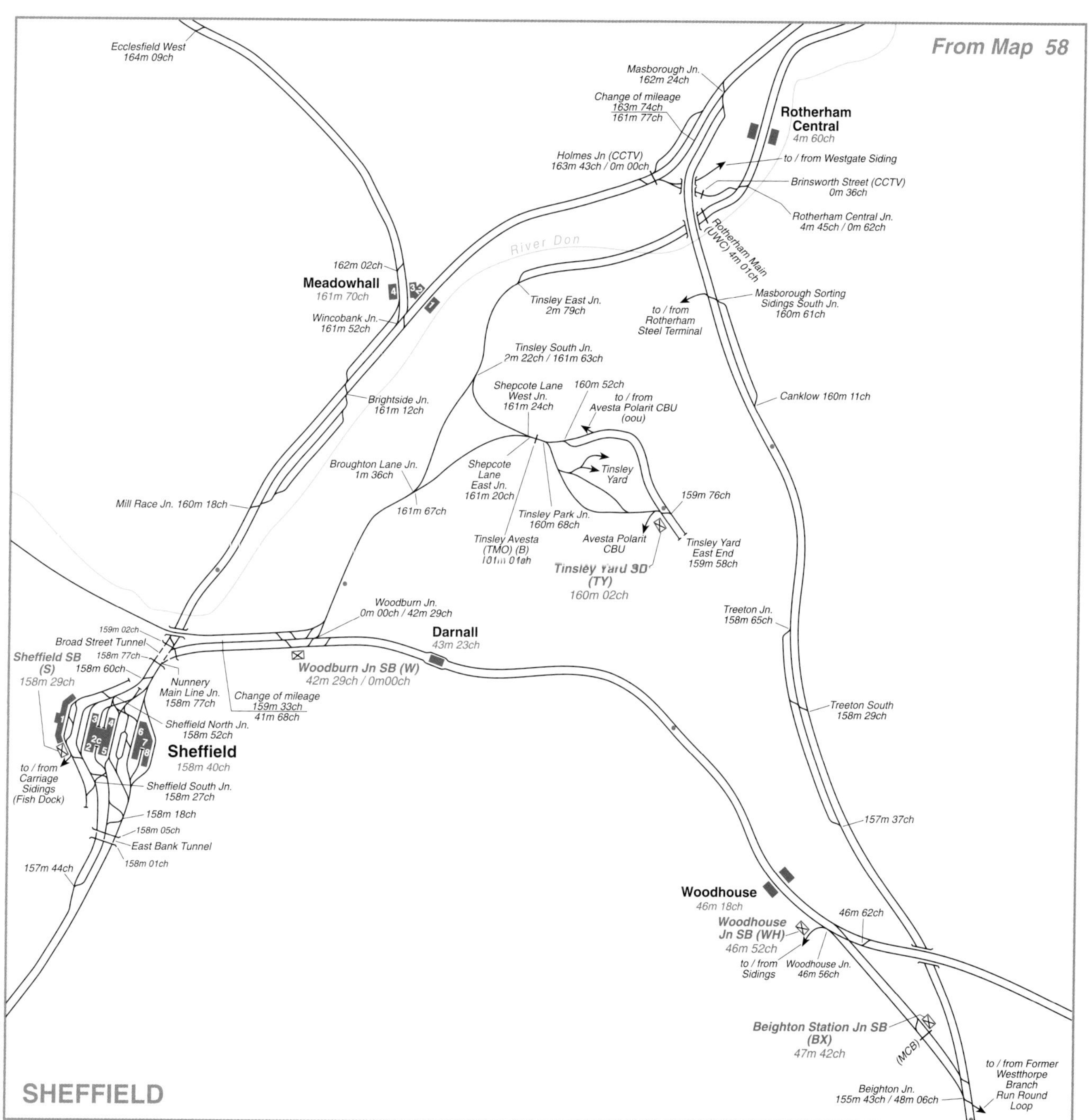
From Map 58
Ecclesfield West 164m 09ch
Masborough Jn. 162m 24ch
Change of mileage 163m 74ch / 161m 77ch
Rotherham Central
4m 60ch
to / from Westgate Siding
Holmes Jn (CCTV) 163m 43ch / 0m 00ch
Brinsworth Street (CCTV) 0m 36ch
Rotherham Central Jn. 4m 45ch / 0m 62ch
Rotherham Main (UWC) 4m 01ch
River Don
162m 02ch
Meadowhall
161m 70ch
Wincobank Jn. 161m 52ch
Tinsley East Jn. 2m 79ch
to / from Rotherham Steel Terminal
Masborough Sorting Sidings South Jn. 160m 61ch
Tinsley South Jn. 2m 22ch / 161m 63ch
Shepcote Lane West Jn. 161m 24ch
160m 52ch
to / from Avesta Polarit CBU (oou)
Brightside Jn. 161m 12ch
Canklow 160m 11ch
Broughton Lane Jn. 1m 36ch
Shepcote Lane East Jn. 161m 20ch
Tinsley Yard
159m 76ch
Mill Race Jn. 160m 18ch
161m 67ch
Tinsley Park Jn. 160m 68ch
Tinsley Avesta (TMO) (B)
Avesta Polarit CBU
Tinsley Yard East End 159m 58ch
Tinsley Yard SB (TY)
160m 02ch
Treeton Jn. 158m 65ch
Woodburn Jn. 0m 00ch / 42m 29ch
159m 02ch
Broad Street Tunnel
Darnall
43m 23ch
158m 77ch
Sheffield SB (S)
158m 29ch
158m 60ch
Woodburn Jn SB (W)
42m 29ch / 0m00ch
Nunnery Main Line Jn. 158m 77ch
Change of mileage 159m 33ch / 41m 68ch
Treeton South 158m 29ch
Sheffield North Jn. 158m 52ch
Sheffield
158m 40ch
to / from Carriage Sidings (Fish Dock)
Sheffield South Jn. 158m 27ch
158m 18ch
158m 05ch
East Bank Tunnel
158m 01ch
157m 37ch
157m 44ch
Woodhouse
46m 18ch
46m 62ch
Woodhouse Jn SB (WH)
46m 52ch
to / from Sidings
Woodhouse Jn. 46m 56ch
Beighton Station Jn SB (BX)
47m 42ch
(MCB)
to / from Former Westthorpe Branch Run Round Loop
Beighton Jn. 155m 43ch / 48m 06ch
SHEFFIELD

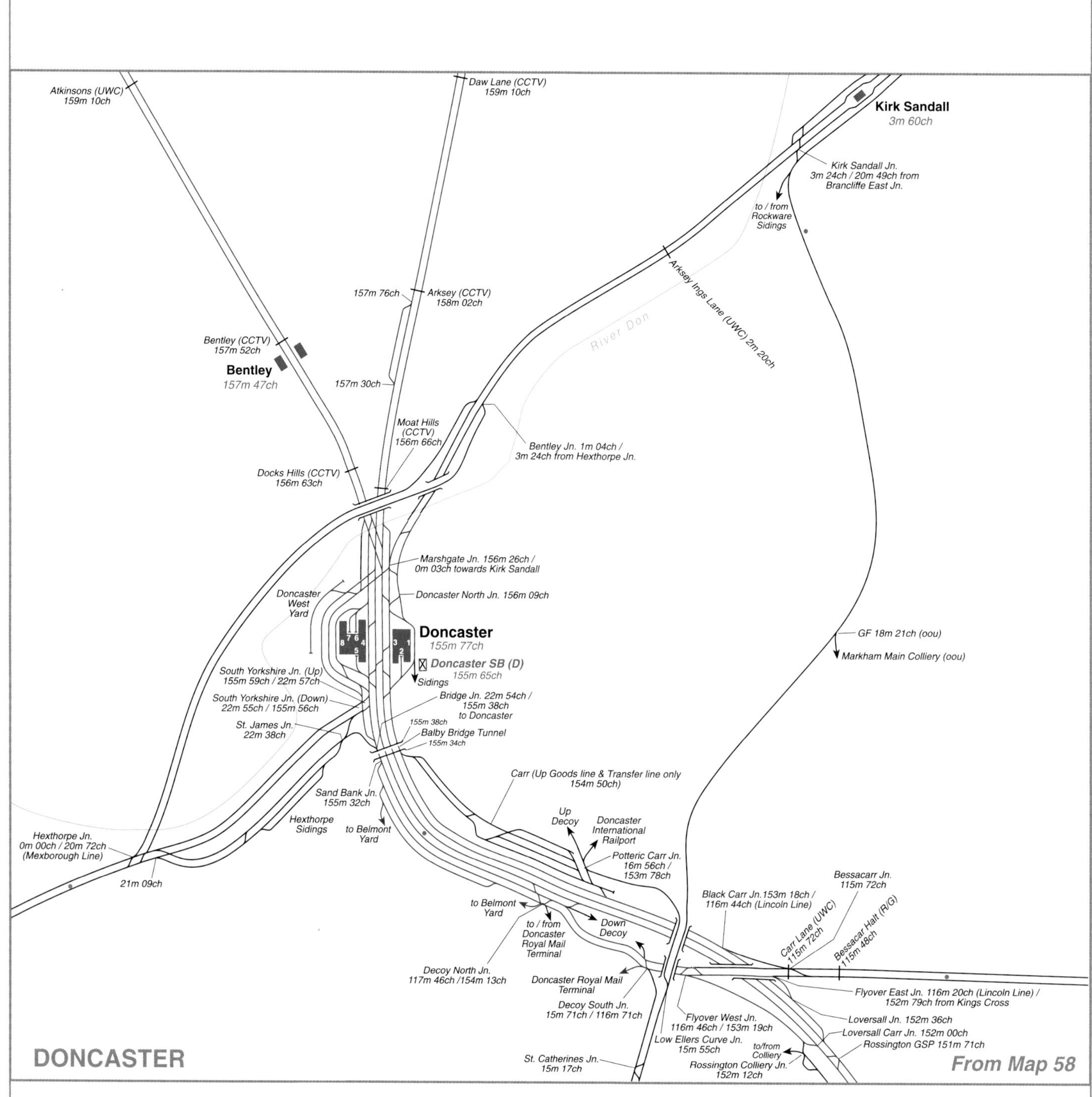

Atkinsons (UWC)
159m 10ch
Daw Lane (CCTV)
159m 10ch
Kirk Sandall
3m 60ch
Kirk Sandall Jn.
3m 24ch / 20m 49ch from
Brancliffe East Jn.
to / from
Rockware
Sidings
157m 76ch
Arksey (CCTV)
158m 02ch
Arksey Ings Lane (UWC) 2m 20ch
River Don
Bentley (CCTV)
157m 52ch
Bentley
157m 47ch
157m 30ch
Moat Hills
(CCTV)
156m 66ch
Bentley Jn. 1m 04ch /
3m 24ch from Hexthorpe Jn.
Docks Hills (CCTV)
156m 63ch
Marshgate Jn. 156m 26ch /
0m 03ch towards Kirk Sandall
Doncaster North Jn. 156m 09ch
Doncaster
West
Yard
Doncaster
155m 77ch
Doncaster SB (D)
155m 65ch
Sidings
GF 18m 21ch (oou)
Markham Main Colliery (oou)
South Yorkshire Jn. (Up)
155m 59ch / 22m 57ch
South Yorkshire Jn. (Down)
22m 55ch / 155m 56ch
Bridge Jn. 22m 54ch /
155m 38ch
to Doncaster
St. James Jn.
22m 38ch
155m 38ch
Balby Bridge Tunnel
155m 34ch
Carr (Up Goods line & Transfer line only
154m 50ch)
Sand Bank Jn.
155m 32ch
Hexthorpe
Sidings
to Belmont
Yard
Up
Decoy
Doncaster
International
Railport
Potteric Carr Jn.
16m 56ch /
153m 78ch
Hexthorpe Jn.
0m 00ch / 20m 72ch
(Mexborough Line)
21m 09ch
Bessacarr Jn.
115m 72ch
Black Carr Jn.153m 18ch /
116m 44ch (Lincoln Line)
Carr Lane (UWC)
115m 72ch
Bessacar Halt (R/G)
115m 48ch
to Belmont
Yard
to / from
Doncaster
Royal Mail
Terminal
Down
Decoy
Decoy North Jn.
117m 46ch /154m 13ch
Doncaster Royal Mail
Terminal
Decoy South Jn.
15m 71ch / 116m 71ch
Flyover East Jn. 116m 20ch (Lincoln Line) /
152m 79ch from Kings Cross
Flyover West Jn.
116m 46ch / 153m 19ch
Loversall Jn. 152m 36ch
Loversall Carr Jn. 152m 00ch
Low Ellers Curve Jn.
15m 55ch
to/from
Colliery
Rossington GSP 151m 71ch
St. Catherines Jn.
15m 17ch
Rossington Colliery Jn.
152m 12ch
DONCASTER
From Map 58

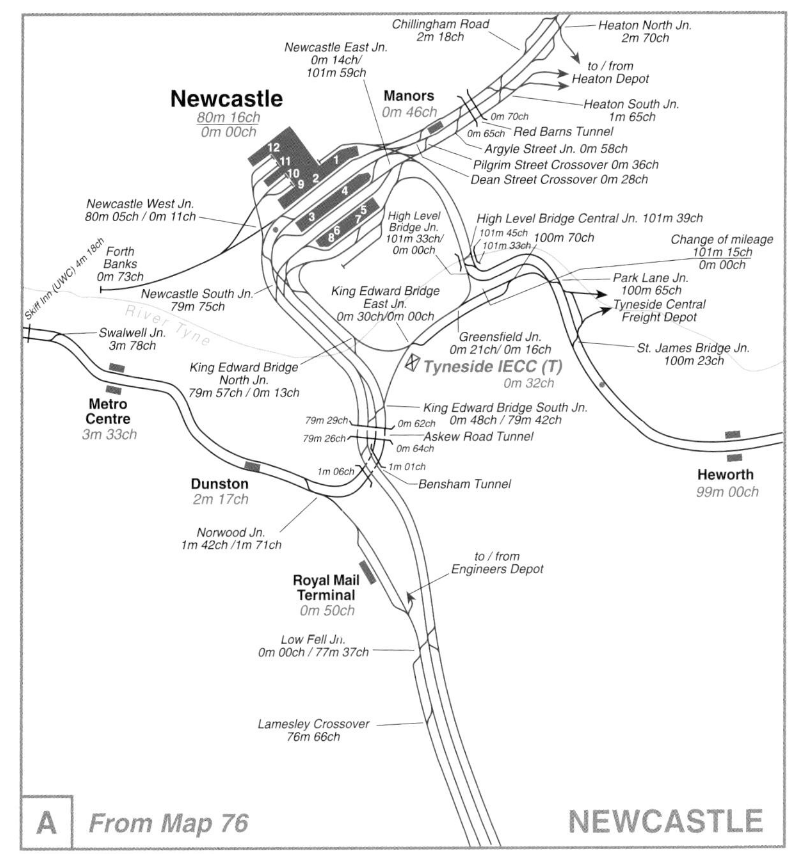

Chillingham Road
2m 18ch
Heaton North Jn.
2m 70ch
Newcastle East Jn.
0m 14ch/
101m 59ch
to / from
Heaton Depot
Newcastle
80m 16ch
0m 00ch
Manors
0m 46ch
Heaton South Jn.
1m 65ch
0m 70ch
0m 65ch
Red Barns Tunnel
Argyle Street Jn. 0m 58ch
Pilgrim Street Crossover 0m 36ch
Dean Street Crossover 0m 28ch
Newcastle West Jn.
80m 05ch / 0m 11ch
High Level Bridge Jn.
101m 33ch/
0m 00ch
High Level Bridge Central Jn. 101m 39ch
101m 45ch
101m 33ch
100m 70ch
Change of mileage
101m 15ch
0m 00ch
Forth Banks
0m 73ch
Skiff Inn (UWC) 4m 18ch
Newcastle South Jn.
79m 75ch
King Edward Bridge
East Jn.
0m 30ch/0m 00ch
Park Lane Jn.
100m 65ch
Tyneside Central
Freight Depot
River Tyne
Swalwell Jn.
3m 78ch
Greensfield Jn.
0m 21ch/ 0m 16ch
St. James Bridge Jn.
100m 23ch
King Edward Bridge
North Jn.
79m 57ch / 0m 13ch
Tyneside IECC (T)
0m 32ch
Metro Centre
3m 33ch
King Edward Bridge South Jn.
0m 48ch / 79m 42ch
79m 29ch
0m 62ch
79m 26ch
Askew Road Tunnel
0m 64ch
1m 01ch
1m 06ch
Dunston
2m 17ch
Bensham Tunnel
Heworth
99m 00ch
Norwood Jn.
1m 42ch /1m 71ch
to / from
Engineers Depot
Royal Mail Terminal
0m 50ch
Low Fell Jn.
0m 00ch / 77m 37ch
Lamesley Crossover
76m 66ch
A
From Map 76
NEWCASTLE

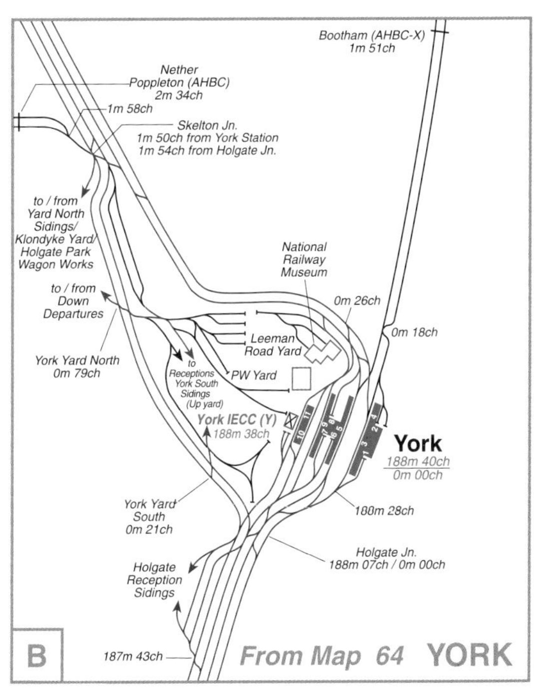

Bootham (AHBC-X)
1m 51ch
Nether
Poppleton (AHBC)
2m 34ch
1m 58ch
Skelton Jn.
1m 50ch from York Station
1m 54ch from Holgate Jn.
to / from
Yard North
Sidings/
Klondyke Yard/
Holgate Park
Wagon Works
National
Railway
Museum
to / from
Down
Departures
0m 26ch
0m 18ch
Leeman
Road Yard
York Yard North
0m 79ch
to
Receptions
York South
Sidings
(Up yard)
PW Yard
York IECC (Y)
188m 38ch
York
188m 40ch
0m 00ch
York Yard
South
0m 21ch
188m 28ch
Holgate Jn.
188m 07ch / 0m 00ch
Holgate
Reception
Sidings
187m 43ch
B
From Map 64
YORK

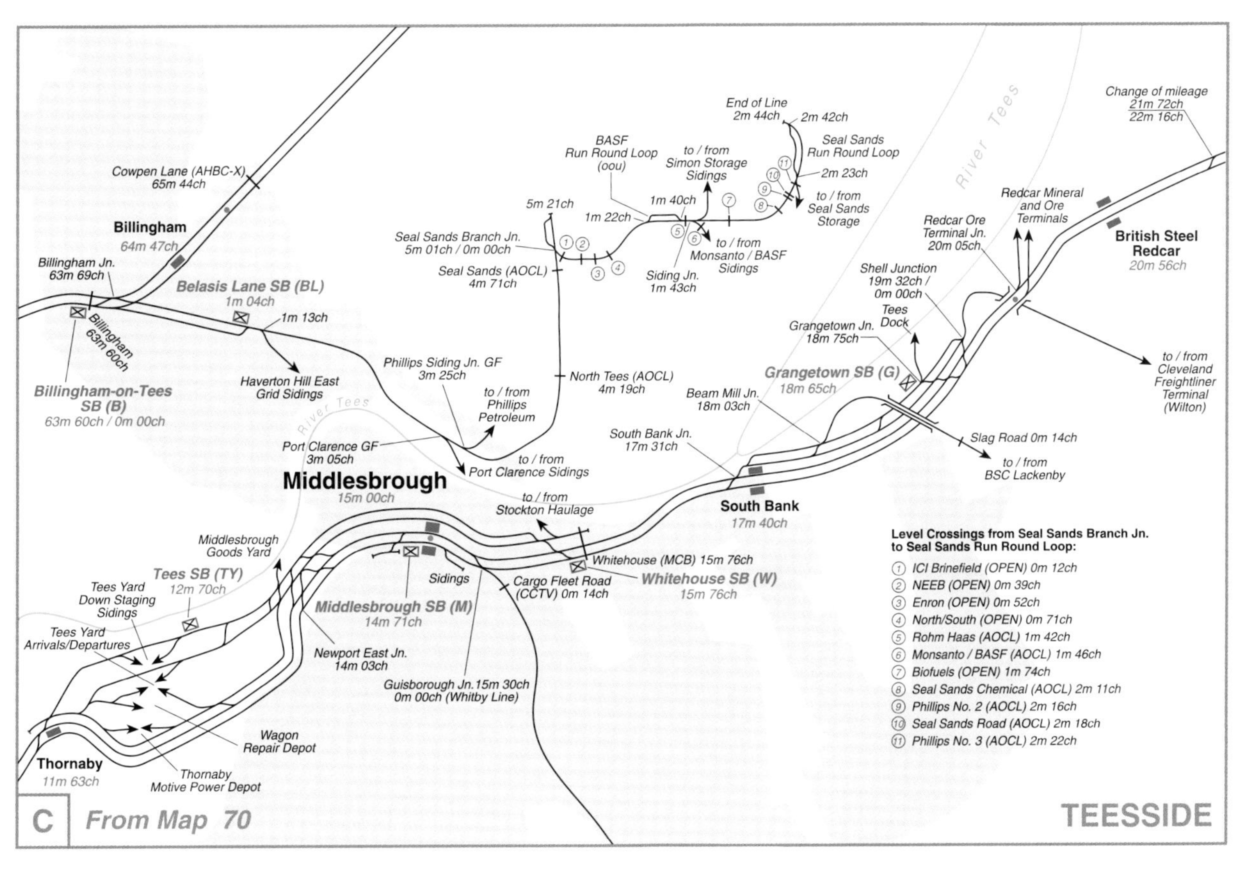

End of Line
2m 44ch
2m 42ch
Change of mileage
21m 72ch
22m 16ch
BASF
Run Round Loop
(oou)
to / from
Simon Storage
Sidings
Seal Sands
Run Round Loop
2m 23ch
River Tees
Cowpen Lane (AHBC-X)
65m 44ch
5m 21ch
1m 40ch
1m 22ch
to / from
Seal Sands
Storage
Redcar Mineral
and Ore
Terminals
Billingham
64m 47ch
Seal Sands Branch Jn.
5m 01ch / 0m 00ch
to / from
Monsanto / BASF
Sidings
Redcar Ore
Terminal Jn.
20m 05ch
British Steel
Redcar
20m 56ch
Billingham Jn.
63m 69ch
Seal Sands (AOCL)
4m 71ch
Siding Jn.
1m 43ch
Shell Junction
19m 32ch /
0m 00ch
Belasis Lane SB (BL)
1m 04ch
1m 13ch
Billingham
63m 60ch
Tees
Dock
Grangetown Jn.
18m 75ch
to / from
Cleveland
Freightliner
Terminal
(Wilton)
Phillips Siding Jn. GF
3m 25ch
North Tees (AOCL)
4m 19ch
Grangetown SB (G)
18m 65ch
Billingham-on-Tees
SB (B)
63m 60ch / 0m 00ch
Haverton Hill East
Grid Sidings
to / from
Phillips
Petroleum
Beam Mill Jn.
18m 03ch
Slag Road 0m 14ch
to / from
BSC Lackenby
Port Clarence GF
3m 05ch
South Bank Jn.
17m 31ch
to / from
Port Clarence Sidings
Middlesbrough
15m 00ch
to / from
Stockton Haulage
South Bank
17m 40ch
Middlesbrough
Goods Yard
Whitehouse (MCB) 15m 76ch
Tees SB (TY)
12m 70ch
Sidings
Cargo Fleet Road
(CCTV) 0m 14ch
Whitehouse SB (W)
15m 76ch
Tees Yard
Down Staging
Sidings
Middlesbrough SB (M)
14m 71ch
Tees Yard
Arrivals/Departures
Newport East Jn.
14m 03ch
Guisborough Jn.15m 30ch
0m 00ch (Whitby Line)
Wagon
Repair Depot
Thornaby
11m 63ch
Thornaby
Motive Power Depot
Level Crossings from Seal Sands Branch Jn.
to Seal Sands Run Round Loop:
1 ICI Brinefield (OPEN) 0m 12ch
2 NEEB (OPEN) 0m 39ch
3 Enron (OPEN) 0m 52ch
4 North/South (OPEN) 0m 71ch
5 Rohm Haas (AOCL) 1m 42ch
6 Monsanto / BASF (AOCL) 1m 46ch
7 Biofuels (OPEN) 1m 74ch
8 Seal Sands Chemical (AOCL) 2m 11ch
9 Phillips No. 2 (AOCL) 2m 16ch
10 Seal Sands Road (AOCL) 2m 18ch
11 Phillips No. 3 (AOCL) 2m 22ch
C
From Map 70
TEESSIDE

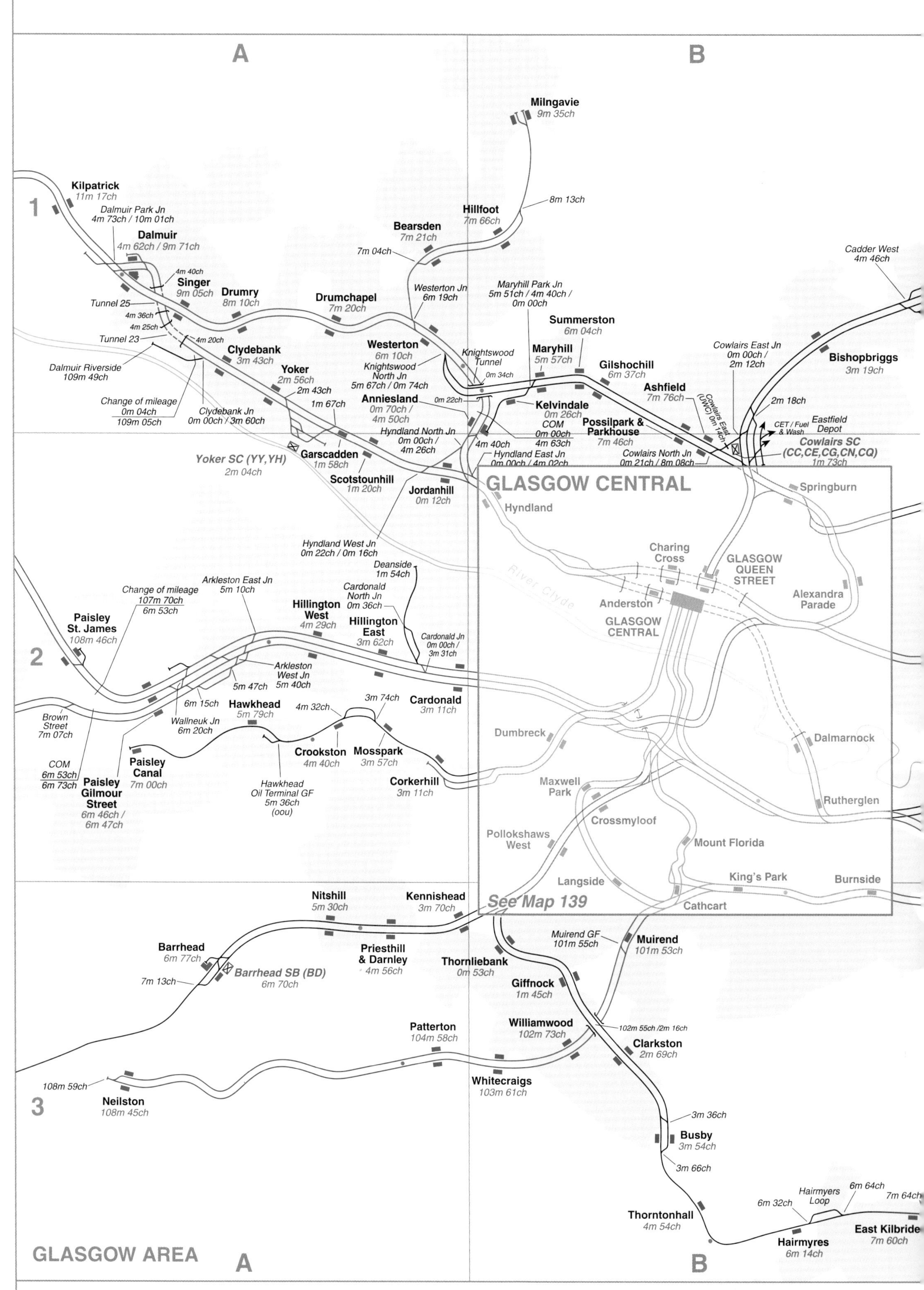
A
B
1
2
3
Milngavie
9m 35ch
8m 13ch
Hillfoot
7m 66ch
Bearsden
7m 21ch
7m 04ch
Kilpatrick
11m 17ch
Dalmuir Park Jn
4m 73ch / 10m 01ch
Dalmuir
4m 62ch / 9m 71ch
4m 40ch
Singer
9m 05ch
Drumry
8m 10ch
Drumchapel
7m 20ch
Westerton Jn
6m 19ch
Maryhill Park Jn
5m 51ch / 4m 40ch / 0m 00ch
Summerston
6m 04ch
Tunnel 25
4m 36ch
4m 25ch
Tunnel 23
4m 20ch
Dalmuir Riverside
109m 49ch
Change of mileage
0m 04ch
109m 05ch
Clydebank Jn
0m 00ch / 3m 60ch
Clydebank
3m 43ch
Yoker
2m 56ch
2m 43ch
1m 67ch
Westerton
6m 10ch
Knightswood North Jn
5m 67ch / 0m 74ch
Knightswood Tunnel
0m 34ch
Maryhill
5m 57ch
Gilshochill
6m 37ch
Anniesland
0m 70ch / 4m 50ch
0m 22ch
Kelvindale
0m 26ch
COM
0m 00ch
4m 63ch
Possilpark & Parkhouse
7m 46ch
Ashfield
7m 76ch
Cowlairs East Jn
0m 00ch / 2m 12ch
Cowlairs East (UWC) 0m 14ch
Cowlairs North Jn
0m 21ch / 8m 08ch
Cadder West
4m 46ch
Bishopbriggs
3m 19ch
2m 18ch
CET / Fuel & Wash
Eastfield Depot
Cowlairs SC (CC,CE,CG,CN,CQ)
1m 73ch
Hyndland North Jn
0m 00ch / 4m 26ch
4m 40ch
Hyndland East Jn
0m 00ch / 4m 02ch
Yoker SC (YY,YH)
2m 04ch
Garscadden
1m 58ch
Scotstounhill
1m 20ch
Jordanhill
0m 12ch
GLASGOW CENTRAL
Hyndland
Springburn
Charing Cross
GLASGOW QUEEN STREET
Alexandra Parade
Anderston
GLASGOW CENTRAL
River Clyde
Hyndland West Jn
0m 22ch / 0m 16ch
Deanside
1m 54ch
Cardonald North Jn
0m 36ch
Change of mileage
107m 70ch
6m 53ch
Arkleston East Jn
5m 10ch
Paisley St. James
108m 46ch
Hillington West
4m 29ch
Hillington East
3m 62ch
Cardonald Jn
0m 00ch / 3m 31ch
Arkleston West Jn
5m 40ch
5m 47ch
6m 15ch
Brown Street
7m 07ch
Wallneuk Jn
6m 20ch
Hawkhead
5m 79ch
4m 32ch
3m 74ch
Cardonald
3m 11ch
Crookston
4m 40ch
Mosspark
3m 57ch
Corkerhill
3m 11ch
COM
6m 53ch
6m 73ch
Paisley Canal
7m 00ch
Paisley Gilmour Street
6m 46ch / 6m 47ch
Hawkhead Oil Terminal GF
5m 36ch
(oou)
Dumbreck
Dalmarnock
Maxwell Park
Crossmyloof
Rutherglen
Pollokshaws West
Mount Florida
Langside
King's Park
Burnside
See Map 139
Cathcart
Nitshill
5m 30ch
Kennishead
3m 70ch
Barrhead
6m 77ch
7m 13ch
Barrhead SB (BD)
6m 70ch
Priesthill & Darnley
4m 56ch
Thornliebank
0m 53ch
Muirend GF
101m 55ch
Muirend
101m 53ch
Giffnock
1m 45ch
Williamwood
102m 73ch
102m 55ch /2m 16ch
Clarkston
2m 69ch
Patterton
104m 58ch
Whitecraigs
103m 61ch
108m 59ch
Neilston
108m 45ch
3m 36ch
Busby
3m 54ch
3m 66ch
Thorntonhall
4m 54ch
Hairmyers Loop
6m 32ch
6m 64ch
7m 64ch
Hairmyres
6m 14ch
East Kilbride
7m 60ch
GLASGOW AREA
A
B

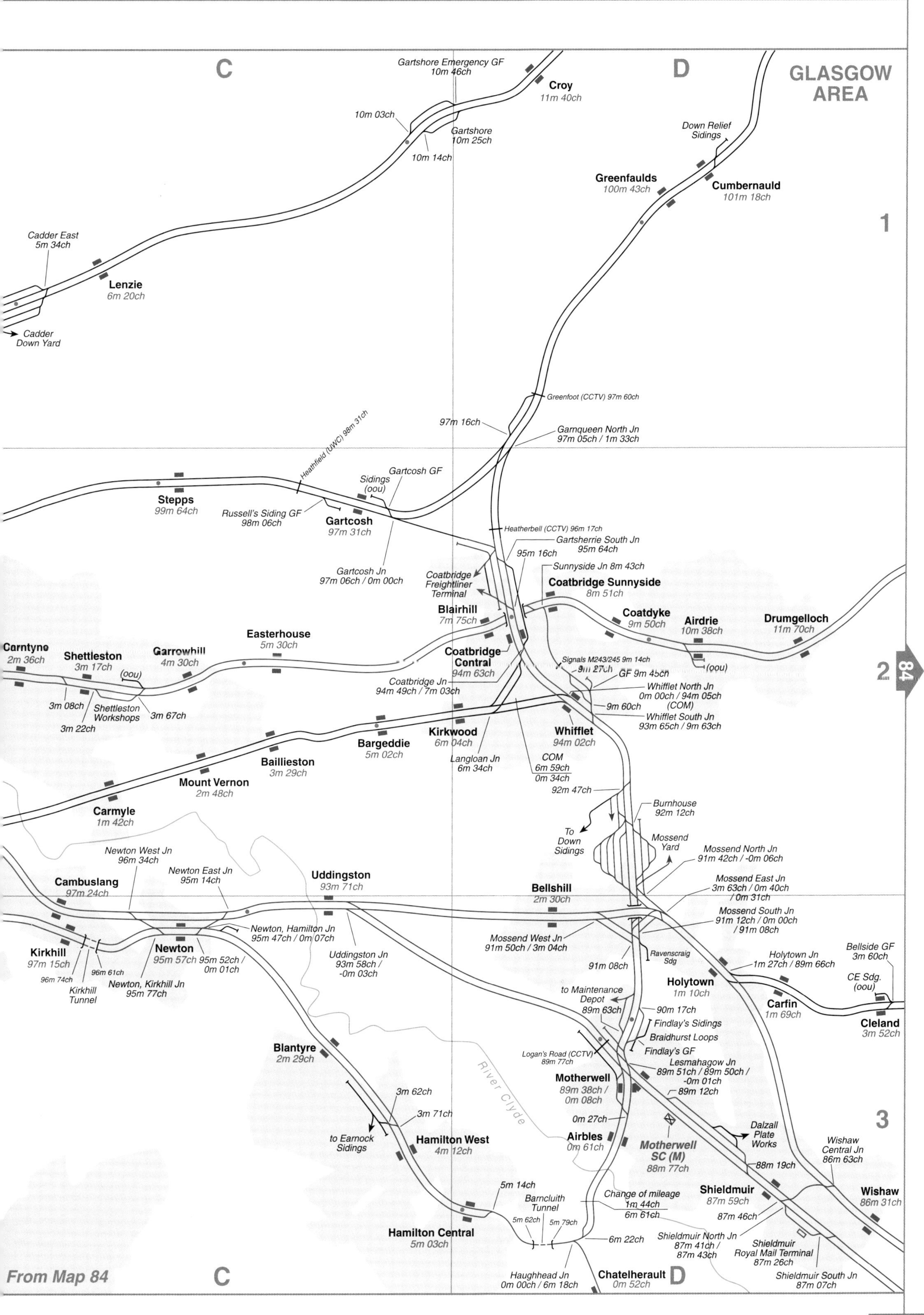

GLASGOW AREA
C
D
1
2
3
84
Gartshore Emergency GF
10m 46ch
Croy
11m 40ch
10m 03ch
Gartshore
10m 25ch
10m 14ch
Down Relief
Sidings
Greenfaulds
100m 43ch
Cumbernauld
101m 18ch
Cadder East
5m 34ch
Lenzie
6m 20ch
Cadder
Down Yard
Greenfoot (CCTV) 97m 60ch
97m 16ch
Garnqueen North Jn
97m 05ch / 1m 33ch
Heathfield (UWC) 98m 31ch
Gartcosh GF
Sidings
(oou)
Stepps
99m 64ch
Russell's Siding GF
98m 06ch
Gartcosh
97m 31ch
Heatherbell (CCTV) 96m 17ch
Gartsherrie South Jn
95m 64ch
95m 16ch
Sunnyside Jn 8m 43ch
Gartcosh Jn
97m 06ch / 0m 00ch
Coatbridge
Freightliner
Terminal
Coatbridge Sunnyside
8m 51ch
Blairhill
7m 75ch
Coatdyke
9m 50ch
Airdrie
10m 38ch
Drumgelloch
11m 70ch
Easterhouse
5m 30ch
Carntyne
2m 36ch
Shettleston
3m 17ch
Garrowhill
4m 30ch
(oou)
Coatbridge
Central
94m 63ch
Signals M243/245 9m 14ch
9m 27ch
GF 9m 45ch
(oou)
Whifflet North Jn
0m 00ch / 94m 05ch
(COM)
9m 60ch
Whifflet South Jn
93m 65ch / 9m 63ch
Coatbridge Jn
94m 49ch / 7m 03ch
3m 08ch
Shettleston
Workshops
3m 67ch
3m 22ch
Kirkwood
6m 04ch
Whifflet
94m 02ch
Bargeddie
5m 02ch
Langloan Jn
6m 34ch
COM
6m 59ch
0m 34ch
Baillieston
3m 29ch
Mount Vernon
2m 48ch
Carmyle
1m 42ch
92m 47ch
Burnhouse
92m 12ch
To
Down
Sidings
Mossend
Yard
Mossend North Jn
91m 42ch / -0m 06ch
Newton West Jn
96m 34ch
Newton East Jn
95m 14ch
Cambuslang
97m 24ch
Uddingston
93m 71ch
Bellshill
2m 30ch
Mossend East Jn
3m 63ch / 0m 40ch
/ 0m 31ch
Mossend South Jn
91m 12ch / 0m 00ch
/ 91m 08ch
Newton, Hamilton Jn
95m 47ch / 0m 07ch
Mossend West Jn
91m 50ch / 3m 04ch
Ravenscraig
Sdg
Kirkhill
97m 15ch
Newton
95m 57ch
95m 52ch /
0m 01ch
Uddingston Jn
93m 58ch /
-0m 03ch
91m 08ch
Holytown Jn
1m 27ch / 89m 66ch
Bellside GF
3m 60ch
96m 74ch
96m 61ch
Kirkhill
Tunnel
Newton, Kirkhill Jn
95m 77ch
Holytown
1m 10ch
CE Sdg.
(oou)
to Maintenance
Depot
89m 63ch
90m 17ch
Carfin
1m 69ch
Cleland
3m 52ch
Findlay's Sidings
Braidhurst Loops
Findlay's GF
Blantyre
2m 29ch
Logan's Road (CCTV)
89m 77ch
Lesmahagow Jn
89m 51ch / 89m 50ch /
-0m 01ch
River Clyde
Motherwell
89m 38ch /
0m 08ch
89m 12ch
3m 62ch
3m 71ch
0m 27ch
Dalzall
Plate
Works
to Earnock
Sidings
Hamilton West
4m 12ch
Airbles
0m 61ch
Motherwell
SC (M)
88m 77ch
Wishaw
Central Jn
86m 63ch
88m 19ch
5m 14ch
Barncluith
Tunnel
Change of mileage
1m 44ch
6m 61ch
Shieldmuir
87m 59ch
Wishaw
86m 31ch
87m 46ch
5m 62ch
5m 79ch
Hamilton Central
5m 03ch
6m 22ch
Shieldmuir North Jn
87m 41ch /
87m 43ch
Shieldmuir
Royal Mail Terminal
87m 26ch
From Map 84
Haughhead Jn
0m 00ch / 6m 18ch
Chatelherault
0m 52ch
Shieldmuir South Jn
87m 07ch

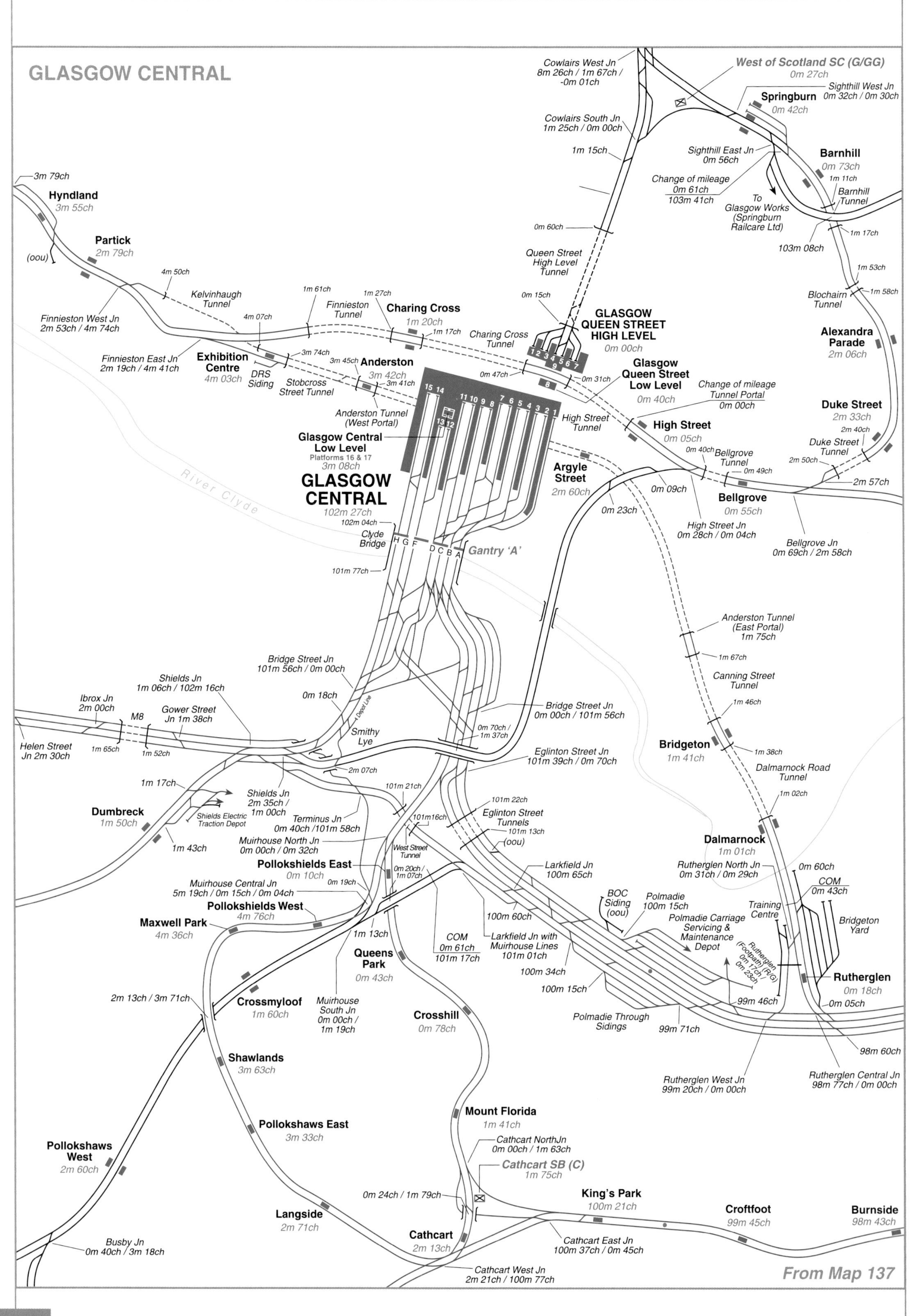

GLASGOW CENTRAL
Cowlairs West Jn
8m 26ch / 1m 67ch / -0m 01ch
West of Scotland SC (G/GG)
0m 27ch
Springburn
0m 42ch
Sighthill West Jn
0m 32ch / 0m 30ch
Cowlairs South Jn
1m 25ch / 0m 00ch
1m 15ch
Sighthill East Jn
0m 56ch
Barnhill
0m 73ch
Change of mileage
0m 61ch
103m 41ch
1m 11ch
Barnhill Tunnel
To Glasgow Works (Springburn Railcare Ltd)
1m 17ch
103m 08ch
0m 60ch
3m 79ch
Hyndland
3m 55ch
Partick
2m 79ch
(oou)
Queen Street High Level Tunnel
1m 53ch
1m 58ch
Blochairn Tunnel
4m 50ch
Kelvinhaugh Tunnel
1m 61ch
1m 27ch
Finnieston Tunnel
Charing Cross
1m 20ch
0m 15ch
GLASGOW QUEEN STREET HIGH LEVEL
0m 00ch
Finnieston West Jn
2m 53ch / 4m 74ch
4m 07ch
1m 17ch
Charing Cross Tunnel
Alexandra Parade
2m 06ch
Finnieston East Jn
2m 19ch / 4m 41ch
Exhibition Centre
4m 03ch
DRS Siding
3m 74ch
3m 45ch
Anderston
3m 42ch
3m 41ch
Stobcross Street Tunnel
0m 47ch
0m 31ch
Glasgow Queen Street Low Level
0m 40ch
Change of mileage
Tunnel Portal
0m 00ch
Duke Street
2m 33ch
Anderston Tunnel (West Portal)
High Street Tunnel
High Street
0m 05ch
2m 40ch
Duke Street Tunnel
Glasgow Central Low Level
Platforms 16 & 17
3m 08ch
0m 40ch
Bellgrove Tunnel
2m 50ch
0m 49ch
2m 57ch
GLASGOW CENTRAL
102m 27ch
Argyle Street
2m 60ch
0m 09ch
Bellgrove
0m 55ch
River Clyde
0m 23ch
102m 04ch
Clyde Bridge
Gantry 'A'
High Street Jn
0m 28ch / 0m 04ch
Bellgrove Jn
0m 69ch / 2m 58ch
101m 77ch
Anderston Tunnel (East Portal)
1m 75ch
1m 67ch
Canning Street Tunnel
Bridge Street Jn
101m 56ch / 0m 00ch
Shields Jn
1m 06ch / 102m 16ch
0m 18ch
1m 46ch
Ibrox Jn
2m 00ch
M8
Gower Street Jn 1m 38ch
Depot Line
Bridge Street Jn
0m 00ch / 101m 56ch
Smithy Lye
0m 70ch / 1m 37ch
Helen Street Jn 2m 30ch
1m 65ch
1m 52ch
Bridgeton
1m 41ch
1m 38ch
Eglinton Street Jn
101m 39ch / 0m 70ch
2m 07ch
Dalmarnock Road Tunnel
1m 17ch
101m 21ch
1m 02ch
Dumbreck
1m 50ch
Shields Electric Traction Depot
Shields Jn
2m 35ch / 1m 00ch
Terminus Jn
0m 40ch /101m 58ch
101m16ch
101m 22ch
Eglinton Street Tunnels
101m 13ch
(oou)
1m 43ch
Muirhouse North Jn
0m 00ch / 0m 32ch
West Street Tunnel
Dalmarnock
1m 01ch
Pollokshields East
0m 10ch
Larkfield Jn
100m 65ch
Rutherglen North Jn
0m 31ch / 0m 29ch
0m 60ch
COM
0m 43ch
Muirhouse Central Jn
5m 19ch / 0m 15ch / 0m 04ch
0m 19ch
0m 20ch / 1m 07ch
BOC Siding (oou)
Polmadie
100m 15ch
Training Centre
Pollokshields West
4m 76ch
Maxwell Park
4m 36ch
100m 60ch
Polmadie Carriage Servicing & Maintenance Depot
Bridgeton Yard
1m 13ch
COM
0m 61ch
101m 17ch
Larkfield Jn with Muirhouse Lines
101m 01ch
Queens Park
0m 43ch
Rutherglen (Footpath) (R/G)
0m 17ch
0m 23ch
100m 34ch
Rutherglen
0m 18ch
100m 15ch
2m 13ch / 3m 71ch
Crossmyloof
1m 60ch
Muirhouse South Jn
0m 00ch / 1m 19ch
99m 46ch
0m 05ch
Crosshill
0m 78ch
Polmadie Through Sidings
99m 71ch
Shawlands
3m 63ch
98m 60ch
Rutherglen West Jn
99m 20ch / 0m 00ch
Rutherglen Central Jn
98m 77ch / 0m 00ch
Mount Florida
1m 41ch
Pollokshaws East
3m 33ch
Cathcart NorthJn
0m 00ch / 1m 63ch
Pollokshaws West
2m 60ch
Cathcart SB (C)
1m 75ch
0m 24ch / 1m 79ch
King's Park
100m 21ch
Croftfoot
99m 45ch
Burnside
98m 43ch
Langside
2m 71ch
Cathcart
2m 13ch
Busby Jn
0m 40ch / 3m 18ch
Cathcart East Jn
100m 37ch / 0m 45ch
Cathcart West Jn
2m 21ch / 100m 77ch
From Map 137

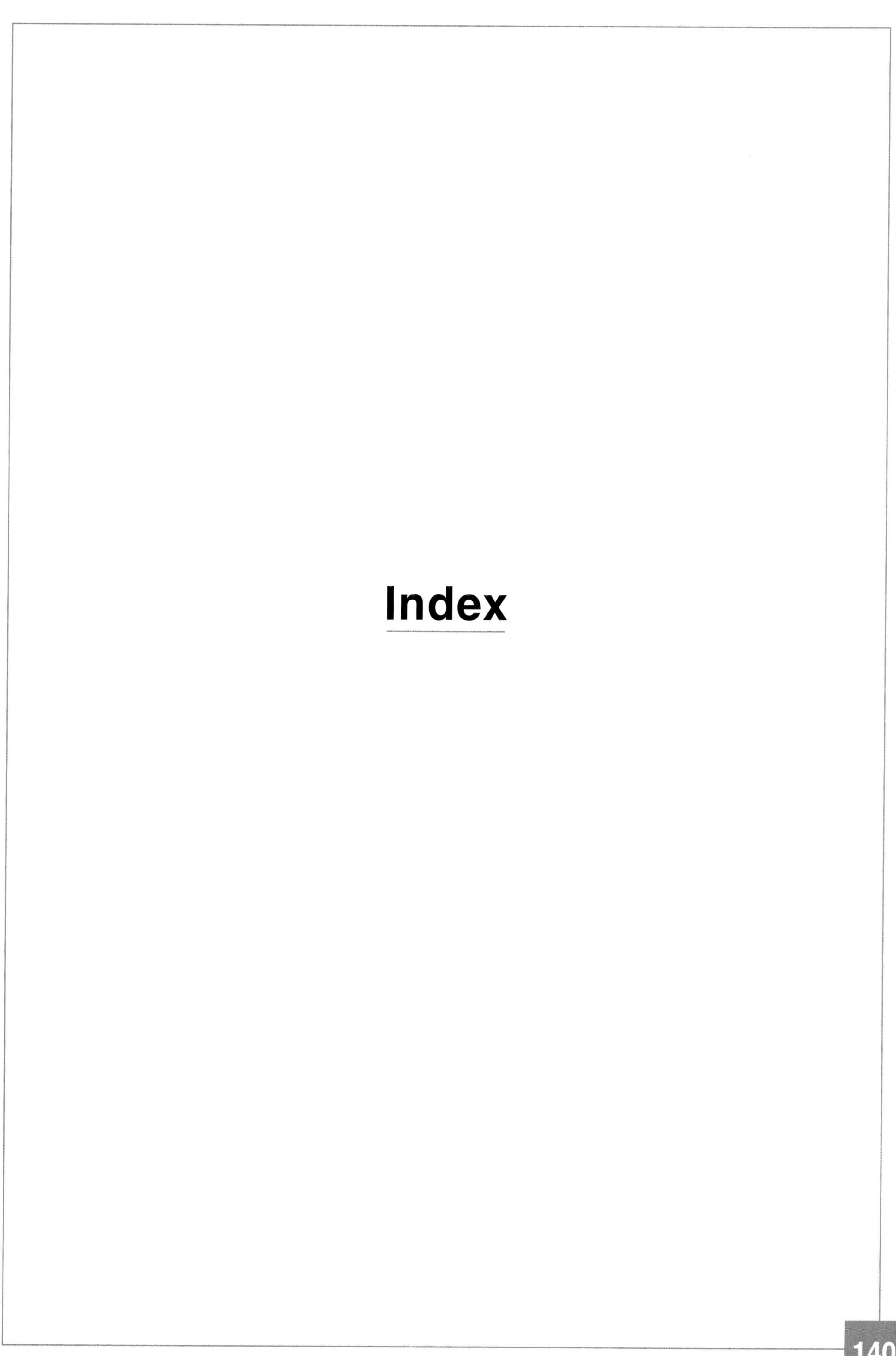

Index

Index

This Index lists the named locations that appear in the main map and inset map sections of the Atlas in alphabetical order. The reference which appears after each name gives the map number and the grid reference of the square in which the name or its related relevant symbol appears. Stations are shown in capital letters. To create an index of manageable size, the full list of names was subjected to reduction against a set of rules. Duplicates were removed and the editor selected the most appropriate reference for such duplicated items. Generally, level crossings, ground frames and sidings were subjected to a greater reduction in the number of entries than junction and tunnels; sidings especially so. Where a location appears within an Area or Inset map, that reference is given in preference to the reference within the main body of maps.
Heritage lines are listed by the commonly used name for the railway and the primary stations only. All names are in capital letters and in italic form.

Astley Bridge Jn. 129 B1
Astley SB (AY) 129 A2
ASTON 126 A
Aston Hall LC 29 A1
Aston Magna No.1 LC 29 A1
Aston North Jn. 126 A
Aston SB (AN) 126 B
Aston South Jn. 126 A
Aston-by-Stone (CCTV) 48 D2
Astral (Shuttlewords) (UWC) 50 B3
Athelney (AHBC) 7 A1
ATHERSTONE 40 C1
ATHERTON 129 A1
Atherton Goods Yard SB (AN) 129 A1
Atkinson Wood Farm(UWC) 71 A1
Atkinsons (UWC) 135 A1
Atmire (UWC) 33 B2
ATTADALE 97 B2
ATTENBOROUGH 128 A
Attenborough (CCTV) 128 A
Attenborough Jn. 128 A
ATTLEBOROUGH 44 C1
Attleborough (MCB-OD) 44 D1
Attleborough North Jn. 40 C1
Attleborough South Jn. 40 C1
Auchencruive GF 78 D1
Auchengray (AHBC-X) 85 A3
Auchinduich No. 1 (UWC) 102 C1
Auchinduich No. 2 (UWC) 102 C1
AUCHINLECK 79 A1
Auchintee (UWC) 97 B2
Auchterarder SB (AR) 91 B2
Auckley (AHBC) 59 A1
AUDLEY END 32 D1
Audley End (Norfolk) (AHBC-X) 44 D2
Audley End Tunnel 32 D1
AUGHTON PARK 55 B1
Aughton Road (CCTV) 61 B3
Auldclune No. 3 (UWC) 93 A2
Austin Crossing 22 C3
Aveley Viaduct 20 D1
AVIEMORE 94 D1
AVIEMORE 94 D1
Aviemore SB (AV) 94 D1
Avon View Farm (UWC) 16 D2
AVONCLIFF 16 C2
Avoncliff Mill (UWC) 16 C2
AVONMOUTH 124
Avonmouth Dock (CCTV) 124
Avonmouth Station (CCTV) 124
Awre (CCTV) 28 C3
Axe (AHBC) 7 A2
AXMINSTER 7 A3
Axminster (CCTV) 7 A3
Axminster East Jn. 7 A3
Axminster West Jn. 7 A3
Aycliffe 70 C1
AYLESBURY 30 D2
Aylesbury Jn. 30 D2
Aylesbury North Goods Loop 30 D2
Aylesbury Vale Jn. 30 D2
AYLESBURY VALE PARKWAY 30 D2
AYLESFORD 21 A3
Aylesford 21 A3
Aylesford SB (AF) 21 A3
Aylesford Village (CCTV) 21 A3
AYLESHAM 22 C3
Aynho Jn. 30 C1
AYR 78 C1
Ayr Harbour 78 C1
Ayr MPD 78 D1

B

B Group GF 127 A
B.A.C. (UWC) 124
BACHE 132 A
Back Settlement (R/G-X) 95 A1
Badcock's Middle (UWC) 6 C2
Badden (UWC) 102 D2
Badgeney Road (AHBC-X) 43 A1
Bagillt (UWC) 55 B3
BAGLAN 122 B
BAGSHOT 19 A3
Bagshot Tunnel 19 A3
Baguley Fold Jn. SB (BF) 128 B
Bagworth Jn. 50 C3
BAILDON 63 B2
Baildon Tunnels 63 B2
Baileyfield GF 86 E
Baileys 42 D1
Baileys (UWC) 49 A2
BAILLIESTON 138 C2
Bainton (AHBC) 51 B3
Bainton Green (AHBC) 51 B3
Baker's 11 B1
Bakers (Accom) 13 B2
Balavil Burn (UWC) 94 D2
Balavil Gates (UWC) 94 D2
Balby Bridge Tunnel 135 A1
Balcathie (UWC) 95 A3
BALCOMBE 12 C1
Balcombe Tunnel 12 C1
Balcombe Tunnel Jn. 12 C1
Balderton (AHBC) 55 B3
Balderton (CCTV) 51 A1
Balderton Tunnel 55 B3
Baldinnies No. 1 (UWC) 91 B2
BALDOCK 32 C1
Baldwin's No. 78 44 D1
Balfour (UWC) 73 B2
BALHAM 112 D2
Balham Jn. 112 D2
Balkeith South (UWC) 102 D2
Ballach (UWC) 103 B1
Ballachladdich Farm (UWC) 99 A1
Ballachraggin (UWC) 99 A1
Ballachroan (UWC) 94 C2
Ballast Hole 13 A1
Ballentoul (UWC) 93 A2
Ballintraid (UWC) 99 B1
BALLOCH 83 B1
BALMOSSIE 92 D1
Balnacra (AOCL) 97 B2
Balnacra No. 2 (UWC) 97 B2
Balne (MCB) 64 D3
Balne Lane 133 A
Balne Low Gate 64 D3
Balshaw Lane Jn. 62 C3
Balspardon (UWC) 99 B2
Balsporran (UWC) 94 C3
Baltic Siding 10 C1
BAMBER BRIDGE 62 C3
Bamber Bridge (MCB) 62 C3
BAMFORD 57 B2
Bamfurlong Jn. 132 C
Bamfurlong Sidings Jn. 132 C
BANAVIE 90 C2
Banavie (R/C) 90 C2
Banavie SC (B) 90 C2
BANBURY 30 C1
Banbury North SB (BN) 29 B1
Banbury Road GF 30 C2
Banbury South SB (BS) 29 B1
Banc-y-Berllan (UWC) 24 D2
BANGOR 53 B2
Bangor SB (BR) 53 B2
Bangor Tunnel 53 B2
BANK HALL 131 A
Bank Head (UWC) 62 C3
Bank House Tunnel 63 B3
Bank Jn 79 A2
Banks Crossing 9 B2
Bannolds (AHBC-X) 43 A3
BANSTEAD 116 C2
BARASSIE 78 C1
Barassie Jn 78 D1
Barcroft (MCG) 58 D1
Bardon Hill GF 50 C3
Bardon Hill Quarries 50 C3
Bardon Hill SB (BH) 50 C3
BARDON MILL 75 A1
Bardon Mill (R/G) 75 A2
Bardon Mill SB (BM) 75 A1
BARE LANE 62 C1
Bare Lane (MCB) 62 C1
Bare Lane Jn.h 62 C1
Bare Lane SB (BL) 62 C1
Barf Farm (UWC) 72 C3
Barford Footpath Crossing (R/G) 8 D1
BARGEDDIE 138 C2
BARGOED 26 C1
Bargoed SB (BD) 26 C1
Bargoed South Jn. 26 C1
Barham Sidings 44 D3
Barhams 45 A3
Barkers (UWC) 49 A2
BARKING 110 D2
Barking Flyover 110 C2
Barking Freightliner Terminal 110 D3
Barking Station Jn. 110 C2
Barking Tilbury Line Jns. 110 C2
BARLASTON 48 D1
Barlaston (CCTV) 48 D1
Barlby (BOCM) (MCB) 65 A2
Barlby North Jn. 65 A2
BARMING 21 A3
Barmoor Clough Tunnel 57 B2
BARMOUTH 35 B2
Barmouth North GF 35 B2
Barmouth South (TMO) 35 B2
Barmouth Swing Bridge 35 B2
Barmouth Tunnel 35 B2
Barn End (UWC) 45 A1
Barnby 45 B2
Barnby (CCTV) 51 A1
Barnby Lane (CCTV) 51 A1
Barnby Moor & Sutton (CCTV) 59 A2
Barncluith Tunnel 138 D3
BARNEHURST 20 D2
BARNES 112 C2
Barnes (UWC) 51 B2
BARNES BRIDGE 112 C2
Barnes Jn. 112 C2
Barnet South Crossovers 32 C3
Barnet Tunnel 108 D1
BARNETBY 60 C1
Barnetby East SB (BE) 60 C1
BARNHAM 11 A3
Barnham SB (BH) 11 A3
BARNHILL 139
Barnhill SB (BH) 92 C2
Barnhill Tunnel 139
Barn's Green (AHBC) 11 B2
BARNSLEY 58 C1
Barnsley (MCB) 58 C1
Barnsley SB (BY) 58 C1
Barnsley Station Jn. 58 C1
BARNSTAPLE 5 A1
Barnstaple GF 5 A1
Barnstone Tunnel 50 C2
BARNT GREEN 39 A2
Barnt Green Jn. 39 A2
Barnwell Jn. 43 A3
Barnwell Siding 43 A3
Barnwood No. 3 GF 28 D2
Baron House (R/G-X) 74 D1
Baron Wood No.1 Tunnel 74 C3
Baron Wood No.2 Tunnel 74 C2
Barrel Lane (R/G) Footpath 59 A3
BARRHEAD 137 A3
Barrhead SB (BD) 137 A3
BARRHILL 77 B2
Barrhill SB (BR) 77 B2
BARROW HAVEN 66 C3
Barrow Haven (OPEN) 66 C3
Barrow Hill North Jn. 58 C2
Barrow Hill South Junction 58 C3
Barrow Road (MCG) 66 C3
Barrow Road Refuse Transfer Stn 125 A
BARROW UPON SOAR 50 C2
Barrowby Lane (Public BW) 64 C3
BARROW-IN-FURNESS 61 A1
Barrow-in-Furness SB (BF) 67 A3
BARRY 25 E
BARRY DOCKS 25 E
Barry Docks 25 E
BARRY ISLAND 25 E
Barry Island Viaduct 25 E
Barry Jn. 25 E
BARRY LINKS 95 A3
Barry SB (B) 25 E
Barry West (CCTV) 95 A3
Barthomley (R/G) 48 C1
Barthomley Jn. 48 C1
Barton & Broughton Loop 62 C2
Barton Hill Depot 125 B
Barton Hill SB BH) 65 A1
Barton Lane (AHBC) 128 A
Barton Mill (MCB) 65 A1
Barton Mill Carriage Sidings 18 D3
Barton North Jn. 49 B3
Barton Road (MCG) 66 C3
Barton South Jn. 49 B3
Barts (Accom) 13 B2
Barway Sidings (AHBC) 43 A2
Baschurch (AHBC-X) 47 A2
Base Green No. 31 44 C3
Basford Chemicals (UWC) 128 A
Basford Hall Jn. 48 E
Basford Hall Jn. SB (BH) 48 E
Basford Hall Sidings 48 E
Basford Wood GF 48 E
BASILDON 21 A1
Basin Road (CCTV) 11 A3
BASINGSTOKE 18 D3
Basingstoke SCC (BE) 18 C3
Basins (UWC) 45 A1
BAT & BALL 20 D3
Bates (UWC) 59 B2
Bath Goods 16 C2
Bath Road Jn. 19 A2
Bath Row Tunnel 126 B
BATH SPA 16 C2
Bath West GF 16 C2
Bathampton Jn. 16 C2
BATHGATE 85 A2
Bathgate East Jn 85 A2
Bathgate LMD 85 A2
Bathing Pool (UWC) 68 C3
Bathley Lane (CCTV) 59 A3
BATLEY 64 C3
Batley SB (B) 64 C3
BATTERSBY 71 A2
Battersby Jn. 71 A2
Battersby Road (AOCL) 71 A1
BATTERSEA PARK 119
Battersea Park Jn. 119
Battersea Pier Jn. 119
Battersea Pier Sdgs. Staff Halt 119
Battersea Railway Bridge 119
BATTLE 13 A2
Battle Road (AHBC) 13 A2
BATTLESBRIDGE 21 A1
Baulkholme (UWC) 65 A3
Bawtry Crossovers GSP 58 D2
Bax Crossing 21 B3
BAYFORD 32 C3
Baylam (AHBC-X) 44 D3
Bayles & Wylies (UWC) 50 C1
Beach Road (RC) 12 D3
Beacon Hill Tunnel 63 B3
BEACONSFIELD 31 A3
Beal (CCTV) 82 D3
Beal Crossovers 82 D3
Beald Drove (AHBC-X) 43 A2
Bealings (ABCL) 34 C1
Beam Mill Jn. 136 C
BEARLEY 39 B3
Bearley Jn. 39 B3
Bears Grove 46 D2
BEARSDEN 137 A1
BEARSTED 21 A3
Bearty Fen (MCG) 52 C2
BEASDALE 89 A2
Beattock North GSP 80 C2
Beattock South 80 C2
Beattock Summit 80 C2
BEAULIEU ROAD 9 B2
BEAULY 99 A2
Beauly Ferry (UWC) 99 A2
Beaumont Hill (UWC) 39 B2
Beaverpool Tunnel 54 C3
Beavers (UWC) 51 B1
Beavers Hill (OC) 23 B3
Beavers Hill Farm (UWC) 23 A3
BEBINGTON 131 A
Bebside (AHBC-X) 76 C1
BECCLES 45 B2
Beccles Bypass (ABCL) 45 B2
BECKENHAM HILL 114 C3
BECKENHAM JUNCTION 113 B3
BeckenhamSpur Jn. 113 B3
Becketts (AHBC) 13 B2
BECKFOOT 67 B2
Beckingham SB (B) 59 A2
BEDALE 70 C2
BEDDGELERT 53 B3
Beddingham (AHBC) 12 C3
Bedford (No. 125) (UWC) 43 A2
BEDFORD MIDLAND 121 A
Bedford North Jn. 121 A
Bedford South Jn. 121 A
BEDFORD ST. JOHNS 121 A
Bedford Station Jn. 121 A
BEDHAMPTON 10 D2
Bedhampton (CCTV) 10 D2
Bedlam Tunnel 16 C3
Bedlington North SB (BN) 81 A3
Bedlington South SB (BS) 81 A3
Bedmill Farm No. 1 Crossing 7 B2
BEDMINSTER 124
BEDWORTH 40 C1
BEDWYN 17 B2
Beech Bank (UWC) 52 C2
Beech Drive (UWC) 17 B2
Beech Hill (AHBC) 59 A1
Beeches Farm (UWC) 55 B3
Beechgrove GF 16 D3
Beechwood Tunnel 40 C2
Beecroft (Wright's) (UWC) 44 D2
BEESTON 128 A
Beeston Castle & Tarporley SB (BC) 56 C3
Beeston Depot GF 128 A
Beeston North Jn. 128 A
Beeston Regis 46 C1
Beeston South Jn. 128 A
Beighton Jn. 134 A1
Beighton Station Jn SB (BX) 134 A1
Bejowan (UWC) 2 C1
BEKESBOURNE 22 C3
Belasis Lane SB (BL) 136 C
Belaugh Lane 46 D2
Belford (CCTV) 81 A1
Belford Burn (Public FP) 82 D3
Belford Crossovers 82 D3
Bell House (UWC) 67 A1
Bellasize (UWC) 65 B3
Belle Isle 109 A3
BELLE VUE 130 C2
Belleport (UWC) 99 B1
BELLGROVE 139
Bellgrove Jn 139
Bellgrove Tunnel 139
BELLINGHAM 113 B3
Bells 45 A1
BELLSHILL 138 D2
Bellside GF 138 D3
Bellwater Jn. SB (BJ) 52 E
BELMONT 116 D2
Belmont (CCTV) 78 D1
Belmont (MCB) 64 C1
Belmont Tunnel 53 B2
BELPER 49 B1
Belper GF 50 C1
Belsize Tunnels 108 C2
BELTRING 13 A1
Beltring (AHBC) 13 A1
BELVEDERE 20 D2
Belvedere Rd Jn. 120 B
BEMPTON 72 D3
Bempton (AHBC) 72 D3
Bempton Sands (UWC) 72 D3
Ben Alder (UWC) 94 C3
BEN RHYDDING 63 B2
Benarth Farm (UWC) 27 A2
BENFLEET 21 A1
Benhar 84 D2
Beningborough Footpath (R/G) 64 D1
Bennar Fawr (AOCL) 35 B2
Bennetland (UWC) 65 A3
Bennetts (UWC) 49 A2
Bensham Tunnel 136 A
BENTHAM 62 D1
Bentley (AHBC-X) 34 C2

C

Canning Street North SB 131 B
Canning Street Tunnel 139
CANNOCK 49 A3
CANNON STREET 120 B
Cannon StreetBridge 120 B
Cannons Mill Lane Footpath 32 D2
CANONBURY 109 A2
Canonbury Tunnel 109 A2
Canonbury West Jn. 109 A2
Canterbury / Wye ACC (EBT) 22 C3
CANTERBURY EAST 22 C3
Canterbury Line Jn. 14 C1
Canterbury Rd Jn. 119
CANTERBURY WEST 22 C3
Canterbury West SB (EDH) 22 C3
CANTLEY 45 A1
Cantley (MWG) 45 A1
Cantley SB (C) 45 A1
Canton Depot 123 A
CAPEL BANGOR 35 F
Capeldrae 92 C3
CAPENHURST 55 B3
CARBIS BAY 1 A2
Carbis Branch Jn. 2 D1
Cardells (R/G) 42 C2
CARDENDEN 92 C3
CARDIFF BAY 123 A
CARDIFF CENTRAL 123 A
Cardiff Docks 123 A
Cardiff East Jn. 123 A
CARDIFF QUEEN STREET 123 A
Cardiff SB (C) 123 A
Cardiff West Jn. 123 A
CARDONALD 137 A2
Cardonald Jn 137 A2
Cardonald North Jn. 137 A2
CARDROSS 83 B1
Cardross (CCTV) 83 A1
CARFIN 138 D3
Cargo Fleet Road (CCTV) 136 C
CARK & CARTMEL 68 C3
Carleton Crossing (CN) SB 61 B2
CARLISLE 127 A
Carlisle Kingmoor Yard 127 A
Carlisle Lane Jn. 119
Carlisle North Jn. 127 A
Carlisle SB (CE) 127 A
Carlisle South Jn 127 A
CARLTON 50 E
Carlton (CCTV) 59 A3
Carlton Lane (UWC) 8.59 55 B2
Carlton Loops 59 A3
Carlton Road Jn. 108 C2
CARLUKE 84 D3
CARMARTHEN 24 C2
Carmarthen Bridge Jn. 24 A2
Carmarthen Jn SB (CJ) 24 C2
Carmont (MCB) 96 C3
Carmont SB (CM) 96 C3
Carmuirs East Jn 84 D1
Carmuirs East SB (CEJ) 84 D1
Carmuirs West Jn. 84 D1
CARMYLE 138 C2
Carnaby (AHBC-X) 66 C1
CARNFORTH 62 C1
CARNFORTH 68 C3
Carnforth North Jn. 62 C1
Carnforth South Jn. 62 C1
Carnforth Station Jn. 68 C3
Carnforth Station Jn. SB (CS) 68 C3
Carno (AHBC) 36 D1
CARNOUSTIE 95 A3
Carnoustie (MCB) 95 A3
Carnoustie SB (CA) 95 A3
CARNTYNE 138 C2
CARPENDERS PARK 31 A3
Carpenters Road North Jn. 109 B2
Carpenters Road South Jn. 110 C2
Carr House Farm (UWC) 72 C3
Carr Lane (UWC) 55 B2
Carr Lane (UWC) 135 A1
CARRBRIDGE 94 D1
Carriage sidings 67 B3
CARROG 47 E
Carrs (UWC) 73 B2
Carsebreck (UWC) 91 A3
Carsgoe No. 1 (UWC) 104 C2
Carsgoe No. 2 (UWC) 104 C2
Carsgoe No. 3 (UWC) 104 C2
Carsgoe No. 4 (UWC) 104 C2
CARSHALTON 116 D1
CARSHALTON BEECHES 116 D2
CARSTAIRS 85 A3
Carstairs East Jn. 85 A3
Carstairs South Jn. 85 A3
Carstairs Station Jn. 85 A3
Cart Lane (UWC) 68 C3
Carterhouse Jn. (UWC) 56 C2
Carters (UWC) 70 D1
Cartsburn Tunnel 83 A2
CARTSDYKE 83 A1
Casefield (UWC) 33 B2
CASTLE BAR PARK 107 B3
Castle Bromwich Jn. 126 A
CASTLE CAEREINION 37 E
CASTLE CARY 7 B1
Castle Cary Jn. 7 B1
Castle Donington 128 A
Castle Farm No. 1 (UWC) 67 B3
Castle Gardens (UWC) 37 A2
Castle Hill East GF 70 D2
Castle Hill Portal 14 D1
Castle Hill Tunnel 63 A3
Castle Hill West GF 70 D2
Castle Hills Jn. 70 D2
Castlefield Jn. 128 B
CASTLEFORD 64 C3
Castleford East Jn. 64 C3
Castleford LC SB (CD) 64 D3
Castleford West Jn. 64 C3
CASTLETON 57 A1
Castleton Central Materials Depot 56 D1
Castleton East Jn. SB (CE) 130 C1
Castleton Farm Crossing 8 C2
Castleton LWR Depot 57 A1
CASTLETON MOOR 71 A1
Castleton North Jn. 130 C1
Castleton South Jn. 130 C1
Cat Lane (UWC) 64 D2
Cater Crossing 22 D3
CATERHAM 117 B3
CATFORD 113 B2
CATFORD BRIDGE 113 B2
Cathan Farm (UWC) 24 D2
CATHAYS 123 A
CATHCART 139
Cathcart East Jn. 139
Cathcart North Jn. 139
Cathcart SB (C) 139
Cathcart West Jn. 139
CATTAL 64 D1
Cattal (MCG) 64 D1
Cattal SB (C) 64 D1
Cattewater Jn. 122 A
Cauldwell Depot 121 A
CAUSELAND 3 A2
Causeway (CCTV) 18 C1
Causewayhead Jn. 91 A3
Cave (MCB) 65 B3
Cavendish Street Tunnel 131 B
Caverswall (MCB) 49 A1
Caverswall SB (CL) 49 A1
Cawdor (AOCL) 25 A1
Cayton (AHBC) 72 C3
Cedar Jn. 108 D3
Cefn Coed (UWC) 25 A1
Cefn Gast Farm No. 2 (UWC) 36 E
Cefn Jn. 25 B2
Cefn Suran (UWC) 37 A2
CEFN-Y-BEDD 47 A1
Cement Depot No. 1 (OC) 127 A
Cement Works Sidings 16 D3
Cemetery (UWC) (Aberdovey) 35 B3
Cemetery Lane (UWC) 16 C2
Cemetery LC (Kingussie) 94 C2
Cemmes Road (R/G) 36 C1
Central OrdnanceDepot GF 30 C2
Central Rivers Depot 49 B2
Central Treviscoe GF 2 C1
Central Tunnel 130 C1
CHADWELL HEATH 110 D2
Chaffers (TMO) 63 A2
CHAFFORD HUNDRED 20 D1
Chain Bridge (AHBC) 42 D1
Chain Bridge (MCB) 76 C2
CHALFONT & LATIMER 31 A3
Chalk Lane (CCTV) LC 66 C3
Chalk Tunnel 19 A3
CHALKWELL 21 B1
Challoch (UWC) 77 B3
Challow 17 B1
Chandlers Crossing 9 E
CHANDLERS FORD 9 B2
Chandlers Ford (UWC) 9 B2
Channelsea Freight Loop 109 B2
Channelsea North Jn. 109 B2
Channelsea South Jn. 110 C2
Chantry (Footpath) 8 D1
Chantry Road (RC) 45 A3
Chapel (AOCL) 2 C1
Chapel Farm 1 (UWC) 2 C1
Chapel Farm 3 (UWC) 2 C1
Chapel L.N.W. Tunnel 57 B2
Chapel Lane (MCG) 56 C1
Chapel Lane GF 26 D2
Chapel Road (MCG) 45 A1
CHAPEL-EN-LE-FRITH 57 B2
Chapel-en-le-Frith SB (CH) 57 A2
CHAPELTON 5 A1
Chapelton (UWC) 99 A2
CHAPELTOWN 58 C1
Chapleton Station (UWC) 5 A1
CHAPPEL & WAKES COLNE 33 B2
Chard Jn. 7 A2
Chard Jn. (CCTV) 7 A2
Chard Jn. Down Sidings GF 7 A2
Charfield Hall Farm (UWC) 16 C1
Charfield Loops 16 C1
CHARING 21 B3
CHARING CROSS 139
Charing Cross Tunnel 139
CHARLBURY 29 B2
Charlbury Jn. 29 B2
Charlemont Road (R/G) 126 A
Charleston (UWC) 91 B1
Charlestown Foundry (TMO) 85 A1
Charlestown Jn. 85 A1
CHARLTON 114 C1
Charlton (UWC) 28 D1
Charlton Jn. 114 C1
Charlton Lane 114 C1
Charlton Tunnel 114 C1
Charlton Tunnel (Bristol) 124
Charlton Tunnel (Dover) 14 D1
Chart Crossing 54m 00ch 13 B1
Chart Leacon Repair Shop 13 B1
CHARTHAM 22 C3
Chartham 22 C3
Chartham Hatch (AHBC) 22 C3
CHASSEN ROAD 129 B2
CHATELHERAULT 138 D3
CHATHAM 21 A3
Chatham Dockyard 21 A2
Chatham Tunnel 21 A2
CHATHILL 81 A1
Chathill (CCTV) 81 A1
Chathill Crossovers 81 A1
Chatterley Valley 48 D1
Chawson Footpath Crossing 38 D3
CHEADLE HULME 130 C3
Cheadle Hulme North Jn. 130 C3
Cheadle Hulme South Jn. 130 C3
Cheadle Jn. 130 C3
Cheadle Wood Farm (UWC) 130 C3
Cheal Road (MCG) 52 C2
CHEAM 116 C2
CHEDDINGTON 31 A2
CHEDDLETON 49 A1
CHELFORD 56 D3
Chelford North Jn. 56 D3
Chelford South Jn. 56 D3
CHELMSFORD 33 A3
CHELSFIELD 118 D1
Chelsfield Tunnel 118 D2
Cheltenham Alstone (MCB) 28 D2
CHELTENHAM RACECOURSE 28 D2
CHELTENHAM SPA 28 D2
Chenson No. 1 (UWC) 5 B2
Chenson No. 2 (UWC) 5 B2
Chenson No. 3 (UWC) 5 B2
CHEPSTOW 27 B3
Chepstow Tunnel 15 B1
Cheriton Jn. 14 D1
Cherry Hinton By-pass (CCTV) 43 A3
Cherry Hinton High Street (CCTV) 43 A3
Cherry Holt (AHBC-X) 52 C2
Cherry Orchard (UWC) 37 A1
CHERRY TREE 62 D3
Cherry Tree (CCTV) 66 C2
Cherry Tree GF 62 D3
Cherry Willingham (AHBC-X) 60 C3
Cherryhalt (UWC) 59 A2
Cherry's No 4 (UWC) 30 C1
CHERTSEY 19 B2
CHESHUNT 32 C3
Cheshunt Jn. 32 C3
CHESSINGTON NORTH 115 B1
CHESSINGTON SOUTH 115 B2
CHESTER 132 A
Chester East Jn. 132 A
Chester Line Jn. 18 E
Chester North Jn. 132 A
Chester PSB (CR) 132 A
CHESTER ROAD 126 A
Chester South Jn. 132 A
Chester Station Yard 132 A
Chester Traincare Centre 132 A
Chester West Jn. 132 A
CHESTERFIELD 58 C3
Chesterfield North Jn. 58 C3
Chesterfield South Jn. 58 C3
CHESTER-LE-STREET 76 C2
Chesterton (CCTV) 43 A3
Chesterton Jn. 43 A3
Chesterton Yard 43 A3
CHESTFIELD &SWALECLIFFE 22 C3
CHETNOLE 7 B2
Chettisham (AHBC-X) 43 A2
Chevington (CCTV) 81 A2
Chevington North Crossovers 81 A2
CHICHESTER 11 A3
Chichester SB (CC) 10 D2
Chicks Farm Crossing 8 C3
Chicks Farm Foot (UWC) 8 C3
CHILHAM 22 C3
Chilham Mill (CCTV) 22 C3
Chilham Road (CCTV) 22 C3
Chillingham Road 136 A
Chiltern Railways LMD 107 B2
Chiltern Railways Servicing Depot 30 D2
Chilvers Coton Jn. 40 C1
CHILWORTH 19 A3
Chilworth (CCTV) 11 A1
CHINGFORD 109 B1
CHINLEY 130 D3
Chinley East Jn. 57 B2
Chinley North Jn. 57 B2
Chinley SB (CY) 57 B2
Chinley South Jn. 57 B2
CHINNOR 30 D3
CHINNOR & PRINCES RISBOROUGH RAILWAY 30 D3
Chinnor Branch Sdg 30 D2
CHIPPENHAM 16 D2
Chippenham Jn. 43 B2
Chippenham Jn. SB (CM) 43 B3
Chipping Campden (CCTV) 29 A1
Chipping Sodbury East GF 16 C1
Chipping Sodbury Tunnel 16 C1
CHIPSTEAD 116 D3
Chipstead Line Jn. 117 A2
CHIRK 47 A2
Chirk Tunnel 47 A2
CHISLEHURST 118 F
Chislehurst Jn. 118 F
Chislehurst Tunnels 114 C3
Chiswells Farm (UWC) 30 C1
CHISWICK 112 C2
Chitts Hill (MCB) 33 B2
Chivers (AHBC-X) 43 B2
CHOLSEY 18 C1
CHOLSEY 18 C1
CHOLSEY & WALLINGFORD RAILWAY 18 C1
Choppington (AHBC) 76 C1
CHORLEY 62 C3
Chorley Tunnel 62 C3
CHORLEYWOOD 31 A3
CHRISTCHURCH 9 A3
Christian Street Jn. 120 B
Christleton Tunnel 56 C3
Christon Bank (CCTV) 81 A1
CHRIST'S HOSPITAL 11 B1
CHURCH & OSWALDTWISTLE 62 D3
Church (Footpath) 13 A1
Church Crossing 39m 30ch 21 A3
Church Dam 12m 76ch 45 B1
Church End Farm (UWC) 11m 20ch 64 D2
Church Farm (UWC) 13m 69ch 65 B3
Church Farm (UWC) 7m 09ch 25 B3
Church Farm No. 1 (UWC) 101m 39ch 16 D2
Church Farm No. 2 (UWC) 102m 10ch 16 D2
CHURCH FENTON 64 D2
Church Fenton North Jn. 64 D2
Church Fenton South Jn. 10m 77ch 64 D2
Church House Farm 33 B2
Church House Farm (UWC) 38 C3
Church Lane 45 B1
Church Lane (CCTV) 33 A1
Church Lane (CCTV) 48 A3
Church Lane (CCTV) 59 A3
Church Lane (CCTV) 71 D2
Church Lane (MCG) 52 C2
Church Path Sidings 20 D2
Church Road 45 A1
Church Road Tunnel 126 A
Church Street (AHBC-X) 33 B3
Church Street (CCTV) 76 D3
Church Street (TMO) 41 A3
CHURCH STRETTON 37 B1
CHURNET VALLEY RAILWAY 49 A1
CHURSTON 4 D2
CILMERI 36 D3
Cilmeri (UWC) 36 D3
Cilmeri Tunnel 36 D3
Cilyrychen (ABCL) 24 D2
City Basin Jn. 6 C3
CITY THAMESLINK 120 B
ClachnaharryCanal Bridge 99 A2
Clacton SB (C) 34 C3
CLACTON-ON-SEA 34 C3
CLANDON 19 B3
CLAPHAM 62 D1
CLAPHAM HIGH STREET 119
CLAPHAM JUNCTION 119
Clapham Yard 119
CLAPTON 109 B2
Clapton GF 109 B2
Clapton Jn. 109 B2
Clapton Tunnel 109 B2
Clara Vale (AHBC-X) 75 B1
CLARBESTON ROAD 23 A2
Clarbeston Road Jn SB (CR) 23 A2
Clarbeston Road Jn. 23 A2
Clarborough Jn. 59 A2
Clarborough Tunnel 59 A2
Clarence Road SB (CR) 76 D3
Clark & Butchers 43 A2
Clarks (UWC) 70 D2
CLARKSTON 137 B3
CLAVERDON 39 B3
Claverton (UWC) 16 C2
Claxby & Usselby (AHBC-X) 60 C2
Claxby Gatehouse (No. 24) (AHBC-X) 60 C1
Clay Cross North Jn. 58 C3
Clay Cross South Jn. (former) 58 C3
Clay Cross Tunnel 58 C3
Clay Lane (AHBC) 10 D2
Clay Mills Jn. 49 B2
Clay Mills Jn. (CCTV) 49 B2
Claydon (AOCL) 30 D1
Claydon (CCTV) 34 C1
Claydon (L&NE) Jn SB (CN) 30 D1
Claydon (L&NE) Jn. 30 D1
Clayfield (AHBC) 29 A1

CLAYGATE 115 B2
Claypole (CCTV) 51 A1
Claypole Down Loop 51 A1
Claypole Up Loop 51 A1
Clayton Bridge (CCTV) 130 C2
Clayton Tunnel 12 C2
Clayton West Jn. 57 B1
Clayway 43 A2
Cleatop (UWC) 62 D1
CLEETHORPES 60 D1
Cleghorn (CCTV) 84 D3
Cleifiog Uchaf (UWC) 53 A2
CLELAND 138 D3
Clements No. 1 (UWC) 59 B3
Clerkenwell No. 1 Tunnel 109 A3
Clerkenwell No. 2 Tunnel 109 A3
Clerkenwell No. 3 Tunnel 109 A3
Clerks Tunnel 4 D1
Cleveland Freightliner Terminal (Wilton) 136 C
Cliff Hill No.1 GF 50 C3
Cliff Hill No.2 GF 50 C3
Cliffe (CCTV) 65 A2
Cliffe Vale Jn. 48 D1
Clifford Farm (Footpath) 21 B1
Cliffsend (AHBC) 22 D3
CLIFTON 129 B1
Clifton (CCTV) 76 C1
Clifton (CCTV) 81 A3
Clifton and Lowther No.1 GF 68 D1
Clifton and Lowther No.2 GF 68 D1
Clifton BridgeNo. 1 Tunnel 124
Clifton BridgeNo. 2 Tunnel 124
CLIFTON DOWN 124
Clifton DownTunnel 124
Cliftonville Tunnel 14 E
Clink Road Jn. 16 C3
Clints Field (UWC) 62 D1
Clints Field (UWC) 68 D3
Clipstone Colliery 58 D3
Clipstone Colliery Jn. 58 D3
Clipstone East Jn. 58 D3
Clipstone SB (C) 58 D3
Clipstone South Jn. 58 D3
Clipstone West Jn. 58 D3
CLITHEROE 62 D2
CLOCK HOUSE 113 B3
Clock House (AHBC) 21 B3
Closglas Farm 1 (UWC) 24 D2
Closglas Farm 3 (UWC) 24 D2
Clover-le-Dale (UWC) 55 B1
Club Lane (UWC) 62 C3
CLUNDERWEN 23 B2
Clunes (UWC) 93 A2
Clyde Bridge 139
CLYDEBANK 137 A1
Clydebank Jn. 137 A1
Clyne (TMO) 25 B1
Clynemilton East (UWC) 103 B3
Coal Road 21 A2
Coaley GF 28 C3
Coalville Jn. 50 C3
Coalville Station (CCTV) 50 C3
COATBRIDGE CENTRAL 138 D2
Coatbridge Freightliner Terminal 138 D2
Coatbridge Jn. 138 D2
COATBRIDGE SUNNYSIDE 138 D2
COATDYKE 138 D2
Coatsworth Farm No.2 (UWC) 81 A3
COBHAM & STOKE D'ABERNON 115 A2
Cockett Tunnel 122 B
Cockett West Jn. 24 D3
Cocklebury Sidings 17 A1
Cockshute Sidings 48 D1
Cockspin Road (AHBC-X) 43 B2
Codford (AHBC) 16 D3
Codnor Park Jn. 50 C1
CODSALL 39 A1
Coed Cae No. 1 (UWC) 36 D1
Coed Ddol (UWC) 36 C1
Coed Farm No. 1 (UWC) 24 C2
Coed Ifan (UWC) 36 E
Coed Moor (UWC) 27 B1
Coed Y Dinas (UWC) 47 A3
Coed-y-Llyn No.1 (UWC) 35 B1
Coffue Drove 43 A2
COGAN 123 A
Cogan Jn. 123 A
Cogan Loops 123 A
Cogan Tunnel 123 A
Cogload Jns 7 A1
COLCHESTER 33 B2
Colchester Jn. 33 B2
Colchester PSB (CO) 33 B2
Colchester Road (CCTV) 34 C2
COLCHESTER TOWN 33 B2
Cold Blow (R/G) 14 D1
Coldham (TMO) 52 D3
Coldham Lane Depot 121 B
Coldham Lane Jn. 121 B
Coldharbour (OPEN) 13 B2
Coleham LM Depot 47 B3
Coleham Sidings 47 B3
Coleman St. Tunnel 114 C1
Colemans (UWC) 8 C3
Coles (UWC) 39 A3
Coleshill East Jn. 39 B1
COLESHILL PARKWAY 39 B1
Coleshill West Jn. 39 B1
COLESLOGGETT HALT 2 D1
Collaton Barton Farm 1 (UWC) 5 B2
College Farm (UWC) 32 C1
Colliers (UWC) 49 A2
COLLINGHAM 59 B3
Collingham (AHBC) 59 B3
COLLINGTON 13 A3
Collins Farm (UWC) 3 B1
Collyhurst Street Sdgs. 128 B
Colnbrook Oil Terminal 19 B2
COLNE 63 A2
Colne Jn. 33 B2
Colthrop (MCB) 18 C2
Coltishall Lane 46 D2
Coltness 84 D3
Colton Jn. 64 D2
Colton North Jn. 64 D2
Colton South Jn. 64 D2
COLWALL 28 C1
ColwallTunnel 28 C1
Colwich Jn. 49 A2
Colwick (CCTV) 50 E
COLWYN BAY 54 D1
COMBE 29 B2
Combe Lane Crossing 19 B3
Common Moor 1 (UWC) 5 B3
Common Moor 2 (UWC) 5 B3
Common Road (MCG) 65 A1
COMMONDALE 71 A1
Compeday (UWC) 18 C2
Concrete Road (UWC) 43 A1
Coney Green Jn. 58 C3
CONGLETON 57 A3
CONISBROUGH 58 D1
Conisbrough Tunnel 58 D1
CONNEL FERRY 87 B1
Connell Park (TMO) 79 A2
Connington North (CCTV) 42 C1
Connington South 42 C1
CONONLEY 63 A2
Cononley (CCTV) 63 A2
CONSALL 49 A1
CONWAY PARK 131 B
CONWY 54 C2
Conwy Tubular Bridge 54 C2
Conwy Tunnel 54 C2
COODEN BEACH 13 A3
Cookes (Accom) 13 B2
COOKHAM 19 A1
Cookham (ABCL) 19 A1
Cooks (CCTV) 115 B2
Cooks 1 (UWC) 28 D1
Cooks 2 (UWC) 28 D1
Cooks Lane (UWC) 40 D1
Cooks No. 1 (Accom) 13 B2
COOKSBRIDGE 12 C2
Cooksbridge (CCTV) 12 C2
Cooksholme (UWC) 28 D1
COOMBE 3 A2
Coombe (UWC) 3 A2
Coombe Jn. 2 D1
Coombe No 1 GF 2 D1
Coombe No 2 GF 2 D1
Coopers 13 B2
Coopers (UWC) 48 D1
Coopers (UWC) 62 C3
Coopers Lane Jn. 32 D2
Coopies Lane (AHBC) 81 A3
Copenhagen Jn. 108 D2
Copenhagen Tunnel 108 D2
Copley Hill East Jn. 133 B
Copley Hill West Jn. 133 B
Copmanthorpe No. 2 (R/G) 64 D2
Copperas (UWC) 34 C2
Coppermill Jn. 109 B2
COPPLESTONE 5 B2
Copy Pit Crossover 63 A3
Copyhold Jn. 12 C2
CORBRIDGE 75 B1
CORBY 41 B1
Corby Automotive Terminal 41 B1
Corby Gates (MCB) 74 C2
Corby Gates SB (CG) 74 C2
Corby North Jn. 41 B1
Corby Tunnel 41 B1
CORFE CASTLE 8 F
CORKERHILL 137 A2
CORKICKLE 67 A1
Corks Farm No. 2 49 B3
Cornard (ABCL) 33 B1
Cornton (AHBC) 91 A3
Cornton No. 2 (R/G) 91 A3
CORPACH 90 C2
Corpach (AOCL) 90 C2
Corporation Rd Tunnel 131 A
Corriebeg Farm No. 1 (UWC) 89 B2
Corriebeg No. 2 (UWC) 89 B2
Corriemoillie No. 1 (UWC) 98 D1
CORROUR 90 D3
Corscombe (UWC) 5 A3
Corton Steps Crossing 16 D3
CORYTON 123 A
Coryton Tunnel 4 D1
COSELEY 39 A1
COSFORD 38 D1
COSHAM 10 C2
Cosham Jn. 10 D2
Coswarth (AOCL) 2 C1
Coswarth 1 & 2 (UWC) 2 C1
Coswarth 3 & 4 (UWC) 2 C1
Coswarth Tunnel 2 C1
Coton Hill Yard 47 B3
Cottage Jn. 118 E
Cottage Lane (AHBC) 59 B3
Cottam Power Station 59 A2
COTTINGHAM 66 C2
Cottingham North (CCTV) 65 B2
COTTINGLEY 64 C3
Cottons (UWC) 128 A
Coulags No. 1 (UWC) 97 B2
Coulags No. 3 (UWC) 97 B2
COULSDON SOUTH 117 A3
COULSDON TOWN 117 A2
Coundon Road (CCTV) 40 C2
COUNTRY PARK HALT 38 D1
COUNTY SCHOOL 44 F
Couplands (UWC) 52 E
Court Farm (UWC) 37 A1
Court Sart Jn. (Down Line) 122 B
Courthill Loop North Jn. 114 E
Courthill Loop South Jn. 114 E
Courtybella 123 B
Cousins (UWC) 72 C3
Cousins No. 2 (UWC) 33 A2
Couston Bottom Crossing 9 A1
Cove (CCTV) 73 B1
COVENTRY 40 C2
Coventry North Jn. 40 C2
Coventry South Jn. 40 C2
Coventry Yard 40 C2
Covered Way 108 D2
Cow Creek (UWC) 44 D3
Cow Crossing 12 C3
Cow Green GSP 44 D3
Cow Lane 11 A3
Cowbridge Rd SB (CR) 25 B3
Cowburn Tunnel 57 B2
COWDEN 12 D1
COWDENBEATH 85 B1
Cowlairs East (UWC) 137 B1
Cowlairs East Jn. 137 B1
Cowlairs North Jn. 137 B2
Cowlairs SC (CC,CE,CG,CN,CQ) 137 B2
Cowlairs South Jn. 139
Cowlairs West Jn. 139
Cowley Bridge Jn. 6 C3
Cowpen Lane (AHBC-X) 136 C
Coxall Farm 1 (UWC) 37 B2
Coxall Farm 2 (UWC) 37 B2
Coxbench (TMO) 50 C1
Coxes Farm 43 A3
Coxmoor (UWC) 5 B3
Cox's Walk (UWC) 51 A2
Coychurch Footpath (R/G-X) 25 B2
Crabley Creek SB (CC) 65 B3
Crabtree (R/G) 56 C1
Cradburns No. 4 (UWC) 51 B1
Cradlehall 99 B2
CRADLEY HEATH 39 A2
Cradley heath (CCTV) 39 A2
Crag Hall SB 71 A1
Crag Hill Farm (UWC) 69 A3
Cragmill (CCTV) 82 D3
Craig (UWC) 97 B2
Craig No. 2 (UWC) 77 B3
Craig Rhymney (UWC) 26 C1
CRAIGENDORAN 83 A1
Craigendoran Jn. 83 A1
Craigentinny 86 E
Craigentinny Depot 86 E
Craigfryn (UWC) 36 D1
Craiglockhart Jn. 85 E
Craigo SB (CO) 95 A1
CRAMLINGTON 76 C1
Crampmoor Crossing 9 B1
Cranbourne (R/G-X) 32 C3
Crane St. Jn. 127 B
Crankley Point (R/G) 51 A1
Cranmore Loop 13 A1
Crannaford (AHBC) 6 C3
Crannel's (UWC) 18 C2
Cranswick (AHBC-X) 66 C2
CRAVEN ARMS 37 B2
Craven Arms (MCB) 37 B2
Craven Arms Jn. 37 B2
Craven Arms SB (CA) 37 B2
Crawford (UWC) 9 B1
Crawford (UWC) 45 B1
CRAWLEY 12 C1
Crawley High Street (CCTV) 12 C1
Crawley New Yard 12 C1
Cray Lane (AHBC) 11 B2
CRAYFORD 20 D2
Crayford Creek Jn. 20 D2
Crayford Spur 'A' Jn. 20 D2
Crayford Spur 'B' Jn. 20 D2
Creag Mhor (UWC) 89 A1
Creag Ruadh LC 94 C2
Creamore Farm (UWC) 47 B2
Creasey's (Rye Meads) (UWC) 32 C3
CREDITON 5 B3
Crediton (MCB) 5 B3
Crediton SB (CN) 5 B3
Crescent Jn. 42 E
Crescent Road (AHBC) 61 B3
CRESSING 33 A2
Cressing (ABCL) 33 A2
CRESSINGTON 131 A
Cresswell (AHBC) 49 A2
CRESWELL 58 D3
Crewe 48 E
Crewe Bank SB (CB) 47 B3
Crewe Coal Yard SB (CY) 48 E
Crewe Electric Maintenance Depot 48 E
Crewe Jn. 47 B3
Crewe Jn. SB (CJ) 47 B3
Crewe North Jn. 48 E
Crewe SB (CE) 48 E
Crewe Sorting Sidings North SB (NH) 48 E
Crewe South Jn. 48 E
Crewe Steel Works SB (SW) 48 E
Crewe Works 48 E
CREWKERNE 7 B2
Crewkerne (AHBC) 7 B2
Crewkerne Tunnel 7 A2
CREWS HILL 32 C3
Creykes (R/G) 65 A3
CRIANLARICH 88 D2
Crianlarich Jn. 88 D2
CRICCIETH 35 A1
Crick Tunnel 40 D2
CRICKLEWOOD 108 C2
Cricklewood Curve Jn. 108 C2
Cricklewood Depot Jn. SB (CD) 108 C2
Cricklewood South Jn. 108 C2
Cridling Stubbs (AHBC) 64 D3
Crinow Farm 2 (UWC) 23 B2
Crismill Crossovers 21 A3
Critchlows (UWC) 49 A2
Croes Newydd North Fork SB (CN) 47 A1
Croft Sidings 40 D1
Croftcarnoch No. 2 (UWC) 94 D2
CROFTFOOT 139
Crofton (R/G) 17 B2
Crofton East Jn. 133 A
Crofton Old Station No.1 (MCG) 133 A
CROFTON PARK 113 B2
Crofton Road Jn. 113 B2
Crofton West Jn 133 A
CROMER 46 C1
Cromer Jn. 46 C1
CROMFORD 49 B1
Cromwell Lane (CCTV) 59 A3
Cromwell Road (ABCL) 45 A2
Crook Wheel (UWC) 68 C3
CROOKSTON 137 A2
Crosby Garrett Tunnel 69 A1
Crosfield's Crossing (CCTV)GF 132 B
Cross Brook Farm (UWC) 38 C3
Cross Common (AHBC-X) 65 A3
Cross Cottage (UWC) 16 C3
CROSS GATES 64 C2
Cross Lane (AHBC) 59 B3
Cross St. Tunnel 114 C1
CROSSFLATTS 63 B2
CROSSHILL 139
Crossing GF 35 A1
CROSSKEYS 26 D2
Crosskeys Jn. 26 D2
Crossley Evans Sidings 63 B2
CROSSMYLOOF 139
CROSTON 62 C3
CROUCH HILL 108 D2
Crouch Hill Tunnel 108 D2
Crow Nest Jn. 56 C1
Crow Nest Jn. SB (CN) 56 D2
CROWBOROUGH 12 D1
Crowborough Jn. 12 D1
Crowborough Tunnel 12 D1
CROWCOMBE HEATHFIELD 6 D1
CROWHURST 13 A2
Crowhurst Bridge (AHBC) 13 A2
CROWLE 59 A1
Crown Point Control Tower 44 E
Crown Point Depot 44 E
Crown St. Tunnel 131 B
CROWTHORNE 18 D2
Croxton (AHBC-X) 44 C2
CROY 138 D1
Cruach Snow Shed 90 D3
Crugwallins Siding 2 C2
Crundale (AHBC) 23 A2
Crundale Mill (UWC) 23 A2
CRYSTAL PALACE 113 B3
Crystal Palace Tunnel 113 A3
Cuaich (UWC) 94 C2
CUDDINGTON 56 C3
CUFFLEY 32 C3
Culcheth Farm (UWC) 129 A2
Culgaith (MCB) 74 D3
Culgaith SB (C) 74 D3
Culgaith Tunnel 74 D3
Culgower No. 1 (UWC) 103 B3
Culgower No. 3 (UWC) 103 B2
CULHAM 30 C3
Culloden 99 B2

DUNLOP 83 B3
Dunnerholme (UWC) 67 B3
Dunragit LC 77 A3
Dunragit SB (DR) 77 A3
DUNROBIN 102 D1
Dunrobin (UWC) 102 D1
Dunrod (UWC) 83 A2
Dunrod Loop East 83 A2
Dunrod Loop West 83 A2
Dunstalls (UWC) 49 B2
DUNSTER 6 E
DUNSTON 136 A
Dunston Flixton Mill Pump (UWC) 45 B1
DUNTON GREEN 20 D3
DURHAM 76 C2
Durham Farm 12 C3
Durn (UWC) 36 C1
Durnsford Road Sdgs. 112 D2
DURRINGTON-ON-SEA 11 B3
Duxford (AHBC) 32 D1
DYCE 96 D2
Dyce SB (DY) 96 D2
Dyffryn (UWC) 35 B3
DYFFRYN ARDUDWY 35 B1
DYMCHURCH 14 C2
Dynevor GF 122 B
Dynevor Jn. 122 B
Dytto Farm (UWC) 54 C2

E

Eagle and Thorpe (AHBC-X) 59 B3
Eagle Barnsdale (AHBC) 59 B3
Eagle Crossing 39 A1
EAGLESCLIFFE 70 D1
Eaglescliffe South Jn. 70 D1
EALING BROADWAY 111 B1
Earfit Lane (R/G) 64 D2
Earles Siding SB (ES) 57 B2
EARLESTOWN 56 C1
Earlestown East Jn. 56 C1
Earlestown South Jn. 56 C2
Earlestown West Jn. 56 C2
EARLEY 18 D2
Earlseat Open Cast 92 C3
EARLSFIELD 112 D2
EARLSWOOD (Surrey) 20 E
EARLSWOOD (W. Midlands) 39 B2
Easington (Public FP) 82 D3
East (Blackwells) (UWC) 70 D1
East Anglian Railway Museum 33 B2
East Bank Tunnel 134 A1
EAST BOLDON 76 D2
East Boldon (CCTV) 76 C2
East Brora Muir No. 1 (UWC) 103 B3
East Brora Muir No. 2 (UWC) 103 B3
East Burton (CCTV) 8 C3
East Chiltington 12 C2
East Claydon (UWC) 104 C2
East Cowick (R/G) 65 A3
East Cowton Crossovers 70 C2
EAST CROYDON 117 A1
East Dean Crossing 9 B1
EAST DIDSBURY 130 C3
EAST DULWICH 113 A2
East End 13 B1
EAST FARLEIGH 21 A3
East Farleigh SB (EF) 21 A3
EAST GARFORTH 64 C2
East Gate (CCTV) 34 C2
East Gate Jn. 34 C2
East Green 45 A3
East Grimstead GF 9 A1
EAST GRINSTEAD 12 C1
EAST GRINSTEAD 12 C1
East Guldeford (AHBC) 13 B2
East Ham Depot 110 C2
East Haugh No. 2 LC 93 B2
East Heslerton (AHBC-X) 72 C3
East Homes Jn. 59 B3
East Junction Viaduct 123 A
East Kent SC (EK) 21 A3
EAST KILBRIDE 137 B3
EAST LANCASHIRE RAILWAY 62 D3
East London Line SCC (EL) 120 B
EAST MALLING 21 A3
East Mendalgief 123 B
EAST MIDLANDS PARKWAY 128 A
East Midlands SC (#) 128 A
East Peckham Engineers Tip Siding 13 A1
EAST PUTNEY 112 C2
East PutneyTunnel 112 C2
East Road (R/G) 31 B1
East Shalford Crossing 11 A1
East Somerset Jn. 16 C3
EAST STAFF HALT 14 D1
East Suffolk Jn. 33 E
EAST TILBURY 21 A2
East Tilbury (CCTV) 21 A2
East Watten (UWC) 104 D2
EAST WORTHING 11 B3
EASTBOURNE 12 D3
EASTBROOK 123 B
Eastburn (UWC) 63 B2
Easter Balgour (UWC) 91 B2
Easter Dalguise No. 1 (UWC) 91 B1
Easter Dalguise No. 2 (UWC) 91 B1
EASTERHOUSE 138 C2
Eastern Access (AOCL) 66 C3
Eastern Jn. 21 B3
Easterton (UWC) 100 C1
Eastfield (UWC) 91 B2
Eastfield Depot 137 B1
Eastfield SB (E) 42 E
Eastfields Road (CCTV) 112 D3
Eastgate 75 B3
EASTHAM RAKE 55 B2
Easthaven (UWC) 95 A3
EASTLEIGH 10 E
Eastleigh SB (E) 10 E
Eastleigh South Jn. 10 E
Eastleigh Works 10 E
Eastriggs 73 B1
EASTRINGTON 65 A3
Eastrington (AHBC-X) 65 A3
Eastville (AHBC-X) 52 E
Eaton (R/G-X) 44 D1
Eaton Lane (Public BW) 59 A2
Eaves Lane (R/G) Bridleway 59 A3
Eaves Tunnel 57 B2
EBBSFLEET 20 D2
Ebbsfleet SR 20 D2
Ebbw Jn. 123 B
EBBW VALE PARKWAY 26 C1
Ebley (UWC) 28 C3
ECCLES 129 B2
Eccles (CCTV) 55 B1
ECCLES ROAD 44 C2
Eccles Road (MCB-OD) 44 C1
Eccles SB (ES) 129 B2
Eccles Station Jn. 129 B2
ECCLESBOURNE VALLEY RAILWAY 49 B1
Ecclesfield West 134 A1
ECCLESTON PARK 56 C2
Eckington 28 D1
ECML Up Sidings 109 A1
EDALE 57 B2
Edale SB (EE) 57 B2
Edderton No. 1 (UWC) 102 D2
EDEN PARK 117 B1
Eden Valley 68 D1
EDENBRIDGE 20 C3
EDENBRIDGE TOWN 12 C1
Edenbridge Tunnel 20 C3
EDGE HILL 131 B
Edge Hill East Jn. 131 A
Edge Hill SB (LE) 131 A
Edge Hill West Jn. 131 B
Edge Lane Jn. 131 A
Edgeley Jn. 130 C3
Edgeley Jn. No.1 130 C3
Edgeley Jn. No.1 SB (EY1) 130 C3
Edgeley Jn. No.2 SB (EY2) 130 C3
EDINBURGH PARK 85 B2
Edinburgh SC (E) 85 E
EDINBURGH WAVERLEY 85 E
EDMONTON GREEN 109 A1
Edstone Hall No. 1 (UWC) 39 B3
Effingham Jn. 115 A3
EFFINGHAM JUNCTION 115 A3
Eggborough Power Station 64 D3
EGGESFORD 5 B2
Eggesford (TMO) 5 B2
Egginton (AHBC) 49 B2
Egginton Jn. SB (EN) 49 B2
EGHAM 19 A2
Egham (CCTV) 19 A2
Egleton (UWC) 51 A3
Eglinton Street Jn. 139
Eglinton Street Tunnels 139
Egmanton (CCTV) 59 A3
EGTON 71 B2
Elderslie 83 B2
Elderslie East 83 B2
Elderslie West 83 B2
Eldrable No. 1 (UWC) 103 B2
ELEPHANT & CASTLE 120 B
Elephant (FP) 32 D2
Elford GF 49 B3
ELGIN 100 D1
Elgin (MCB) 100 D1
Elgin East Freight Yard 100 D1
Elgin SB (EL) 100 D1
Eliots (UWC) 128 A
Elland Tunnel 63 B3
Ellands No. 3 (UWC) 100 C1
ELLESMERE PORT 55 B2
Ellesmere Port SB (EP) 55 B3
Ellingers 34 D1
Elm Grove (UWC) 9.00 55 B2
Elm Road (AHBC) 42 D1
Elm Tree Farm (UWC) 71 B3
ELMERS END 117 B1
Elmley Lovett GF 39 A2
Elmside (UWC) 73 B1
Elmsley (OPEN) 58 D3
ELMSTEAD WOODS 114 C3
ELMSWELL 44 C3
Elmswell (CCTV) 44 C3
Elmton & Creswell Jn. SB (EC) 58 D3
ELSECAR 58 C1
ELSENHAM 32 D2
Elsenham (MCG) 32 D2
Elsenham GSP 32 D2
Elsham (MCB) 60 C1
Elsham SB (EM) 59 B1
ELSTREE & BOREHAMWOOD 31 B3
Elstree Tunnels 107 B1
ELTHAM 114 C2
ELTON & ORSTON 50 D1
Elworth Jn. 56 D3
ELY 43 A2
Ely Dock Jn. 43 A2
Ely Freight Terminal 43 A2
Ely North Jn. 43 A2
Ely Station North 43 A2
Ely West Curve No. 1 43 A2
Ely West Curve No. 2 43 A2
Ely West Jn. 43 A2
EMBSAY 63 B1
EMBSAY & BOLTON ABBEY RAILWAY 63 B1
EMERSON PARK 20 D1
EMSWORTH 10 D2
Endon (AOCL) 48 D1
ENFIELD CHASE 109 A1
ENFIELD LOCK 109 B1
Enfield Lock (CCTV) 109 B1
ENFIELD TOWN 109 A1
Engine Shed Jn. 65 A3
Engine Shed Jn. 133 B
Engine Sidings No. 1 GF 128 A
England Lane (MCG) 64 D3
England Springs (UWC) 66 C2
Englemere 19 A2
English Bridge Jn. 47 B3
Enham (UWC) 17 B3
ENTWISTLE 62 D3
EPSOM 116 C2
EPSOM DOWNS 116 C2
Eptons (UWC) 52 E
ERDINGTON 126 A
ERIDGE 12 D1
ERIDGE 12 D1
ERITH 20 D2
Ernesettle North GF 122 A
Ernesettle South GF 122 A
Errol LC (MCB) 92 C2
Errol SB (ER) 92 C2
Erwbeilli Farm (UWC) 36 E
ESHER 115 B1
Esholt Jn. 63 B2
Esholt Tunnel 63 B2
Essex Portals 20 D2
ESSEX ROAD 109 A3
Esso (UWC) 20 D2
Esso Siding (Cambridge) 121 B
Esso Siding (Cambridge) 121 B
Estrea (AHBC-X) 42 D1
ETCHINGHAM 13 A2
Etchingham (CCTV) 13 A2
Eton College Road 8 D1
Euro Terminal (Swindon) 17 A1
Euro Terminal (Willesden) 120 A
Eurofreight Terminal 129 B1
EUROTUNNEL TERMINAL 14 C1
EUSTON 108 D3
EUXTON BALSHAW LANE 62 C3
Euxton Jn. 62 C3
Evelench (UWC) 39 A3
Evershot Tunnel 7 B2
Everton (CCTV) 42 C3
EVESHAM 28 D1
Evesham SB (E) 28 D1
Evesham West Jn. 28 D1
EWELL EAST 116 C2
EWELL WEST 116 C2
EXETER CENTRAL 6 C3
Exeter Central Goods Jn. 6 C3
Exeter SB (E) 5 B3
EXETER ST DAVIDS 5 B3
Exeter St Davids Jn. 5 B3
EXETER ST THOMAS 6 C3
EXHIBITION CENTRE 139
EXMOUTH 4 D1
Exmouth Jn. 6 C3
Exmouth Jn. SB (EJ) 6 C3
EXTON 6 C3
Extons Road (CCTV) 43 B1
Eye Court Farm (UWC) 38 C3
Eyhorne Tunnel 21 A3
EYNSFORD 20 D2
Eynsford Tunnel 20 D2
Eyton (AHBC-X) 47 B2

F

Factory Jn. 119
FAIRBOURNE 35 B2
Fairbourne (AOCL) 35 B2
Fairburn Tunnel 64 D3
FAIRFIELD 130 C2
Fairfield (UWC) 17 B2
Fairfields 43 A1
Fairheads (R/G) (UWC) 32 D1
Fairheads No. 39 (UWC) 33 A2
FAIRLIE 83 A3
Fairlie High Siding GF 83 A3
Fairlie Tunnel 83 A3
FAIRWATER (TYLLGOED) 123 A
Fairwood Jn. 16 C3
Falcon Jn. 119
FALCONWOOD 114 D2
FALKIRK GRAHAMSTON 84 D1
FALKIRK HIGH 84 D1
Falkirk Tunnel 84 D1
Falkland 78 D1
Fallodon (CCTV) 81 A1
FALLS OF CRUACHAN 87 B2
FALMER 12 C3
Falmer Tunnel 12 C2
FALMOUTH 1 B3
FALMOUTH DOCKS 2 C3
Falmouth No 1 GF 1 B3
Falmouth No 2 GF 2 C3
FALMOUTH TOWN 1 B3
Fancy (UWC) 2 C1
FAREHAM 10 C2
Fareham East Jn. 10 C2
Fareham No.1 Tunnel 10 C2
Fareham No.2 Tunnel 10 C2
Farington Curve Jn. 62 E
Farington Jn. 62 E
Farleigh Lane 21 A3
Farlington Jn. 10 D2
Farm (UWC) 40 C3
Farmer Johnson's (UWC) 56 C3
Farmers (UWC) 25 B3
Farmstead Rise (UWC) 64 D1
FARNBOROUGH 19 A3
FARNBOROUGH NORTH 19 A3
Farnborough North Footpath (R/G) 18 D3
FARNCOMBE 11 A1
Farncombe East (CCTV) 11 A1
Farncombe SB (WZ) 11 A1
FARNHAM 11 A1
Farnham (CCTV) 18 D3
Farnham Depot 18 D3
Farnham Road 19 A1
Farnham SB (FN) 11 A1
FARNINGHAM ROAD 20 D2
Farningham Road GF 20 D2
FARNWORTH 129 B1
Farnworth Tunnels 129 B1
FARRINGDON 109 A3
Farringdon Jn. 109 A3
Fassfern No. 1 (UWC) 89 B2
Fassfern No. 2 (UWC) 89 B2
FAULDHOUSE 84 D2
FAVERSHAM 21 B3
Fawkham Jn. 20 D2
Fawley Oil Refinery 10 C2
FAYGATE 11 B1
FAZAKERLEY 131 A
Fazakerley (UWC) 131 A
FEARN 102 D3
Fearn GF 102 D3
FEATHERSTONE 64 C3
Featherstone (CCTV) 64 C3
Feeder Bridge Jn. 125 B
Felixstowe Beach (CCTV) 34 D2
Felixstowe Beach Jn. 34 D2
Felixstowe Dock Jn. 34 D2
Felixstowe North Quay Terminal 34 C2
FELIXSTOWE TOWN 34 D2
FELLGATE 76 C1
FELTHAM 111 A2
Feltham Jn. 111 A2
Feltham SB (F) 111 A2
Feltham West (CCTV) 111 A2
Felton Lane (CCTV) 81 A2
Feltons (UWC) 38 C2
Fen Crossing (UWC) 51 B1
FENCHURCH STREET 120 B
Fenham Hill (Public FP) 82 D3
Fenham Low Moor (CCTV) 82 D3
FENITON 6 D3
Feniton (CCTV) 6 D3
Fenny Compton Jn. 40 A3
FENNY STRATFORD 31 A1
Fenny Stratford (CCTV) 31 A1
Fenton Manor Tunnel 48 D1
Fenwick (MCG) 64 D3
FERNHILL 26 C1
Ferrett Lane (UWC) 56 C1
FERRIBY 65 B3
Ferring (CCTV) 11 B3
Ferry Lane (AOCL) 34 D1
FERRY MEADOWS 42 C1
Ferry Road (CCTV) 13 B2
Ferry Toll Tunnel 85 A1
Ferrybridge North Jn. 64 D3
Ferrybridge Power Station 64 D3
Ferrybridge SB (FE) 64 D3
Ferrybridge South Jn. 64 D3
Ferryhill 76 C3
Ferryhill Jn. 96 D2
Ferryhill SB (F) 76 C3
Ferryhill South Jn. 76 C3
FERRYSIDE 24 C2
Ferryside (MCB) 24 C2
Ferryside SB (F) 24 C2
Fersit Tunnel 90 D2
FFAIRFACH 24 D2
Ffairfach (AOCL) 24 D2
FFESTINIOG RAILWAY 54 C3
Ffestiniog Tunnel 54 C3

Ffos Fach Isaf (UWC) 24 D3
Ffynnongain (R/G) 24 C2
Fiddlers Ferry (UWC) 56 C2
Fiddlers Ferry Power Stn. 56 C2
Fiddlers Ferry SB 56 C1
Field (UWC) 6 C3
Field House Farm (UWC) 56 D3
Field Lane (AOCL) 64 D3
Fieldens 45 A3
Fielders (OPEN) 13 B2
Fields Farm (UWC) 48 C1
Fifoots Power Station / Uskmouth 123 B
Filbert Grove (UWC) 65 A3
FILEY 72 C3
Filey (CCTV) 72 C3
Filey Junction 72 C3
Fillan 88 D2
FILTON ABBEY WOOD 124
Filton Jn. No. 2 124
Filton Jn. No.1 124
Filton South Jn. 124
Filton Tip (AOCL) 124
Filton West Jn. No. 1 124
Filton West Jn. No. 2 124
FINCHLEY ROAD & FROGNAL 108 C2
Findern (AHBC) 128 A
Findlay's GF 138 D3
Fine Lane (MCG) 49 B3
FINGHALL 70 C2
Finney Lane (UWC) 62 C3
Finnieston East Jn. 139
Finnieston Tunnel 139
Finnieston West Jn. 139
Finningley (MCB) 59 A1
FINSBURY PARK 109 A2
Finsbury Park Jn. 109 A2
FINSTOCK 29 B2
Fir Tree Farm No.3 (UWC) 45 B2
Firbeck Jn. 58 D2
Firle 12 D3
Firsby East Jn. (former) 52 E
Firtree Farm No.2 (UWC) 45 B2
Fish Dock Road (CCTV) 60 D1
FISHBOURNE 10 D2
Fishbourne Footpath (R/G) 10 D2
Fishergate Tunnel 62 E
FISHERGROUND HALT 67 B2
Fisherman Path (UWC) 55 B1
Fishers (Footpath) 13 A1
Fisher's (UWC) 16 C2
FISHERSGATE 11 B3
Fisherton Tunnel 9 A1
FISHGUARD & GOODWICK 23 A1
FISHGUARD HARBOUR 23 A1
Fishguard Harbour (AOCL) 23 A1
Fishley (UWC) 5 A1
Fishpool Farm (UWC) 55 A2
FISKERTON 50 D1
Fiskerton (MCG) 50 D1
Fiskerton Jn SB (FJ) 50 D1
FITZWILLIAM 64 C3
FIVE WAYS 126 B
Flamborough (AHBC) 72 D3
Flax BourtonTunnel 124
Flax Mill (MCG) 52 C2
Flaxby Grange (UWC) 64 C1
Flaxton (AHBC-X) 65 A1
FLEET 18 D3
Flemingate (RC) 66 C2
Fletton Jn. 42 E
FLIMBY 73 A3
FLINT 55 B3
FLITWICK 31 A1
Flitwick Jn. 31 A1
Flixborough Wharf 65 B3
FLIXTON 129 B2
Flockton Sidings 133 A
Flordon GSP 44 D1
FLOWERY FIELD 130 D2
Flyover East Jn. 135 A1
Flyover Jn 31 A1
Flyover West Jn. 135 A1
Fobbing (AHBC-X) 21 A1
Foley Crossing SB (FY) 48 D1
FOLKESTONE 14 C1
FOLKESTONE CENTRAL 14 C1
Folkestone East SB (YE) 14 D1
FOLKESTONE HARBOUR 14 C1
Folkestone Harbour SB (EBB) 14 D1
FOLKESTONE WEST 14 C1
Folly Bank (AHBC) 52 C3
Folly Farm (UWC) 22 C3
Folly Lane (UWC) 60 C1
Folly Road (CCTV) 14 D1
Foot crossing (WL) 17 A1
Football Field (UWC) 36 D1
FORD 11 A3
Ford (CCTV) 11 A3
Ford Bridge (UWC) 38 C3
Ford Crossing 11 B1
Ford Jn. 11 A3
Ford River Bridge 11 A3
Ford Siding GF 25 B3
Forden (AOCL) 37 A1
Fordgate 7 A1
Fordham (AHBC-X) 43 A3
Fords Jn. 25 B3
Foreign Ore Branch Jn. 59 B1
FOREST GATE 110 C2
Forest Gate Jn. 110 C2
FOREST HILL 113 B2
Forest House Farm (UWC) 56 C3
Forestry 43 B2
Forestry (UWC) 16 C3
Forgandenny Ford (UWC) 91 B2
Forge Farm (R/G) 12 D1
FORMBY 55 B1
FORRES 100 C1
Forres LC (MCB) 100 C1
Forres SB (FS) 100 C1
FORSINARD 103 B1
Forsinard (AOCL) 103 B1
FORT MATILDA 83 A1
Fort Pitt Tunnel 21 A2
FORT WILLIAM 90 C2
Fort William Jn SB (FW) 90 C2
Forteviot (AHBC-X) 91 B2
Forteviot Farm (UWC) 91 B2
Forth Banks 136 A
Forth Railway Bridge 85 A1
Forty Steps (F/P) 11 B2
Forty Steps (UWC) 18 C3
Fosseway (AHB) 49 A3
Foster Yeoman Sidings 17 A1
Foster Yeoman Terminal (Thamesport) 21 B2
Fouldubs Jn SB (FD) 85 A1
Foulis (R/G) 99 A1
Fountain (AOCL) 25 B2
Four Ashes Up Goods Loop 48 D3
Four Lane Ends (R/G) 56 C1
FOUR OAKS 39 B1
Fowey Dock 2 D2
Fowlers (UWC) 32 C2
Fowlers Lane 44 C1
Fox Grove Crossing 21 B3
Fox Street Sdg. 50 D3
Foxes (UWC) 59 B2
FOXFIELD 67 B3
Foxfield (UWC) 67 B3
Foxfield SB 67 B3
Foxhall Jn. 18 E
Foxhills Tunnel 19 A3
Foxlow Jn. 58 C2
FOXTON 32 C1
Foxton (MCB) 42 D3
Frame Park Sdgs. (Down Yard) 108 D1
Frampton (UWC) 28 D3
FRANT 12 D1
Frating (CCTV) 34 C2
FRATTON 10 F
Fratton East 10 F
Fratton Traincare Depot 10 F
Freeby (UWC) 50 B2
Freemans Lane (UWC) 59 A2
Freemans SB (F) 81 B3
FRESHFIELD 55 B1
Freshfield (CCTV) 55 B1
FRESHFORD 16 C2
Freshford (UWC) 16 C2
Friargate (CCTV) 60 D1
Friars Jn. 112 C1
Friday Street GF 62 C3
FRIMLEY 19 A3
Frimley Crossing 19 A3
Frinkley Lane (AHBC-X) 51 A1
Frinkley Lane (Public BW) 51 A1
Frinton (CCTV) 34 C2
FRINTON-ON-SEA 34 C2
Friog Cutting 35 B2
Frisby (MCB) 50 B3
Frisby SB (FY) 50 D2
FRIZINGHALL 63 B2
FRODSHAM 56 C2
Frodsham Jn. 56 C2
Frodsham Jn. SB (FJ) 56 C2
Frodsham Tunnel 56 C2
Frog Lane (UWC) 56 C1
Frogmore 2 (UWC) 29 A2
Frognall Farm 21 B3
FROME 16 C3
Frome Avoiding Line 16 C3
Frome North Jn. 16 C3
Fron (UWC) 37 A1
Frost Lane (AHBC) 9 B2
FROSTERLEY 75 B3
Frying Pan Farm (UWC) 16 D2
Fulbourn (AHBC) 43 A3
Fullers End Footpath (R/G-X) 32 D2
Fullerton (UWC) 96 C1
FULWELL 111 B3
Fulwell Jn. 111 B3
Fulwell Tunnel 111 B3
Fulwood Tunnel 131 A
Funthams Lane (CCTV) 42 C1
Funtington (AHBC) 10 D2
Furness Abbey (UWC) 67 B3
Furness Abbey Tunnel 67 B3
FURNESS VALE 130 D3
Furness Vale SB 130 D3
FURZE PLATT 19 A1
Furze Platt (ABCL) 19 A2
Furzebrook GF 8 D3

G

Gaer Jn. 123 B
Gaer Tunnel 123 B
Gaerwen (MCB) 53 B2
Gaerwen GF 53 B2
Gaerwen Isa (UWC) 53 B2
Gaerwen Jn. 53 B2
Gaerwen SB (GN) 53 B2
Gaerwen Uchaf No. 1 (UWC) 53 B2
Gaerwen Uchaf No. 2 (UWC) 53 B2
Gailes (AHBC-X) 78 C1
GAINSBOROUGH CENTRAL 59 A2
Gainsborough Central SB (GC) 59 B2
GAINSBOROUGH LEA ROAD 59 B2
Gainsborough Lea Road SB (LR) 59 B2
Gainsborough Trent Jn. SB (TJ) 59 B2
Gallachers (UWC) 17 B3
Galton Jn. 126 A
Galton Tunnel 126 A
Gambols (UWC) 17 A1
Gamekeepers 45 B2
Gannow Jn. 62 D2
Ganton (AHBC-X) 72 C3
Ganton Hall (UWC) 72 C3
Gantry 'A' 139
Garclaugh No. 1 (UWC) 79 A2
Garclaugh No. 2 (UWC) 79 A2
Garclaugh No. 3 (UWC) 79 A2
Garden Street (CCTV) 60 D1
Gardners (CCTV) 21 A1
GARELOCHHEAD 83 A1
GARFORTH 64 C2
GARGRAVE 63 A1
Garlands No. 1 (UWC) 27 B3
Garnant Branch (TMO) 24 D2
Garnqueen North Jn. 138 D1
Garola House (UWC) 66 C3
Garriongill Jn. 84 D3
GARROWHILL 138 C2
GARSCADDEN 137 A2
GARSDALE 69 A2
Garsdale SB (G) 69 A2
Garstang & Catterall Emergency GFs 62 C2
GARSTON 31 B3
Garston (UWC) 131 A
Garston Jn. 131 A
GARSWOOD 56 C1
GARTCOSH 138 C2
Gartcosh GF 138 C2
Gartcosh Jn. 138 C2
GARTH (Maesteg) 25 B2
GARTH (Powys) 36 D3
Gartly (AHBC) 101 B3
Gartsherrie South Jn. 138 D2
Gartshore 138 C1
Gartshore Emergency GF 138 C1
GARVE 98 D1
Garve (AOCL) 98 D1
Gas Factory Jn. 109 B3
Gas House Lane (UWC) 15 A2
Gascoigne Wood SB (GW) 64 D3
Gascoyne Crossing 21 B3
Gasworks Tunnel 108 D3
Gatehead (AHBC) 78 D1
GATHURST 56 C1
GATLEY 130 C3
GATWICK AIRPORT 12 C1
Geilston (UWC) 83 A1
Geilston Farm (UWC) 83 A1
GELERT'S FARM 35 C1
Gelshfield (UWC) 104 C2
Gelshfield Gates (UWC) 104 C2
Gelynis (R/G-X) 123 A
Genwen Jn. 24 D3
George IV Tunnel 114 C1
GEORGEMAS JUNCTION 104 C2
Georgemas No. 1 GF 104 C2
Gerards Bridge Jn. GF 56 C1
GERRARDS CROSS 19 A1
Gibbet Hill Jn. 40 A2
Gibbons (UWC) 26 C2
GIDEA PARK 20 D1
Giffen (RNAD Beith) 83 A3
GIFFNOCK 137 B3
Gifford Street Portals 109 A3
GIGGLESWICK 62 D1
GILBERDYKE 65 B3
Gilberdyke (UWC) 65 B3
Gilberdyke Jn. 65 B3
Gilberdyke Jn. SB (G) 65 B2
Giles (UWC) 34 C3
GILFACH FARGOED 26 C1
Gilfach Farm 3 (UWC) 36 E
Gillets (UWC) 62 C3
Gillingham 21 A3
GILLINGHAM (Dorset) 8 C1
GILLINGHAM (Kent) 21 A2
Gillingham GF 8 C1
Gillingham SB (ET) (Kent) 21 A3
Gillingham Tunnel (Dorset) 8 C1
Gillingham Tunnel (Kent) 21 A2
GILSHOCHILL 137 B1
GIPSY HILL 113 A3
Girlings (R/G-X) 44 E
GIRVAN 77 B1
Girvan SB (GV) 77 B1
Gisburn Tunnel 63 A2
Gishbourne (UWC) 28 D1
Gislingham 44 D3
Gissing No. 2 (AHBC-X) 44 D2
Glackton (UWC) 100 C2
GLAISDALE 71 B2
GLAN CONWY 54 C2
Glan Conwy (AHBC) 54 C2
Glanhafren (UWC) 37 A1
Glanirfon (UWC) 36 E
Glanrhyd (OC) 36 E
Glanrhyd Bridge 36 E
Glanrhyd Saeson Farm 1 (UWC) 36 E
Glantowy (UWC) 36 E
Glan-y-Mor Elias (UWC) 54 C2
GLANYRAFON 35 F
Glanyrynys Farm 36 E
GLASGOW CENTRAL LOW LEVEL 139
GLASGOW QUEEN STREET HIGH LEVEL 139
GLASGOW QUEEN STREET LOW LEVEL 139
GLASSHOUGHTON 64 C3
Glass's 16 C2
Glaston Tunnel 51 A3
GLAZEBROOK 129 A2
Glazebrook East Jn. SB (GE) 129 A2
Glebe St. Jn. 48 D1
Gledholt Nth & Sth Tunnels 63 B3
Glen Douglas 88 C3
Glen Parva GF 40 D1
Glen Parva Jn. 50 C3
Glenamman Footpath (R/G) 25 A1
GLENEAGLES 91 B3
GLENFINNAN 89 B2
Glenfinnan Viaduct 89 B2
GLENGARNOCK 83 A3
GLENROTHESWITH THORNTON 92 C3
Glenwhilly SB (GW) 77 B2
GLOSSOP 130 D2
GLOUCESTER 28 C2
Gloucester Foot crossing (WL) 28 C2
Gloucester New Yard 28 C2
Gloucester Road Jn. 118 E
Gloucester SB (G) 28 C2
Gloucester West 28 C2
Gloucester Yard Jn. 28 C2
Gloucester Yard No. 2 GF 28 C2
GloucesterBarnwood Jn. 28 D2
GLOUCESTERSHIRE WARWICKSHIRE RAILWAY 28 D2
GLYNDE 12 D3
GLYNDYFRDWY 47 E
Glynisw (UWC) 24 C2
Glyn-y-Mul (UWC) 122 B
GOATLAND 71 B2
GOBOWEN 47 A2
Gobowen North SB (GN) 47 A2
Gobowen South GF 47 A2
Gobwen North (MCB) 47 A2
GODALMING 11 A1
GODLEY 130 D2
Godnow Bridge (MCG) 59 A1
Godregarreg Farm 1 36 E
GODSTONE 20 C3
Godstone Tip 12 C1
Golant (OC) 2 D2
Goldborne Jn. 56 C1
Golden High Hedges (MCG) 52 C2
Goldies (UWC) 45 A1
Goldmire Quarry (UWC) 67 B3
GOLDTHORPE 58 C1
Golf Course Bridleway 75 B1
GOLF STREET 95 A3
GOLFA 37 E
GOLSPIE 102 D1
GOMSHALL 19 B3
Gonalston (AHBC) 50 D1
Gonerby Tunnel 51 A2
Gooderhams (UWC) 44 C3
Gooderham's No.41 44 C2
Gooderham's No.42 44 C2
GOODMAYES 110 D2
Goodridges (UWC) 51 A3
GOODRINGTON SANDS 4 D2
Goodrington Yard Sidings 4 D2
Goods Yard Sidings 67 B3
GOOLE 65 A3
Goole Bridge SB (GB) 65 A3
Goole Docks 65 A3
Goole SB (G) 65 A3
Goonbarrow Jn. SB (G) 2 C1
GOOSTREY 56 D3
Goostrey Jn. 56 D3
GORDON HILL 109 A1
Gore Top 22 D3
Gorgie Jn 85 E
GORING & STREATLEY 18 C1
GORING-BY-SEA 11 B3
Goring-by-Sea (CCTV) 11 B3
Gorsecoch (UWC) 24 C2
Gorsey Lane (UWC) 50 D1
Gorshwen No. 2 (UWC) 35 B2
Gorstan (UWC) 98 D1
Gorton 88 D1

H

Haughs (UWC) 101 B2
Hauxton (AHBC) 32 C1
HAVANT 10 D2
Havant ASC (HT) 10 D2
Havant Jn. 10 D2
Havant New Lane (MCB)(CCTV) 10 D2
HAVEN STREET 10 C3
HAVENHOUSE 52 E
Havenhouse (AHBC-X) 52 E
Havensmouth 13 A3
HAVERFORDWEST 23 A2
Haverigg (AHBC) 67 B3
Haw Bank Tunnel 63 A2
Haw Lane (UWC) 63 A1
HAWARDEN 55 B3
HAWARDEN BRIDGE 55 B3
Hawkeridge Jn. 16 D2
Hawkes Point Foot Crossing 1 A2
Hawkesbury Lane (CCTV) 40 C1
Hawkesbury Lane Sidings GF 40 C1
HAWKHEAD 137 A2
Hawkhead Oil Terminal GF 137 A2
Hawks 43 A2
Hawkwood Jn. 118 F
HAWORTH 63 A2
Hawthorn Bank (CCTV) 52 C3
Haxby Road (CCTV) 64 D1
Haxby Station (CCTV) 64 D1
Haxey (CCTV) 59 A1
Hay Crossing 7 A2
Haydock Branch Jn. 56 C1
HAYDON BRIDGE 75 A1
Haydon Bridge (MCB) 75 A2
Haydon Bridge SB (HB) 75 A1
HAYDONS ROAD 112 D3
HAYES 118 A1
HAYES & HARLINGTON 111 A1
Hayes (UWC) 42 D3
Hayes Up Goods Loop 111 A1
Hayfields (UWC) 59 A1
HAYLE 1 A2
Hayle Footpath (R/G) 1 A2
HAYMARKET 85 E
Haymarket Central Jn. 85 E
Haymarket East Jn. 85 E
Haymarket North & South Tunnels 85 E
Haymarket Sprinter Depot 85 E
Haymarket Tunnel 131 B
Haymarket West Jn. 85 E
Hayside (UWC) 49 B2
Haystacks 13 B2
Haywards (AOCL) 34 D1
HAYWARDS HEATH 12 C2
Haywards Heath Tunnel 12 C2
Haywood (CCTV) 58 D1
HAZEL GROVE 130 C3
Hazel Grove East Jn. 130 C3
Hazel Grove High Level Jn. 130 C3
Hazel Grove SB (HG) 130 C3
Hazel Grove West Jn. 130 C3
HEADCORN 13 A1
HEADINGLEY 64 C2
Headingley Tunnel 64 C2
Heads of Lochiel (UWC) 89 B2
HEADSTONE LANE 107 A1
HEALD GREEN 130 C3
Heald Green N. Jn. 130 C3
Heald Green S. Jn. 130 C3
Heald Green W. Jn. 129 B3
Healey Mills SB (HM) 64 C3
Healey Mills Yard 64 C3
HEALING 60 C1
Healing (R/G) 60 C1
Healy Mills A Jn. 63 B3
Healy Mills B Jn. 63 B3
Heartlands Park GF 126 A
Heartlands Park Sdgs. 126 A
Heath Farm 46 D2
Heath Farm (UWC) 37 B2
HEATH HIGH LEVEL 123 A
Heath Jn. 123 A
Heath Junction SB (HJ) 123 A
HEATH LOW LEVEL 123 A
Heath No. 59 44 C2
Heatherbell (CCTV) 138 D2
Heatherinch (UWC) 92 C3
Heathfield 4 D1
Heathfield (UWC) (Aspatria) 73 B3
Heathfield (UWC) (Gartcosh) 138 C2
Heathfield Branch Jn. 4 D1
Heathrow Airport Jn. 111 A1
HEATHROW CENTRAL (TERMINALS 1, 2 & 3) 111 A1
Heathrow Express Depot 108 C3
HEATHROW TERMINAL 4 111 A1
HEATHROW TERMINAL 5 111 A2
Heathrow Tunnel Jn. 111 A1
HEATON CHAPEL 130 C2
Heaton Depot 136 A
Heaton Lodge Jns. 63 B3
Heaton Norris Jn. 130 C3
Heaton Norris Jn. SB (HN) 130 C3
Heaton North Jn. 136 A
Heaton South Jn. 136 A
HEBDEN BRIDGE 63 A3
Hebden Bridge SB (HB) 63 A3
Heck GF 64 D3
Heck Ings (MCG) 64 D3
Heck Lane (MCG) 64 D3
HECKINGTON 52 C1
Heckington (MCG) 52 C1
Heckington SB (HN) 51 B1
HEDGE END 10 C2
HEDNESFORD 49 A3
Hednesford SB (HD) 49 A3
Heeley 58 C2
HEIGHINGTON 70 C1
Heighington (MCB) 70 C1
Heighington SB (H) 70 C1
Hele & Bradninch 6 C2
Helen Street Jn. 139
HELENSBURGH CENTRAL 83 A1
HELENSBURGH UPPER 83 A1
HELLIFIELD 63 A1
Hellifield SB (H) 62 D1
Helm Tunnel 68 D1
HELMSDALE 103 B2
Helpston (MCB) 51 B3
Helpston Jn. 51 B3
HELSBY 56 C3
Helsby Jn. 56 C3
Helsby Jn. SB (HY) 56 C3
Helston Farm No. 1 3 B1
HEMEL HEMPSTEAD 31 A3
Hemerdon GF 3 B2
Hemingfield Tunnel 58 C1
Hempstead 12 D2
Hemsworth 64 C3
Henblas (UWC) 35 B2
Hendon 76 D2
HENDON 107 B2
Hendre Waelod No.1 54 C2
Hendre Waelod No.2 (UWC) 54 C2
Hendrewen (UWC) 23 A1
Hendrewen Farm 1 (UWC) 24 D3
Hendrewen Farm 3 (UWC) 24 D2
Hendy Jn. 24 D3
Hendy Sewage Works(UWC) 24 D3
Henfaes Farm (UWC) 54 C2
HENGOED 26 C2
HENIARTH 37 E
Henley Branch Jn. 18 D1
HENLEY-IN-ARDEN 39 B3
HENLEY-ON-THAMES 18 D1
HENSALL 64 D3
Hensall (MCB) 64 D3
Hensall SB (H) 64 D3
Henwick (MCB) 38 D3
Henwick Hall (MCB) 64 D3
Henwick SB (HK) 38 D3
Heol-y-Deliaid (UWC) 25 A2
Hepscott (AHBC) 81 A3
Hepscott Jn. 81 A3
Herbrandston Jn. 23 A2
HEREFORD 27 B1
Hereford SB (H) 27 B1
Hereford Storage Sdg. 48 C3
Hereford Yard Jn. 27 B1
HERNE BAY 22 A2
HERNE HILL 113 A2
Herne Hill North Jn. 113 A2
Herne Hill South Jn. 113 A2
Herringe 14 C1
HERSHAM 115 A1
HERSTON HALT 8 F
HERTFORD EAST 32 C2
HERTFORD NORTH 32 C3
Hesketh (UWC) 55 B1
Heslerton Station (AHBC-X) 71 B3
Hessay 64 D2
Hessay WD GF 64 D1
Hessel East Jn. 66 C3
HESSLE 66 C3
Hessle Road SB (HR) 65 B3
Hessle Road South Jn. 66 C3
Hest Bank (MCB) 62 C1
Hest Bank Jn. 62 C1
Hestham Hall (UWC) 67 B3
HESWALL 55 B2
Hethersett GF 44 D1
Hethfelton No. 1 Crossing 8 D3
Hethfelton No. 2 Crossing 8 D3
Hett Mill (CCTV) 76 C3
HEVER 12 D1
Hever Jn. 12 C1
Hewish (AHBC) 7 A2
HEWORTH 136 A
HEXHAM 75 A1
Hexham SB (H) 75 B2
Hexthorpe Jn. 135 A1
HEYFORD 30 C2
Heyope 1 (UWC) 37 A2
Heyope 2 (UWC) 37 A2
HEYSHAM PORT 61 B1
Heysham Power Station 62 C1
HEYWOOD 56 D1
Heywood Road Jn. 16 D3
Heyworth (MCG) 64 D3
Hiams Fen House No.20 43 B2
Hibaldstow (AHBC-X) 59 B1
Hicks Lodge GF 49 B3
High Balernoch (UWC) 83 A1
HIGH BROOMS 12 D1
High Brooms GF 12 D1
High Eggborough (MCG) 64 D3
High Ferry (AHBC) 52 C1
High Ferry Lane (AHBC) 52 C1
High Hall (UWC) 27 B3
High Lane (UWC) 62 C3
High Level Bridge Central Jn. 136 A
High Level Bridge Jn. 136 A
High Marnham 59 A3
High Meads Jn. 109 B2
High Meads Loop Enclosure 110 C2
High Oaks Jn. 40 D2
HIGH ROCKS 12 D1
High Scampston (AHBC-X) 71 B3
HIGH STREET 139
High Street Crossover (Cheltenham) 28 D2
High Street Jn. 139
High Street Tunnel 139
High Tor No 1 Tunnel 58 C3
High Tor No 1A Tunnel 58 C3
High Tor No 2 Tunnel 58 C3
HIGH WYCOMBE 30 D3
HIGHAM 21 A2
Higham Tunnel 21 A2
HIGHAMS PARK 109 B1
HIGHBRIDGE & BURNHAM 15 A3
Highbridge West 15 A3
HIGHBURY & ISLINGTON 109 A2
Highbury Vale Jn. 109 A2
Highdyke 51 A2
Higher Barn (UWC) 8 C3
Higher Delacre (UWC) 56 D3
Higher Doomsford (UWC) 5 A2
Higher Town Tunnel 1 B2
Highfields (UWC) 49 A2
HIGHLEY 38 D2
Highover Farm (UWC) 31 B2
HIGHTOWN 55 B1
Hightown Farm (UWC) 74 D1
Highworth GF 17 A1
Highworth Jn. 17 A1
HILDENBOROUGH 20 D3
Hilgay (AHBC) 43 A1
Hillam Gates (CCTV) 64 D3
HILLFOOT 137 B1
Hillhouse No.3 GF 61 B2
Hillhouse No.4 GF 61 B2
Hillhouse No.5 GF 61 B2
Hilling (UWC) 23 B3
Hillings Road 45 B2
HILLINGTON EAST 137 A2
HILLINGTON WEST 137 A2
Hillmorton Jn. 40 E
HILLSIDE 55 B1
Hilly Laid (TMO) 61 B2
HILSEA 10 D2
Hilton (MCG) 49 B2
Hilton Jn SB (HJ) 91 B2
Hilton Jn. 91 B2
Hilton Mills (UWC) 102 D2
HINCHLEY WOOD 115 B1
HINCKLEY 40 C1
Hinderton Field Tunnel 131 B
HINDLEY 56 C1
Hindlow No. 1 GF 57 B3
Hindlow No. 2 GF 57 B3
Hindlow Tunnel 57 B3
Hindon Road Crossing 8 D1
Hinds (UWB) 40 D1
Hinksey North 30 C3
Hinksey Reception GF 30 C3
Hinksey South 30 C3
HINTON ADMIRAL 9 A3
Hinxton (AHBC) 32 D1
Hipperholme Tunnel 63 B3
Hirst Lane (MCG) 81 A3
Hirwaun (TMO) 25 B1
Hirwaun Pond 25 B1
HITCHIN 31 B2
Hitchin 'A' GF 32 C2
Hitchin Grade Separation 31 B2
Hitch's No. 12 43 A3
HITHER GREEN 114 C2
Hither Green TMD 114 C2
Hither Green Yard 114 C2
Hives Farm (UWC) 50 B3
Hobhole Bank (UWC) 52 C1
Hobhole Bank Bridleway 52 C1
Hochkings (UWC) 59 B3
Hockham Road 44 C2
HOCKLEY 21 B1
Hockley (CCTV) 49 A2
Hockley No. 2 Tunnel 126 B
Hockley No.1 Tunnel 126 B
Hoe Street Tunnel 109 B1
Hoghton (AHBC) 62 C3
Hogs Croft 32 D2
Hogwell 21 A1
Holbeck Depot 133 B
Holbeck Depot Jn. 133 B
Holbeck Jn. 133 B
Holbrook (TMO) 50 C1
Holdingham Lane (UWC) 51 B1
Holehouse Jn GF 78 D2
Holesmouth Jn. 124
Holgate Jn. 136 B
Hollands (Streehay) 49 B3
Holliday St. Tunnel 126 B
HOLLINGBOURNE 21 B3
Holloway 108 D2
Holly Moor (UWC) 7 A1
Holme (CCTV) 42 C1
Holme (No. 31) (AHBC) 8 D3
Holme Green (R/G) 42 C3
Holme Lode (CCTV) 42 C1
Holme Road 43 B1
Holme Tunnel 63 A3
HOLMES CHAPEL 56 D3
Holmes Jn (CCTV) 134 A1
Holmes Summit 84 D2
HOLMWOOD 11 B1
HOLT 46 C1
Holt Lane Tunnel 58 C3
Holton Gatehouse (AHBC-X) 60 C1
HOLTON HEATH 8 D3
Holton Heath GF 8 D3
Holton-le-Moor SB (H) 60 C1
Holts (UWC) 40 D1
Holybourne Oil Terminal Sidings 18 D3
HOLYHEAD 53 A1
Holyhead SB (HD) 53 A1
HOLYTOWN 138 D3
Holytown Jn. 138 D3
Holywell (ABCL) 76 C1
Holywell (UWC) 2 C1
HOLYWELL HALT 63 B1
Holywell Junction SB (HJ) 55 A2
Holywell Viaduct 109 A3
Holywood LC 73 A1
Holywood SB (HW) 80 C3
Home Farm (UWC) 48 D1
Homedown (UWC) 28 D1
HOMERTON 109 B2
HONEYBOURNE 29 A1
Honeybourne GF 29 A1
Honington (AHBC-X) 51 A1
HONITON 6 D3
Honiton Tunnel 6 D3
HONLEY 57 B1
HONOR OAK PARK 113 B2
Hoo Jn. 21 A2
HOO JN. STAFF HALT 21 A2
Hoods Mill (UWC) 51 B3
HOOK 18 D3
Hook Moor Farm (UWC) 65 A3
HOOTON 55 B2
Hooton North Jn. 55 B2
Hooton SB (HN) 55 B2
Hooton South Jn. 55 B3
HOPE 57 B2
Hope Exchange (UWC) 55 B3
HOPE/YR HOB 47 A1
Hopetown Jn. 70 C1
Hoplands Farm 22 C3
Hopperton Grange (UWC) 64 C1
Hopperton Old Station (UWC) 64 C1
Hoppity Tunnel 116 C3
HOPTON HEATH 37 B2
Horbury Jn GF 133 A
Horbury Jn SB (HJ) 133 A
Horbury Jn. 133 A
Horbury Station Jn. 64 C3
HORLEY 12 C1
HORNBEAM PARK 64 C1
Horninglow Bridge Jn. 49 B2
HORNSEY 108 D1
Horrocks (UWC) 74 C2
Horrocksford Jn. SB (H) 62 D2
Horse Fen 43 A3
Horsemoor (AHBC-X) 43 A1
Horsfall Tunnel 63 A3
HORSFORTH 64 C2
HORSHAM 11 B1
Horsham Jn. 11 B1
Horsham Road (CCTV) 12 C1
HORSLEY 115 A3
HORSTED KEYNES 12 C1
Horton (UWC) 22 C3
Horton Road (MCB) 28 C2
Horton Road Jn. 28 C2
HORTON-IN-RIBBLESDALE 69 A3
HORWICH PARKWAY 129 A1
HOSCAR 56 C1
Hoscar (AHBC-X) 56 C1
Hosegood's (UWC) 6 C3
Hospital (CCTV) 62 C3
Hospital Mill (R/G) 92 D3
Hothfield Sidings 13 B1
Hoton House (AHBC-X) 65 A3
HOUGH GREEN 56 C2
Hough Lane (AHBC-X) 51 A1
Hough Lane (Public BW) 51 A1
HOUNSLOW 111 A2
Hounslow Jn. 111 A2
Houstry (UWC) 104 C2
HOVE 14 E
Hove Jn. 14 E
Hove Tunnel 14 E
HOVETON & WROXHAM 46 D2
How Mill (AHBC-X) 74 C2
HOW WOOD 31 B3

Llanfair SB 53 B2
LLANFAIRFECHAN 54 C2
LLANFAIRPWLL 53 B2
LLANGADOG 36 E
Llangadog (AOCL) 36 E
LLANGAMMARCH 36 E
Llangammarch Tunnel 36 E
LLANGENNECH 24 D3
Llangennech (UWC) 24 D3
Llanglan Fechan No. 2 (UWC) 36 C1
Llanglan Fechan No. 4 (UWC) 36 C1
LLANGOLLEN 47 E
LLANGOLLEN RAILWAY 47 A1
Llangyfelach Tunnel 122 B
LLANGYNLLO 37 A2
Llangynllo Tunnel 37 A2
LLANHARAN 25 B2
Llanharan (UWC) 26 C2
LLANHILLETH 26 D1
Llanidloes Road (CCTV) 36 D1
Llanion (OC) 23 A3
LLANISHEN 26 C2
Llanlliwe Farm (UWC) 23 B2
LLANRWST 54 C2
LLANRWST NORTH 54 C2
Llanrwst SB 54 C2
Llanrwst Tunnel 54 C2
LLANSAMLET 122 B
Llanstephan Footpath (R/G) 24 C2
Llantrisant West (CCTV) 26 C2
LLANTWIT MAJOR 25 B3
Llanwern Exchange Sidings 15 A1
Llanwern West Jn. 15 A1
Llanwern Works East Connection 15 A1
Llanwern Works West Connection 15 A1
LLANWRDA 36 E
Llanwrda (OC) 36 E
LLANWRTYD 36 E
Llechryd (UWC) 35 B2
Lloyds & Martin (UWC) 43 A2
LLWYFAN 24 C2
Llwyn Cadwgan (UWC) 35 B1
Llwyn Jack Farm (UWC) 36 E
Llwyndrys (UWC) 23 B2
Llwyndyrys (UWC) 23 B2
LLWYNGWRIL 35 B2
Llwyngwyddil 2 (UWC) 23 B2
Llwynllanc Farm 1 (UWC) 25 A1
Llwynpener 2 (UWC) 23 B2
Llwynpiod No. 1 (UWC) 36 D3
Llwynpiod No. 2 (UWC) 36 D3
LLWYNYPIA 25 B2
Llynfi Jn. 25 B2
Llynmellin Farm (UWC) 37 A3
Llysfaen Emergency GF 54 D2
LOCH AWE 87 B2
LOCH EILOUTWARD BOUND 90 C2
Loch View Farm (UWC) 104 C2
LOCHAILORT 89 A2
LOCHEILSIDE 89 B2
LOCHGELLY 92 C3
Lochiel OB (UWC) 90 C2
Lochinver Farm (UWC) 100 D1
LOCHLUICHART 98 D1
Lochluichart Lodge (UWC) 98 D1
Lochluichart Station (UWC) 98 D1
Lochridge Jn. 83 B3
Lochside (UWC) 103 B1
LOCHWINNOCH 83 A2
Lochwinnoch Crossovers 83 B2
Lock Lane (MCB) 128 A
LOCKERBIE 80 D3
Lockerbie North GSP 80 D3
Lockerbie South GSP 73 B1
Lockington (AHBC-X) 66 C2
LOCKWOOD 63 B3
Lockwood Tunnel 63 B3
Loco Yard Jn. 17 A1
Lodge Farm (OC) 3 A2
Lodge Farm (UWC) 51 A1
Logan's Road (CCTV) 138 D3
Lolham (CCTV) 51 B3
LONDON BRIDGE 120 B
London Bridge SB (L) 120 B
LONDON CHARING CROSS 120 B
LONDON FIELDS 109 B3
LONDON PADDINGTON 112 C1
London Road (ABCL) 45 A2
London Road (AHBC) 52 C3
LONDON ROAD (Brighton) 14 E
LONDON ROAD (Guildford) 19 A3
London Road (TMO) 30 C2
London Road Jn. 127 A
London Road Jn. 128 A
London Road Viaduct 14 E
London Road Yard 127 A
London Tunnel 1 109 B2
London Tunnel 2 110 C2
LONDON VICTORIA 119
LONDON WATERLOO 119
Long Ashes (UWC) 74 C3
LONG BUCKBY 40 D2
Long Byre (AHBC-X) 74 D1
Long Dyke Jn. 123 A
LONG EATON 128 A
Long Eaton (CCTV) 128 A
Long Green FP (R/G-X) 33 B2
Long Lane (CCTV) 70 D2
Long Lawford Jn. 40 D2
Long Lye, Kilmarnock 83 B3
Long Marsh (UWC) 67 B3
Long Plantation (UWC) 72 C3
LONG PRESTON 62 D1
Long Rock (CCTV) 1 A3
Long Salts 22 D3
Long Valley Sidings 18 D3
Longannet No. 4 (UWC) 85 A1
Longannet Power Station 85 A1
Longannet SB(LG) 85 A1
LONGBECK 71 A1
Longbeck (MCB) 71 A1
Longbeck SB (L) 71 A1
LONGBRIDGE 39 A2
LONGCROSS 19 A2
LONGFIELD 20 D2
Longforgan (MCB) 92 C2
Longforgan SB (LF) 92 C2
Longhedge Jn. 119
Longhirst (CCTV) 81 A3
Longlands Jn. 70 D2
Longlands Tunnel 70 D2
Longley (UWC) 100 C1
LONGNIDDRY 86 C1
LONGPORT 48 D1
Longport Jn. 48 D1
Longsight Depot Jn. 128 B
Longsight South Jn. 128 B
Longsight Traincare & TMD 128 B
LONGTON 48 D1
Lonlas Tunnel 122 B
LOOE 3 A2
Lookout (UWC) 24 C3
Loover Barn 12 D2
LOSTOCK 129 A1
LOSTOCK GRALAM 56 D3
LOSTOCK HALL 62 E
Lostock Hall Depot Sdg 62 E
Lostock Hall Jn. 62 E
Lostock Jn. 129 A1
Lostock Works 56 D3
LOSTWITHIEL 2 D1
Lostwithiel (MCB) 2 D1
Lostwithiel Jn. 2 D1
Lostwithiel SB (LL) 2 D1
Lothbeg (UWC) 103 B3
LOUGHBOROUGH 50 C2
LOUGHBOROUGH CENTRAL 50 C3
LOUGHBOROUGH JN. 113 A2
Loughborough Jn. 113 A2
Loughborough North Jn. 50 C2
Loughborough South Jn 50 C2
Loughor Viaduct 24 D3
Lounge Jn. 50 C2
Lovers Walk Depot 14 E
Loversall Carr Jn. 135 A1
Loversall Jn. 135 A1
Low Eggborough (UWC) 64 D3
Low Ellers Curve Jn. 135 A1
Low Farm (UWC) 45 B2
Low Fell Jn. 136 A
Low Gates (MCB) 70 D2
Low Gates SB (LG) 70 D2
Low House (MCB) 74 C2
Low House Crossing SB (LH) 74 C2
Low Mill (R/G) 74 C2
Low Moor (CCTV) 62 D2
Low Moor Farm (UWC) 64 C1
Low Row (MCB) 74 D2
Low Row SB (LR) 74 D2
Low Scampston (AHBC-X) 71 B3
Low Street (CCTV) 21 A2
LOWDHAM 50 D1
Lowdham GF 50 D1
Lowdham SB (LM) 50 D1
Lower Barn Farm (UWC) 28 C2
Lower Barn No. 1 12 D2
Lower Burton Farm (UWC) 38 C3
Lower Crianlarich 88 D2
Lower Cullernie (UWC) 99 B2
Lower Hall (UWC) 37 A2
Lower House Farm (UWC) 37 A2
Lower Radway Green (UWC) 48 D1
Lower Shakespeare Cliff Shaft 14 D1
Lower Stannage Farm (UWC) 37 B2
LOWER SYDENHAM 113 B3
Lower Trenowin (UWC) 1 A2
LOWESTOFT 45 B2
Lowestoft SB (L) 45 B2
Lowfield (UWC) 64 D3
Lowthorpe (AHBC-X) 66 C1
Lowton Jn. 56 C1
Loxley Lane (AHBC-X) 49 A2
Lucker (CCTV) 81 A1
Lucks Road (AHBC-X) 52 C3
Ludgate GW Jn. 119
LUDLOW 38 C2
Ludlow Tunnel 38 C2
Luffenham (CCTV) 51 A3
Lugton SB LU) 83 B3
Lund Lane (UWC) 65 A3
Lune Viaduct 62 C1
LUTON 31 B2
LUTON AIRPORT PARKWAY 31 B2
Luton North Jn. 31 A2
Luton South Jn. 31 B2
LUXULYAN 2 D1
Luxulyan Tunnel 2 D1
Lybster No. 1 (UWC) 104 C2
Lybster No. 2 (UWC) 104 C2
Lydd Town (TMO) 14 A2
Lydden Tunnel 14 D1
Lyde Court (UWC) 27 B1
LYDNEY 27 B3
Lydney (CCTV) 27 B3
LYDNEY JUNCTION 27 B3
LYDNEY TOWN 27 B3
LYE 39 A2
LYMINGTON PIER 9 B3
LYMINGTON TOWN 9 B3
Lymn Bank (AOCL-X) 52 E
LYMPSTONE COMMANDO 6 C3
LYMPSTONE VILLAGE 6 C3
Lynchat (UWC) 94 D2
Lyneham (UWC) 29 B2
Lynwilg No. 1 (UWC) 94 D1
Lyon Crossing 37 B2
Lyons Wood Farm (UWC) 47 B2
LYTHAM 61 B3

M

MacBeaths (UWC) 103 B3
Macclesfield 57 A3
Macclesfield SB (MD) 57 A3
Macclesfield Tunnel 57 A3
Machen Fach Farm (UWC) 123 B
Machen Quarry 123 B
MACHYNLLETH 36 C1
Machynlleth SC (MH) 36 C1
Macleans (UWC) 89 A1
Macraes (UWC) 94 C2
Madeley Chord Jn 48 D1
Madeley Jn (Telford) 48 C3
Madeley Jn. SB (MJ) 48 C3
Madeley South Jn. 48 C3
Madryn Farm (UWC) 54 C2
Maes (ABCL) 35 A1
MAESTEG 25 B2
MAESTEG (EWENNY ROAD) 25 B2
Maesteg (UWC) 55 A2
Maes-y-Coed Farm (UWC) 36 E
Magdalen Road SB (MR) 43 B1
MAGHULL 55 B1
Maghull (CCTV) 55 B1
Magor 15 A1
MAIDEN NEWTON 7 B3
Maidendale 70 C1
MAIDENHEAD 19 A2
Maidenhead East 19 A2
MAIDSTONE BARRACKS 21 A3
Maidstone East 21 A3
Maidstone East SB (ME) 21 A3
MAIDSTONE WEST 21 A3
MaidstoneWest SB (MS) 21 A3
Maindee East Jn. 123 B
Maindee North Jn. 123 B
Maindee West Jn. 123 B
Maindy Bach (UWC) 26 C3
Mair No. 2 (Accom) 13 B2
Malden (CCTV) 112 C3
MALDEN MANOR 116 C1
MALLAIG 89 A1
Mallaig GF 89 A1
Malt House (UWC) 47 A3
Maltby Colliery SB (M) 58 D2
Malting Lane (AHBC-X) 52 C2
Maltings (UWC) 32 C3
MALTON 71 B3
Malton (MCB) 71 B3
Malton SB (M) 71 B3
MALVERN LINK 28 C1
Malvern Wells SB (MW) 28 C1
MANCHESTER AIRPORT 129 B3
Manchester East SCC (AS) (GB) (SE) 130 C3
Manchester International Depot 128 B
Manchester North SB (MN) 128 B
MANCHESTER PICCADILLY 128 B
Manchester Piccadilly East Jn. 128 B
Manchester Piccadilly SB (MP) 128 B
Manchester South SB (MS) 130 C3
MANCHESTER VICTORIA 128 B
Manchester Victoria East Jn. 128 B
Manchester Victoria West Jn. 128 B
MANEA 43 A1
Manea (MCB) 43 A2
Manea SB (M) 43 A1
Mann Island Jn. 131 B
Manning Upper House (UWC) 27 B1
MANNINGTREE 34 C2
Manningtree East Jn. 34 C2
Manningtree North Jn. 34 C2
Manningtree South Jn. (CCTV) 34 C2
Manor Farm (UWC) (Langworth) 60 C3
Manor Farm (UWC) (Oxford) 30 C3
Manor Farm (UWC) (Saltmarche) 65 A3
Manor Farm 2 (UWC) (Bucknell) 37 B2
Manor Farm 3 (UWC) (Bucknell) 37 B2
Manor Farm No 1 (UWC) (Banbury) 30 C1
Manor Neuk (UWC) 91 A3
MANOR PARK 110 C2
Manor Powis (UWC) 91 A3
MANOR ROAD 55 B2
Manor Way (CCTV) 20 D1
MANORBIER 23 B3
Manorbier Newton (OC) 23 A3
Manorbier Station (AOCL) 23 B3
MANORS 136 A
Manse (UWC) 102 D2
MANSFIELD 58 D3
Mansfield Jn. 50 E
Mansfield Road (CCTV) 59 A3
Mansfield Viaduct 58 D3
MANSFIELD WOODHOUSE 58 D3
Mansfield Woodhouse Jn. 58 D3
Manson (R/G) 64 C2
Mantle Lane SB (ML) 50 C3
Manton Jn. 51 A3
Manton Jn. GF 51 A3
Manton Jn. SB (MJ) 51 A3
Manton North Jn. 51 A3
Manton Tunnel 51 A3
Manton Wood 58 D2
Manuells Farm 2 (UWC) 1 C1
MARCH 42 D1
March East (MCB) 42 D1
March East Jn. 42 D1
March East Jn. SB (ME) 43 A1
March South (MCB) 42 D1
March South Jn. SB (MS) 43 A1
March West Jn. 42 D1
Marchey's House (MCB) 81 A3
Marchey's House Jn. 81 A3
Marchey's House SB (MH) 81 A3
Marchington Old Station (UWC) 49 A2
MARCHWOOD 9 B2
Marchwood 9 B2
Marchwood SB (MW) 9 B2
MARDEN 13 A1
Mare Brook (UWC) 29 A1
Mares (UWC) 28 D1
Mares Close (UWC) 76 C1
Margam Abbey Works 25 A2
Margam Depot 25 A2
Margam East 25 A2
Margam Middle Jn 122 B
Margam Moors Jn. 25 A2
Margaretting FP (R/G) 33 A3
MARGATE 22 D2
Marina (ABCL) 19 A1
Maritime Freightliner Terminal 9 E
Mark Beech Tunnel 12 D1
Markdhu No. 1 (UWC) 77 B2
MARKET HARBOROUGH 41 A1
Market Harborough Jn. 41 A1
Market Harborough Sdgs. 41 A1
MARKET RASEN 60 C2
Market Rasen Footpath (R/G) 60 C2
MARKINCH 92 C3
Marklach No. 1 (UWC) 77 B2
Marklach No. 3 (UWC) 77 B2
Markle (AHBC) 86 D1
MARKS TEY 33 B2
Marks Tey GF 33 B2
Marks Tey Jn. 33 B2
Marlborough Road (BW) 128 A
Marley Green (UWC) 47 B1
Marley Lane (CCTV) 13 A2
Marley Tunnels 4 C2
MARLOW 19 A1
MARPLE 130 D3
Marple North Tunnel 130 D3
Marple South Tunnel 130 D3
Marple Wharf Jn. 130 D3
Marr House Farm (UWC) 65 B3
Marrel (ABCL) 103 B2
Marriots (UWC) 50 D1
MARSDEN 57 B1
Marsh Brook (MCB) 37 B1
Marsh Brook SB (MB) 37 B1
Marsh Farm (OPEN) 66 C3
Marsh House (CCTV) 56 C2
Marsh Junction SB (M) 60 D1
Marsh Lane (ABCL) 30 D2
Marsh Lane (AHBC) 66 C3
Marsh Lane (Liverpool) 131 A
Marsh Lane (UWC) 44 C3
Marsh Lane Jn. (Leeds) 133 B
Marsh Mills 122 A
Marsh West Jn. 60 C1
Marshalls (UWC) 63 A2
Marshgate Jn. 135 A1
Marshmoor 32 C3
Marshwood Farm No. 2 Crossing 7 A2
MARSKE 71 A1
Marston (AHBC-X) 31 A1
MARSTON GREEN 126 A
Marston Moor 64 D1
Marston on Dove (AHBC) 49 B2
Marston Vale SCC (MV) 31 A1
Martello Tunnel 14 D1
MARTIN MILL 14 D1
Martin Road (UWC) 60 C3
Martins 43 A1
MARTINS HERON 19 A2
Martins Lane (R/G) 61 B3

MARTON 71 A1
Marton Lane (ABCL) 71 A1
Maryburgh (UWC) 99 A2
MARYHILL 137 B1
Maryhill Park Jn 137 B1
MARYLAND 110 C2
Maryland East Crossovers 110 C2
MARYLEBONE 108 D3
Marylebone IECC (ME) 108 D3
MARYPORT 73 A3
Maryport (CCTV) 73 A3
Maryport Station SB (MS) 73 A3
Masborough Jn. 134 A1
Masborough Sorting Sidings South Jn. 134 A1
Masons 1 (UWC) 23 B2
Masters (UWC) 16 C3
MATLOCK 58 C3
MATLOCK 58 C3
MATLOCK BATH 49 B1
Matlock GF 58 C3
Matt Pitts Lane (AOCL-X) 52 E
Mauchline SB (MM) 78 D1
Maud Foster (AHBC) 52 C1
MAULDETH ROAD 130 C2
Maxey (CCTV) 51 B3
MAXWELL PARK 139
Maxwelltown Goods Jn 73 A1
MAYBOLE 78 C2
Maylord (UWC) 37 A2
Maypole Rasen (UWC) 60 C2
Mays (CCTV) 19 A2
MAZE HILL 114 C1
McDougall Crossing 21 A3
McIvors (UWC) 103 B3
Mckenzies (UWC) 58 D3
McNicols (UWC) 102 C2
Mead Lane FP (R/G-X) 32 C3
Meadow Croft Farm (UWC) 66 C3
Meadow Lane (CCTV) 128 A
Meadow Lane Jn. 128 A
MEADOWHALL 134 A1
Meads (R/G-X) 7 A1
Meads Farm Crossing 7 B2
Meads Lane (UWC) 72 C3
Meaford Crossing (CCTV) 48 D2
Meardsall Lane (UWC) 59 B3
Meddiant Isaf (UWC) 54 C2
Meddiant No.3 (UWC) 54 C2
Meddiant No.4 (UWC) 54 C2
Meddiant No.5 (UWC) 54 C2
Medge Hall (MCG) 59 A1
Medhurst Row 20 D3
Medhurst Row (R/G) 12 D1
MEDSTEAD & FOUR MARKS 10 D1
Medway Viaduct 21 A3
MEILLIONEN 53 B3
Meir Tunnel 48 D1
Melbourne Jn. 128 A
MELDON QUARRY 5 A3
Meldon Quarry (oou) 5 A3
MELDRETH 32 C1
Meldreth Road (AHBC) 32 C1
MELKSHAM 16 D2
Mellaig No. 1 (UWC) 102 D1
Melling Tunnel 62 C1
Mellis (AHBC-X) 44 D2
Melrose Avenue (UWC) 55 B2
MELTON 34 D1
Melton (AOCL) 34 D1
Melton Jn. & GF 50 B3
Melton Lane (MCB) 65 B3
Melton Lane SB (ML) 65 B3
MELTON MOWBRAY 50 B3
Melton Station SB (MM) 50 D2
Menadue (UWC) 2 D1
Menai Bridge North Jn. 53 B2
Menai Bridge South Jn. 53 B2
MENHENIOT 3 A2
MENSTON 63 B2
MEOLS 55 B2
MEOLS COP 61 B3
MEOPHAM 20 D2
Mercer's (UWC) 129 B3
Merchants Quay (UWC) 73 A3
Merehead Quarry 16 C3
Merehead Quarry Jn. 16 C3
Merllyn (MCG) 35 A1
Merrick No. 2 (Accom) 13 B2
Merrick No. 3 (Accom) 13 B2
Merrings (UWC) 21 A1
Merryhill (TMO) 85 A1
MERRYTON 84 C3
Mersey Tunnel 131 B
Merseyrail SB (ML) 131 A
Mersham Tunnel 14 C1
MERSTHAM 20 C3
Merstham Tunnel 116 D3
MERTHYR TYDFIL 26 C3
MERTHYR VALE 26 C1
METHERINGHAM 60 C3
Methil Dock 92 D3
Methley Jn. 64 C3
Methley North (R/G) 64 C3
METRO CENTRE 136 A
Metropolitan Jn. 120 B
Meusydd Mill (UWC) 24 D2

MEXBOROUGH 58 D1
Mexborough Jn. 58 D1
MICHELDEVER 18 C3
Mickle Trafford SB (MT) 56 C3
MICKLEFIELD 64 D3
Micklefield Jn. 64 D2
Mickleham Tunnel 115 B3
Micklewood No. 2 (UWC) 38 C1
Mickley (R/G) 75 B2
Mid Cannock Colliery GF 49 A3
Mid Fearn (UWC) 102 C2
MID HANTS RAILWAY 10 D1
MID NORFOLK RAILWAY 44 D1
Midcalder Jn 85 A2
Middle 22 D3
Middle Drove (R/G) 42 D1
Middle Hill Tunnel 16 C2
Middle Jn. 21 B3
Middle Road Crossing 8 D1
Middle Siding East GF 125 B
Middle Stoke 21 B2
Middlemere (AHBC) 43 A2
MIDDLESBROUGH 136 C
Middlesbrough Goods Yard 136 C
Middlesbrough SB (M) 136 C
Middleton (ABCL) 45 A3
Middleton Place (UWC) 67 A2
Middleton Towers 43 B1
Middleton Towers No. 6 (TMO) 43 B1
Middleway (CCTV) 2 D1
Middlewich Loop 56 D3
Middlewich Loop East Jn. 56 D3
MIDDLEWOOD 130 D3
Middlewood Tunnel 130 D3
Midge Hall SB (MH) 62 C3
MIDGHAM 18 C2
Midgham (CCTV) 18 C2
MIDLAND RAILWAY CENTRE (BUTTERLEY) 50 C1
Midland Road Jn. 108 D3
Midland Yard Jn. 40 C1
Midley (OPEN) 13 B2
Milburn Jn. 99 B2
Mildearie No. 2 (UWC) 101 A2
Mile Drove 22 D3
Mile End (AHBC-X) 43 B2
Miles Platting Jn. 128 B
MILFORD 11 A1
Milford (AHBC) 11 A1
MILFORD HAVEN 23 A3
Milford Jn. 48 D2
Milford Jn. 64 D3
Milford SB (M) 64 D3
Milford Tunnel 49 B1
Mill Dam (UWC) 67 B3
Mill Deeping (UWC) 50 B3
Mill Drove (AHBC-X) 43 A2
Mill Green (MCB) 52 C2
Mill Green LC SB (MG) 52 C2
Mill Hall 21 A3
MILL HILL 62 D3
MILL HILL BROADWAY 107 B1
Mill Lane (UWC) 65 B3
Mill Lane Jn. 41 A2
Mill Lane Jn. SB (M) 63 B3
Mill Path (Footpath) 8 D1
Mill Race Jn. 134 A1
Mill St (TMO) 62 E
MILLBROOK (Beds) 31 A1
MILLBROOK (S/hampton) 9 E
Millbrook Freightliner Terminal 9 E
Millburn Grange (UWC) 40 C2
Millens (UWC) 90 D2
Millerhill East Jn. 86 E
Millerhill SB (M) 86 E
Millerhill South Jn. 86 E
Millerhill West Jn. 86 E
Millerhill Yard 86 E
Millers (UWC) 67 B3
Millfield Farm (UWC) 65 A2
MILLIKEN PARK 83 B2
MILLOM 67 B3
Millom SB 67 B3
MILLS HILL 130 C1
Millstead Crossing 21 A3
Milltown Viaduct 2 D1
Millwood Tunnel 63 A3
Milner Royd Jn. 63 B3
Milner Royd Jn. SB (MR) 63 B3
MILNGAVIE 137 B1
Milnthorpe GFs 68 C3
Milton 18 C1
Milton Court Crossing 19 B3
Milton Fen (AHBC) 43 A3
Milton Keynes Central 30 D1
Milton Keynes Nth Jn. 31 A1
Milton Keynes Sth Jn. 31 A1
Milton No. 1 (UWC) 104 D2
Milton No. 3 (UWC) 104 D2
Milton of Gollanfield (UWC) 99 B2
Milton of Larg No. 1 (UWC) 77 B3
Milton of Larg No. 2 (UWC) 77 B3
Milton Siding 18 E
Milton Village (MCB) 74 C2
Miltonise (UWC) 77 B2
Milverton Jn. 40 A3

MINEHEAD 6 E
Minety (MCG) 16 D1
MINFFORDD 35 B1
MINFFORDD 35 B1
MINSTER 22 D3
Minster (R/G) 22 D3
Minster East Jn. 22 D3
Minster South Jn. 22 D3
Minster West SB (EBE) 22 D3
Mintholme (CCTV) 62 C3
MIRFIELD 63 B3
Mirfield East Jn. 63 B3
MISTLEY 34 C2
Mistley (Footpath) (R/G) 34 C2
MITCHAM EASTFIELDS 112 D3
MITCHAM JUNCTION 116 D1
MITESIDE HALT 67 B2
Mitre Bridge Jn. 120 A
Moat Farm No 1 (UWC) 30 D2
Moat Hills (CCTV) 135 A1
MOBBERLEY 56 D2
Mobberley (MCB) 56 D2
Mobberley SB (MY) 56 D2
MOD Annan 73 B1
MOD Ashchurch 28 D1
MOD Bicester 30 C2
MOD Caerwent 15 B1
MOD Ernesettle 122 A
MOD Kineton 40 C3
MOD Long Marston 39 B3
MOD Longtown 74 C1
MOD Ludgershall 17 B3
MOD Marchwood 9 B2
MOD Pigs Bay 21 B2
MOD Sidings (Chippenham) 16 D2
MOD Sidings (Redmire) 69 B2
MOD Smalmstown 74 C1
Moel Llys Tunnel 54 C2
Moira West SB (MW) 49 B3
Molewood Tunnel 32 C2
Molinnis (AOCL) 2 C1
Mona (UWC) 54 C2
MoncrieffeTunnel 91 B2
MONIFIETH 92 D1
Monk Bretton Loop 58 C1
MONKS RISBOROUGH 30 D2
Monk's Siding (MCB) 132 B
Monkton Court 22 D3
Monkton GF 78 C1
Monktonhall Jn 86 E
Monkwearmouth Jn. 76 D2
Monsanto GF 123 B
Monsanto/BASF (AOCL) 136 C
MONTPELIER 124
Montpelier Jn. 14 E
Montpelier Tunnel 124
MONTROSE 95 B2
Montrose North SB (MN) 95 A2
Montrose South Jn. 95 A2
Moor Farm (UWC) 128 A
Moor Farm 1 (UWC) 67 B3
Moor Lane (UWC) 50 C3
Moor Lane (UWC) 59 B1
Moor Lane Tunnels 129 A1
MOOR PARK 31 A3
Moorcock Tunnel 69 A2
Moorends Farm (UWC) 65 A3
Moores 43 A3
MOORFIELDS (DEEP LEVEL) 131 B
MOORFIELDS (LOW LEVEL) 131 B
MOORGATE 109 A3
MOORSIDE 129 B1
Moorswater 3 A1
Moorswater (OC) 3 A2
Moorswater Viaduct 3 A2
MOORTHORPE 58 C1
Moorthorpe Jn. 58 C1
Moortown (AHBC-X) 60 C1
Morangie (UWC) 102 D2
MORAR 89 A1
Morar (AOCL) 89 A1
MORCHARD ROAD 5 B2
MORDEN SOUTH 116 D1
MORECAMBE 61 B1
Morecambe Jn. GF 61 B1
Morecambe South Jn. 62 C1
Moreton (AHBC) 8 C3
MORETON (Dorset) 8 C3
Moreton (UWC) 23 B3
MORETON (Wirral) 131 A
Moreton Cutting 18 E
MORETON-IN-MARSH 29 A1
Moreton-in-Marsh SB (MM) 29 A1
Moreton-on-Lugg 27 B1
Moreton-on-Lugg SB (ML) 27 B1
Morfa Main (UWC) 24 C3
MORFA MAWDDACH 35 B2
Morfa No.1 (UWC) 35 B1
Morfa-Rhydd-y-Pwll (UWC) 54 C2
Morlais Jn. 24 D3
Morlanga (UWC) 26 C3
MORLEY 64 C3
Morley Tunnel 64 C3
MORPETH 81 A3
Morpeth Electrification Depot 81 A3
Morpeth Jn. 81 A3

Morpeth North Jn. 81 A3
Morpeth SB (M) 81 A3
Morris Cowley GF 30 C3
Morris Farm No. 2 8 D1
Morris Hill (CCTV) 28 D2
Morris Motors GF 30 C3
Morse Gorse (UWC) 49 A3
Morston Hall (AHBC) 34 D2
MORTIMER 18 D2
MORTLAKE 112 C2
Mortlake (CCTV) 112 C2
Morton 58 C3
Morton Carr (AOCL) 71 A1
Morton Grange Farm No. 4 (UWC) 71 A1
Morvich No. 1 (UWC) 102 D1
Morvich No. 3 (UWC) 102 D1
Morvich No. 4 (UWC) 102 D1
Morvich No. 6 (UWC) 102 D1
Morvich No. 7 (UWC) 102 D1
Morvich No. 8 (UWC) 102 D1
Moseley Tunnel 126 A
MOSES GATE 129 B1
Moses Gate Jn. 129 B1
Mosley St GF 49 B2
Moss (MCB) 64 D3
Moss (Tip) 67 A3
Moss Lane (UWC) 130 C2
Moss Road (AHBC-X) 83 A1
MOSS SIDE 61 B3
Moss Side (ABCL) 61 B3
Mossband Jn. 74 C1
Mossend Jns. 138 D3
Mossend Yard 138 D2
Mosset Park (UWC) 100 C1
Mossgiel Tunnel 78 D1
MOSSLEY 130 D1
MOSSLEY HILL 131 A
MOSSPARK 137 A2
MOSTON 130 C1
Mostyn Dock 55 A2
Mostyn SB (MN) 55 A2
MOTHERWELL 138 D3
Motherwell SC (M) 138 D3
MOTSPUR PARK 116 C1
Motspur Park Jn. 116 C1
MOTTINGHAM 114 C2
MOTTISFONT & DUNBRIDGE 9 B1
Motts Lane FP (R/G-X) 33 B3
MOULDSWORTH 56 C3
Mouldsworth GF 56 C3
Moulin (UWC) 93 A2
Moulinearn (R/G) 93 B2
MOULSECOOMB 14 E
Moulton (AHBC-X) 44 D2
Mound Tunnels 85 E
Mount Bures (ABCL) 33 B2
MOUNT FLORIDA 139
Mount Gould Jn. 122 A
Mount Pleasant (CCTV) 9 E
Mount Pleasant Tunnel (Hastings) 13 B3
Mount Pleasant Tunnel (Liverpool) 131 B
Mount Street Tunnel 114 C1
MOUNT VERNON 138 C2
Mountain (OPEN) 13 B2
MOUNTAIN ASH 26 C1
Mountains No. 29 (UWC) 51 B1
Mountfield Sidings GF 13 A2
Mountfield Tunnel 13 A2
Mountnessing Jn. 21 A1
Mountsorrel LC 50 B3
Mountsorrel Sdgs. 50 C3
Mow Cop (CCTV) 48 D1
Moy loop 99 B3
MPV Depot 118 E
Mucking (AHBC) 21 A2
Mucky Lane (UWC) 50 B3
Mud Lane (UWC) 15 A2
MUIR OF ORD 99 A2
MUIREND 137 B3
Muirend GF 137 B3
Muirhouse Central Jn. 139
Muirhouse Farm (UWC) 73 B1
Muirhouse North Jn. 139
Muirhouse South Jn. 139
MUNCASTER MILL 67 B2
Munceys 43 A3
Munllyn (UWC) 37 A1
Murdercombe Tunnel 16 C3
Murie (AHBC) 92 C2
Murthly (AHBC) 91 B1
MUSSELBURGH 86 E
Muston (AHBC) 72 C3
Mutley Tunnel 122 A
Myremill (UWC) 78 C2
MYTHOLMROYD 63 A3
Mywars No. 2 (UWC) 36 C1

N

Naas (AHBC) 27 B3
Nadins Swadlincote GF 49 B2
NAFFERTON 66 C1
NAILSEA & BACKWELL 15 B2
NAIRN 100 C2
Nairn East 100 C1
Nairn SB (NA) 100 C1
Nairn West 100 C1

Nairns (No. 117) (UWC) 43 A2
NANT GWERNOL 35 B2
Nantmawr Branch Jn. (former) 47 A2
NANTMOR 53 B3
NANTWICH 48 C1
Nantwich (MCB) 48 C1
Nantwich Emergency GF 48 C1
Nantwich SB (NH) 48 C1
Nant-y-Cefn (UWC) 25 B1
Nantyci No. 2 (UWC) 24 C2
NANTYRONEN 35 E
NAPPERS HALT 4 C3
NARBERTH 23 B2
Narberth Tunnel 23 B2
NARBOROUGH 40 D1
Narborough (MCB) (CCTV) 40 D1
Narroways Hill Jn. 124
Nashenden Crossovers 21 A3
Nash's (UWC) 49 B2
National Railway Museum 136 B
Navarino Road Jn. 109 A2
NAVIGATION ROAD 129 B3
Navigation Road (CCTV) 129 B3
Naworth (AHBC-X) 74 D2
Naylors (UWC) 51 A3
Neasden Jn. SB (NJ) 120 A
Neasden South Jn. 107 B2
NEATH 122 B
Neath and Brecon Jn. 122 B
Neath and Brecon Jn. SB (NB) 122 B
NEEDHAM MARKET 34 C1
NEILSTON 137 A3
NELSON 63 A2
Nene Carriage Sidings 42 E
NENE VALLEY RAILWAY 42 C1
NESTON 55 B2
Nether Lane (AHBC-X) 66 C1
Nether Poppleton (AHBC) 136 B
NETHERFIELD 50 E
Netherfield Jn. 50 E
Netherfield Jn. SB (NJ) 50 E
NETHERTOWN 67 A1
NETLEY 10 C2
Neudd Farm 2 (UWC) 36 D3
Neville Hill Depot 133 B
Neville Hill East Jn. 133 B
Neville Hill West Jn. 133 B
NEW BARNET 108 D1
New Barnetby (MCG) 60 C1
NEW BECKENHAM 113 B3
New Bilton 40 E
New Bridge (UWC) 43 B2
New Bridge Lane (Footpath) 13 B1
NEW BRIGHTON 131 A
NEW CLEE 60 D1
NEW CROSS 120 B
NEW CROSS GATE 112 C2
NEW CROSS GATE 120 B
New Cross Gate Depot 120 B
New Cross Gate Down Jn. 120 B
New Cross Gate North Jn. 120 B
New Cross Gate Up Jn. 120 B
NEW CUMNOCK 79 A2
New Cumnock SB (NC) 79 A2
New Cut (UWC) 45 B1
New Earswick (UWC) 64 D1
NEW ELTHAM 114 D2
New England North 42 E
New England Sidings(West Yard) 42 E
New Fishbourne (AHBC) 10 D2
New Furnace Tunnel 63 B3
NEW HOLLAND 66 C3
New Holland Bulk Terminal 66 C3
New House Farm (UWC) 38 C1
New House Farm No. 43 33 A2
NEW HYTHE 21 A3
New Inn (OPEN) 66 C3
New Kew Jn. 112 C1
NEW LANE 62 C3
New Lane (AHBC) 56 C1
NEW MALDEN 112 C3
New Middle Sdgs. (Crewe) 48 E
NEW MILLS CENTRAL 130 D3
New Mills Central SB (NM) 130 D3
NEW MILLS NEWTOWN 130 D3
New Mills South Jn. 130 D3
New Mills South Jn. SB (NS) 130 D3
New MillsTunnel 130 D3
NEW MILTON 9 A3
New Oak Farm (UWC) 65 A3
New Place (F/P) 11 B2
NEW PUDSEY 63 B2
NEW ROMNEY 14 C2
NEW SOUTHGATE 108 D1
New St. North Tunnel 126 B
New St. South Tunnel 126 B
New York Farm (UWC) 64 C2
NEWARK CASTLE 51 A1
Newark Castle LC SB (NC) 51 A1
Newark Crossing 59 A3
Newark Crossing East Jn. 59 A3
Newark Crossing South Jn. 51 A1
NEWARK NORTH GATE 51 A1
Newark South Jn. 51 A1
Newbold Jn. 40 D2
NEWBRIDGE 26 D2
Newbridge Jn. 85 A2
NEWBURY 18 C2
NEWBURY RACECOURSE 18 C2
NEWCASTLE 136 A
Newcastle East Jn. 136 A
Newcastle Jn. 48 D1
Newcastle Rd (AHBC-X) 48 C1
Newcastle South Jn. 136 A
Newcastle West Jn. 136 A
Newcombes (UWC) 5 A3
NEWCRAIGHALL 86 E
Newfeild (UWC) 104 D2
Newham (CCTV) 81 A1
NEWHAVEN HARBOUR 12 C3
Newhaven Harbour RC 12 C3
Newhaven Harbour SB (NH) 12 C3
NEWHAVEN MARINE 12 D3
Newhaven Town 12 C3
NEWHAVEN TOWN 12 D3
Newhaven Town SB (CCO) 12 D3
NEWINGTON 21 B3
Newland East SB (NE) 28 C1
Newlands East (MCB) 38 D3
Newmans (UWC) 21 B1
NEWMARKET 43 A3
Newnham Barton Farm (UWC) 5 A2
Newnham Tunnel 28 C3
NEWPORT (Cambs) 32 D2
NEWPORT (Gwent) 123 B
Newport Docks 123 B
Newport East Jn. 136 C
Newport SB (N) 123 B
Newport Tunnel (New) 123 B
Newport Tunnel (Old) 123 B
NEWQUAY 1 B1
Newsham (MCB) 76 C1
Newsham North Jn. 76 C1
Newsham Road (TMO) 76 C1
Newsham SB (N) 76 C1
NEWSTEAD 50 C1
Newstead Tilford Road (AHBC) 50 D1
NEWTON (Glasgow) 138 C3
NEWTON (Manchester) 130 D2
Newton (UWC) 49 A2
NEWTON ABBOT 4 D1
Newton Abbot East Jn. 4 D1
Newton Abbot West Jn. 4 D1
NEWTON AYCLIFFE 70 C1
Newton East Jn. 138 C2
Newton Flotman (AHBC-X) 44 D1
Newton Heath TMD 130 C1
Newton Jn. (Birmingham) 126 A
Newton Jn. (Prestwick) 78 D1
Newton Lodge (UWC) 23 A3
Newton of Struthers (UWC) 100 D1
NEWTON ST CYRES 6 C3
Newton Street Tunnel 83 A1
Newton West Jn. 138 C2
Newton, Hamilton Jn. 138 C3
Newton, Kirkhill Jn. 138 C3
NEWTONDALE 71 B2
Newtonhill SB (NH) 96 D3
NEWTON-LE- WILLOWS 56 C2
Newton-le-Willows Jn. 56 C2
NEWTONMORE 94 C2
Newtonmore Station LC 94 C2
NEWTON-ON-AYR 78 C1
NEWTOWN 37 A1
Newtown GF 37 A1
Newtown Tunnel 130 D3
Newtown West 123 A
Niddrie South Jn. 86 E
Niddrie West Jn. 86 E
Niffany (UWC) 63 A2
Nigg (AHBC) 102 D3
Nine Elms Jn. 119
NINIAN PARK 123 A
NITSHILL 137 A3
No.94 Water Drove (MCG) 52 C2
Noblethorpe (MCG) 59 A1
Noose Lane (CCTV) 127 B
NORBITON 111 B3
NORBURY 113 A3
Norbury Hollow (MCG) 130 C3
NORCHARD 27 B3
Norchard Farm 1 (UWC) 23 B3
Nordans Farm (UWC) 38 C3
NORDEN 8 F
Normanby Park GF 65 B3
NORMANS BAY 13 A3
NORMANTON 64 C3
Normanton (AHBC-X) 50 D2
North (FP) 81 A3
North (MCB) 81 A3
NORTH BERWICK 86 C1
NORTH CAMP 19 A3
North Carr (MCG) 59 A2
North Cove 45 B2
North DownsTunnel 21 A3
NORTH DULWICH 113 A2
NORTH ELMHAM 44 F
North Erewash (CCTV) 128 A
North Fen (AHBC-X) 43 A2
North GF 35 B2
North Green (AOCL) 45 A3
North Halling 21 A3
North Jn. (Southport) 61 B3
North Jn. (Sunderland) 76 C2
North Kelsey (AHBC-X) 60 C1
North Kent East Jn. 120 B
North Lincoln Jn. 59 B1
North London Tunnel 2 Portal 110 D3
NORTH NORFOLK RAILWAY 46 C1
North Pole Depot 112 C1
NORTH QUEENSFERRY 85 A1
North Queensferry Tunnel 85 A1
NORTH ROAD 70 C1
North Seaton (MCB) 81 A3
NORTH SHEEN 111 B2
North Somerset Jn. 125 B
North Stafford Jn. (Crewe) 48 E
North Stafford Jn. (Derby) 128 A
North Stoke Tunnel 11 A2
North Tees (AOCL) 136 C
North Tunnel 32 C2
NORTH WALSHAM 46 D2
NORTH WEMBLEY 107 B2
North Wembley Jn. 107 B2
NORTH YORKSHIRE MOORS RAILWAY 71 B2
NORTHALLERTON 70 D2
Northallerton East Jn. 70 D2
Northallerton High Jn. 70 D2
Northam Jn. 9 E
Northam Traincare Centre 9 E
NORTHAMPTON 41 A2
Northampton North Jn. 41 A2
Northampton South Jn. 41 A3
Northchurch Tunnels 31 A2
Northcote 2 Crossing 6 D3
Northenden Jn. 130 C3
Northenden Jn. SB (NN) 129 B3
NORTHFAMBRIDGE 21 B1
NORTHFIELD 126 A
NORTHFLEET 20 D2
Northgate Street Tunnels 132 A
NORTHIAM 13 B2
NORTHOLT PARK 107 A2
Northolt Park Jn. 107 A2
Northorpe SB (N) 59 B2
NORTHUMBERLAND PARK 109 B1
Northwall (R/G) 22 D3
Northway (AHBC) 28 D1
NORTHWICH 56 D3
Northwich East Jn. 56 D2
Northwich South Jn. 56 D3
Northwich Station Jn. 56 D3
Northwich West Jn. 56 D3
NORTHWOOD HALT 38 D2
Norton (MCB) 64 D3
Norton (R/G) 56 C2
NORTON BRIDGE 48 D2
Norton Bridge Jn. 48 D2
Norton Bridge North Jn. 48 D2
Norton Bridge South Jn. 48 D2
Norton Farm 1 (UWC) 6 C3
Norton Farm 2 (UWC) 6 C3
NORTON FITZWARREN 6 D1
Norton Fitzwarren Jn. 6 D1
Norton Junction SB (NJ) 28 C1
Norton Marsh (UWC) 45 B1
Norton SB (NN) 56 C2
Norton West (MCB) 70 D1
Norton-onTees East Jn. 70 D1
Norton-on-Tees East SB (NE) 70 D1
Norton-on-Tees South Jn. 70 D1
Norton-on-Tees South SB (NS) 70 D1
Norton-onTees West Jn. 70 D1
Norton-on-Tees West SB (NW) 70 D1
Norway Crossing 11 B3
Norwell Lane (CCTV) 59 A3
NORWICH 44 E
Norwich Road (AHBC-X) 45 A1
Norwood (MCG) 58 D3
Norwood Fork Jn. 118 E
Norwood Jn. 136 A
Norwood Road (AHBC) 42 D1
NORWOODJUNCTION 118 E
NOTTINGHAM 50 E
Nottingham Branch Jn. 51 A2
Nottingham East Jn. 50 E
Nottingham West Jn. 50 E
NUNEATON 40 C1
Nuneaton North Chord 40 C1
Nuneaton North Jn. 40 C1
Nuneaton South Jn. 40 C1
NUNHEAD 113 B2
Nunhead Jn. 113 B2
Nunnery Main Line Jn. 134 A1
NUNTHORPE 71 A1
Nunthorpe SB (N) 70 D1
Nursery (UWC) 34 C2
Nutborne (AHBC) 10 D2
NUTBOURNE 10 D2
NUTFIELD 20 C3

O

OAKAMOOR 49 A1
OAKENGATES 48 C3
Oakengates Tunnel 48 C3
Oakenshaw Farm (UWC) 133 A
Oakenshaw Jn. 133 A
Oakenshaw South Jn. 133 A
OAKHAM 51 A3
Oakham SB (OA) 51 A3
Oakham Station (MCB) 51 A3
OAKLEIGH PARK 108 D1
Oakmead 19 B2
Oakwood Farm (R/G) 64 C1
OAKWORTH 63 A2
OBAN 87 A2
OCC LC 20 D3
OCKENDON 20 D1
OCKLEY 11 B1
Odd Lane (UWC) 18 A2
Oddingley (MCG) 39 A3
Oddington (AOCL) 30 C2
Oerffrwyd (UWC) 36 D1
Offham Crossing 11 A2
Offord (CCTV) 42 C2
Ogmore House Farm (UWC) 23 B2
Oil Sidings GF 61 B2
Oil Terminal (AOCL) 85 A1
OKEHAMPTON 5 A3
OKEHAMPTON 5 A3
Okeltor (OC) 3 B1
Old Castle (UWC) 103 B3
Old Ends (CCTV) 28 C3
Old Hall Mill (UWC) 45 B1
Old Hay 13 A1
OLD HILL 39 A2
Old Hill Tunnel 39 A2
Old Junction (UWC) 60 C1
Old Kent Road Jn. 120 B
Old Kew Jn. 111 B1
Old Leake (AHBC-X) 52 C1
Old Mill (UWC) (Llwynypia) 26 C2
Old Mill (UWC) (Marsh Brook) 37 B1
Old Oak Common Depot 108 C3
Old Oak Common East Jn. 108 C3
Old Oak Common West Jn. 108 C3
Old Oak Sidings 120 A
OLD ROAN 131 A
OLD STREET 109 A3
Oldbury Branch (oou) 39 A1
OLDFIELD PARK 16 C2
Olive Mount Jn. 131 A
Olive Mount Tunnel 131 A
OLTON 126 A
Olympic Park Jn. 110 C2
Onibury (MCB) 37 B2
Onibury SB (OY) 37 B2
Onllwyn Washery 25 B1
Oram's No. 1 (UWC) 62 C3
Oram's No. 3 (UWC) 62 C3
Ord Mains (UWC) 99 B1
Ord Works GF 123 B
Ordsall Lane Jn. 128 B
ORE 13 B3
Ore GF 13 A2
Ore Tunnel 13 B2
Oriel Road Tunnel 131 A
Orient Way Sidings 109 B2
Orival (UWC) 90 C2
ORMSKIRK 56 C1
ORPINGTON 118 D1
ORRELL 56 C1
ORRELL PARK 131 A
Orston Lane (MCBR) 50 D2
ORTON MERE 42 C1
Orton Mere 42 E
Osterfen (CCTV) 51 A1
Oswestry Branch Jn. 47 A2
OTFORD 20 D3
Otford Jn. 20 D3
Oubeck 62 C1
Oulmsdale Burn (UWC) 103 B2
OULTON BROAD NORTH 45 B1
Oulton Broad North 45 B2
Oulton Broad North Jn. 45 B1
Oulton Broad North SB (OB) 45 B2
OULTON BROAD SOUTH 45 B2
Oulton Broad Swing Bridge 45 B2
Ouse Valley Viaduct 12 C2
Ouston Crossovers 76 C2
OUTWOOD 64 C3
Over Jn. 28 C2
Overbury St. Tunnel 131 B
OVERPOOL 55 B3
OVERTON 18 C3
Oving Crossing 11 A3
Ox Pasture Farm 1 (UWC) 38 C3
Ox Pasture Lane (Public BW) 60 C3
OXENHOLME 68 D2
Oxenholme Emergency GF 68 C2
Oxenholme Jn. 68 C2
OXENHOPE 63 A2
OXFORD 30 C3
Oxford Car Terminal 30 C3
Oxford North Jn. 30 C3
OXFORD ROAD 128 B
Oxford Road Jn. 125 A
Oxford SB (OX) 30 C3
Oxford Station LC 30 C3
Oxley (Stafford Road) Jn. 127 B
Oxley SB (OY) 127 B
Oxmardyke 65 B3
Oxmarsh Crossing SB (OM) 66 C3

P

Post Office No. 2 (UWC) 36 D1
Potteric Carr Jn. 135 A1
POTTERS BAR 32 C3
Potters Bar Tunnel 32 C3
Potters Grange Jn. 65 A3
Potters Lock No. 1 (UWC) 128 A
Poulters (UWC) 64 D2
Poulton Jn. 61 B2
Poulton SB (PT) 61 B2
POULTON-LE-FYLDE 61 B2
Pound Lane Crossing 7 B2
Poundbury Tunnel 8 C3
Pouparts Jn. 119
Powderhall Branch Jn 86 E
Powderham (UWC) 6 C3
Powell (UWC) 27 A2
POYNTON 130 C3
Pratts (UWC) 45 B1
Pratts Lower (R/G-X) 11 B2
PREES 47 B2
Prees (MCB) 47 B2
Prees SB (PS) 47 B2
PRESCOT 56 C2
PRESTATYN 55 A2
Prestatyn SB (PN) 55 A2
PRESTBURY 57 A2
Prestbury Tunnel 57 A2
PRESTON 62 E
Preston Brook Tunnel 56 C2
Preston Docks 62 E
Preston Fylde Jn. 62 E
Preston North Jn. 62 E
PRESTON PARK 14 E
Preston Park Jn. 14 E
Preston Ribble Jn. 62 E
Preston SB (PN) 62 E
Preston South Jn. 62 E
PRESTONPANS 86 C2
PRESTWICK INTERNATIONAL AIRPORT 78 D1
PRESTWICK TOWN 78 C1
Price Church Farm (UWC) 27 A2
PRIESTHILL & DARNLEY 137 A3
PRIMROSE HILL 108 D3
Primrose Hill Jn. 108 D3
Primrose Hill Tunnels 108 C3
Prince of Wales (UWC) 47 B3
Prince of Wales Colliery 64 D3
Prince of Wales SB (P) 64 D3
Princes Bridge (AHBC) 10 D1
PRINCES RISBOROUGH 30 D2
Princes Risborough Jn 30 D2
Princes St (AOCL) 83 A3
Princess Royal Distribution Centre 120 A
PRITTLEWELL 21 B1
Prologis Park Siding 40 C1
Proof House Jn 126 B
PRUDHOE 75 B2
Prudhoe (MCB) 75 B2
Prudhoe SB(PE) 75 B1
PULBOROUGH 11 A2
Pulborough SB (PH) 11 B2
Pulford (AHBC) 47 B1
Pulford (AHBC) 55 B3
Pump House (UWC) 67 A1
Pumpfield Farm (R/G) 9 B2
Pumphouse No. 122 (UWC) 52 C2
PURFLEET 20 D2
Purfleet (CCTV) 20 D2
Purfleet Rifle Range (UWC) 20 D2
PURLEY 117 A2
PURLEY OAKS 117 A2
Purton Collins Lane (AHBC) 17 A1
Purton Common (UWC) 17 A1
PUTNEY 112 C2
Puxton & Worle (MCB) 15 A2
PWLLHELI 35 A1
Pwllheli Goods (ABCL) 35 A1
Pye Road (UWC) 92 C2
Pyewipe Jn. 59 B3
Pyewipe Road SB (P) 60 D1
PYLE 25 B2
Pylle Hill GF 125 B

Q

Quadring (AHBC-X) 52 C2
Quainton Road (site of) 30 D2
QUAKERS YARD 26 C2
Quarrington (CCTV) 51 B1
Quarry Burn (UWC) 98 C1
Quarry Hill Jn. 133 B
Quarry Tunnel 117 A3
Quay Sdg. (Conwy) 54 C2
Quay Viaduct 10 C2
Quay Ward No. 1 (UWC) 36 C1
Quay Ward No. 2 (UWC) 36 C1
Quay Ward No. 3 (UWC) 36 C1
Quay Ward No. 4 (UWC) 36 C1
Queen Adelaide (AHBC-X) 43 A2
Queen St. North Jn. 123 A
Queen St. South Jn. 123 A
Queen Street High Level Tunnel 139
QUEENBOROUGH 21 B2
Queenborough Yard 21 B2
QUEENS PARK (Glasgow) 139
QUEENS PARK (London) 108 C3
QUEENS ROAD PECKHAM 112 C2
Queens Road Tunnel 109 B2
QUEENSTOWN ROAD (BATTERSEA) 119
Quintinshill EGF 73 B1
QUINTREL DOWNS 2 C1
Quintrel Downs (ABCL) 2 C1
Quoiggs No. 1 (UWC) 91 A3
QUORN & WOODHOUSE 50 C3

R

Rabber Farm (UWC) 37 A3
Racecourse Sidings GF 18 A2
Rackheath Road (AHBC-X) 45 A1
RADCLIFFE 50 E
Radford Jn. GF 128 A
RADLETT 31 B3
Radlett Jn. 31 B3
RADLEY 30 C3
Radway Green (CCTV) 48 D1
RADYR 123 A
Radyr Branch Jn. 123 A
Radyr Jn. 123 A
Radyr Junction SB (VR) 123 A
Raigmore (CCTV) 99 B2
Raikes (UWC) 36 D3
Rail Innovation & Development Centre 59 A3
Rainbow Hill Jn.(former) 38 D3
Rainbow Hill Tunnel 38 D3
RAINFORD 56 C1
Rainford Jn. SB 56 C1
Rainham (CCTV) (Essex) 20 D1
RAINHAM (Essex) 20 D1
RAINHAM (Kent) 21 A3
Rainham SB (EU) (Kent) 21 B2
RAINHILL 56 C2
Raiths Farm FreightTerminal 96 D2
Rallt (UWC) 36 D1
Rampart Lane (UWC) 64 D3
RAMSBOTTOM 62 D3
Ramsey Road (AHBC) 42 D1
RAMSGATE 22 D3
RAMSGREAVE & WILPSHIRE 62 D3
RANNOCH 90 D3
Ranskill (MCB) 59 A2
Ranskill Loops 59 A2
Rat Hole No. 80 (UWC) 59 A2
Ratcliffe Jn. 128 A
Ratcliffe North Jn. 128 A
Ratcliffe Power Station 128 A
Ratcliffe South Jn. (OOU) 128 A
RAUCEBY 51 B1
Rauceby SB (RY) 51 B1
Raven (AOCL) 26 A1
RAVENGLASS 67 A2
RAVENGLASS & ESKDALE RAILWAY 67 A2
Ravenhead Jn. 56 C2
RAVENSBOURNE 114 C3
Ravensbourne Jn. 114 C3
RAVENSTHORPE 63 B3
Ravenstruther 84 D3
RAWCLIFFE 65 A3
Rawcliffe (AHBC) 65 A3
Rawcliffe Branch (UWC) 65 A3
Raws (UWC) 63 A2
RAWTENSTALL 62 D3
RAYLEIGH 21 A1
RAYNES PARK 112 C3
READING 125 A
Reading Depot 125 A
Reading East Jn. (future) 125 A
Reading Lane Jn. 109 B2
Reading New Jn. 125 A
Reading SB (R) 125 A
Reading Spur Jn 18 D2
Reading Triangle Sidings 125 A
READING WEST 125 A
Reading West Jn. 125 A
Reads GF 26 C1
Rearsby (AHBC) 50 B3
Reasby Manor (UWC) 60 C2
Recreation 21 B2
Rectory Farm (UWC) 42 D3
Rectory Jn. 50 E
Rectory Jn. SB (RJ) 50 E
RECTORY ROAD 109 A2
Rectory Road (AHBC-X) 44 D2
Red Barns Tunnel 136 A
Red Cap Lane (ABCL) 52 C1
Red Cow (CCTV) 6 C3
Red Hill Tunnel (Hereford) 27 B1
Red Hill Tunnels (E. Midlands) 128 A
Red House (Kelsale) 45 A3
Red House (UWC) 37 A1
Red House Farm (Suffolk) 34 D1
Red House Farm (UWC) (Blyth) 76 C1
Red House Farm No. 1 36 D1
Red Lane (MCG) 64 C3
Red Van (UWC) 94 C3
REDBRIDGE 9 E
Redcar (MCB) 71 A1
REDCAR CENTRAL 71 A1
REDCAR EAST 71 A1
Redcar Ore Terminal Jn. 136 C
Redcar SB(R) 71 A1
REDDISH NORTH 130 C2
REDDISH SOUTH 130 C2
REDDITCH 39 A2
Redford Jn 92 C3
REDHILL 20 E
Redhill Tunnel 20 E
REDLAND 124
Redland GF 43 B3
Redmire 69 B2
REDMIRE 71 B2
Redmoor (AOCL) 43 A1
Rednal Farm (UWC) 47 A2
REDRUTH 1 B2
Redruth Tunnel 1 B2
REEDHAM (Norfolk) 45 A1
REEDHAM (Surrey) 117 A2
Reedham Jn. 45 B1
Reedham Jn. SB (RJ) 45 A1
Reedham Swing Bridge SB (RB) 45 A1
Reeds Farm (UWC) 48 C1
Reepham (CCTV) 60 C3
Regent Road (AOCL) 131 A
Regent Street (CCTV) 44 C3
Regents Canal Jn. 108 D3
REIGATE 20 C3
Reigate SB (RG) 20 C3
Renishaw Park 58 C2
RENTON 83 B1
Renwick Road Jn 110 D3
Reston GSP 82 C2
RETFORD 59 A2
Retford North 59 A2
Retford South Jn. 59 A2
Retford Western Jn. 59 A2
REVENGLASS 67 A2
RHEIDOL FALLS 35 F
RHIWBINA 123 A
Rhiwderin (AOCL) 123 B
RHIWFRON 35 F
Rhiwlas Hall No. 4 (UWC) 36 C1
Rhohm Haas (AOCL) 136 C
RHOOSE 26 C3
Rhosfach (UWC) 36 C1
Rhosferig Tunnel 36 D3
RHOSNEIGR 53 A2
Rhowniar (UWC) 35 B3
RHYD DDU 53 B3
Rhydllyn 2 (UWC) 37 A3
Rhydwhimen (R/G) 37 A1
Rhyd-y-Fynnon Farm (UWC) 24 D2
RHYDYRONEN 35 B2
RHYL 54 D1
Rhyl SB (RL) 54 D1
RHYMNEY 26 C1
RIBBLE STEAM RAILWAY 62 F
RIBBLEHEAD 69 A3
Ribblehead GF 68 D3
Riccarton 78 D1
RICE LANE 131 A
Richborough (AHBC) 22 D3
Richborough Castle 22 D3
RICHMOND 111 B2
Richmond Hill Tunnel 133 B
Richmond SB (GB) 111 B2
Rickerscote 48 D2
RICKMANSWORTH 31 A3
Ridden's Lane Crossing 12 C2
RIDDLESDOWN 117 A2
Riddlesdown Tunnel 117 A2
RIDGMONT 31 A1
RIDING MILL 75 B2
Rigg (UWC) 73 B1
Rigton SB (RN) 64 C2
Rillington (AHBC-X) 71 B3
Rimmell's (UWC) 27 B1
Ripe (AHBC) 12 D3
Rippings (UWC) 50 B2
Rippins Main (UWC) 50 B3
Ripple Lane Yard 110 D3
RISCA & PONTYMISTER 123 B
Risca South Jn. 123 B
Risehill Tunnel 69 A2
RISHTON 62 D2
Rishton Tunnel 62 D2
Ritchies (UWC) 58 D1
River Bank (No. 305) 59 B3
River Board 45 B1
River Bridge 25 B2
Riverford (UWC) 99 A2
Riverhall (CCTV) 13 A2
Riverside Sdgs. 41 A2
Riverside Yard (Exeter) 6 C3
Robarts I (UWG) 32 D3
Robarts III (UWG) 32 D3
ROBERTSBRIDGE 13 A2
Robertsbridge 13 A2
Robertsbridge SB (RB) 13 A2
Robertstown (TMO) 25 B1
Robeston Elf Sidings 23 A3
Robin Hood Tunnel 57 B1
Robin's Bottom Plantation (UWC) 72 C3
Robinsons (UWC) (Metheringham) 60 C3
Robinsons (UWC) (Ulceby) 66 C3
ROBY 131 A
ROCHDALE 57 A1
Rochdale East Jn. 57 A1
Rochdale West SB (TH) 130 C1
ROCHE 2 C1
ROCHESTER 21 A3
Rochester Bridge Jn. 21 A3
Rochester SB (ER) 21 A2
ROCHFORD 21 B1
ROCK FERRY 131 B
Rock Ferry North Jn 131 B
Rock Ferry South Jn. 131 B
Rockcliffe Hall SB (RH) 55 B3
Rockliffe Hall Tunnel 55 B3
Rodidge (MCG) 49 B3
Roffey Road (AHBC) 11 B1
ROGART 102 D1
Rogart (OPEN) 102 D1
ROGERSTONE 123 B
Rogie (UWC) 98 D1
ROLLESTON 50 D1
Rolleston (MCG) 50 D1
Rolleston Mill (UWC) 50 D1
Rolt Street Jn. 120 B
ROLVENDEN 13 B1
ROMAN BRIDGE 54 C3
Roman BridgeTunnel 54 C3
Romanby Road (CCTV) 70 D2
Rome Street Jn. 127 A
ROMFORD 20 D1
ROMILEY 130 D3
Romiley Jn. 130 D3
Romiley Jn. SB (RJ) 130 C3
ROMNEY SANDS 14 C2
ROMNEY WARREN 14 C2
ROMNEY, HYTHE & DYMCHURCH RAILWAY 14 C1
ROMSEY 9 B1
Roodee Jn. 132 A
Rookery No.1 21 B1
ROOSE 61 B1
ROPLEY 10 C1
Rosarie (AOCR) 101 A2
Rose Bank (UWC) 98 C1
Rose Farm (UWC)(oou) 30 D2
ROSE GROVE 62 D2
Rose Grove West Jn. 62 D3
ROSE HILL 130 D3
Rose Lane 44 D1
Rose Street (CCTV) 99 B2
Roseisle GF 100 D1
Rosewain (R/G) 73 B2
Rosier No. 2 Crossing 11 B2
Roskear Jn. (MCB) 1 B2
Roskear Jn. SB (R) 1 B2
Rossal No. 2 (UWC) 102 D1
Rossett (R/G) 47 B1
Rossington (CCTV) 58 D1
Rossington Colliery Jn. 135 A1
Rossington GSP 135 A1
ROSYTH 85 A1
Rosyth Dockyard 85 A1
ROTHERHAM CENTRAL 134 A1
Rotherham Central Jn. 134 A1
Rotherham Main (UWC) 134 A1
Rotherham Steel Terminal 134 A1
ROTHERHITHE 120 B
Rothery (UWC) 73 B2
ROTHLEY 50 C3
Roudham Hall Road 44 C2
Rough Crossing 22 D3
ROUGHTON ROAD 46 C1
Round Oak Sdgs 39 A2
Roundabout Drove 43 A2
Roundham (R/G-X) 30 C2
Roundstone (CCTV) 11 B3
Roundwood 58 C1
Rounton Gates (AHBC-X) 70 D2
Routes No.8 (R/G) 34 C1
Rover Sidings 17 A1
Rovie (AOCL) 102 D1
Rovie (UWC) 102 D1
Rowes (UWC) 49 B2
Rowland Hall (AHBC-X) 65 A3
ROWLANDS CASTLE 10 D2
ROWLEY REGIS 39 A2
ROWSLEY SOUTH 58 C3
Rowston (MCG) 51 B1
Roxby 65 B3
Roxton Sidings (MCG) 60 C1
Roxton Sidings SB (RS) 60 C1
ROY BRIDGE 90 D2
Royal Albert Bridge 122 A
Royal Oak 112 C1
Royal Oak (AHBC-X) 72 C3
Royal Portbury Dock 15 B1
ROYDON 32 C3
Roydon Lock (UWC) 32 C3
ROYSTON 32 C1
RR Loop (oou) 23 A3
RUABON 47 A1
Ruabon Road Tunnel 47 A1
Ruckholt Road Jn. 109 B2
Ruddle Bridge (UWC) 28 C2
RUFFORD 56 C1
Rufford (MCB) 62 C3
Rufford Colliery 58 D3
Rufford Colliery Jn. 58 D3
Rufford No.1 Coal Stacking Site 58 D3
Rufford SB (RD) 62 C3
RUGBY 40 D2

Rugby North Jn 40 E
Rugby SCC (#) 40 E
Rugby South Jn. 40 E
Rugby Trent Valley Jn. 40 E
Rugeley North Jn. 49 A2
Rugeley Power Station 49 A3
RUGELEY TOWN 49 A3
RUGELEY TRENT VALLEY 49 A2
Ruislip Gardens Jn. 107 A2
Rumney River Bridge Jn. 123 B
RUNCORN 56 C2
RUNCORN EAST 56 C2
Runcorn SB (RN) 56 C2
Ruscombe 18 D1
Rusham (AHBC) 19 A2
RUSHCLIFFE HALT 50 C2
Rushey Moor (UWC) 58 D1
Rushey Platt 17 A1
Rushey Sidings (AHBC-X) 59 A3
Rushton No. 2 (R/G) 8 D3
RUSKINGTON 51 B1
Rusper Road (MCG) 11 B1
Russell Street Tunnel 131 B
Russell's Siding GF 138 C2
Rustons Tip (R/G) 59 B3
RUSWARP 71 B1
Ruswarp (ABCL) 71 B1
RUTHERGLEN 139
Rutherglen (Footpath) (R/G) 139
Rutherglen Central Jn. 139
Rutherglen East Jn. 139
Rutherglen North Jn. 139
Rutherglen West Jn. 139
Ryans Disposal Point 25 B1
RYDE ESPLANADE 10 C3
Ryde Esplanade Tunnel 10 C3
RYDE PIER HEAD 10 C3
RYDE ST JOHNS ROAD 10 C3
Ryde St. Johns Road SB (WFP) 10 C3
Ryde Train Care Depot 10 C3
RYDER BROW 130 C2
RYE 13 B2
Rye Hill Farm (UWC) 66 C3
RYE HOUSE 32 C3
Rye SB (RY) 13 B2
Ryecroft Jn. 127 B
Ryhope Grange Jn. 76 D2
Ryhope Grange SB (RG) 76 D2
Rylstone (TMO) 63 A1

S

Sack Lane 11 A3
Sadler's (UWC) 39 A3
ST. ALBANS 31 B3
ST. ALBANS ABBEY 31 B3
St. Andrews Jn. 126 B
ST. ANDREWS ROAD 124
St. Andrews SB (SA) 124
St. Annes Crossing 21 B3
St. Annes Park No. 3 (Foxes Wood) Tunnel 124
St. Annes Park Tunnel No. 2 124
ST. ANNES-ON-THE-SEA 61 B3
ST. AUSTELL 2 C2
St. Austell GF 2 C2
ST. BEES 67 A1
St. Bees (MCB) 67 A1
St. Bees SB (SB) 67 A1
St. Blazey Bridge (CCTV) 2 D1
St. Blazey Jn. 2 D2
St. Blazey SB (SB) 2 D1
ST. BUDEAUX FERRY ROAD 122 A
St. Budeaux Jn. 122 A
ST. BUDEAUX VICTORIA RD 122 A
St. Catherines Jn. 135 A1
St. Catherine's Tunnel 19 A3
St. Clears (CCTV) 24 C2
St. Clears GF 24 C2
ST. COLUMB ROAD 2 C1
St. Cross Tunnel 10 C1
St. David's Golf Club (UWC) 35 B1
St. Davids Tunnel 6 C3
St. Dennis GF 2 C1
ST. DENYS 9 E
St. Dunstan's (CCTV) 22 C3
ST. ERTH 1 A2
St. Erth Jn. 1 A2
St. Erth SB (SE) 1 A2
St. Fagans (MCB) 123 A
St. George's (CCTV) 26 C3
St. George's Church (UWC) 26 C3
St. Georges Road (CCTV) 66 C3
St. Germains (CCTV) 86 C1
ST. GERMANS 3 A2
St. Germans (AHBC) 43 B1
ST. HELENS CENTRAL 56 C1
ST. HELENS JUNCTION 56 C2
St. Helens Station SB (SH) 56 C1
ST. HELIER 116 D1
ST. IVES 1 A2
St. James Bridge Jn. 136 A
St. James Deeping SB (SD) 52 C3
St. James Jn. 135 A1
St. James No.1 Tunnel 131 B
St. James No.2 Tunnel 131 B
St. James No.3 Tunnel 131 B
St. James No.4 Tunnel 131 B
ST. JAMES PARK 6 C3
ST. JAMES STREET 109 B1
ST. JOHNS 114 E
St. Johns Wood Tunnel 108 C3
St. Johns Wood Tunnel 108 C3
ST. KEYNE WISHING WELL 3 A2
ST. LEONARDS WARRIOR SQUARE 13 A3
St. Margarets (CCTV) 32 C3
ST. MARGARETS (Essex) 32 C3
ST. MARGARETS (Middx) 111 B2
ST. MARY CRAY 118 D1
St. Mary Cray Down Jn. 118 F
St. Mary Cray Up Jn. 118 F
St. Mary's (MCG) 28 D3
ST. MARY'S BAY 14 C2
ST. MARY'S HALT 27 B3
St. Marys North Jn. 128 A
ST. MICHAEL'S 131 B
St. Michael's Tunnel 131 B
ST. NEOTS 42 C2
St. Neots North Jn. 42 C2
St. Neots South Jn. 42 C2
ST. PANCRAS INTERNATIONAL 108 D3
St. Pancras SR 109 A3
St. Pauls Bridge 120 B
ST. PETER'S 76 D2
St. Phillips Marsh GF 125 B
St. Phillips Marsh HST Depot 125 B
St. Pinnock Viaduct East 2 D1
St. Stephen's (CCTV) 22 C3
SALFORD CENTRAL 128 B
SALFORD CRESCENT 128 B
SALFORDS 12 C1
SALHOUSE 46 D2
SALISBURY 9 A1
Salisbury Depot 9 A1
Salisbury SB (SY) 9 A1
Salisbury Tunnel Jn. 9 A1
Salmon Pool (AOCL) 5 B3
Salop Goods Jn. 48 E
Salop Goods Jn. SB (SG) 48 E
Salop Goods Loop Jn. 48 E
SALTAIRE 63 B2
SALTASH 122 A
SALTBURN 71 A1
Saltburn Riding School (UWC) 71 A1
Saltburn West Jn. 71 A1
SALTCOATS 83 A3
Saltcoats (MCG) 67 A2
Saltend and BP Chemicals 66 C3
Salterhebble Tunnels 63 B3
Saltford Tunnel 16 C2
Salthouse Jn. 61 B1
Salthouse Jn. GF 61 B1
Salthouse No. 1 (UWC) 67 B3
Salthouse No. 3 (UWC) 67 B3
Saltley PSB (SY) 126 B
SALTMARSHE 65 A3
Saltmarshe SB (SA) 65 A3
Saltmoor (UWC) 38 C2
Saltney Jn. 132 A
Saltwood Jn. 14 C1
Saltwood Tunnel 14 C1
Salvation Army Colony No. 2 21 A1
SALWICK 62 C2
Salwick SB (SK) 62 C2
SAMPFORD COURTENAY 5 A3
Sand Bank Jn. 135 A1
Sand Lane (UWC) 71 B3
SANDAL & AGBRIGG 133 A
SANDBACH 56 D3
Sandbach North Jn. 56 D3
Sandbach South Jn. 56 D3
SANDERSTEAD 117 A2
Sandhill (AHBC-X) 43 A2
Sandhill Lane (MCG) 64 D3
SANDHILLS 131 A
Sandhills Jn. 131 A
SANDHURST 18 D2
Sandilands (ABCL) 35 B3
SANDLING 14 C1
Sandling Tunnel 14 C1
SANDOWN 10 C3
SANDPLACE 3 A2
Sandringham Avenue (UWC) 55 B2
Sandscale / British Cellophane (AOCL) 67 B3
Sandstone Tunnel 124
Sandways (OC) 3 B1
SandwayTunnel 21 B3
SANDWELL AND DUDLEY 39 A1
SANDWICH 22 D3
Sandwich 22 D3
Sandwich SB (SW) 22 D3
SANDY 42 C3
Sandy Lane 44 C1
Sandy Lane (AHBC-X) 30 C2
Sandy Lane (UWC) (Exeter) 6 C3
Sandy Lane (UWC) (Prestatyn) 55 A2
Sandy North Jn. 42 C3
Sandy South Jn. 42 C3
SANKEY 56 C2
SANQUHAR 79 B2
Santon (AHBC-X) 43 B2
Santon Ore Mining (UWC) 59 B1
Sapperton Tunnels 28 D3
SARN 25 B2
Sarn (UWC) 36 D1
Sarnau (CCTV) 24 C2
Sarnau Lower (UWC) 24 C2
Sarnlas (UWC) 23 B2
Sarre Bridge 22 D3
SAUNDERSFOOT 23 B3
SAUNDERTON 30 D3
Saunderton TunnelUp Line only 30 D3
Savernake GF 17 A2
SAWBRIDGEWORTH 32 C2
Sawbridgeworth (CCTV) 32 D3
Sawley (CCTV) 128 A
Sawston (CCTV) 32 D1
Sawston (UWC) 32 D1
SAXILBY 59 B3
Saxilby SB 59 B2
SAXMUNDHAM 45 A3
Saxmundham (MCB) 45 A3
Saxmundham Jn. 45 A3
Saxmundham Road (TMO) 45 A3
Saxmundham SB (SM) 45 A3
Saxondale (UWC) 50 D1
Scalm Lane (R/G) 64 D3
SCARBOROUGH 72 C2
Scarrington Lane (AHBC-X) 50 D1
Scholes Tunnel 56 C2
School Lane (Public BW) 59 A2
School Road 9 B2
Schoolhill Tunnel 96 D2
Scoop 1 (UWC) 5 A2
Scopwick SB (SK) 51 B1
Scorborough (AHBC-X) 66 C2
Scotby (UWC) 74 C2
Scothearn (AHBC-X) 60 C2
SCOTSCALDER 104 C2
SCOTSTOUNHILL 137 A2
Scours Lane 125 A
Scout Tunnel 130 D2
Screener (UWC) 59 B2
Scremerston (CCTV) 82 D3
Scrooby (UWC) 59 A2
Scropton (MCG) 49 B2
Scropton Mill Lane (UWC) 49 B2
Scropton SB 49 B2
SCUNTHORPE 59 B1
Scunthorpe SB (S) 59 B1
Scunthorpe West Jn. 59 B1
SEA MILLS 124
Sea Mills (UWC) 124
SEABURN 76 D2
Seacroft (AOCL-X) 52 E
Seafield (TMO) 86 C3
SEAFORD 12 D3
SEAFORTH & LITHERLAND 131 A
SEAHAM 76 D2
Seaham LC 76 D2
Seal 20 D3
Seal Sands (AOCL) 136 C
Seal Sands Branch Jn. 136 C
Seal Sands Chemical (AOCL) 136 C
Seal Sands Road (AOCL) 136 C
Seal Sands Run Round Loop 136 C
Sealand Road Escape Shaft 111 A1
SEAMER 72 C3
Seamer Carr Farm 72 C3
Seamer SB (SR) 72 C3
Seamer South Jn. 72 C3
Seamer West Jn. 72 C3
Seapark No. 2 (UWC) 100 D1
SEASCALE 67 A2
SEATON CAREW 76 D3
Seaton Carew Jn. 76 D3
Seaton Snook Jn. 76 D3
Seaton Tunnel 41 B1
Seaton-on-Tees 76 D3
Second Drove (AHBC-X) 43 A2
SEER GREEN & JORDANS 31 A3
Seggiehill (UWC) 92 D2
Seghill North (AHBC) 76 C1
SELBY 64 D2
SELBY 65 A2
Selby Mine Sdgs 64 D3
Selby Potter Sidings 65 A2
Selby Road (AHBC) 58 D1
Selby SB (S) 64 D2
Selby South Jn. 65 A3
Selby Swing Bridge 65 A2
Selby West Jn. 64 D2
SELHURST 118 E
Selhurst Depot 118 E
Selhurst Depot SB (S) 118 E
Selhurst Jn. 118 E
SELLAFIELD 67 A2
Sellafield SB (SD) 67 A2
SELLING 22 C3
Selling Tunnel 22 C3
SELLY OAK 126 A
Selmeston (AHBC) 12 D3
Selsdon Road Jn. 117 A2
Sergeants (UWC) 49 A2
SETTLE 63 A1
Settle Junction SB (SJ) 63 A1
SEVEN KINGS 110 D2
SEVEN SISTERS 109 A1
Seven Sisters Jn. 109 A1
SEVENOAKS 20 D3
Sevenoaks Tunnel 20 D3
Sevenscore (AHBC) 22 D3
SEVERN BEACH 124
Severn Bridge Jn. (Shresbury) 47 B3
Severn Bridge Jn. SB (SBJ) (Shrewsbury) 47 B3
Severn Tunnel 15 B1
Severn Tunnel East 124
Severn Tunnel Jn. 15 B1
SEVERN TUNNEL JUNCTION 15 B1
SEVERN VALLEY RAILWAY 38 D2
Sewage Farm (UWC) 21 A2
Sewage Works Lane (UWC) 33 B1
Sewerage Works (UWC) 24 D2
Sewerby (AHBC) 66 C1
Sewstern Lane (R/G) 51 A2
Seymour Jn. 58 D3
Seymour Jn. SB (SE) 58 D2
Shacklegate Jn 111 B3
SHADWELL 120 B
Shadwell LC 44 C2
Shady Lane (UWC) 63 A2
Shaftholme Jn. 58 D1
Shakehole Bridge 21 A3
Shakespeare Tunnel 14 D1
SHALFORD 19 A3
Shalford Jn. 11 A1
Shalfords (UWC) 33 B1
Shalmsford Street (UWC) 22 C3
SHANKLIN 10 C3
Shap Hardendale Quarry 68 D1
Shap Summit GSP 68 D1
Sharnbrook Jn. 41 B2
Sharnbrook Tunnel 41 B2
Sharpes (UWC) 15 B1
Sharpness Docks 28 C3
SHAWFORD 10 C1
SHAWLANDS 139
Shaw's (R/G) 56 C1
Shearing No.2 (UWC) 45 B1
Sheen Hill No. 1 (UWC) 29 A1
Sheen Hill No. 2 (UWC) 29 A1
Sheerness Docks 21 B2
Sheerness Dockyard 21 B2
SHEERNESS-ON-SEA 21 B2
Sheet (AHBC) 10 D1
Sheet Stores Jn. 128 A
SHEFFIELD 134 A1
Sheffield North Jn. 134 A1
SHEFFIELD PARK 12 C2
Sheffield SB (S) 134 A1
Sheffield South Jn. 134 A1
SHELFORD 32 D1
Shelford (CCTV) 32 D1
Shell Junction 136 C
Shell Mex (OPEN) 66 C3
Shelwick Green (UWC) 27 B1
Shelwick Jn. 27 B1
SHENFIELD 20 D1
Shenfield Jn. 20 D1
SHENSTONE 49 A3
Shepcote Lane East Jn. 134 A1
Shepcote Lane West Jn. 134 A1
SHEPHERD'S BUSH 112 C1
Shepherds Lane Jn. 119
SHEPHERDS WELL 14 C1
Shepiston Lane Escape Shaft 111 A1
SHEPLEY 57 B1
SHEPPERTON 19 B2
SHEPRETH 42 D3
Shepreth (AHBC) 42 D3
Shepreth Branch Jn. 42 D3
SHERBORNE 7 B2
Sherborne (CCTV) 7 B2
Sherburn Jn. 64 D2
SHERBURN-IN-ELMET 64 D2
Shere Heath Crossing 11 B1
Sheriffs Court 22 D3
SHERINGHAM 46 C1
SHERINGHAM 46 C1
Sheringham East LC 46 C1
Sherrington Crossing 16 D3
Sherrington Footpath Crossing 8 D1
Sherrington Footpath Crossing 16 D3
SHETTLESTON 138 C2
Shettleston Workshops 138 C2
Shevock (UWC) 101 B3
Shewalton Moss GF 78 D1
SHIELDMUIR 138 D3
Shieldmuir North Jn. 138 D3
Shieldmuir South Jn. 138 D3
Shields Electric Traction Depot 139
Shields Jn. 139
SHIFNAL 48 C3
SHILDON 70 C1
Shildon SB (S) 76 C3
Shildon Tunnel 76 C3
SHIPLAKE 18 D1
Shiplake (AOCL) 18 D1
SHIPLEY 63 B2
Shipley East Jn. 63 B2
Shipley South Jn. 63 B2
Shipley Tunnel 63 B2
Shipley West Jn. 63 B2
SHIPPEA HILL 43 A2

Shippea Hill (MCB-OD) 43 B2
Ships (UWC) 28 D1
SHIPTON 29 B2
SHIREBROOK 58 D3
Shirebrook East Jn. 58 D3
Shirebrook Jn. 58 D3
Shirebrook Jn. SB (SJ) 58 D3
Shirebrook South Jn. 58 D3
SHIREHAMPTON 124
SHIREOAKS 58 D2
Shireoaks East Jn. 58 D2
Shireoaks Station (CCTV) 58 D2
Shireoaks West Jn. 58 D2
SHIRLEY 39 B2
Shoals Hook (UWC) 23 A2
SHOEBURYNESS 21 B1
SHOLING 10 C2
Shore House Farm (UWC) 62 D3
SHOREDITCH HIGH STREET 109 A3
SHOREHAM 20 D3
Shoreham by Sea (CCTV) 11 B3
Shoreham East (CCTV) 11 B3
SHOREHAM-BY-SEA 11 B3
Shornmead 21 A2
Short Drove 43 A2
SHORTLANDS 114 C3
Shortlands Jn. 114 C3
Shortridge Farm 2 (UWC) 5 A1
Shortridge Farm 3 (UWC) 5 A1
Shotlock Hill Tunnel 69 A2
SHOTTLE 49 B1
SHOTTON (HIGH LEVEL) 55 B3
SHOTTON (LOW LEVEL) 55 B3
SHOTTS 84 D2
Shotwick GF 55 B3
Shrewbridge Rd (AHBC-X) 48 C1
SHREWSBURY 47 B3
Shrub Hill Jn. 38 D3
Shugborough Tunnel 49 A2
Shuttlewords Top (UWC) 50 C2
Sibsey SB (S) 52 C1
Sibster Buolter (UWC) 104 C2
Sibster Moss (UWC) 104 C2
Sibster No. 1 (UWC) 104 D2
Sibsterburn (UWC) 104 C2
SIDCUP 114 D2
Sideway Jn. 48 D1
Sighthill East Jn. 139
Sighthill West Jn. 139
SILEBY 50 B3
Sileby Jn. 50 B3
SILECROFT 67 B3
Silecroft SB (ST) 67 A3
SILKSTONE COMMON 58 C1
Silkstream Jn. 107 B1
Silt Drove (T) 43 A1
SILVER STREET 109 A1
SILVERDALE 68 C3
Silverdale (AHBC) 68 C3
Silverdale Tunnel 48 D1
Silwood Jn. 120 B
Simmon House (AHBC-X) 52 D1
Simpsons (UWC) 52 C1
Sin Green (UWC) 22 D3
Sincil Bank (CCTV) 59 B3
SINFIN CENTRAL 128 A
SINFIN NORTH 128 A
SINGER 137 A1
Singlewell Crossover 21 A2
Singlewell Loops 21 A2
Singlewell SR 21 A2
Sipson Farm Escape Shaft 111 A1
Sitting Mill (UWC) 58 C2
SITTINGBOURNE 21 B3
Sittingbourne SB (EV) 21 B3
Six Mile Bottom (AHBC) 43 A3
Sizewell (TMO) 45 B3
Sizewell Power Station 45 B3
Skeffington Rd (TMO) 62 E
SKEGNESS 52 E
Skegness SB 52 E
Skellow Jn. 58 D1
Skelly Crag (MCG) 67 B3
Skelton Bridge Jn. 64 D1
Skelton Jn. (Manchester) 129 B3
Skelton Jn. (York) 136 B
Skew Bridge Jn. 62 E
Skewbridge Tip (UWC) 59 B3
SKEWEN 122 B
Skiff Inn (UWC) 136 A
SKIPTON 63 A1
Skipton Middle Jn. 63 A2
Skipton North Jn. 63 A2
Skipton South Jn. 63 A1
SLADE GREEN 20 D2
Slade Green Jn. 20 D2
Slade Green T & RSM depot 20 D2
Slade Lane Jn. 130 C2
Slag Road 136 C
SLAITHWAITE 63 B3
SLATEFORD 85 E
Slateford Jn. 85 E
SLEAFORD 51 B1
Sleaford East Jn. 51 B1
Sleaford East LC SB (SE) 51 B1
Sleaford North Jn SB (SN) 51 B1
Sleaford North Jn. 51 B1
Sleaford Sidings 52 C1
Sleaford South Jn. 51 B1
Sleaford South SB (SS) 51 B1
Sleaford West Jn. 51 B1
Sleaford West LC SB (SW) 51 B1
SLEIGHTS 71 B2
Sleights (CCTV) 50 C1
Slipe Lane (UWC) 32 C3
Slochd loop 99 B3
Sloley Church Lane (AHBC) 46 D2
SLOUGH 19 A2
Slough New SB (SN) 19 A1
Slough SB (S) 19 A1
Slough West 19 A2
Slutchers Lane (FP) 132 B
Slymlakes (UWC) 7 A3
SMALL HEATH 126 A
Small Heath South Jn. 126 A
Smallbrook Jn. 10 C3
SMALLBROOK JUNCTION 10 C3
Smeafield (CCTV) 82 D3
SMETHWICK GALTON BRIDGE 126 A
Smethwick Jn. 126 A
SMETHWICK ROLFE ST. 126 A
Smithdown Lane Tunnel 131 B
Smithfield Road (AHBC-X) 60 C1
Smithfield Sidings 109 A3
Smiths 18 D2
Smiths Lower Cefn (UWC) 47 A3
SMITHY BRIDGE 63 A3
Smithy Bridge (MCB) 63 A3
Snailham (Accom) 13 B2
SNAITH 65 A3
Snaith (AOCL) 65 A3
Snaith and Pontefract Highway (AHBC-X) 64 D3
Snaith East (UWC) 65 A3
Snaith Road (AHBC) 65 A3
Snape 45 A3
Snargate 13 B2
Sneinton Footpath (MCB) 50 E
Snelland (AHBC-X) 60 C2
SNODLAND 21 A3
Snodland (MCB) 21 A3
Snodland LC SB (EDM) 21 A3
Snow Hill Tunnel (Birmingham) 126 B
Snow Hill Tunnel (North Portal) 109 A3
SNOWDON RANGER 53 B3
SNOWDOWN 22 C3
Sod Hall (UWC) 62 C3
Soham Common 43 A2
Soham Station 43 A2
Soho East Jn. 126 A
Soho LMD 126 A
Soho North Jn. 126 A
Soho South Jn. 126 A
SOLE STREET 21 A3
SOLIHULL 126 A
Solomans 1 2 C1
Solomans 2 2 C1
Solomans 3 2 C1
Somerhill Tunnel 12 D1
SOMERLEYTON 45 B1
Somerleyton Swing Bridge 45 B1
Somerleyton Swing Bridge SB (SB) 45 A1
Somerton (UWC) 30 C1
Somerton GF 7 B1
Somerton Tunnel 7 B1
Songar Grange Farm (UWC) 39 B3
Sordale No. 1 (UWC) 104 C2
Sordale No. 2 (UWC) 104 C2
Sordale No. 4 (UWC) 104 C2
Sough Tunnel 62 D3
SOUTH ACTON 112 C1
South Acton Jn. 112 C1
SOUTH BANK 136 C
South Bank Jn. 136 C
SOUTH BERMONDSEY 120 B
South Bermondsey Jn. 120 B
SOUTH CROYDON 117 A1
South Croydon Jn. 117 A2
SOUTH DEVON RAILWAY 4 C2
South Drove (AHBC-X) 52 C3
SOUTH ELMSALL 58 D1
South Emsall (UWC) 58 D1
South Farm No.1 (UWC) 58 D1
South Farm No.2 (UWC) 58 D1
SOUTH GREENFORD 107 B3
SOUTH GYLE 85 E
SOUTH HAMPSTEAD 108 C3
South Hampstead Tunnels 108 C3
South Harrow Tunnel 107 B2
South Ings (UWC) 52 C2
South Jn. (Inverkeithing) 85 B1
South Jn. (Southport) 55 B1
South Jn. GF 128 A
SOUTH KENTON 107 B2
South Kirby Jn. 58 C1
South Marsh 21 B2
South Marston 17 A1
SOUTH MERTON 112 C3
South Milford Footpath (R/G) 64 D3
SOUTH MILLFORD 64 D3
SOUTH RUISLIP 107 A2
South Scarle (AHBC) 59 B3
South Stoke Crossing 11 A2
SOUTH TOTTENHAM 109 A1
South Tottenham East Jn. 109 A1
South Tottenham Station Jn. SB (S) 109 A1
South Tottenham West Jn. 109 A1
South Wales SCC (NT) 123 A
SOUTH WIGSTON 50 C3
SOUTH WOODHAM FERRERS 21 A1
South Yorkshire Jn. 135 A1
SOUTHALL 111 A1
Southall Depot 111 B1
Southall East Jn. 111 A1
Southall West Jn. 111 A1
SOUTHAMPTON AIRPORT (PARKWAY) 10 C2
SOUTHAMPTON CENTRAL 9 E
Southampton Docks 9 E
Southampton Tunnel 9 E
Southborne (CCTV) 10 D2
SOUTHBOURNE 10 D2
SOUTHBURY 109 A1
Southcote Jn. 18 D2
SOUTHEASE 12 C3
SOUTHEND AIRPORT 21 B1
SOUTHEND CENTRAL 21 B1
SOUTHEND EAST 21 B1
SOUTHEND VICTORIA 21 B1
Southerham Crossing 12 C3
Southerham Jn. 12 C3
Southfield Lane (UWC) 64 D3
SOUTHFIELDS 112 C2
Southfleet Crossover 21 A2
Southfleet Jn. 21 A2
Southfleet Tunnel 21 A2
SOUTHMINSTER 21 B1
SOUTHPORT 61 B3
Southwark Bridge Jn. 120 B
SOUTHWICK 12 C3
SOWERBY BRIDGE 63 B3
Sowerby Bridge Tunnel 63 B3
Sowters (UWC) 50 C1
Spa Road 120 B
Spa Street (UWC) 59 B3
SPA VALLEY RAILWAY 12 D1
Spadeoak Crossing (UWC) 19 A1
SPALDING 52 C2
Spalding (MCB) 52 C2
Spalding SB (S) 52 C2
Sparnick Tunnel 1 B2
Sparrow's Bridge (UWC) 14 C1
Spatham Lane (AHBC) 12 C2
SPEAN BRIDGE 90 C2
Spean Bridge Loop 90 C2
Specklies (UWC) 51 A2
Speedway (AOCL) 122 A
Speedway Jn. 122 A
Speeton (AHBC) 72 D3
Speke Jn. GF 131 A
Speke Jn. SB (SE) 131 A
Spellbrook (CCTV) 32 C2
Spellow No.1 Tunnel 131 A
Spellow No.2 Tunnel 131 A
Spencers (UWC) 62 D1
Spetchley 39 A3
Spey Viaduct 101 A2
Spinks Lane (R/G-X) 44 D1
SPITAL 131 A
Spital Jn. 42 E
Spittal (R/G) 82 D2
Spittal Tunnel 23 A2
Spon End Viaduct 40 C2
SPONDON 128 A
SPOONER ROW 44 D1
Spooner Row (MCB-OD) 44 D1
Sportsfield (UWC) 64 C3
Spratts Water 45 B2
Spring Grove 13 B1
Spring Lodge (AHBC) 64 D3
SPRING ROAD 126 A
Springbank North Jn. 65 B2
Springbank South Jn. 65 B3
SPRINGBURN 139
SPRINGFIELD 92 D2
Springfield No. 1 (UWC) 100 C1
Springhead Rd Jn. (Down) 21 A2
Springs Branch Jn. 132 C
Springs Jn. 63 B2
Springs Tunnel 63 B2
Springwell Lane (AHBC) 70 D2
Springwood Jn. 57 B1
Spronces (AHBC-X) 44 D1
Spurriers No. 2 (UWC) 49 B2
SQUIRES GATE 61 B3
STADIUM OF LIGHT 76 D2
Staff Halt (Wimbledon) 111 B3
Staff LC 111 A2
STAFFORD 48 D2
Stafford No. 4 SB (SD4) 48 D2
Stafford No. 5 SB (SD5) 48 D2
Stafford Trent Valley Jn. No 1. 48 D2
Stafford's Bridge (UWC) 6 C3
STAINES 19 B2
Stainforth Jn. 59 A1
Stainforth Road (AHBC) 59 A1
Stainforth Tunnel 63 A1
Stainton (AHBC-X) 60 C2
Stainton Jn. 127 A
STALLINGBOROUGH 60 C1
Stallingborough (MCB) 60 C1
Stallingborough SB (ST) 60 C1
Stallington (CCTV) 49 A2
STALYBRIDGE 130 D2
Stalybridge Jn. 130 D2
Stalybridge Tunnel 130 D2
Stalybridge Tunnel Jn. 130 D2
STAMFORD 51 B3
Stamford (CCTV) 81 A1
STAMFORD HILL 109 A2
Stamford Tunnel 51 B3
Standedge Tunnel 57 A1
Standish Jn. 28 C3
Stanfield Farm (UWC) 73 B1
STANFORD-LE-HOPE 21 A2
Stanford-Le-Hope (CCTV) 21 A1
Stangrah Farm (UWC) 67 A3
STANHOPE 75 B3
STANHOPE 75 B3
Stanley Jn SB (SJ) 91 B1
Stanley Jn. 91 B1
STANLOW & THORNTON 56 C3
Stanningley Tunnel 63 B2
Stanningley Tunnel 64 C2
Stannington (CCTV) 76 C1
Stannington (CCTV) 81 A3
STANSTED AIRPORT 32 D2
Stansted Airport Tunnel 32 D2
Stansted East Jn. 32 D2
STANSTED MOUNTFITCHET 32 D2
Stansted North Jn. 32 D2
Stansted South Jn. 32 D2
STAPLEHURST 13 A1
STAPLETON ROAD 124
Star (AHBC) 13 B2
Star Lane (CCTV) 18 D2
STARBECK 64 C1
Starbeck (MCB) 64 C1
Starbeck SB (SB) 64 C1
STARCROSS 4 D1
Stathams (UWC) 49 A2
Station GF 25 B1
Station Jn. 56 C1
Station Road (Billingshurt) 11 B2
Station Road (Rainham, Kent) 21 A3
STAVELEY 68 C2
Staveley (AHBC) 68 C2
STAVERTON 4 C2
Staverton Farm (UWC) 16 C2
Staythorpe SB (SK) 50 D1
STECHFORD 126 A
Stechford North Jn. 126 A
Stechford South Jn. 126 A
STEETON & SILSDEN 63 B2
Stenson Jn. 128 A
Stenson Raynors (UWC) 128 A
Stenton GSP 86 D1
STEPPS 138 C2
Stepps Cottage (UWC) 90 C2
STEVENAGE 32 C2
STEVENSTON 83 A3
Stevenston (CCTV) 83 A3
Steventon 18 C1
STEWARTBY 31 A1
Stewartby Brickworks (CCTV) 31 A1
STEWARTON 83 B3
Stewarts Lane Depot 119
Stewarts Lane Jn. 119
Steynton (UWC) 23 A2
Stilton Fen 42 C1
STIRLING 84 D1
Stirling Middle SB (SM) 84 D1
Stirling North SB (SN) 91 A3
Stoats Nest Jn. 117 A2
Stobcross Street Tunnel 139
Stockbridge Road (CCTV) 11 A3
Stocking Farm (UWC) 38 D1
Stockley Bridge Jn. 111 A1
Stockley Flyover 111 A1
STOCKPORT 130 C3
Stockport No.1 SB (ST1) 130 C3
Stockport No.2 SB (ST2) 130 C3
Stocks Lane (CCTV) 18 C1
STOCKSFIELD 75 B2
STOCKSMOOR 57 B1
Stocksmoor Jn. 57 B1
STOCKTON 70 D1
Stockton Brook Tunnel 48 D1
Stockton Crossing 8 D1
Stockton Cut Jn. 70 D1
Stockyard (UWC) 50 C1
STOGUMBER 6 D1
Stoke 8 D3
Stoke 51 A2
Stoke (ABCL) 21 B2
Stoke Canon (CCTV) 6 C3
Stoke Creek 21 A2
Stoke Edith (AHBC) 27 B1
Stoke Gifford East Jn. 124
Stoke Gifford Jn.No. 1 124
Stoke Gifford Jn.No. 2 124
Stoke Gifford West Jn. 124
Stoke Jn. 48 D1
Stoke Lane (AHBC-X) 50 E

Location	Page	Grid
Tinsley Avesta (TMO)	134	A1
Tinsley East Jn.	134	A1
Tinsley Green Jn.	12	C1
Tinsley Park Jn.	134	A1
Tinsley South Jn.	134	A1
Tinsley Yard	134	A1
Tinsley Yard SB (TY)	134	A1
Tinsleys (Campains Lane) (R/G)	52	C3
Tinwell (UWC)	51	B3
TIPTON	39	A1
Tir Isaf (UWC)	122	A
Tir-Allen Farm 1 (UWC)	36	E
Tir-Allen Farm 2 (UWC)	36	E
Tir-Allen Farm 3 (UWC)	36	E
TIR-PHIL	26	C1
Tirydail (ABCL)	24	D2
TISBURY	8	D1
Tisbury Quarry (R/G)	8	D1
Tisbury West (AHBC)	8	D1
Tiverton Loops	6	C2
TIVERTON PARKWAY	6	C2
Tivetshall (AHBC-X)	44	D2
Toadmoor Tunnel	50	C1
TODDINGTON	*29*	*A1*
Toddington (AHBC)	11	B3
Todholes No. 1 (UWC)	104	C2
Todholes No. 2 (UWC)	104	C2
Todholes No. 3 (UWC)	104	C2
TODMORDEN	63	A3
Tofthill (UWC)	92	C2
Tolans (UWC)	55	B2
Toll of Cults (UWC)	101	B3
TOLWORTH	116	C1
Tomatin Loop	99	B3
Tomich No. 1 (UWC)	102	C1
Tomich No. 2 (UWC)	102	C1
Tomlinsons (UWC)	59	B3
TON PENTRE	25	B2
TONBRIDGE	12	D1
Tonbridge East Jn.	12	C1
Tonbridge SB (PE)	20	D3
TONDU	25	B2
Tondu Jn.	25	B2
Tondu SB (TU)	25	B2
TONFANAU	35	B2
Tonfanau (UWC)	35	B3
Tongside No. 1 (UWC)	104	C2
Tongside No. 2 (UWC)	104	C2
TONYPANDY	25	B2
TOOTING	112	D3
Topley Pike GF	57	B3
TOPSHAM	6	C3
Topsham (CCTV)	6	C3
Torcoed 2 (UWC)	25	B2
Torness Siding GSP	82	C1
Torphin (UWC)	85	A2
TORQUAY	4	D2
TORRE	4	D1
Torworth (CCTV)	59	A2
Totley Tunnel	58	C2
Totley Tunnel East SB (TE)	58	C2
TOTNES	4	C2
Totnes East	4	C2
TOTNES LITTLEHEMPSTON	*4*	*C2*
Toton Down Marshalling Yard	128	A
Toton Jn.	128	A
Toton TMD	128	A
TOTTENHAM HALE	109	B1
Tottenham North Curve Tunnels	108	D2
Tottenham South Jn.	109	B1
TOTTON	9	B2
Totton (CCTV)	9	B2
Totton West Jn.	9	B2
Totton Yard	9	B2
Towan (UWC)	1	A2
Towers Farm (UWC)	130	C3
TOWN GREEN	55	B1
Towneley (MCB)	63	A3
Towneley Tunnel	63	A3
Towney (UWC)	18	C2
Townhill Jn.	85	A1
Townsend Crossing	8	D1
Trabboch (UWC)	78	D1
Trabboch (UWC)	78	D1
Traeth Mawr (ABCL)	35	B1
TRAFFORD PARK	129	B2
Trafford Park East Jn.	128	B
Trafford Park Reversing Line	128	B
Trafford Park West Jn.	129	B2
Tram Inn (MCB)	27	B1
Tram Inn SB (TI)	27	B1
Travel's Lane (AHBC)	9	B2
Tredington (AHBC)	28	D2
Trees (CCTV)	32	D2
Treeton Jn.	134	A1
Treeton South	134	A1
Treffeddian (UWC)	35	B3
Treffoliad Farm 1 (UWC)	36	E
TREFFOREST	26	C2
TREFFOREST ESTATE	26	C2
Tregoss Moor (AOCL)	2	C1
TREHAFOD	26	C2
TREHERBERT	25	B2
Trelavour Sidings	2	C1
Tremains DPL	25	B3
Tremorfa Works GF	123	A
Trencreek (AOCL)	1	B1
Trent East Jn. (Gainsborough)	59	A1
Trent East Jn. (Toton Yard)	128	A
Trent Jn.	65	B3
Trent Lane (UWC)	50	D1
Trent Lane Footpath (R/G)	50	E
Trent SB (TT)	128	A
Trent South Jn.	128	A
Trent West Jn.	59	A2
Trent Yard	128	A
TREORCHY	25	B2
Tresithney 2 (UWC)	2	C1
Treverrin Tunnel	2	D1
Treviscoe Sidings	2	C1
Trewern Farm (UWC)	23	B2
Trewern Mill (UWC)	23	B2
TRIMLEY	34	D2
Trimley (CCTV)	34	D2
TRING	31	A2
Tring North Jn.	31	A2
Tring South Jn.	31	A2
Trinity Lane (MCG)	32	C3
Troad yr Rhin (UWC)	54	C2
TROED-Y-RHIW	26	C1
Troed-y-Rhiw Fedwen (UWC)	37	A2
Troed-Y-Rhiw South Jn.	26	C1
TROON	78	C1
Trotts Lane (AHBC)	9	B2
TROWBRIDGE	16	C2
Trowell South Jn.	128	A
Trowse Jn.	45	A1
Trowse Swing Bridge	44	E
Trowse Swing Bridge SB (TSB)	44	E
Truro	1	B2
TRURO	2	C2
Truro (MCB)	1	B2
Truro SB (T)	1	B2
Tucker's (UWC)	16	C2
Tuckwells (UWC)	30	C3
Tuebrook Sidings	131	A
Tuffley	28	C2
TULLOCH	90	D2
TULSE HILL	112	C2
TUNBRIDGE WELLS	12	D1
TUNBRIDGE WELLS WEST	*12*	*D1*
Tunnel 23	137	A1
Tunnel 25	137	A1
Tunnel Road Tunnel	131	B
Tunnicliffs No. 1 (UWC)	49	A2
Tunstead Church Lane	46	D2
Tunstead Market Street (AHBC)	46	D2
Turf Lock (UWC)	6	C3
TURKEY STREET	109	A1
Turnchapel Branch Jn.	122	A
Turners Lane Jn.	133	A
Turton (AOCL)	62	D3
TUTBURY & HATTON	49	B2
Tutbury (MCB)	49	B2
Tutbury Crossing SB	49	B2
Tuxford GSP	59	A3
Tuxford No. 2 GF	59	A3
Tuxford No.1 GF	59	A3
Tuxford West J. GF	59	A3
Tweedmouth Crossover	82	D2
Tweedmouth SB (TW)	82	D2
Twerton Long Tunnel	16	C2
Twerton Short Tunnel	16	C2
TWICKENHAM	111	B2
Twickenham Jn	111	B2
Two Mile Bottom (AHBC-X)	44	C2
TWYFORD	18	D2
Twyford East	18	D2
Twyford West	18	D2
TY CROES	53	A2
Ty Croes (MCG)	53	A2
TY GLAS	123	A
Ty Mawr Farm (UWC)	36	D1
Ty Pella (UWC)	36	C1
Ty-Ddu (UWC)	37	A3
Tyddyn Morfa (UWC)	53	B2
Tyddyn-y-pwll (UWC)	36	D1
Tye Green Jn.	32	D2
TYGWYN	35	B1
Tygwyn (ABCL)	35	B1
Tyminster (CCTV)	11	A3
Ty'n Llan 1 (UWC)	35	B2
Ty'n Llan 3 (UWC)	35	B2
Tyn-Ddol (UWC)	54	C2
TYNDRUM LOWER	88	D2
Tyndrum Lower (UWC)	88	D2
Tyneside Central Freight Depot	136	A
Tyneside IECC (T)	136	A
Tynewydd 2	23	B2
Tyning	8	D1
Tynycerig 1 (UWC)	24	D3
Tynycerig 2 (UWC)	24	D3
Tynycerig 3 (UWC)	24	D3
Tynycynllwyn (UWC)	24	D3
Tyn-y-Maes (UWC)	36	E
Tyn-y-Morfa (MCG)	55	A2
Ty'n-yr-Wtre No. 2 (UWC)	36	D1
TYSELEY	126	A
Tyseley Depot	126	A
Tyseley No.1 SB (TY1)	126	A
Tyseley North Jn.	126	A
Tyseley South Jn.	126	A
Tytherington Tunnel	16	C1
Ty-Uchaf (AOCL)	25	A1
TYWYN	35	B3
TYWYN	*35*	*B3*
Tywyn GF	35	B3
TYWYN WHARF	*35*	*B3*
U		
UCKFIELD	12	D2
UDDINGSTON	138	C2
Uddingston Jn	138	C3
Uffington & Barnack (MCG)	51	B3
Uffington (Swindon)	17	B1
Uffington SB (UN) (Lincs)	51	B3
Uffold	34	D1
Ufford (ABCL)	34	D1
Ufton (AHBC)	18	C2
Ugley Lane (UWC)	32	D2
ULCEBY	66	C3
Ulceby Jn. (MCB)	66	C3
Ulceby Jn. SB (UJ)	66	C3
Ulceby North Jn.	66	C3
Ulceby South Jn.	66	C3
Ulgham Grange (CCTV)	81	A3
Ulgham Lane (CCTV)	81	A3
ULLESKELF	64	D2
ULVERSTON	67	B3
Ulverston SB (UN)	67	B3
UMBERLEIGH	5	A1
Umberleigh (AOCL)	5	A1
Underhill (UWC)	67	B3
Union Street Jn.	120	B
UNITED F.C. HALT	129	B2
Universal Grinding Wheel Works	48	D2
UNIVERSITY (Birmingham)	126	A
Unthank (TMO)	75	B3
UPHALL	85	A2
Uphill Jn.	15	A2
UPHOLLAND	56	C1
Upholland Tunnel	56	C1
UPMINSTER	20	D1
Upminster East Jn.	20	D1
Upminster IECC (UR)	20	D1
Upminster LUL Depot	20	D1
Upney Jn.	110	D3
Upper Cellws (UWC)	37	A3
Upper Chapel Hill Farm (UWC)	23	B2
Upper Denton (AHBC-X)	74	D1
Upper Denton West (UWC)	74	D1
UPPER HALLIFORD	111	A3
UPPER HOLLOWAY	108	D2
Upper Holloway SB (UH)	108	D2
Upper Leigh (AHBC-X)	40	A2
Upper Llegodig (UWC)	37	A1
Upper Portland (AHBC)	50	C1
Upper Trenowin (UWC)	1	A2
UPPER TYNDRUM	88	D2
UPPER WARLINGHAM	117	B3
Upperby Bridge Jn.	127	A
Upperby Jn.	127	A
Upperby Yard	127	A
Upperby Yard GF	127	A
UPTON	131	A
Upton Lovell (AHBC)	16	D3
UPWEY	8	E
Urlay Nook (MCB)	70	D1
Urlay Nook SB (UN)	70	D1
URMSTON	129	B2
Urrard No. 1 (UWC)	93	A2
Urrard No. 2 (UWC)	93	A2
Usan Jn	95	B2
UTTOXETER	49	A2
Uttoxeter Racecourse (UWC)	49	A2
Uttoxeter SB (UR)	49	A2
V		
Valance Road Portal	109	A3
VALE OF RHEIDOL RAILWAY	*35*	*E*
Vale Viaduct	12	C2
Vale Wood Crossing	11	B2
VALLEY	53	A1
Valley (MCB)	53	A2
Valley SB (VY)	53	A2
Valleyfield Colliery (UWC)	85	A1
Vange Wharf (CCTV)	21	A1
Vaseys (UWC)	70	D2
VAUXHALL	119	
Vauxhall Jn.	126	B
Veals Lane (AHBC)	9	B2
Viaduct Maintenance (UWC)	32	C3
Victoria ASC (VC) (VS)	119	
Victoria Road (CCTV)	45	B2
Victory (AHBC)	6	D1
Vine Road (CCTV)	112	C2
Vineyard Farm No. 1 (UWC)	27	A2
Vineyard Farm No. 2 (UWC)	27	A2
VIRGINIA WATER	19	A2
Virtual Quarry GF	127	A
Vitriol Works SB (VW)	130	C1
Voltaire Jn.	119	
W		
Wadborough (AHBC)	28	D1
WADDON	117	A1
WADHURST	12	D1
Wadhurst GF	12	D1
Wadhurst Tunnel	12	D1
Wagon Lane (AHBC)	13	A1
Wagon Repairs GF	127	A
WAINFLEET	52	E
Wainfleet (UWC)	52	E
Wainfleet Bypass (AHBC-X)	52	E
Wainfleet SB (W)	52	E
WAINHILL HALT	*30*	*D3*
Wakefield Europort	64	C3
WAKEFIELD KIRKGATE	133	A
Wakefield Kirkgate East Jn.	133	A
Wakefield Kirkgate SB (K)	133	A
Wakefield Kirkgate West Jn.	133	A
Wakefield Road Tunnel	63	B3
WAKEFIELD WESTGATE	133	A
Wakefield Westgate South Jn.	133	A
Waldersea (TMO)	52	D3
Walesby (AHBC-X)	60	C2
WALKDEN	129	B1
Walkden SB (WN)	129	B1
Walkers (UWC)	70	D2
Walkers No. 63 (UWC)	59	B3
Wall End	22	D3
WALLASEY GROVE ROAD	131	A
WALLASEY VILLAGE	131	A
Wallers Ash	10	C1
Wallers Ash Tunnel	10	C1
WALLINGFORD	*18*	*C1*
WALLINGTON	116	D2
Wallington Viaduct	10	C2
Wallneuk Jn.	137	A2
Wallsend (CCTV)	13	A3
WALLYFORD	86	C2
WALMER	22	D3
Walnut Grove (UWC)	92	C2
Walpole (CCTV)	46	D2
WALSALL	127	B
Walsall North Jn.	127	B
Walsall Pleck Jn.	127	B
Walsall PSB (WL)	127	B
Walsall South Jn.	127	B
WALSDEN	63	A3
Walters Hall	22	D3
WALTHAM CROSS	32	C3
Waltham Nurseries (UWC)	67	B3
WALTHAMSTOW CENTRAL	109	B1
WALTHAMSTOW QUEENS ROAD	109	B1
WALTON	131	A
Walton Common	21	A2
Walton Jn. (Aintree Line)	131	A
Walton Old Jn.	132	B
Walton Street (CCTV)	66	C3
Walton Street Jn.	66	C3
WALTON-ON-THAMES	115	A1
WALTON-ON-THE-NAZE	34	D2
Wamphray GSP	80	D3
WANBOROUGH	19	A3
WANDSWORTH COMMON	112	D2
WANDSWORTH ROAD	119	
WANDSWORTH TOWN	112	D2
Wansdyke (UWC)	17	B2
WANSFORD	*42*	*C1*
Wansford Road (CCTV)	66	C1
WANSTEAD PARK	110	C2
Wantage Road	17	B1
WAPPING	120	B
WARBLINGTON	10	D2
Warblington (CCTV)	10	D2
Warden (AHBC-X)	75	A1
Wards (Kent)	13	A1
Wards (UWC) (Lancs)	68	D3
Wards (UWC) (Lincs)	51	B3
Wards (UWC) (Moray)	100	D1
Wards Dyke (UWC)	52	C1
Wards Sidings GF	51	B3
WARE	32	C2
Ware (CCTV)	32	C3
WAREHAM	8	D3
Wareham SB (WR)	8	D3
Warehorne (AHBC)	13	B1
WARGRAVE	18	D1
Warkworth (CCTV)	81	A2
WARMINSTER	16	C3
WARNHAM	11	B1
Warnham Station Road (R/G)	11	B1
Warren Hill Tunnel	43	B3
Warren House	21	A3
Warren House (MWL)	50	D1
Warrenhill (UWC)	73	A1
WARRINGTON BANK QUAY	132	B
WARRINGTON CENTRAL	132	B
Warrington Central SB (WC)	132	B
Warrington North Jn.	132	B
Warrington SB (WN)	132	B
Warrington South Jn.	132	B
Warsop Jn.	58	D3
WARWICK	40	C3
WARWICK PARKWAY	40	C3
Warwickshire Oil Sidings	40	C1
WASHFORD	*6*	*E*
Washstones (R/G) (UWC)	50	B3
Washwood Heath East Jn.	126	A
Washwood Heath West Sdgs.	126	A
Wassicks (AHBC-X)	44	C3
Waste Bank Tunnel	74	D3
Waste Recycling (Claydon)	30	D1

WATCHET 6 E
Water Eaton (UWC) 30 C2
Water Lane (UWC) 6 C3
WATER ORTON 126 A
Water Orton East Jn. 39 B1
Water Orton West Jn. 126 A
Water Pit Lane 46 D2
WATERBEACH 43 A3
Waterbeach (AHBC) 43 A3
Waterfields No. 1 (UWC) 64 D3
Waterford (RC) 100 C1
WATERINGBURY 21 A3
Wateringbury SB (WB) 21 A3
WATERLOO 131 A
Waterloo (AHBC) 18 D2
Waterloo (CCTV) 131 A
Waterloo (Llandrindod) 36 D2
WATERLOO EAST 120 B
Waterloo Yard (Exeter) 6 C3
Waterside 78 D2
Waterside (CCTV) 91 A3
Waterslack Quarry (UWC) 68 C3
Waterton (AOCL) 25 B3
Waterworks (UWC) 49 B3
Watery Lane S.F. (WL) 39 A1
Watery Road GF 47 A1
WATFORD HIGH STREET 31 A3
Watford Jn PSB (WJ) 31 B3
WATFORD JUNCTION 31 B3
Watford Lodge Tunnel 40 D2
WATFORD NORTH 31 B3
Watford North Jn. 31 B3
Watford South Jn. 31 B3
Watford Tunnel (Fast) 31 A3
Watford Tunnel (Slow) 31 B3
Wath (UWC) 73 A1
WATLINGTON 43 B1
Watlington Road (CCTV) 43 B1
Watson (UWC) 29 A1
Watsons (UWC) 50 C3
Watten (AOCL) 104 D2
Watton (AHBC-X) 66 C2
WATTON-AT-STONE 32 C2
WAUNFAWR 53 B3
WAUN-GRON PARK 123 A
Waverley (East End) 85 E
Waverley (West End) 85 E
Wavertree Junction 131 A
WAVERTREE TECH. PARK 131 A
Way & Works Jn. 128 A
WEARDALE RAILWAY 76 C3
Weasel Hall Tunnel 63 A3
Weasenham Lane (TMO) 43 A1
Weaver Jn. (Dn) 56 C3
Weaver Jn. (Up) 56 C2
Weaverthorpe (MCG) 72 C3
Weaverthorpe SB (W) 72 C3
Websters 32 D1
WEDGWOOD 48 D1
Wedgwood (CCTV) 48 D1
Wednesbury Town Jn. 39 A1
Wednesfield Heath Tunnel 127 B
Week Street Tunnel 21 A3
WEELEY 34 C2
Weer Lane (UWC) 49 B2
WEETON 64 C2
Weig Lane (AOCL) 36 D1
Weish Highland Railway (Flat Crossing) 35 B1
Welbeck Colliery 58 D3
Welbeck Colliery Jn. 58 D3
Welbury (AHBC-X) 70 D2
WELHAM GREEN 31 B3
Welland Bank (UWC) 52 C3
Wellhouse (UWC) 99 A2
Wellhouse Tunnel 57 B1
WELLING 114 D2
WELLINGBOROUGH 41 B2
Wellingborough North Jn. 41 B2
Wellington (AHBC) (Hereford) 27 B1
WELLINGTON (Shrops) 48 C3
Wellington Crossovers (Taunton) 6 D1
Wellowgate (CCTV) 60 D1
Wellpark Tunnel 83 A1
Wells Engine 43 A2
Wells Tunnel 12 D1
Welney Road (AHBC-X) 43 A2
WELSH HIGHLAND RAILWAY 53 B3
WELSHPOOL 47 A3
WELSHPOOL & LLANFAIR RAILWAY 47 A3
WELSHPOOL (RAVEN SQUARE) 37 E
Welshpool GF 47 A3
Welsh's Bridge Jn. 99 B2
Welton (MCG) 65 B3
Welton Crossover 60 C3
WELWYN GARDEN CITY 31 B3
WELWYN NORTH 32 C2
Welwyn South Tunnel 32 C2
WEM 47 B2
Wem (MCB) 47 B2
Wem Emergency Crossover GFs 47 B2
Wem SB (WM) 47 B2
WEMBLEY CENTRAL 120 A
Wembley Central GF 120 A
Wembley Mainline SCC (WM) 120 A
WEMBLEY STADIUM 107 B2
Wembley Yard 120 A
Wembley Yard SB (WY) 120 A
Wembley Yard South Jn. 120 A
WEMYSS BAY 83 A2
Wemyss Bay Jn. 83 A2
WENDOVER 31 A3
Wenhaston (AOCL) 45 A2
WENNINGTON 68 D3
Wennington Crossover 20 D1
WENSLEYDALE RAILWAY 70 C2
Wensum Jn. 44 E
Wentloog Freight Terminal East Jn. 123 B
Wentloog Freight Terminal West Jn. 123 B
Werneth Tunnel 130 C1
Wescoehill Tunnel 64 C2
West (OPEN) 76 D3
WEST ALLERTON 131 A
West Bank Hall (AHBC-X) 65 A3
West Bank Terminal (Ipswich) 33 E
West Barnes(CCTV) 116 C1
WEST BROMPTON 112 C1
West Burton East Jn. 59 A2
West Burton Power Station 59 A2
West Burton SB (WB) 59 A2
West Burton West Jn. 59 A2
WEST BYFLEET 19 A3
WEST CALDER 85 A2
West Calder Goods GF 85 A2
West Cowick (R/G) 65 A3
WEST CROYDON 117 A1
WEST DRAYTON 19 B1
West Drayton (TMO) 19 A1
West Drayton East 19 A1
West Drayton Jn. 19 A1
WEST DULWICH 113 A2
WEST EALING 111 B1
West Ealing Jn. 111 B1
West Grimstead Crossing 9 A1
WEST HAM 110 C3
West Hampstead Jn. North 108 C2
West Hampstead Jn. South 108 C2
West Hampstead PSB (WH) 108 C2
WEST HAMPSTEAD THAMESLINK 108 C2
WEST HAMSTEAD 108 C2
West Heslerton (AHBC-X) 71 B3
West Holmes Jn. 59 B3
WEST HORNDON 20 D1
West House (TMO) 45 A3
WEST KILBRIDE 83 A3
West Kinnauld No. 2 (UWC) 102 D1
West Kinnauld No. 3 (UWC) 102 D1
WEST KIRBY 55 A2
West Lodge (UWC) 74 D2
West London Jn. (Clapham) 119
West London Jn. (Willesden) 120 A
West London Waste Transfer Stn 111 B1
WEST MALLING 21 A3
West Midlands SC (#) 126 B
WEST NORWOOD 113 A3
West Norwood Jn. 113 A3
West of Scotland SC (G/GG) 139
West Parade North Jn. 66 C2
West River (R/G-X) 43 A2
WEST RUISLIP 19 B1
WEST RUNTON 46 C1
West Sleekburn Jn. 81 A3
WEST SOMERSET RAILWAY 6 D1
WEST ST. LEONARD'S 13 A3
West Street (AHBC) 9 B2
West Street Jn. SB (WS) 52 C1
WEST SUTTON 116 D1
West Thurrock Jn. 20 D1
WEST WICKHAM 117 B1
WEST WORTHING 11 B3
West Worthing (CCTV) 11 B3
Westborough (Public BW) 51 A1
Westbrecks (AHBC) 59 A2
Westbrook Lane (R/G) 59 B3
WESTBURY 16 C3
Westbury (AHBC) 28 C2
Westbury (AHBC) 47 A3
Westbury East Loop Jn. 16 D3
Westbury Line Jn. 125 A
Westbury North Jn. 16 C2
Westbury SB (W) 16 D3
Westbury South Jn. 16 C3
WESTCLIFF-ON-SEA 21 B1
WESTCOMBE PARK 114 C1
WESTENHANGER 14 C1
Westenhanger Crossovers 14 C1
Westenhanger SR 14 C1
Wester Fearn (UWC) 102 C2
Wester Fearn No. 1 (UWC) 102 C2
WESTER HALLES 85 E
WESTERFIELD 33 E
Westerfield Jn. 33 E
Westerfield Station (AHBC) 33 E
Westerleigh Jn. 16 C1
Westerleigh Yard 16 C1
Western Entrance (CCTV) 66 C3
Western Growers 1 A2
Western Jn. 21 B3
Western Sidings 48 E
Westernhanger SR 14 C1
WESTERTON 137 A1
Westerton Jn. 137 A1
Westfield Disposal Point 92 C3
Westford (Footpath) (R/G) 6 D2
Westford (UWC) 7 A2
WESTGATE-ON-SEA 22 D2
Westhall (ABCL) 45 A2
WESTHOUGHTON 129 A1
Westley Road (R/G) 43 A3
Westminster Tunnel 131 A
Weston 18 C3
Weston (AOCL) 45 A2
WESTON MILTON 15 A2
Weston Rhyn (AHBC) 47 A2
Westons (UWC) 49 A2
Weston-S-M Up GF 15 A2
WESTON-SUPER-MARE 15 A2
Wests Bridge Farm (UWC) 27 A2
Westwood Sidings 42 E
WETHERAL 74 C2
WEYBOURNE 46 C1
WEYBRIDGE 19 B3
WEYMOUTH 8 E
Weymouth Jn. 8 E
WEYMOUTH QUAY 8 E
WHALEY BRIDGE 57 A2
WHALLEY 62 D2
Wharf Road (AHBC-X) 32 C3
Whatley Quarry 16 C3
WHATSTANDWELL 49 B1
Whatstandwell (UWC) 49 B1
Whatstandwell Tunnel 49 B1
Wheeler Street Tunnel 21 A3
Wheler Street Jn. 109 A3
WHIFFLET 138 D2
Whifflet North Jn. 138 D2
Whifflet South Jn. 138 D2
Whiley Hill (AHBC) 70 C1
WHIMPLE 6 C3
WHINHILL 83 A2
Whipps Farm 20 D1
Whisby Quarry (UWC) 59 B3
Whissendine SB 51 A3
WHISTON 56 C2
Whitacre East Jn. 39 B1
Whitacre West Jn. 39 B1
Whitbeck (AOCL) 67 A3
WHITBY 71 B1
Whitchester Tunnel 74 D1
WHITCHURCH (Cardiff) 123 A
WHITCHURCH (Hants) 18 C3
WHITCHURCH (Shrops) 47 B2
White Hart (CCTV) 112 C2
WHITE HART LANE 109 A1
White Hart Lane GF 109 A1
White House Mill (UWC) 23 B2
WHITE NOTLEY 33 A2
White Notley (ABCL) 33 A2
Whiteball Tunnel 6 D2
Whitebridge (UWC) 94 C3
WHITECHAPEL 109 A3
WHITECRAIGS 137 B3
WHITECROFT 27 B2
Whitehall (R/G) 22 C3
Whitehall East Jn. 133 B
Whitehall West Jn. 133 B
WHITEHAVEN 67 A1
Whitehaven Tunnel 67 A1
Whitehills (UWC) 100 D1
Whitehouse (MCB) 136 C
Whitehouse Farm (UWC) 52 C1
Whitehouse Jn. 48 D2
Whitehouse Lane (R/G) Footpath 51 A1
Whitehouse SB (W) 136 C
WhitehouseTunnel 30 D3
Whitehurst (UWC) 47 A2
Whitehurst Tunnel 47 A2
Whitelaw Footpath (R/G) 85 B2
Whitemoor Drove (AHBC) 42 D1
Whitemoor Jn. 42 D1
Whitemoor Yard 42 D1
Whitemoss (AHBC-X) 91 B2
Whites (UWC) (Long Eaton) 128 A
Whites (UWC) (Oxon) 40 C3
Whites (UWC) (Wilts) 16 C3
White's Farm (UWC) 28 D1
WHITLAND 23 B2
Whitland (MCB) 23 B2
Whitland Jn. 23 B2
Whitland SB (W) 23 B2
Whitland Tunnel 23 B2
WHITLEY BRIDGE 64 D3
Whitley Bridge (CCTV) 64 D3
Whitley Bridge Jn. 64 D3
Whitlingham Jn. 44 E
Whitlingham Lane 44 E
WHITLOCKS END 39 B2
WHITSTABLE 22 C3
Whitterleys Farm (UWC) 37 B2
Whittington (AHBC) 47 A2
Whittle International GF 62 E
WHITTLESEA 42 D1
Whittlesea (MCB) 42 D1
Whittlesea SB (W) 42 D1
WHITTLESFORD PARKWAY 32 D1
WHITTON 111 B2
Whitton Jn. 111 B2
WHITWELL 58 D3
Whitwell Tunnel 58 D3
Whitwood Jn. 64 D3
Whixley (MCG) 64 C1
Whyke Road (CCTV) 11 A3
WHYTELEAFE 117 A3
Whyteleafe (CCTV) 117 A3
WHYTELEAFE SOUTH 117 A3
Whyteleafe South (CCTV) 117 B3
Wichnor Jn. 49 B3
WICK 104 D2
Wickenby (MCG) 60 C2
Wickenby SB (W) 60 C2
WICKFORD 21 A1
Wickford Jn. 21 A1
Wickham Knights (UWC) 18 C2
WICKHAM MARKET 34 D1
Wickwar Tunnel 16 C1
WIDDRINGTON 81 A3
Widdrington (CCTV) 81 A3
Widdrington Sidings Crossover 81 A3
WIDNES 56 C2
WIDNEY MANOR 39 B2
Wig (UWC) 54 C2
Wig Farm (UWC) 54 C2
Wigan North Jn. 132 C
WIGAN NORTH WESTERN 132 C
Wigan South Jn. 132 C
Wigan Station Jn. 132 C
WIGAN WALLGATE 132 C
Wigan Wallgate Jn. 132 C
Wigan Wallgate SB (WW) 132 C
Wigston North Jn. 50 B3
Wigston South Jn. 50 B3
WIGTON 73 B2
Wigton GF 73 B2
Wigton SB (WN) 73 B2
WILDMILL 25 B2
Wilkinsons (UWC) 71 B3
Willaston (CCTV) 48 E
Willerby Carr (UWC) 72 C3
Willersley Tunnel 49 B1
Willesden Carriage Shed North SB (CN) 120 A
Willesden Carriage Shed South SB (CS) 120 A
Willesden Jn. (Acton Branch) 120 A
WILLESDEN JUNCTION (H.L) 120 A
WILLESDEN JUNCTION (L.L) 120 A
Willesden North Jn. 120 A
Willesden Suburban Jn. 120 A
Willesden Suburban SB (WS) 120 A
Willesden Turnback Siding 120 A
WILLIAMWOOD 137 B3
Willingdon Jn. 12 D3
WILLINGTON 128 A
Willington (AHBC) 49 B2
WILLITON 6 E
Willoughby Road (AHBC) 52 C1
Willow Gap (UWC) 75 A2
Willow Walk (FPW) 21 A3
Willows Lane (AHBC) 52 C1
Willows Marsh (AOCL) 45 A3
Willox Bridge 1 (UWC) 27 B1
WILMCOTE 39 B3
Wilmington (AHBC) 12 D3
Wilmington Green (Footpath) 12 D3
WILMSLOW 57 A2
Wilmslow North Jn. 130 C3
Wilmslow South Jn. 57 A2
WILNECOTE 49 B3
Wilpshire Tunnel 62 D2
Wilsford (AHBC-X) 51 B1
Wilson's Crossing 21 A3
Wilstrop (MCG) 64 D1
Wilton Jn. 9 A1
Wilton South 9 A1
Wiltshires (UWC) 49 B2
WIMBLEDON 112 C3
Wimbledon ASC (W) 112 C3
WIMBLEDON CHASE 112 C3
Wimbledon East Depot 111 B2
WIMBLEDON PARK 111 B2
Wimbledon Park Sdgs. 112 C3
Wimbledon Traincare Depot 112 C3
Wimbledon West Jn. 112 C3
Wincanton Sdgs. (Corby) 41 B1
Winchburgh Jn. 85 A1
Winchburgh Tunnel 85 A2
WINCHCOMBE 28 D2
WINCHELSEA 13 B2
Winchelsea (AOCL) 13 B2
WINCHESTER 10 C1
WINCHFIELD 18 D3
WINCHMORE HILL 109 A1
Wincobank Jn. 134 A1
WINDERMERE 68 C2
Windmill Bridge Jn. 118 E
Windmill Lane (CCTV) 32 C3
Windmill Lane (UWC) 59 A2
Windmill Lane Tunnel 132 A
Windridge (UWC) 40 C1
WINDSOR & ETON CENTRAL 19 A2
WINDSOR & ETON RIVERSIDE 19 A2
Windsor (UWC) 59 A1
Windsor Bridge North Jn. 128 B
Windsor Bridge South Jn. 128 B
Winfrith (UWC) 8 C3

Chains to Yards Conversion Table

80 chains = 1 mile

Chains	Yards	Km equivalent	Chains	Yards	Km equivalent	Chains	Yards	Km equivalent
1	22	0.020	28	616	0.563	55	1,210	1.106
2	44	0.040	29	638	0.583	56	1,232	1.127
3	66	0.060	30	660	0.604	57	1,254	1.147
4	88	0.080	31	682	0.624	58	1,276	1.167
5	110	0.101	32	704	0.644	59	1,298	1.187
6	132	0.121	33	726	0.664	60	1,320	1.207
7	154	0.141	34	748	0.684	61	1,342	1.227
8	176	0.161	35	770	0.704	62	1,364	1.247
9	198	0.181	36	792	0.724	63	1,386	1.267
10	220	0.201	37	814	0.744	64	1,408	1.287
11	242	0.221	38	836	0.764	65	1,430	1.308
12	264	0.241	39	858	0.785	66	1,452	1.328
13	286	0.262	40	880	0.805	67	1,474	1.348
14	308	0.282	41	902	0.825	68	1,496	1.368
15	330	0.302	42	924	0.845	69	1,518	1.388
16	352	0.322	43	946	0.865	70	1,540	1.408
17	374	0.342	44	968	0.885	71	1,562	1.428
18	396	0.362	45	990	0.905	72	1,584	1.448
19	418	0.382	46	1,012	0.925	73	1,606	1.469
20	440	0.402	47	1,034	0.945	74	1,628	1.489
21	462	0.422	48	1,056	0.966	75	1,650	1.509
22	484	0.443	49	1,078	0.986	76	1,672	1.529
23	506	0.463	50	1,100	1.006	77	1,694	1.549
24	528	0.483	51	1,122	1.026	78	1,716	1.569
25	550	0.503	52	1,144	1.046	79	1,738	1.589
26	572	0.523	53	1,166	1.066	80	1,760	1.609
27	594	0.543	54	1,188	1.086			

Notes